T0331763

Stochastic Processes

This comprehensive guide to stochastic processes gives a complete overview of the theory and addresses the most important applications. Pitched at a level accessible to beginning graduate students and researchers from applied disciplines, it is both a course book and a rich resource for individual readers. Subjects covered include Brownian motion, stochastic calculus, stochastic differential equations, Markov processes, weak convergence of processes, and semigroup theory. Applications include the Black–Scholes formula for the pricing of derivatives in financial mathematics, the Kalman–Bucy filter used in the US space program, and also theoretical applications to partial differential equations and analysis. Short, readable chapters aim for clarity rather than for full generality. More than 350 exercises are included to help readers put their new-found knowledge to the test and to prepare them for tackling the research literature.

RICHARD F. BASS is Board of Trustees Distinguished Professor in the Department of Mathematics at the University of Connecticut.

Stochastic Processes

Richard F. Bass

University of Connecticut

CAMBRIDGE
UNIVERSITY PRESS

University Printing House, Cambridge CB2 8BS, United Kingdom

Cambridge University Press is part of the University of Cambridge.

It furthers the University's mission by disseminating knowledge in the pursuit of
education, learning and research at the highest international levels of excellence.

www.cambridge.org
Information on this title: www.cambridge.org/9781107008007

© R. F. Bass 2011

First published 2011
Reprinted 2013

A catalogue record for this publication is available from the British Library

Library of Congress Cataloguing in Publication data
Bass, Richard F.
Stochastic processes / Richard F. Bass.
p.   cm. – (Cambridge series in statistical and probabilistic mathematics ; 33)
Includes index.
ISBN 978-1-107-00800-7 (hardback)
1. Stochastic analysis.   I. Title.
QA274.2.B375   2011
519.2′32 – dc23   2011023024

ISBN 978-1-107-00800-7 Hardback

To Meredith, as always

Contents

Preface

Why study stochastic processes? This branch of probability theory offers sophisticated theorems and proofs, such as the existence of Brownian motion, the Doob–Meyer decomposition, and the Kolmogorov continuity criterion. At the same time stochastic processes also have far-reaching applications: the explosive growth in options and derivatives in financial markets throughout the world derives from the Black–Scholes formula, while NASA relies on the Kalman–Bucy method to filter signals from satellites and probes sent into outer space.

A graduate student taking a year-long course in probability theory first learns about sequences of random variables and topics such as laws of large numbers, central limit theorems, and discrete time martingales. In the second half of the course, the student will then turn to stochastic processes, which is the subject of this text. Topics covered here are Brownian motion, stochastic integrals, stochastic differential equations, Markov processes, the Black–Scholes formula of financial mathematics, the Kalman–Bucy filter, as well as many more.

The 42 chapters of this book can be grouped into seven parts. The first part consists of Chapters 1–8, where some of the basic processes and ideas are introduced, including Brownian motion. The next group of chapters, Chapters 9–15, introduce the theory of stochastic calculus, including stochastic integrals and Itô's formula. Chapters 16–18 explore jump processes. This requires a study of the foundations of stochastic processes, which is also known as the general theory of processes. Next we take up Markov processes in Chapters 19–23. A formidable obstacle to the study of Markov processes is the notation, and I have attempted to make this as accessible as possible. Chapters 24–29 involve stochastic differential equations. Two very important applications, to financial mathematics and to filtering, appear in Chapters 28 and 29, respectively. Probability measures on metric spaces and the weak convergence of random variables taking values in a metric space prove to be relevant to the study of stochastic processes. These and related topics are treated in Chapters 30–35. We then return to Markov processes, namely, their construction and some important examples, in Chapters 36–42. Tools used in the construction include infinitesimal generators, Dirichlet forms, and solutions to stochastic differential equations, while two important examples that we consider are diffusions on the real line and Lévy processes.

The prerequisites to this book are a sound knowledge of basic measure theory and a course in the classical aspects of probability. The probability topics needed are provided (with proofs) in an appendix.

There is far too much material in this book to cover in a single semester, and even too much for a full year. I recommend that as a minimum the following chapters be studied: Chapters 1–5, Chapters 9–13, Chapters 19–21, and Chapter 24. If possible, include either

Chapter 28 or Chapter 29. In Chapter 11, the statement and corollaries of Itô's formula are very important, but the proof of Itô's formula may be omitted.

I would like to thank the many students who patiently sat through my lectures, pointed out errors, and made suggestions. I especially would like to thank my colleague Sasha Teplyaev who taught a course from a preliminary version of this book and made a great number of useful suggestions.

Frequently used notation

Here are some notational conventions we will use. We use the letter c, either with or without subscripts, to denote a finite positive constant whose exact value is unimportant and which may change from line to line. We use $B(x, r)$ to denote the open Euclidean ball centered at x with radius r. $a \wedge b$ is the minimum of a and b, while $a \vee b$ is the maximum of a and b. $x^+ = x \vee 0$ and $x^- = (-x) \vee 0$. The symbol $\exists$ is used in a few formulas and means "there exists." $\mathbb{Q}$, $\mathbb{Q}_+$, $\mathbb{N}$, and $\mathbb{Z}$ denote the rationals, the positive rationals, the natural numbers, and the integers, respectively. If C is a matrix, C^{T} is the transpose of C.

For a set A, we use A^c for the complement of A. If A is a subset of a topological space, $\overline{A}$, A^0, and ∂A denote the closure, interior, and boundary of A, respectively.

Given a topological space $\mathcal{S}$, we use $C(\mathcal{S})$ for the space of continuous functions on $\mathcal{S}$, where we use the supremum norm. If $\mathcal{S}$ is a domain in $\mathbb{R}^d$, $C^k(\mathcal{S})$ refers to the set of continuous functions with domain $\mathcal{S}$ whose partial derivatives up to order k are continuous. C^∞ functions are those that are infinitely differentiable.

We will on a few occasions use the Fourier transform, which we define by

$$\widehat{f}(u) = \int e^{iu \cdot x} f(x) \, dx$$

for f integrable. This agrees with the convention in Rudin (1987).

If X is a stochastic process whose paths are right continuous with left limits, then $X_{t-} = \lim_{s < t, s \to t} X_s$ and $\Delta X_t = X_t - X_{t-}$.

1

Basic notions

In a first course on probability one typically works with a sequence of random variables $X_1, X_2, \ldots$ For stochastic processes, instead of indexing the random variables by the positive integers, we index them by $t \in [0, \infty)$ and we think of X_t as being the value at time t. The random variable could be the location of a particle on the real line, the strength of a signal, the price of a stock, and many other possibilities as well.

We will also work with increasing families of σ-fields $\{\mathcal{F}_t\}$, known as filtrations. The σ-field $\mathcal{F}_t$ is supposed to represent what we know up to time t.

1.1 Processes and σ-fields

Let $(\Omega, \mathcal{F}, \mathbb{P})$ be a probability space. A real-valued *stochastic process* (or simply a process) is a map X from $[0, \infty) \times \Omega$ to the reals. We write $X_t = X_t(\omega) = X(t, \omega)$. We will impose stronger measurability conditions shortly, but for now we require that the random variables X_t be measurable with respect to $\mathcal{F}$ for each $t \geq 0$.

A collection of σ-fields $\mathcal{F}_t$ such that $\mathcal{F}_t \subset \mathcal{F}$ for each t and $\mathcal{F}_s \subset \mathcal{F}_t$ if $s \leq t$ is called a *filtration*. Define $\mathcal{F}_{t+} = \cap_{\varepsilon > 0} \mathcal{F}_{t+\varepsilon}$. A filtration is *right continuous* if $\mathcal{F}_{t+} = \mathcal{F}_t$ for all $t \geq 0$. The σ-field $\mathcal{F}_{t+}$ is supposed to represent what one knows if one looks ahead an infinitesimal amount. Most of the filtrations we will come across will be right continuous, but see Exercise 1.1.

A *null set N* is one that has outer probability 0. This means that

$$\inf\{\mathbb{P}(A) : N \subset A, A \in \mathcal{F}\} = 0.$$

A filtration is *complete* if each $\mathcal{F}_t$ contains every null set. A filtration that is right continuous and complete is said to satisfy the *usual conditions*.

Given a filtration $\{\mathcal{F}_t\}$, whether or not it satisfies the usual conditions, we define $\mathcal{F}_\infty$ to be the σ-field generated by $\cup_{t \geq 0} \mathcal{F}_t$, that is, the smallest σ-field containing $\cup_{t \geq 0} \mathcal{F}_t$, and we write

$$\mathcal{F}_\infty = \bigvee_{t \geq 0} \mathcal{F}_t.$$

Recall that the arbitrary intersection of σ-fields is a σ-field, but the union of even two σ-fields need not be a σ-field.

We say that a stochastic process X is *adapted* to a filtration $\{\mathcal{F}_t\}$ if X_t is $\mathcal{F}_t$ measurable for each t. Often one starts with a stochastic process X and wants to define a filtration with respect to which X is adapted.

The simplest way to do this is to let $\mathcal{F}_t$ be the σ-field generated by the random variables $\{X_s, s \leq t\}$. More often one wants to have a slightly larger filtration than the one generated by X.

We define the *minimal augmented filtration* generated by X to be the smallest filtration that is right continuous and complete and with respect to which the process X is adapted. For each t, $\mathcal{F}_t$ is in general strictly larger than the smallest σ-field with respect to which $\{X_s : s \leq t\}$ is measurable because of the inclusion of the null sets. It is important to include the null sets; see Exercise 1.5. There is no widely accepted name for what we call the minimal augmented filtration; I like this nomenclature because it is descriptive and sufficiently different from "filtration generated by X" to avoid confusion.

The minimal augmented filtration generated by the process X_t can be constructed in three steps. First, let $\{\mathcal{F}_t^{00}\}$ be the smallest filtration with respect to which X is adapted, that is,

$$\mathcal{F}_t^{00} = \sigma(X_s; s \leq t). \tag{1.1}$$

Let $\mathbb{P}^*$ be the outer probability corresponding to $\mathbb{P}$: for $A \subset \Omega$,

$$\mathbb{P}^*(A) = \inf\{\mathbb{P}(B) : B \in \mathcal{F}, A \subset B\}.$$

Let $\mathcal{N}$ be the collection of null sets, so that $\mathcal{N} = \{A \subset \Omega : \mathbb{P}^*(A) = 0\}$. The second step is to let $\mathcal{F}_t^0$ be the smallest σ-field containing $\mathcal{F}_t^{00}$ and $\mathcal{N}$, or

$$\mathcal{F}_t^0 = \sigma(\mathcal{F}_t^{00} \cup \mathcal{N}). \tag{1.2}$$

The third step is to let

$$\mathcal{F}_t = \cap_{\varepsilon > 0} \mathcal{F}_{t+\varepsilon}^0. \tag{1.3}$$

Exercise 1.2 asks you to check that $\{\mathcal{F}_t\}$ is the minimal augmented filtration generated by X. We will refer to $\{\mathcal{F}_t^{00}\}$ as the *filtration generated by X*.

Two stochastic processes X and Y are said to be *indistinguishable* if $\mathbb{P}(X_t \neq Y_t$ for some $t \geq 0) = 0$. X and Y are *versions* of each other if for each $t \geq 0$, we have $\mathbb{P}(X_t \neq Y_t) = 0$. An example of two processes that are versions of each other but are not indistinguishable is to let $\Omega = [0, 1]$, $\mathcal{F}$ the Borel σ-field on $[0, 1]$, $\mathbb{P}$ Lebesgue measure on $[0, 1]$, $X(t, \omega) = 0$ for all t and ω, and $Y(t, \omega)$ equal to 1 if $t = \omega$ and 0 otherwise. Note that the functions $t \to X(t, \omega)$ are continuous for each ω, but the functions $t \to Y(t, \omega)$ are not continuous for any ω.

If X is a stochastic process, the functions $t \to X(t, \omega)$ are called the *paths* or *trajectories* of X. There will be one path for each ω. If the paths of X are continuous functions, except for a set of ω's in a null set, then X is called a *continuous process*, or is said to be continuous. We similarly define right continuous process, left continuous process, etc.

A function $f(t)$ is *right continuous with left limits* if $\lim_{h>0, h\downarrow 0} f(t+h) = f(t)$ for all t and $\lim_{h<0, h\uparrow 0} f(t+h)$ exists for all $t > 0$. Almost all our stochastic processes will have the property that except for a null set of ω's the function $t \to X(t, \omega)$ is right continuous and has left limits. One often sees *cadlag* to refer to paths that are right continuous with left limits; this abbreviates the French "continue à droite, limite à gauche."

1.2 Laws and state spaces

Let S be a topological space. The Borel σ-field on S is defined to be the σ-field generated by the open sets of S. A function $f\colon S \to \mathbb{R}$ is Borel measurable if $f^{-1}(G)$ is in the Borel σ-field of S whenever G is an open subset of $\mathbb{R}$. A random variable $Y\colon \Omega \to S$ is measurable with respect to a σ-field $\mathcal{F}$ of subsets of Ω if $\{\omega \in \Omega : Y(\omega) \in A\}$ is in $\mathcal{F}$ whenever A is in the Borel σ-field on S.

A stochastic process taking values in a topological space S is a map $X\colon [0, \infty) \times \Omega \to S$, where for each t, the random variable X_t is measurable with respect to $\mathcal{F}$.

Recall that if we have a probability space $(\Omega, \mathcal{F}, \mathbb{P})$ and $Y\colon \Omega \to \mathbb{R}$ is a random variable, then the *law* of Y is the probability measure $\mathbb{P}_Y$ on the Borel subsets of $\mathbb{R}$ defined by $\mathbb{P}_Y(A) = \mathbb{P}(Y \in A)$. Similarly, if $Y\colon \Omega \to \mathbb{R}^d$ is a d-dimensional random vector, then the law of Y is the probability measure $\mathbb{P}_Y$ on the Borel subsets of $\mathbb{R}^d$ defined by $\mathbb{P}_Y(A) = \mathbb{P}(Y \in A)$. We extend this definition to random variables Y taking values in a topological space S. In this case $\mathbb{P}_Y$ is a probability measure on the Borel subsets of S with the same definition: $\mathbb{P}_Y(A) = \mathbb{P}(Y \in A)$. In particular, if Y and Z are two random variables with the same state space S, then Y and Z will have the same law if $\mathbb{P}(Y \in A) = \mathbb{P}(Z \in A)$ for all Borel subsets A of S.

The relevance of the preceding paragraph to stochastic processes is this. Suppose X and Y are stochastic processes with continuous paths. Let $S = C[0, \infty)$ be the collection of real-valued continuous functions on $[0, \infty)$ together with the usual metric defined in terms of the supremum norm:

$$d(f, g) = \sup_{0 \le t} |f(t) - g(t)|.$$

(Strictly speaking, we should write $C([0, \infty))$, but we follow the usual convention and drop the outside parentheses.) Let the random variable $\overline{X}$ taking values in S be defined by setting $\overline{X}(\omega)$ to be the continuous function $t \to X(t, \omega)$, and define $\overline{Y}$ similarly. More precisely, $\overline{X}\colon \Omega \to S$ with

$$\overline{X}(\omega)(t) = X(t, \omega), \qquad t \ge 0.$$

Then $\overline{X}$ and $\overline{Y}$ are random variables taking values in the metric space S, and saying that $\overline{X}$ and $\overline{Y}$ have the same law means that $\mathbb{P}(\overline{X} \in A) = \mathbb{P}(\overline{Y} \in A)$ for all Borel subsets A of S. When this happens, we also say that the stochastic processes X and Y have the same law.

Two stochastic processes X and Y have the same *finite-dimensional distributions* if for every $n \ge 1$ and every $t_1 < \cdots < t_n$, the laws of $(X_{t_1}, \ldots, X_{t_n})$ and $(Y_{t_1}, \ldots, Y_{t_n})$ are equal.

Most often the topological spaces we will consider will also be metric spaces, but there will be a few occasions when we want to consider topological spaces that are not metric spaces. Suppose $S = \mathbb{R}^{[0, \infty)}$. We furnish S with the product topology. S can be identified with the collection of real-valued functions on $[0, \infty)$, but the topology is not given by the supremum norm nor by any other metric. We use f for elements of S, where $f(t)$ is the tth coordinate of f. We call a subset A of S a *cylindrical set* if there exist $n \ge 1$, non-negative reals $t_1, t_2, \ldots, t_n$, and a Borel subset B of $\mathbb{R}^n$ such that

$$A = \{f \in S : (f(t_1), \ldots, f(t_n)) \in B\}.$$

The appropriate σ-field to use on $\mathcal{S}$ is the one generated by the collection of cylindrical sets.

We want to generalize this notion slightly by allowing more general index sets and by allowing for the possibility of considering only a subset of the product space.

Definition 1.1 Let $\mathcal{U}$ be a topological space, T an arbitrary index set, and B a subset of $\mathcal{U}^T$, the collection of functions from T into $\mathcal{U}$. We say a set C is a *cylindrical subset* of B if there exist $n \geq 1, t_1, \ldots, t_n \in T$, and a Borel subset A of $\mathbb{R}^n$ such that

$$C = \{f \in B : (f(t_1), \ldots, f(t_n)) \in A\}.$$

Exercises

1.1   This exercise gives an example where $\{\mathcal{F}_t^{00}\}$ defined by (1.1) is not right continuous. Let $\Omega = \{a, b\}$, let $\mathcal{F}$ be the collection of all subsets of Ω, and let $\mathbb{P}(\{a\}) = \mathbb{P}(\{b\}) = \frac{1}{2}$. Define

$$X_t(\omega) = \begin{cases} 0, & t \leq 1; \\ 0, & t > 1 \text{ and } \omega = a; \\ t - 1, & t > 1 \text{ and } \omega = b. \end{cases}$$

Calculate $\mathcal{F}_t^{00} = \sigma(X_s; s \leq t)$ and show $\{\mathcal{F}_t^{00}\}$ is not right continuous.

1.2   If X is a stochastic process, let $\mathcal{F}_t^{00}, \mathcal{F}_t^0$, and $\mathcal{F}_t$ be defined by (1.1), (1.2), and (1.3), respectively. Show that $\{\mathcal{F}_t\}$ is the minimal augmented filtration generated by X.

1.3   Let $\{\mathcal{F}_t\}$ be a filtration satisfying the usual conditions and let $\mathcal{B}[0, t]$ be the Borel σ-field on $[0, t]$. A real-valued stochastic process X is *progressively measurable* if for each $t \geq 0$, the map $(s, \omega) \to X(s, \omega)$ from $[0, t] \times \Omega$ to $\mathbb{R}$ is measurable with respect to the product σ-field $\mathcal{B}[0, t] \times \mathcal{F}_t$.
      (1) If X is adapted to $\{\mathcal{F}_t\}$ and we define

$$X_t^{(n)}(\omega) = \sum_{k=0}^{\infty} X_{k/2^n}(\omega) 1_{[k/2^n, (k+1)/2^n)}(t),$$

show that $X^{(n)}$ is progressively measurable for each $n \geq 1$.
      (2) Use (1) to show that if X is adapted to $\{\mathcal{F}_t\}$ and has left continuous paths, then X is progressively measurable.
      (3) If X is adapted to $\{\mathcal{F}_t\}$ and we define

$$Y_t^{(n)}(\omega) = \sum_{k=0}^{\infty} X_{(k+1)/2^n}(\omega) 1_{[k/2^n, (k+1)/2^n)}(t),$$

show that for each $t \geq 0$, the map $(s, \omega) \to Y^{(n)}(s, \omega)$ from $[0, t] \times \Omega$ to $\mathbb{R}$ is measurable with respect to $\mathcal{B}[0, t] \times \mathcal{F}_{t+2^{-n}}$.
      (4) Show that if X is adapted to $\{\mathcal{F}_t\}$ and has right continuous paths, then X is progressively measurable.

1.4   Let $\mathcal{S} = \mathbb{R}^{[0,1]}$, the set of functions from $[0, 1]$ to $\mathbb{R}$, and let $\mathcal{F}$ be the σ-field generated by the cylindrical sets. The purpose of this exercise is to show that the elements of $\mathcal{F}$ depend on only countably many coordinates.

Let $\mathcal{S}_0 = \{(x_1, x_2, \ldots)\}$, the set of sequences taking values in $\mathbb{R}$. Let $\mathcal{F}_0$ be the σ-field generated by the cylindrical subsets of $\mathbb{R}^{\mathbb{N}}$, where $\mathbb{N} = \{1, 2, \ldots\}$.

Show that $B \in \mathcal{F}$ if and only if there exist $t_1, t_2, \ldots$ in $[0, 1]$ and a set $C \in \mathcal{F}_0$ such that

$$B = \{f \in \mathcal{S} : (f(t_1), f(t_2), \ldots) \in C\}.$$

1.5   Null sets are sometimes important! Let $\mathcal{S}$ and $\mathcal{F}$ be as in Exercise 1.4. Show that $D \notin \mathcal{F}$, where

$$D = \{f \in \mathcal{S} : f \text{ is a continuous function on } [0, 1]\}.$$

1.6   Suppose X is a stochastic process, $\{\mathcal{F}_t\}$ its minimal augmented filtration, and $\mathcal{F}_\infty = \vee_{t \geq 0} \mathcal{F}_t$. Suppose with probability one, the paths of X are right continuous with left limits. Let $X_{t-} = \lim_{s < t, s \to t} X_s$, the left-hand limit at time t, and $\Delta X_t = X_t - X_{t-}$, the size of the jump at time t. If

$$A = \{\exists t \geq 0 : \Delta X_t > 1\},$$

prove $A \in \mathcal{F}_\infty$.

1.7   Suppose X is a stochastic process, $\{\mathcal{F}_t\}$ is the minimal augmented filtration for X, and $\mathcal{F}_\infty = \vee_{t \geq 0} \mathcal{F}_t$. If the paths of X are right continuous with left limits with probability one, show that the event

$$A = \{X \text{ has continuous paths}\}$$

is in $\mathcal{F}_\infty$.

Notes

The older literature sometimes uses the notion of a separable stochastic process, but this is rarely seen nowadays. For much more on measurability, see Chapter 16. For the complete story on the foundations of stochastic processes, see Dellacherie and Meyer (1978).

2

Brownian motion

Brownian motion is by far the most important stochastic process. It is the archetype of Gaussian processes, of continuous time martingales, and of Markov processes. It is basic to the study of stochastic differential equations, financial mathematics, and filtering, to name only a few of its applications.

In this chapter we define Brownian motion and consider some of its elementary aspects. Later chapters will take up the construction of Brownian motion and properties of Brownian motion paths.

2.1 Definition and basic properties

Let $(\Omega, \mathcal{F}, \mathbb{P})$ be a probability space and let $\{\mathcal{F}_t\}$ be a filtration, not necessarily satisfying the usual conditions.

Definition 2.1 $W_t = W_t(\omega)$ is a one-dimensional *Brownian motion* with respect to $\{\mathcal{F}_t\}$ and the probability measure $\mathbb{P}$, started at 0, if

(1) W_t is $\mathcal{F}_t$ measurable for each $t \geq 0$.
(2) $W_0 = 0$, a.s.
(3) $W_t - W_s$ is a normal random variable with mean 0 and variance $t - s$ whenever $s < t$.
(4) $W_t - W_s$ is independent of $\mathcal{F}_s$ whenever $s < t$.
(5) W_t has continuous paths.

If instead of (2) we have $W_0 = x$, we say we have a Brownian motion started at x. Definition 2.1(4) is referred to as the *independent increments* property of Brownian motion. The fact that $W_t - W_s$ has the same law as W_{t-s}, which follows from Definition 2.1(3), is called the *stationary increments* property. When no filtration is specified, we assume the filtration is the filtration generated by W, i.e., $\mathcal{F}_t = \sigma(W_s; s \leq t)$. Sometimes a one-dimensional Brownian motion started at 0 is called a *standard Brownian motion*.

Figure 2.1 is a simulation of a typical Brownian motion path.

We define d-dimensional Brownian motion with respect to a filtration $\{\mathcal{F}_t\}$ and started at $x = (x_1, \ldots, x_d)$ to be $(W_t^{(1)}, \ldots, W_t^{(d)})$, where the $W^{(i)}$ are each one-dimensional Brownian motions with respect to $\{\mathcal{F}_t\}$ started at x_i, respectively, and $W^{(1)}, \ldots, W^{(n)}$ are all independent.

The law of a Brownian motion is called *Wiener measure*. More precisely, given a Brownian motion W, we can view it as a random variable taking values in $C[0, \infty)$, the space of real-valued continuous functions on $[0, \infty)$. The law of W is the measure $\mathbb{P}_W$ on

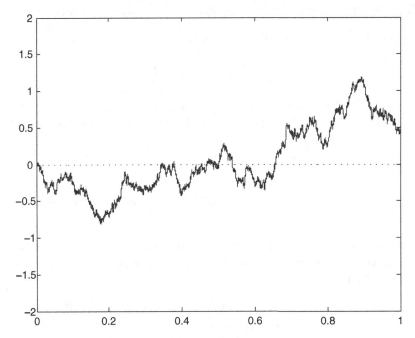

Figure 2.1 Simulation of a typical Brownian motion path.

$C[0, \infty)$ defined by $\mathbb{P}_W(A) = \mathbb{P}(W \in A)$ for all Borel subsets A of $C[0, \infty)$. The measure $\mathbb{P}_W$ is Wiener measure.

There are a number of transformations one can perform on a Brownian motion that yield a new Brownian motion. The first one is called the *scaling property of Brownian motion*, or simply *scaling*.

Proposition 2.2 *If W is a Brownian motion started at 0, $a > 0$, and $Y_t = aW_{t/a^2}$, then Y_t is a Brownian motion started at 0.*

Proof We use $\mathcal{G}_t = \mathcal{F}_{t/a^2}$ for the filtration for Y. Clearly Y_t has continuous paths, $Y_0 = 0$, a.s., and Y_t is $\mathcal{G}_t$ measurable. If $s < t$,

$$Y_t - Y_s = a(W_{t/a^2} - W_{s/a^2})$$

is independent of $\mathcal{F}_{s/a^2}$, hence is independent of $\mathcal{G}_s$. Finally, if $s < t$, and if $s < t$, then $Y_t - Y_s$ will be a normal random variable with mean zero and

$$\text{Var}\,(Y_t - Y_s) = a^2 \text{Var}\,(W_{t/a^2} - W_{s/a^2}) = a^2\left(\frac{t}{a^2} - \frac{s}{a^2}\right) = t - s.$$

This suffices to give our result. $\square$

For some other transformations, see Exercises 2.3 and 2.5.

Recall what it means for a finite collection of random variables to be jointly normal; see (A.29). A stochastic process X is *Gaussian* or *jointly normal* if all its finite-dimensional distributions are jointly normal, that is, if for each $n \geq 1$ and $t_1 < \cdots < t_n$, the collection of random variables $X_{t_1}, \ldots, X_{t_n}$ is a jointly normal collection.

Proposition 2.3 *If W is a Brownian motion, then W is a Gaussian process.*

Proof   Suppose W is a Brownian motion and let $0 = t_0 < t_1 < \cdots < t_n$. Define

$$Z_i = \frac{W_{t_i} - W_{t_{i-1}}}{\sqrt{t_i - t_{i-1}}}, \qquad i = 1, 2, \ldots, n.$$

By Definition 2.1(4), Z_i is independent of $\mathcal{F}_{t_{i-1}}$, and hence independent of $Z_1, \ldots, Z_{j-1}$. By Definition 2.1(3), Z_i is a mean-zero random variable with variance one. We can write

$$W_{t_j} = \sum_{i=1}^{j} (t_i - t_{i-1})^{1/2} Z_i, \qquad j = 1, \ldots, n,$$

and so $(W_{t_1}, \ldots, W_{t_n})$ is jointly normal. It follows that Brownian motion is a Gaussian process.   □

Since the law of a finite collection of jointly normal random variables is determined by their means and covariances, let's calculate the covariance of W_s and W_t when W is a Brownian motion. If $s \leq t$, then

$$t - s = \text{Var}\,(W_t - W_s) = \text{Var}\,W_t + \text{Var}\,W_s - 2\,\text{Cov}\,(W_s, W_t)$$
$$= t + s - 2\,\text{Cov}\,(W_s, W_t)$$

from Definition 2.1(2) and (3). Hence $\text{Cov}\,(W_s, W_t) = s$ if $s \leq t$. This is frequently written as

$$\text{Cov}\,(W_s, W_t) = s \wedge t. \tag{2.1}$$

We have the following converse.

Theorem 2.4 *If W is a process such that all the finite-dimensional distributions are jointly normal, $\mathbb{E}\,W_s = 0$ for all s, $\text{Cov}\,(W_s, W_t) = s$ when $s \leq t$, and the paths of W_t are continuous, then W is a Brownian motion.*

Proof   For $\mathcal{F}_t$ we take the filtration generated by W. If we take $s = t$, then $\text{Var}\,W_t = \text{Cov}\,(W_t, W_t) = t$. In particular, $\text{Var}\,W_0 = 0$, and since $\mathbb{E}\,W_0 = 0$, then $W_0 = 0$, a.s. We have

$$\text{Var}\,(W_t - W_s) = \text{Var}\,W_t - 2\,\text{Cov}\,(W_s, W_t) + \text{Var}\,W_t$$
$$= t - 2s + s = t - s.$$

We have thus established all the parts of Definition 2.1 except for the independence of $W_t - W_s$ from $\mathcal{F}_s$.

If $r \leq s < t$, then

$$\text{Cov}\,(W_t - W_s, W_r) = \text{Cov}\,(W_t, W_r) - \text{Cov}\,(W_s, W_r) = r - r = 0,$$

and so $W_t - W_s$ is independent of W_r by Proposition A.55. This shows that $W_t - W_s$ is independent of $\mathcal{F}_s$.   □

We now look at two results that are more technical. These should only be skimmed on the first reading of the book: read the statements, but not the proofs. The first result says that if W is a Brownian motion with respect to the filtration generated by W, then it is also a Brownian motion with respect to the minimal augmented filtration.

Proposition 2.5 *Let W_t be a Brownian motion with respect to $\{\mathcal{F}_t^{00}\}$, where $\mathcal{F}_t^{00} = \sigma(W_s; s \leq t)$. Let $\mathcal{N}$ be the collection of null sets, $\mathcal{F}_t^0 = \sigma(\mathcal{F}_t^{00} \cup \mathcal{N})$, and $\mathcal{F}_t = \cap_{\varepsilon>0}\mathcal{F}_{t+\varepsilon}^0$.*

(1) W is a Brownian motion with respect to the filtration $\{\mathcal{F}_t\}$.

(2) $\mathcal{F}_t = \mathcal{F}_t^0$ for each t.

Proof (1) The only property we need to check is Definition 2.1(4). If f is a continuous bounded function on $\mathbb{R}$, $A \in \mathcal{F}_s^{00}$, and $s < t$, then because W is a Brownian motion with respect to $\{\mathcal{F}_t^{00}\}$, the independent increments property shows that

$$\mathbb{E}[f(W_t - W_s); A] = \mathbb{E}[f(W_t - W_s)]\mathbb{P}(A). \tag{2.2}$$

If A is such that $A \backslash B$ and $B \backslash A$ are null sets for some $B \in \mathcal{F}_s^{00}$, it is easy to see that (2.2) continues to hold. By linearity, it also holds if A is a finite disjoint union of such sets. If $\mathcal{C}_1$ is the collection of subsets of $\mathcal{F}_s^0$ that are finite disjoint unions of such sets, then $\mathcal{C}_1$ is an algebra of subsets of $\mathcal{F}_s^0$. Let $\mathcal{M}_1$ be the collection of subsets of $\mathcal{F}_s^0$ for which (2.2) holds. It is readily checked that $\mathcal{M}_1$ is a monotone class. By the monotone class theorem (Theorem B.2), $\mathcal{M}_1$ is equal to the smallest σ-field containing $\mathcal{C}_1$, which is $\mathcal{F}_s^0$. Therefore (2.2) holds for all $A \in \mathcal{F}_s^0$.

Now suppose $A \in \mathcal{F}_s = \mathcal{F}_{s+}^0$. Then for each $\varepsilon > 0$, $A \in \mathcal{F}_{s+\varepsilon}^0$, and so using (2.2) with s replaced by $s + \varepsilon$ and t replaced by $t + \varepsilon$, we have

$$\mathbb{E}[f(W_{t+\varepsilon} - W_{s+\varepsilon}); A] = \mathbb{E}[f(W_{t+\varepsilon} - W_{s+\varepsilon})]\mathbb{P}(A). \tag{2.3}$$

Letting $\varepsilon \to 0$ and using the facts that f is bounded and continuous and W has continuous paths, the dominated convergence theorem implies that

$$\mathbb{E}[f(W_t - W_s); A] = \mathbb{E}[f(W_t - W_s)]\mathbb{P}(A). \tag{2.4}$$

This equation holds whenever f is continuous and $A \in \mathcal{F}_s$. By a limit argument, (2.4) holds whenever f is the indicator of a Borel subset of $\mathbb{R}$. That says that $W_t - W_s$ and $\mathcal{F}_s$ are independent.

(2) Fix t and choose $t_0 > t$. Let $\mathcal{M}_2$ be the collection of subsets of $\mathcal{F}_{t_0}^{00}$ whose conditional expectation with respect to $\mathcal{F}_t$ is $\mathcal{F}_t^0$ measurable, that is, $A \in \mathcal{M}_2$ if $A \in \mathcal{F}_{t_0}^{00}$ and $\mathbb{E}[1_A \mid \mathcal{F}_t]$ is $\mathcal{F}_t^0$ measurable. Let $\mathcal{C}_2$ be the collection of events A for which there exist $n \geq 1$, $0 \leq s_0 < s_1 < \cdots < s_n \leq t_0$ with t equal to one of the s_i, and Borel subsets $B_1, \ldots, B_n$ of $\mathbb{R}$ such that

$$A = (W_{s_1} - W_{s_0} \in B_1, \ldots, W_{s_n} - W_{s_{n-1}} \in B_n).$$

Suppose A is of this form, and suppose $t = s_i$. Then by the independence result that we proved in (1),

$$\mathbb{E}[1_A \mid \mathcal{F}_t] = 1_{(W_{s_1} - W_{s_0} \in B_1, \ldots, W_{s_i} - W_{s_{i-1}} \in B_i)}$$
$$\times \mathbb{P}(W_{s_{i+1}} - W_{s_i} \in B_{i+1}, \ldots, W_{s_n} - W_{s_{n-1}} \in B_n),$$

which is $\mathcal{F}_t^0$ measurable. Thus $\mathcal{C}_2 \subset \mathcal{M}_2$. Finite unions of sets in $\mathcal{C}_2$ form an algebra of subsets of $\mathcal{F}_t^{00}$ that generate $\mathcal{F}_t^{00}$. It is easy to check that $\mathcal{M}_2$ is a monotone class, so by the monotone class theorem, $\mathcal{M}_2$ equals $\mathcal{F}_t^{00}$. By linearity and taking monotone limits, if Y is non-negative and $\mathcal{F}_t^{00}$ measurable, then $\mathbb{E}[Y \mid \mathcal{F}_t]$ is $\mathcal{F}_t^0$ measurable.

To finish, suppose $A \in \mathcal{F}_t$. Then since $t < t_0$, we see that $A \in \mathcal{F}_{t_0}^0$. By Exercise 2.7, there exists $Y \in \mathcal{F}_{t_0}^{00}$ such that $1_A = Y$, a.s. Then $\mathbb{E}[Y \mid \mathcal{F}_t]$ is $\mathcal{F}_t^0$ measurable. Since $\mathcal{F}_t^0$ contains all the null sets, $1_A = \mathbb{E}[1_A \mid \mathcal{F}_t]$ is also $\mathcal{F}_t^0$ measurable, or $A \in \mathcal{F}_t^0$. This proves (2). $\square$

The final item we consider in this chapter is a subtle one. The question is this: if W and W' are both Brownian motions, do they have all the same properties? To illustrate this issue, let's revisit the example of Chapter 1 where $\Omega = [0, 1]$, $\mathcal{F}$ is the Borel σ-field on $[0, 1]$, $\mathbb{P}$ is Lebesgue measure on $[0, 1]$, $X(t, \omega) = 0$ for all t and ω, and $Y(t, \omega)$ is 1 if $t = \omega$ and 0 otherwise. For each t, $\mathbb{P}(X_t = Y_t) = 1$, so X and Y have the same finite-dimensional distributions. However, if

$$A = \{f : f \text{ is not a continuous function on } [0, 1]\},$$

then $(X \in A)$ is a null set but $(Y \in A)$ is not. Even though X and Y have the same finite-dimensional distributions, X has continuous paths but Y does not.

To rephrase our question, is it true that $\mathbb{P}(W \in A) = \mathbb{P}(W' \in A)$ for every Borel subset A of $C[0, \infty)$? We know W and W' have the same finite-dimensional distributions because each is jointly normal with zero means and $\operatorname{Cov}(W_s, W_t) = s \wedge t = \operatorname{Cov}(W_s', W_t')$. The fact that the answer to our question is yes then comes from the following theorem. We look at $C[0, t_0]$ instead of $C[0, \infty)$ for the sake of simplicity.

Theorem 2.6 *Let $t_0 > 0$ and let X, Y be random variables taking values in $C[0, t_0]$ which have the same finite-dimensional distributions. Then the laws of X and Y are equal.*

Proof Let $\mathcal{M}$ be the collection of Borel subsets A of $C[0, t_0]$ for which $\mathbb{P}(X \in A)$ equals $\mathbb{P}(Y \in A)$. We will show that $\mathcal{M}$ is a monotone class and then use the monotone class theorem to show that $\mathcal{M}$ is equal to the Borel σ-field on $C[0, t_0]$.

First, let $\mathcal{C}$ be the collection of all cylindrical subsets of $C[0, t_0]$ (defined by Definition 1.1). Since the finite-dimensional distributions of X and Y are equal, then $\mathcal{M}$ contains $\mathcal{C}$. It is easy to check that $\mathcal{C}$ is an algebra of subsets of $C[0, t_0]$. If $A_1 \supset A_2 \supset \cdots$ are elements of $\mathcal{M}$, then

$$\mathbb{P}(X \in \cap_n A_n) = \lim_n \mathbb{P}(X \in A_n) = \lim_n \mathbb{P}(Y \in A_n) = \mathbb{P}(Y \in \cap_n A_n)$$

since $\mathbb{P}$ is a finite measure. Therefore $\cap_n A_n \in \mathcal{M}$. A very similar argument shows that if $A_1 \subset A_2 \subset \cdots$ are elements of $\mathcal{M}$, then $\cup_n A_n \in \mathcal{M}$. Therefore $\mathcal{M}$ is a monotone class. By the monotone class theorem, $\mathcal{M}$ contains the smallest σ-field containing $\mathcal{C}$. We will show that $\mathcal{M}$ contains all the open sets; then $\mathcal{M}$ will contain the smallest σ-field containing the open sets, and we will be done.

Since $C[0, t_0]$ is separable, every open set is the countable union of open balls. Because $\mathcal{M}$ is a σ-field, it suffices to show that $\mathcal{M}$ contains the open balls in $C[0, t_0]$, that is, all sets of the form

$$B(f_0, r) = \{f \in C[0, t_0] : \sup_{0 \le t \le t_0} |f(t) - f_0(t)| < r\}$$

where $r > 0$ and $f_0 \in C[0, t_0]$. For each m and n,

$$\{f \in C[0, t_0] : \sup_{0 \le k \le 2^n t_0} |f(k/2^n) - f_0(k/2^n)| \le r - (1/m)\}$$

is a set in $\mathcal{C}$, and so is in $\mathcal{M}$. As $n \to \infty$, these sets decrease to

$$D_m = \{f \in C[0, t_0] : \sup_{0 \le t \le t_0} |f(t) - f_0(t)| \le r - (1/m)\},$$

since all the functions we are considering are continuous. Finally, D_m increases to $B(f_0, r)$ as $m \to \infty$, so $B(f_0, r)$ is in $\mathcal{M}$ as desired. □

Exercises

2.1 Suppose W is a Brownian motion on $[0, 1]$. Let

$$Y_t = W_{1-t} - W_1.$$

Show that Y_t is a Brownian motion on $[0, 1]$.

2.2 This exercise shows that the projection of a d-dimensional Brownian motion onto a hyperplane yields a one-dimensional Brownian motion. Suppose $(W_t^{(1)}, \ldots, W_t^{(d)})$ is a d-dimensional Brownian motion started from 0 and $\lambda_1, \ldots, \lambda_d \in \mathbb{R}$ with $\sum_{i=1}^d \lambda_i^2 = 1$. Show that $X_t = \sum_{i=1}^d \lambda_i W_t^{(i)}$ is a one-dimensional Brownian motion started from 0.

2.3 This exercise shows that rotating a Brownian motion about the origin yields another Brownian motion. Let W be a d-dimensional Brownian motion started at 0 and let A be a $d \times d$ orthogonal matrix, that is, $A^{-1} = A^{\mathrm{T}}$. Show that $Y_t = AW_t$ is again a d-dimensional Brownian motion.

2.4 Here is a converse to Exercise 2.2: roughly speaking, if all the projections of a d-dimensional process X onto hyperplanes are one-dimensional Brownian motions, then X is a d-dimensional Brownian motion.

Suppose $(X_t^1, \ldots, X_t^d)$ is a d-dimensional continuous process, i.e., one taking values in $\mathbb{R}^d$. Let $\{\mathcal{F}_t\}$ be the minimal augmented filtration generated by X. Suppose that whenever $\lambda_1, \ldots, \lambda_d \in \mathbb{R}$ with $\sum_{i=1}^d \lambda_i^2 = 1$, then $\sum_{i=1}^d \lambda_i X_t^i$ is a one-dimensional Brownian motion started at 0 with respect to the filtration $\{\mathcal{F}_t\}$.

(1) If $u = (u_1, \ldots, u_d)$, let $\|u\| = (\sum u_j^2)^{1/2}$ and let $\lambda_j = u_j / \|u\|$. Calculate

$$\mathbb{E} \exp \left(i \sum_{j=1}^d u_j X_t^j \right) = \mathbb{E} \exp \left(i \|u\| \sum_{j=1}^d \lambda_j X_t^j \right),$$

the joint characteristic function of X_t.

(2) If $t_0 < t_1 < \cdots < t_n$, use independence and (1) to calculate

$$\mathbb{E} \exp \left(i \sum_{k=0}^{n-1} \sum_{j=1}^d u_j^k (X_{t_{k+1}}^j - X_{t_k}^j) \right).$$

(3) Prove that $(X_t^1, \ldots, X_t^d)$ is a d-dimensional Brownian motion started from 0.

(Some care is needed with the filtrations. If we only know that $Y^\lambda = \sum_i \lambda_i X^i$ is a Brownian motion with respect to the filtration generated by Y^λ for each $\lambda = (\lambda_1, \ldots, \lambda_d)$, the assertion is not true. See Revuz and Yor (1999), Exercise I.1.19.)

2.5 Let W_t be a Brownian motion and suppose

$$\lim_{t \to \infty} W_t/t = 0, \qquad \text{a.s.} \tag{2.5}$$

Let $Z_t = tW_{1/t}$ if $t > 0$ and set $Z_0 = 0$. (This is called *time inversion.*) Show that Z is a Brownian motion. (We will see later that the assumption (2.5) is superfluous; see Theorem 7.2.)

2.6   Let X and Y be two independent Brownian motions started at 0 and let $t_0 > 0$. Let

$$Z_t = \begin{cases} X_t, & t \leq t_0, \\ X_{t_0} + Y_{t-t_0}, & t > t_0. \end{cases}$$

Prove that Z is also a Brownian motion.

2.7   Let $\mathcal{F}_t^{00}$ and $\mathcal{F}_t^0$ be defined as in (1.1) and (1.2). Prove that if X is $\mathcal{F}_t^0$ measurable, there exists Z such that Z is $\mathcal{F}_t^{00}$ measurable and $Y = Z$, a.s.

2.8   Let $\mathcal{F}_t^{00}$ and $\mathcal{F}_t^0$ be defined as in (1.1) and (1.2). The *symmetric difference* of two sets A and B is defined by $A \,\Delta\, B = (A \setminus B) \cup (B \setminus A)$. Prove that

$$\mathcal{F}_t^0 = \{A \subset \Omega : A \,\Delta\, B \in \mathcal{N} \text{ for some } B \in \mathcal{F}_t^{00}\}.$$

Notes

Brownian motion is named for Robert Brown, a botanist who observed the erratic motion of colloidal particles in suspension in the 1820s. Brownian motion was used by Bachelier in 1900 in his PhD thesis to model stock prices and was the subject of an important paper by Einstein in 1905. The rigorous mathematical foundations for Brownian motion were first given by Wiener in 1923.

3

Martingales

Although discrete-time martingales are useful in a first course on probability, they are nowhere near as useful as continuous-time martingales are in the study of stochastic processes. The whole theory of stochastic integrals and stochastic differential equations is based on martingales indexed by times $t \in [0, \infty)$. After giving the definition and some examples, we extend Doob's inequalities, the optional stopping theorem, and the martingale convergence theorem to continuous-time martingales. We then derive some estimates for Brownian motion using martingale techniques.

3.1 Definition and examples

We define continuous-time martingales. Let $\{\mathcal{F}_t\}$ be a filtration, not necessarily satisfying the usual conditions.

Definition 3.1 M_t is a continuous-time *martingale* with respect to the filtration $\{\mathcal{F}_t\}$ and the probability measure $\mathbb{P}$ if

(1) $\mathbb{E} |M_t| < \infty$ for each t;
(2) M_t is $\mathcal{F}_t$ measurable for each t;
(3) $\mathbb{E}[M_t \mid \mathcal{F}_s] = M_s$, a.s., if $s < t$.

Part (2) of the definition can be rephrased as saying M_t is adapted to $\mathcal{F}_t$. If in part (3) "=" is replaced by "$\geq$," then M_t is a *submartingale*, and if it is replaced by "$\leq$," then we have a *supermartingale*.

Taking expectations in Definition 3.1(3), we see that if $s < t$, then $\mathbb{E} M_s \leq \mathbb{E} M_t$ is M is a submartingale and $\mathbb{E} M_s \geq \mathbb{E} M_t$ if M is a supermartingale. Thus submartingales tend to increase, on average, and supermartingales tend to decrease, on average.

There are many martingales associated with Brownian motion. Here are three examples.

Example 3.2 Let $M_t = W_t$, where W_t is a Brownian motion. Then M_t is a martingale. To verify Definition 3.1(3), we write

$$\mathbb{E}[M_t \mid \mathcal{F}_s] = M_s + \mathbb{E}[W_t - W_s \mid \mathcal{F}_s] = M_s + \mathbb{E}[W_t - W_s] = M_s,$$

using the independent increments property of Brownian motion and the fact that $\mathbb{E}[W_t - W_s] = 0$.

Example 3.3 Let $M_t = W_t^2 - t$, where W_t is a Brownian motion. To show M_t is a martingale, we write

$$
\begin{aligned}
\mathbb{E}\left[M_t \mid \mathcal{F}_s\right] &= \mathbb{E}\left[(W_t - W_s + W_s)^2 \mid \mathcal{F}_s\right] - t \\
&= W_s^2 + \mathbb{E}\left[(W_t - W_s)^2 \mid \mathcal{F}_s\right] + 2\mathbb{E}\left[W_s(W_t - W_s) \mid \mathcal{F}_s\right] - t \\
&= W_s^2 + \mathbb{E}\left[(W_t - W_s)^2\right] + 2W_s\mathbb{E}\left[W_t - W_s \mid \mathcal{F}_s\right] - t \\
&= W_s^2 + \mathbb{E}\left[(W_t - W_s)^2\right] + 2W_s\mathbb{E}\left[W_t - W_s\right] - t \\
&= W_s^2 + (t - s) - t = M_s.
\end{aligned}
$$

We used the facts that W_s is $\mathcal{F}_s$ measurable and that $W_t - W_s$ is independent of $\mathcal{F}_s$.

Example 3.4 Again let W_t be a Brownian motion, let $a \in \mathbb{R}$, and let $M_t = e^{aW_t - a^2 t/2}$. Since $W_t - W_s$ is normal with mean zero and variance $t - s$, we know $\mathbb{E}\, e^{a(W_t - W_s)} = e^{a^2(t-s)/2}$; see (A.6). Then

$$
\begin{aligned}
\mathbb{E}\left[M_t \mid \mathcal{F}_s\right] &= e^{-a^2 t/2} e^{aW_s} \mathbb{E}\left[e^{a(W_t - W_s)} \mid \mathcal{F}_s\right] \\
&= e^{-a^2 t/2} e^{aW_s} \mathbb{E}\left[e^{a(W_t - W_s)}\right] \\
&= e^{-a^2 t/2} e^{aW_s} e^{a^2(t-s)/2} = M_s.
\end{aligned}
$$

We give one more example of a martingale, although not one derived from Brownian motion.

Example 3.5 Recall that given a filtration $\{\mathcal{F}_t\}$, each $\mathcal{F}_t$ is contained in $\mathcal{F}$, where $(\Omega, \mathcal{F}, \mathbb{P})$ is our probability space. Let X be an integrable $\mathcal{F}$ measurable random variable, and let $M_t = \mathbb{E}[X \mid \mathcal{F}_t]$. Then

$$
\mathbb{E}\left[M_t \mid \mathcal{F}_s\right] = \mathbb{E}\left[\mathbb{E}[X \mid \mathcal{F}_t] \mid \mathcal{F}_s\right] = \mathbb{E}[X \mid \mathcal{F}_s] = M_s,
$$

and M is a martingale.

3.2 Doob's inequalities

We derive the analogs of Doob's inequalities in the stochastic process context.

Theorem 3.6 *Suppose M_t is a martingale or non-negative submartingale with paths that are right continuous with left limits. Then*
(1)

$$
\mathbb{P}(\sup_{s \leq t} |M_s| \geq \lambda) \leq \mathbb{E}|M_t|/\lambda.
$$

(2) If $1 < p < \infty$, then

$$
\mathbb{E}[\sup_{s \leq t} |M_s|]^p \leq \left(\frac{p}{p-1}\right)^p \mathbb{E}|M_t|^p.
$$

Proof We will do the case where M_t is a martingale, the submartingale case being nearly identical. Let $\mathcal{D}_n = \{kt/2^n : 0 \leq k \leq 2^n\}$. If we set $N_k^{(n)} = M_{kt/2^n}$ and $\mathcal{G}_k^{(n)} = \mathcal{F}_{kt/2^n}$, it is clear that $\{N_k^{(n)}\}$ is a discrete-time martingale with respect to $\{\mathcal{G}_k^{(n)}\}$. Let

$$
A_n = \{ \sup_{s \leq t, s \in \mathcal{D}_n} |M_s| > \lambda \}.
$$

By Doob's inequality for discrete-time martingales (see Theorem A.32),

$$\mathbb{P}(A_n) = \mathbb{P}(\max_{k \leq 2^n} |N_k^{(n)}| > \lambda) \leq \frac{\mathbb{E}\, |N_{2^n}^{(n)}|}{\lambda} = \frac{\mathbb{E}\, |M_t|}{\lambda}.$$

Note that the A_n are increasing, and since M_t is right continuous,

$$\cup_n A_n = \{\sup_{s \leq t} |M_s| > \lambda\}.$$

Then

$$\mathbb{P}(\sup_{s \leq t} |M_s| > \lambda) = \mathbb{P}(\cup_n A_n) = \lim_{n \to \infty} \mathbb{P}(A_n) \leq \mathbb{E}\, |M_t|/\lambda.$$

If we apply this with λ replaced by $\lambda - \varepsilon$ and let $\varepsilon \to 0$, we obtain (1).

The proof of (2) is similar. By Doob's inequality for discrete-time martingales (see Theorem A.33),

$$\mathbb{E}\, [\sup_{k \leq 2^n} |N_k^{(n)}|^p] \leq \left(\frac{p}{p-1}\right)^p \mathbb{E}\, |N_{2^n}^{(n)}|^p = \left(\frac{p}{p-1}\right)^p \mathbb{E}\, |M_t|^p.$$

Since $\sup_{k \leq 2^n} |N_k^{(n)}|^p$ increases to $\sup_{s \leq t} |M_s|^p$ by the right continuity of M, (2) follows by Fatou's lemma. $\qquad\square$

3.3 Stopping times

Throughout this section we suppose we have a filtration $\{\mathcal{F}_t\}$ satisfying the usual conditions.

Definition 3.7 A random variable $T : \Omega \to [0, \infty]$ is a *stopping time* if for all t, $(T < t) \in \mathcal{F}_t$. We say T is a finite stopping time if $T < \infty$, a.s. We say T is a bounded stopping time if there exists $K \in [0, \infty)$ such that $T \leq K$, a.s.

Note that T can take the value infinity. Stopping times are also known as *optional times*.

Given a stochastic process X, we define $X_T(\omega)$ to be equal to $X(T(\omega), \omega)$; that is, for each ω we evaluate $t = T(\omega)$ and then look at $X(\cdot, \omega)$ at this time.

Proposition 3.8 *Suppose $\mathcal{F}_t$ satisfies the usual conditions. Then*
(1) T is a stopping time if and only if $(T \leq t) \in \mathcal{F}_t$ for all t.
(2) If $T = t$, a.s., then T is a stopping time.
(3) If S and T are stopping times, then so are $S \vee T$ and $S \wedge T$.
(4) If T_n, $n = 1, 2, \ldots$, are stopping times with $T_1 \leq T_2 \leq \cdots$, then so is $\sup_n T_n$.
(5) If T_n, $n = 1, 2, \ldots$, are stopping times with $T_1 \geq T_2 \geq \cdots$, then so is $\inf_n T_n$.
(6) If $s \geq 0$ and S is a stopping time, then so is $S + s$.

Proof We will just prove part of (1), leaving the rest as Exercise 3.4. Note $(T \leq t) = \cap_{n \geq N}(T < t + 1/n) \in \mathcal{F}_{t+1/N}$ for each N. Thus $(T \leq t) \in \cap_N \mathcal{F}_{t+1/N} \subset \mathcal{F}_{t+} = \mathcal{F}_t$. $\qquad\square$

For a Borel measurable set A, let

$$T_A = \inf\{t > 0 : X_t \in A\}. \tag{3.1}$$

Proposition 3.9 *Suppose $\mathcal{F}_t$ satisfies the usual conditions and X_t has continuous paths.*
(1) If A is open, then T_A is a stopping time.
(2) If A is closed, then T_A is a stopping time.

Proof (1) $(T_A < t) = \cap_{q \in \mathbb{Q}_+, q < t}(X_q \in A)$, where $\mathbb{Q}_+$ denotes the set of non-negative rationals. Since $(X_q \in A) \in \mathcal{F}_q \subset \mathcal{F}_t$, then $(T_A < t) \in \mathcal{F}_t$.

(2) Let $A_n = \{x : \text{dist}(x, A) < 1/n\}$, the set of points within a distance $1/n$ from A. Each A_n is open and thus by (1), T_{A_n} is a stopping time. Moreover, the A_n decrease, so the T_{A_n} increase. Let $T = \sup_n T_{A_n}$, a stopping time by Proposition 3.8(4). Since $A \subset A_n$, then $T_A \geq T_{A_n}$, so $T_A \geq T$. Because X has continuous paths, on $(T < \infty)$, $X_T = \lim_n X_{T_{A_n}}$. If $n \geq m$, then $X(T_{A_n}) \in \overline{A_n} \subset \overline{A_m}$. Therefore $X_T \in \overline{A_m}$ for each m. Since $A = \cap_m \overline{A_m}$, then $X_T \in A$. Therefore $T_A \leq T$, and hence $T = T_A$. $\square$

It is true that under the hypotheses of the preceding proposition, T_A is a stopping time for every Borel set A, but that is much harder to prove; see Section 16.2.

It is often useful to be able to approximate stopping times from the right. If T is a finite stopping time, that is, $T < \infty$, a.s., define

$$T_n(\omega) = (k+1)/2^n \qquad \text{if } k/2^n \leq T(\omega) < (k+1)/2^n. \tag{3.2}$$

Exercise 3.5 asks you to prove that the T_n are stopping times decreasing to T.
  Define

$$\mathcal{F}_T = \{A \in \mathcal{F} : \text{ for each } t > 0, \ A \cap (T \leq t) \in \mathcal{F}_t\}. \tag{3.3}$$

This definition of $\mathcal{F}_T$, which is supposed to be the collection of events that are "known" by time T, is not very intuitive. But it turns out that this definition works well in applications. Exercise 3.6 gives an equivalent definition that is more appealing but not as useful.

Proposition 3.10 *Suppose $\{\mathcal{F}_t\}$ is a filtration satisfying the usual conditions.*
(1) $\mathcal{F}_T$ is a σ-field.
(2) If $S \leq T$, then $\mathcal{F}_S \subset \mathcal{F}_T$.
(3) If $\mathcal{F}_{T+} = \cap_{\varepsilon > 0} \mathcal{F}_{T+\varepsilon}$, then $\mathcal{F}_{T+} = \mathcal{F}_T$.
(4) If X_t has right-continuous paths, then X_T is $\mathcal{F}_T$ measurable.

Proof If $A \in \mathcal{F}_T$, then $A^c \cap (T \leq t) = (T \leq t) \setminus [A \cap (T \leq t)] \in \mathcal{F}_t$, so $A^c \in \mathcal{F}_T$. The rest of the proof of (1) is easy.

Suppose $A \in \mathcal{F}_S$ and $S \leq T$. Then $A \cap (T \leq t) = [A \cap (S \leq t)] \cap (T \leq t)$. We have $A \cap (S \leq t) \in \mathcal{F}_t$ because $A \in \mathcal{F}_S$, while $(T \leq t) \in \mathcal{F}_t$ because T is a stopping time. Therefore $A \cap (T \leq t) \in \mathcal{F}_t$, which proves (2).

For (3), if $A \in \mathcal{F}_{T+}$, then $A \in \mathcal{F}_{T+\varepsilon}$ for every ε, and so $A \cap (T + \varepsilon \leq t) \in \mathcal{F}_t$ for all t. Hence $A \cap (T \leq t - \varepsilon) \in \mathcal{F}_t$ for all t, or equivalently $A \cap (T \leq t) \in \mathcal{F}_{t+\varepsilon}$ for all t. This is true for all ε, so $A \cap (T \leq t) \in \mathcal{F}_{t+} = \mathcal{F}_t$. This says $A \in \mathcal{F}_T$.

(4) Define T_n by (3.2). Note

$$(X_{T_n} \in B) \cap (T_n = k/2^n) = (X_{k/2^n} \in B) \cap (T_n = k/2^n) \in \mathcal{F}_{k/2^n}.$$

Since T_n only takes values in $\{k/2^n : k \geq 0\}$, we conclude $(X_{T_n} \in B) \cap (T_n \leq t) \in \mathcal{F}_t$ and so $(X_{T_n} \in B) \in \mathcal{F}_{T_n} \subset \mathcal{F}_{T+1/2^n}$.

Hence X_{T_n} is $\mathcal{F}_{T+1/2^n}$ measurable. If $n \geq m$, then X_{T_n} is measurable with respect to $\mathcal{F}_{T+1/2^n} \subset \mathcal{F}_{T+1/2^m}$. Since $X_{T_n} \to X_T$, then X_T is $\mathcal{F}_{T+1/2^m}$ measurable for each m. Therefore X_T is measurable with respect to $\mathcal{F}_{T+} = \mathcal{F}_T$. $\qquad\square$

3.4 The optional stopping theorem

We will need Doob's optional stopping theorem for continuous-time martingales. An example to keep in mind is $M_t = W_{t \wedge t_0}$, where W is a Brownian motion and t_0 is some fixed time. Exercise 3.12 is a version of the optional stopping time with slightly weaker hypotheses that is often useful.

Theorem 3.11 *Let $\{\mathcal{F}_t\}$ be a filtration satisfying the usual conditions. If M_t is a martingale or non-negative submartingale whose paths are right continuous, $\sup_{t \geq 0} \mathbb{E}\, M_t^2 < \infty$, and T is a finite stopping time, then $\mathbb{E}\, M_T \geq \mathbb{E}\, M_0$.*

Proof We do the submartingale case, the martingale case being very similar. By Doob's inequality (Theorem 3.6(1)),

$$\mathbb{E}\,[\sup_{s \leq t} M_s^2] \leq 4\mathbb{E}\, M_t^2.$$

Letting $t \to \infty$, we have $\mathbb{E}\,[\sup_{t \geq 0} M_t^2] < \infty$ by Fatou's lemma.

Let us first suppose that $T < K$, a.s., for some real number K. Define T_n by (3.2). Let $N_k^{(n)} = M_{k/2^n}$, $\mathcal{G}_k^{(n)} = \mathcal{F}_{k/2^n}$, and $S_n = 2^n T_n$. By Doob's optional stopping theorem applied to the submartingale $N_k^{(n)}$, we have

$$\mathbb{E}\, M_0 = \mathbb{E}\, N_0^{(n)} \leq \mathbb{E}\, N_{S_n}^{(n)} = \mathbb{E}\, M_{T_n}.$$

Since M is right continuous, $M_{T_n} \to M_T$, a.s. The random variables $|M_{T_n}|$ are bounded by $1 + \sup_{t \geq 0} M_t^2$, so by dominated convergence, $\mathbb{E}\, M_{T_n} \to \mathbb{E}\, M_T$.

We apply the above to the stopping time $T \wedge K$ to get $\mathbb{E}\, M_{T \wedge K} \geq \mathbb{E}\, M_0$. The random variables $M_{T \wedge K}$ are bounded by $1 + \sup_{t \geq 0} M_t^2$, so by dominated convergence, we get $\mathbb{E}\, M_T \geq \mathbb{E}\, M_0$ when we let $K \to \infty$. $\qquad\square$

3.5 Convergence and regularity

We present the continuous-time version of Doob's martingale convergence theorem. We will see that not only do we get limits as $t \to \infty$, but also a regularity result.

Let $\mathcal{D}_n = \{k/2^n : k \geq 0\}$, $\mathcal{D} = \cup_n \mathcal{D}_n$.

Theorem 3.12 *Let $\{M_t : t \in \mathcal{D}\}$ be either a martingale, a submartingale, or a supermartingale with respect to $\{\mathcal{F}_t : t \in \mathcal{D}\}$ and suppose $\sup_{t \in \mathcal{D}} \mathbb{E}\,|M_t| < \infty$. Then*

(1) $\lim_{t \to \infty} M_t$ exists, a.s.

(2) With probability one M_t has left and right limits along $\mathcal{D}$.

The second conclusion says that except for a null set, if $t_0 \in [0, \infty)$, then both $\lim_{t \in \mathcal{D}, t \uparrow t_0} M_t$ and $\lim_{t \in \mathcal{D}, t \downarrow t_0} M_t$ exist and are finite. The null set does not depend on t_0.

Proof Martingales are also submartingales and if M_t is a supermartingale, then $-M_t$ is a submartingale, so we may without loss of generality restrict our attention to submartingales.

By Doob's inequality (Theorem 3.6(1)),

$$\mathbb{P}(\sup_{t \in \mathcal{D}_n, t \leq n} |M_t| > \lambda) \leq \frac{1}{\lambda} \mathbb{E}\, |M_n|.$$

Letting $n \to \infty$ and using Fatou's lemma,

$$\mathbb{P}(\sup_{t \in \mathcal{D}} |M_t| > \lambda) \leq \frac{1}{\lambda} \sup_t \mathbb{E}\, |M_t|.$$

This is true for all λ, so with probability one, $\{|M_t| : t \in \mathcal{D}\}$ is a bounded set.

Therefore the only way either (1) or (2) can fail is that if for some pair of rationals $a < b$ the number of upcrossings of $[a, b]$ by $\{M_t : t \in \mathcal{D}\}$ is infinite. Recall that we define upcrossings as follows.

Given an interval $[a, b]$ and a submartingale M, if $S_1 = \inf\{t : M_t \leq a\}$, $T_i = \inf\{t > S_i : M_t \geq b\}$, and $S_{i+1} = \inf\{t > T_i : M_t \leq a\}$, then the number of upcrossings up to time u is $\sup\{k : T_k \leq u\}$.

Doob's upcrossing lemma (Theorem A.34) tells us that if V_n is the number of upcrossings by $\{M_t : t \in \mathcal{D}_n \cap [0, n]\}$, then

$$\mathbb{E}\, V_n \leq \frac{\mathbb{E}\, |M_n|}{b - a}.$$

Letting $n \to \infty$ and using Fatou's lemma, the number of upcrossings of $[a, b]$ by $\{M_t : t \in \mathcal{D}\}$ has finite expectation, hence is finite, a.s. If $N_{a,b}$ is the null set where the number of upcrossings of $[a, b]$ by $\{M_t : t \in \mathcal{D}\}$ is infinite and $N = \cup_{a<b, a, b \in \mathbb{Q}_+} N_{a,b}$, where $\mathbb{Q}_+$ is the collection of non-negative rationals, then $\mathbb{P}(N) = 0$. If $\omega \notin N$, then (1) and (2) hold.   □

As a corollary we have

Corollary 3.13 *Let $\{\mathcal{F}_t\}$ be a filtration satisfying the usual conditions, and let M_t be a martingale with respect to $\{\mathcal{F}_t\}$. Then M has a version that is also a martingale and that in addition has paths that are right continuous with left limits.*

Proof Let $\mathcal{D}$ be as in the above proof. For each integer $N \geq 1$, $\mathbb{E}\, |M_t| \leq \mathbb{E}\, |M_N| < \infty$ for $t \leq N$ since $|M_t|$ is a submartingale by the conditional expectation form of Jensen's inequality (Proposition A.21). Therefore $M_{t \wedge N}$ has left and right limits when taking limits along $t \in \mathcal{D}$. Since N is arbitrary, M_t has left and right limits when taking limits along $t \in \mathcal{D}$, except for a set of ω's that form a null set. Let

$$\widetilde{M}_t = \lim_{u \in \mathcal{D}, u > t, u \to t} M_u.$$

It is clear that $\widetilde{M}$ has paths that are right continuous with left limits. Since $\mathcal{F}_{t+} = \mathcal{F}_t$ and $\widetilde{M}_t$ is $\mathcal{F}_{t+}$ measurable, then $\widetilde{M}_t$ is $\mathcal{F}_t$ measurable.

Let N be fixed. We will show $\{M_t; t \leq N\}$ is a uniformly integrable family of random variables; see Section A.4. Let $\varepsilon > 0$. Since M_N is integrable, there exists δ such that if $\mathbb{P}(A) < \delta$, then $\mathbb{E}\,[|M_N|; A] < \varepsilon$. If L is large enough, $\mathbb{P}(|M_t| > L) \leq \mathbb{E}\, |M_t|/L \leq \mathbb{E}\, |M_N|/L < \delta$. Then

$$\mathbb{E}\,[|M_t|; |M_t| > L] \leq \mathbb{E}\,[|M_N|; |M_t| > L] < \varepsilon,$$

since $|M_t|$ is a submartingale and $(|M_t| > L) \in \mathcal{F}_t$. Uniform integrability is proved.

Now let $t < N$. If $B \in \mathcal{F}_t$,

$$\mathbb{E}\left[\widetilde{M}_t; B\right] = \lim_{u \in \mathcal{D}, u > t, u \to t} \mathbb{E}\left[M_u; B\right] = \mathbb{E}\left[M_t; B\right].$$

Here we used the Vitali convergence theorem (Theorem A.19) and the fact that M_t is a martingale. Since $\widetilde{M}_t$ is $\mathcal{F}_t$ measurable, this proves that $\widetilde{M}_t = M_t$, a.s. Since N was arbitrary, we have this for all t. We thus have found a version of M that has paths that are right continuous with left limits. That $\widetilde{M}_t$ is a martingale is easy. $\qquad\square$

The following technical result will be used several times in this book. A function f is increasing if $s < t$ implies $f(s) \leq f(t)$. A process A_t has increasing paths if the function $t \to A_t(\omega)$ is increasing for almost every ω.

Proposition 3.14 *Suppose $\{\mathcal{F}_t\}$ is a filtration satisfying the usual conditions and suppose A_t is an adapted process with paths that are increasing, are right continuous with left limits, and $A_\infty = \lim_{t \to \infty} A_t$ exists, a.s. Suppose X is a non-negative integrable random variable, and M_t is a version of the martingale $\mathbb{E}[X \mid \mathcal{F}_t]$ which has paths that are right continuous with left limits. Suppose $\mathbb{E}[XA_\infty] < \infty$. Then*

$$\mathbb{E} \int_0^\infty X\, dA_s = \mathbb{E} \int_0^\infty M_s\, dA_s. \tag{3.4}$$

Proof First suppose X and A are bounded. Let $n > 1$ and write $\mathbb{E} \int_0^\infty X\, dA_s$ as

$$\sum_{k=1}^\infty \mathbb{E}\left[X\left(A_{k/2^n} - A_{(k-1)/2^n}\right)\right].$$

Conditioning the kth summand on $\mathcal{F}_{k/2^n}$, this is equal to

$$\mathbb{E}\left[\sum_{k=1}^\infty \mathbb{E}[X \mid \mathcal{F}_{k/2^n}]\left(A_{k/2^n} - A_{(k-1)/2^n}\right)\right].$$

Given s and n, define s_n to be that value of $k/2^n$ such that $(k-1)/2^n < s \leq k/2^n$. We then have

$$\mathbb{E} \int_0^\infty X\, dA_s = \mathbb{E} \int_0^\infty M_{s_n}\, dA_s. \tag{3.5}$$

For any value of s, $s_n \downarrow s$ as $n \to \infty$, and since M has right-continuous paths, $M_{s_n} \to M_s$. Since X is bounded, so is M. By dominated convergence, the right-hand side of (3.5) converges to

$$\mathbb{E} \int_0^\infty M_s\, dA_s.$$

This completes the proof when X and A are bounded. We apply this to $X \wedge N$ and $A \wedge N$, let $N \to \infty$, and use monotone convergence for the general case. $\qquad\square$

The only reason we assume X is non-negative is so that the integrals make sense. The equation (3.4) can be rewritten as

$$\mathbb{E} \int_0^\infty X\, dA_s = \mathbb{E} \int_0^\infty \mathbb{E}[X \mid \mathcal{F}_s]\, dA_s. \tag{3.6}$$

We also have

$$\mathbb{E} \int_0^t X \, dA_s = \mathbb{E} \int_0^t \mathbb{E}[X \mid \mathcal{F}_s] \, dA_s \qquad (3.7)$$

for each t. This follows either by following the above proof or by applying Proposition 3.14 to $A_{s \wedge t}$.

3.6 Some applications of martingales

The following estimates are very useful.

Proposition 3.15 *If W_t is a Brownian motion, then*

$$\mathbb{P}(\sup_{s \le t} W_s \ge \lambda) \le e^{-\lambda^2/2t}, \qquad \lambda > 0, \qquad (3.8)$$

and

$$\mathbb{P}(\sup_{s \le t} |W_s| \ge \lambda) \le 2e^{-\lambda^2/2t}, \qquad \lambda > 0. \qquad (3.9)$$

Proof For any a the process $\{e^{aW_t}\}$ is a submartingale. To see this, since $x \to e^{ax}$ is convex, the conditional expectation form of Jensen's inequality (Proposition A.21) implies

$$\mathbb{E}[e^{aW_t} \mid \mathcal{F}_s] \ge e^{a\mathbb{E}[W_t \mid \mathcal{F}_s]} = e^{aW_s}.$$

By Doob's inequality (Theorem 3.6(1)),

$$\mathbb{P}(\sup_{s \le t} W_s \ge \lambda) = \mathbb{P}(\sup_{s \le t} e^{aW_s} \ge e^{a\lambda}) \le \frac{\mathbb{E} e^{aW_t}}{e^{a\lambda}}. \qquad (3.10)$$

Since $\mathbb{E} e^{aY} = e^{a^2 \operatorname{Var} Y/2}$ if Y is Gaussian with mean 0 by (A.6), it follows that the right side of (3.10) is bounded by $e^{-a\lambda} e^{a^2 t/2}$. If we now set $a = \lambda/t$, we obtain (3.8). Inequality (3.9) follows by applying (3.8) to W and to $-W$ and adding. $\qquad \square$

Let us use martingales to calculate some probabilities. Let us suppose $a, b > 0$ and set $T = \inf\{t > 0 : W_t = -a \text{ or } W_t = b\}$, the first time Brownian motion exits the interval $[-a, b]$. By Proposition 3.9, T is a stopping time.

We have

Proposition 3.16 *Let W be a Brownian motion, let $T = \inf\{t > 0 : W_t \notin [-a, b]\}$, and let $a, b > 0$. Then*

$$\mathbb{P}(W_T = -a) = \frac{b}{a+b}, \qquad \mathbb{P}(W_T = b) = \frac{a}{a+b}, \qquad (3.11)$$

and

$$\mathbb{E} T = ab. \qquad (3.12)$$

Proof Since $W_t^2 - t$ is a martingale with $W_0 = 0$, it is easy to check that for each u, $W_{t \wedge u}^2 - (t \wedge u)$ is also a martingale. Applying Theorem 3.11, we see that $\mathbb{E} W_{u \wedge T}^2 = \mathbb{E}[u \wedge T]$. As $u \to \infty$, the right-hand side tends to $\mathbb{E} T$ by monotone convergence. $|W_{u \wedge T}|^2$ is bounded by $(a + b)^2$, so by dominated convergence the left-hand side tends to $\mathbb{E}\,W_T^2 \leq (a + b)^2$ as $u \to \infty$. Therefore

$$\mathbb{E}\,T = \mathbb{E}\,W_T^2. \tag{3.13}$$

In particular, $\mathbb{E}\,T < \infty$, so we know $T < \infty$, a.s.

We use that T is finite, a.s., to conclude that $\mathbb{P}(W_T \in \{-a, b\}) = 1$, or

$$1 = \mathbb{P}(W_T = -a) + \mathbb{P}(W_T = b). \tag{3.14}$$

Since W_t is a martingale, then so is $W_{t \wedge u}$ for each u, and therefore $\mathbb{E}\,W_{u \wedge T} = 0$. Letting $u \to \infty$ and using dominated convergence (noting $|W_{u \wedge T}|$ is bounded by $a + b$), we have $\mathbb{E}\,W_T = 0$, or

$$0 = (-a)\mathbb{P}(W_T = -a) + b\mathbb{P}(W_T = b). \tag{3.15}$$

We get (3.11) by solving (3.14) and (3.15) for the unknowns $\mathbb{P}(W_T = -a)$ and $\mathbb{P}(W_T = b)$.

We get (3.12) by (3.13), writing

$$\mathbb{E}\,T = \mathbb{E}\,W_T^2 = (-a)^2 \mathbb{P}(W_T = -a) + b^2 \mathbb{P}(W_T = b),$$

and substituting the values from (3.11). □

In proving Proposition 3.16, we used the fact that $W_{t \wedge T}$ is a martingale and $\mathbb{P}(T < \infty) = 1$. The same proof shows

Corollary 3.17 *Suppose M_t is a martingale with continuous paths and with $M_0 = 0$, a.s., $T = \inf\{t \geq 0 : M_t \notin [-a, b]\}$, and $T < \infty$, a.s. Then*

$$\mathbb{P}(M_T = -a) = \frac{b}{a + b}, \qquad \mathbb{P}(M_T = b) = \frac{a}{a + b}.$$

We can also use martingales to get more subtle results. Suppose $r > 0$. Since $e^{rW_t - r^2 t/2}$ is a martingale, as above

$$\mathbb{E}\,e^{rW_{T \wedge t} - r^2(T \wedge t)/2} = 1.$$

The exponent is bounded by rb if $r > 0$, so we can let $t \to \infty$ and use dominated convergence to get

$$\mathbb{E}\,e^{rW_T - r^2 T/2} = 1.$$

This can be written as

$$e^{-ra}\mathbb{E}\,[e^{-r^2 T/2}; W_T = -a] + e^{rb}\mathbb{E}\,[e^{-r^2 T/2}; W_T = b] = 1.$$

Since $e^{-rW_t - r^2 t/2}$ is also a martingale, similar reasoning gives us

$$e^{ra}\mathbb{E}\,[e^{-r^2 T/2}; W_T = -a] + e^{-rb}\mathbb{E}\,[e^{-r^2 T/2}; W_T = b] = 1.$$

We can solve those two equations to obtain

$$\mathbb{E}\left[e^{-r^2 T/2}; W_T = -a\right] = \frac{e^{rb} - e^{-rb}}{e^{r(a+b)} - e^{-r(a+b)}} \tag{3.16}$$

and

$$\mathbb{E}\left[e^{-r^2 T/2}; W_T = b\right] = \frac{e^{ra} - e^{-ra}}{e^{r(a+b)} - e^{-r(a+b)}}. \tag{3.17}$$

The left-hand sides of (3.16) and (3.17) are the Laplace transforms of the quantities $\mathbb{P}(T \in dt; W_T = -a)/dt$ and $\mathbb{P}(T \in dt; W_T = b)/dt$, respectively, and finding the inverse Laplace transforms of the right-hand sides of (3.16) and (3.17) gives us formulas for $\mathbb{P}(T \in dt; W_T = -a)/dt$ and $\mathbb{P}(T \in dt; W_T = b)/dt$. If we add the two formulas, we get an expression for $\mathbb{P}(T \in dt)/dt$, and integrating over t from 0 to t_0 gives an expression for $\mathbb{P}(T \leq t_0)$.

We sketch how to invert the Laplace transform and leave the detailed calculations and justification for inverting a Laplace transform term by term to the interested reader. See also Karatzas and Shreve (1991), Section 2.8. The right-hand side of (3.16) is equal to

$$\frac{e^{-ra} - e^{-ra-2rb}}{1 - e^{-2r(a+b)}}.$$

Since $e^{-2r(a+b)} < 1$, we can use

$$(1 - x)^{-1} = \sum_{n=0}^{\infty} x^n$$

to expand the denominator as a power series; if we set $\lambda = r^2/2$, then

$$E\left[e^{-\lambda T}; W_T = -a\right] \tag{3.18}$$

$$= \sum_{n=0}^{\infty} \left(e^{-(2n+1)\sqrt{2\lambda}a - 2n\sqrt{2\lambda}b} - e^{-(2n+1)\sqrt{2\lambda}a - (2n+2)\sqrt{2\lambda}b}\right).$$

We then use the fact that the Laplace transform of

$$\frac{k}{2\sqrt{\pi t^3}} e^{-k^2/4t}$$

is $e^{-k\sqrt{\lambda}}$ to find the inverse Laplace transform of the right-hand side of (3.18) by inverting term by term.

Similarly (see Exercises 3.15 and 3.16), if $b > 0$, W is a Brownian motion, and $S = \inf\{t > 0 : W_t = b\}$, then $\mathbb{E}\, e^{-\lambda S} = e^{-\sqrt{2\lambda}b}$. Inverting the Laplace transform,

$$\mathbb{P}(S \in dt) = \frac{b}{\sqrt{2\pi t^3}} e^{-b^2/2t}, \qquad t \geq 0. \tag{3.19}$$

Exercises

3.1   If W is a Brownian motion, show that

$$W_t^3 - 3\int_0^t W_s\, ds$$

is a martingale.

3.2   Suppose $\{\mathcal{F}_t\}$ is a filtration satisfying the usual conditions. Show that if M_t is a submartingale and $\mathbb{E}\, M_t = \mathbb{E}\, M_0$ for all t, then M is a martingale.

3.3   Let X be a submartingale. Show that $\sup_{t \geq 0} \mathbb{E}\, |X_t| < \infty$ if and only if $\sup_{t \geq 0} \mathbb{E}\, X_t^+ < \infty$.

3.4   Prove all parts of Proposition 3.8.

3.5 If T_n is defined by (3.2), show T_n is a stopping time for each n and $T_n \downarrow T$.

3.6 This exercise gives an alternate definition of $\mathcal{F}_T$ which is more appealing, but not as useful. Suppose that $\{\mathcal{F}_t\}$ satisfies the usual conditions. Show that $\mathcal{F}_T$ is equal to the σ-field generated by the collection of random variables Y_T such that Y is a bounded process with paths that are right continuous with left limits and Y is adapted to the filtration $\{\mathcal{F}_t\}$.

3.7 Suppose $\{\mathcal{F}_t\}$ is a filtration satisfying the usual conditions. Show that if T is a stopping time, then T is $\mathcal{F}_T$ measurable.

3.8 Suppose $\{\mathcal{F}_t\}$ is a filtration satisfying the usual conditions and T is a stopping time. Show that if S is a $\mathcal{F}_T$ measurable random variable with $S \geq T$, then S is a stopping time.

3.9 This exercise demonstrates that the conclusion of Corollary 3.13 cannot be extended to submartingales. Find a filtration $\{\mathcal{F}_t\}$ satisfying the usual conditions and a submartingale X with respect to $\{\mathcal{F}_t\}$ such that X does not have a version with paths that are right continuous with left limits.

3.10 Suppose $\{\mathcal{F}_t\}$ is a filtration satisfying the usual conditions. Show that if S and T are stopping times and X is a bounded $\mathcal{F}_\infty$ measurable random variable, then

$$\mathbb{E}\left[\mathbb{E}\left[X \mid \mathcal{F}_S\right] \mid \mathcal{F}_T\right] = \mathbb{E}\left[X \mid \mathcal{F}_{S \wedge T}\right].$$

Hint: Let $Y_t = \mathbb{E}[X \mid \mathcal{F}_t]$ and $Z_t = Y_{t \wedge S}$. Show the left-hand side is equal to $Y_{S \wedge T}$.

3.11 A martingale or submartingale M_t is uniformly integrable if the family $\{M_t : t \geq 0\}$ is a uniformly integrable family of random variables. Show that if M_t is a uniformly integrable martingale with paths that are right continuous with left limits, then $\{M_T; T$ a finite stopping time$\}$ is a uniformly integrable family of random variables. Show this also holds if M_t is a non-negative submartingale with paths that are right continuous with left limits.

3.12 This exercise weakens the conditions on the optional stopping theorem. Show that if M_t is a uniformly integrable martingale that is right continuous with left limits and T is a finite stopping time, then $\mathbb{E}\, M_T = \mathbb{E}\, M_0$.

3.13 Let W be a Brownian motion and let T be a stopping time with $\mathbb{E}\, T < \infty$. Prove that $\mathbb{E}\, W_T = 0$ and $\mathbb{E}\, W_T^2 = \mathbb{E}\, T$. This is not an easy application of the optional stopping theorem because we do not know that $W_{t \wedge T}$ is necessarily a uniformly integrable martingale.

3.14 Suppose that $(W_t^1, \ldots, W_t^d)$ is a d-dimensional Brownian motion. Show that if $i \neq j$, then $W_t^i W_t^j$ is a martingale.

3.15 Let W_t be a Brownian motion, $b > 0$, and $T = \inf\{t > 0 : W_t = b\}$. Show $T < \infty$, a.s. Show $\mathbb{E}\, T = \infty$.
   Hint: Take a limit in (3.11).

3.16 Suppose W is a Brownian motion and $b > 0$. If $S = \inf\{t > 0 : W_t = b\}$, show that the Laplace transform of the density of S is given by

$$\mathbb{E}\, e^{-\lambda S} = e^{-\sqrt{2\lambda} b}.$$

3.17 Let W_t be a Brownian motion. Show that if $\alpha > 1/2$, then

$$\lim_{t \to \infty} \frac{W_t}{t^\alpha} = 0, \qquad \text{a.s.}$$

Hint: Let $\alpha_0 \in (1/2, \alpha)$, estimate

$$\mathbb{P}(\sup_{2^n \leq s \leq 2^{n+1}} |W_s| \geq (2^n)^{\alpha_0})$$

using (3.9), and then use the Borel–Cantelli lemma.

3.18 Let W_t be a one-dimensional Brownian motion and $\alpha \in (0, 1/2]$. Prove that

$$\limsup_{t \to \infty} \frac{|W_t|}{t^\alpha} > 0, \qquad \text{a.s.}$$

3.19 If W is a Brownian motion and b is a constant, then the process $X_t = W_t + bt$ is a *Brownian motion with drift*. Prove that if $b > 0$, then

$$\lim_{t \to \infty} X_t = \infty, \qquad \text{a.s.}$$

4

Markov properties of Brownian motion

In later chapters we will discuss extensively the Markov property and strong Markov property. The Brownian motion case is much simpler, and we do that now.

4.1 Markov properties

Let us begin with the Markov property.

Theorem 4.1 *Let* $\{\mathcal{F}_t\}$ *be a filtration, not necessarily satisfying the usual conditions, and let* W *be a Brownian motion with respect to* $\{\mathcal{F}_t\}$. *If* u *is a fixed time, then* $Y_t = W_{t+u} - W_u$ *is a Brownian motion independent of* $\mathcal{F}_u$.

Proof Let $\mathcal{G}_t = \mathcal{F}_{t+u}$. It is clear that Y has continuous paths, is zero at time 0, and is adapted to $\{\mathcal{G}_t\}$. Since $Y_t - Y_s = W_{t+u} - W_{s+u}$, then $Y_t - Y_s$ is a mean zero normal random variable with variance $(t + u) - (s + u) = t - s$ that is independent of $\mathcal{F}_{s+u} = \mathcal{G}_s$. $\qquad\square$

The strong Markov property is the Markov property extended by replacing fixed times u by finite stopping times.

Theorem 4.2 *Let* $\{\mathcal{F}_t\}$ *be a filtration, not necessarily satisfying the usual conditions, and let* W *be a Brownian motion adapted to* $\{\mathcal{F}_t\}$. *If* T *is a finite stopping time, then* $Y_t = W_{T+t} - W_T$ *is a Brownian motion independent of* $\mathcal{F}_T$.

Proof We will first show that whenever $m \geq 1$, $t_1 < \cdots < t_m$, f is a bounded continuous function on $\mathbb{R}^m$, and $A \in \mathcal{F}_T$, then

$$\mathbb{E}\left[f(Y_{t_1}, \ldots, Y_{t_m}); A\right] = \mathbb{E}\left[f(W_{t_1}, \ldots, W_{t_m})\right] \mathbb{P}(A). \tag{4.1}$$

Once we have done this, we will then show how (4.1) implies our theorem.

To prove (4.1), define T_n by (3.2). We have

$$\mathbb{E}\left[f(W_{T_n+t_1} - W_{T_n}, \ldots, W_{T_n+t_m} - W_{T_n}); A\right] \tag{4.2}$$

$$= \sum_{k=1}^{\infty} \mathbb{E}\left[f(W_{T_n+t_1} - W_{T_n}, \ldots, W_{T_n+t_m} - W_{T_n}); A, T_n = k/2^n\right]$$

$$= \sum_{k=1}^{\infty} \mathbb{E}\left[f(W_{t_1+k/2^n} - W_{k/2^n}, \ldots, W_{t_m+k/2^n} - W_{k/2^n}); A, T_n = k/2^n\right].$$

Following the usual practice in probability that "," means "and," we use the notation "$\mathbb{E}\left[\cdots; A, T_n = k/2^n\right]$" as an abbreviation for "$\mathbb{E}\left[\cdots; A \cap (T_n = k/2^n)\right]$." Since $A \in \mathcal{F}_T$, then $A \cap (T_n = k/2^n) = A \cap ((T < k/2^n) \setminus (T < (k-1)/2^n)) \in \mathcal{F}_{k/2^n}$. We use the independent increments property of Brownian motion and the fact that $W_t - W_s$ has the same law as W_{t-s} to see that the sum in the last line of (4.2) is equal to

$$\sum_{k=1}^{\infty} \mathbb{E}\,[f(W_{t_1+k/2^n} - W_{k/2^n}, \ldots, W_{t_m+k/2^n} - W_{k/2^n})]\,\mathbb{P}(A, T_n = k/2^n)$$

$$= \sum_{k=1}^{\infty} \mathbb{E}\,[f(W_{t_1}, \ldots, W_{t_m})]\,\mathbb{P}(A, T_n = k/2^n)$$

$$= \mathbb{E}\,[f(W_{t_1}, \ldots, W_{t_m})]\,\mathbb{P}(A),$$

which is the right-hand side of (4.1). Thus

$$\mathbb{E}\,[f(W_{T_n+t_1} - W_{T_n}, \ldots W_{T_n+t_m} - W_{T_n}); A] = \mathbb{E}\,[f(W_{t_1}, \ldots W_{t_m})]\,\mathbb{P}(A). \qquad (4.3)$$

Now let $n \to \infty$. By the right continuity of the paths of W, the boundedness and continuity of f, and the dominated convergence theorem, the left-hand side of (4.3) converges to the left-hand side of (4.1).

If we take $A = \Omega$ in (4.1), we obtain

$$\mathbb{E}\,[f(Y_{t_1}, \ldots, Y_{t_m})] = \mathbb{E}\,[f(W_{t_1}, \ldots, W_{t_m})]$$

whenever $m \geq 1$, $t_1, \ldots, t_m \in [0, \infty)$, and f is a bounded continuous function on $\mathbb{R}^m$. This implies that the finite-dimensional distributions of Y and W are the same. Since Y has continuous paths, Y is a Brownian motion.

Next take $A \in \mathcal{F}_T$. By using a limit argument, (4.1) holds whenever f is the indicator of a Borel subset B of $\mathbb{R}^d$, or in other words,

$$\mathbb{P}(Y \in B, A) = \mathbb{P}(Y \in B)\mathbb{P}(A) \qquad (4.4)$$

whenever B is a cylindrical set. Let $\mathcal{M}$ be the collection of all Borel subsets B of $C[0, \infty)$ for which (4.4) holds. Let $\mathcal{C}$ be the collection of all cylindrical subsets of $C[0, \infty)$. Then we observe that $\mathcal{M}$ is a monotone class containing $\mathcal{C}$ and $\mathcal{C}$ is an algebra of subsets of $C[0, \infty)$ generating the Borel σ-field of $C[0, \infty)$. By the monotone class theorem (Theorem B.2), $\mathcal{M}$ is equal to the Borel σ-field on $C[0, \infty)$, and since (4.4) holds for all sets $B \in \mathcal{M}$, this establishes the independence of Y and $\mathcal{F}_T$. $\qquad \square$

In the future, we will not put in the details for the arguments using the monotone class theorem.

Observe that what is needed for the above proof to work is not that W be a Brownian motion, but that the process W have right continuous paths and that $W_t - W_s$ be independent of $\mathcal{F}_s$ and have the same distribution as W_{t-s}. We therefore have the following corollary.

Corollary 4.3 *Let $\{\mathcal{F}_t\}$ be a filtration, not necessarily satisfying the usual conditions, and let X be a process adapted to $\{\mathcal{F}_t\}$. Suppose X has paths that are right continuous with left limits and suppose $X_t - X_s$ is independent of $\mathcal{F}_s$ and has the same law as X_{t-s} whenever $s < t$. If T is a finite stopping time, then $Y_t = X_{T+t} - X_T$ is a process that is independent of $\mathcal{F}_T$ and X and Y have the same law.*

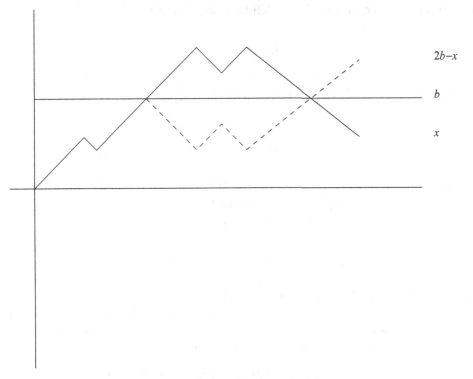

Figure 4.1 The reflection principle.

4.2 Applications

The first application is known as the reflection principle and allows us to get control of the maximum of a Brownian motion. The idea is the following. Suppose that W_t is a Brownian motion and for some path, the Brownian motion goes above a level b before time t but that at time t the value of W_t is less than x, where $x < b$. We could take the graph of this path and reflect it across the horizontal line at level b the first time the path crosses the level b (Figure 4.1). This will give us a new path that ends up above $2b-x$. Thus there is a one-to-one correspondence between paths where the maximum up to time t is above b and W_t is below x and the paths where W_t is above $2b - x$.

More precisely, we have the following.

Theorem 4.4 *Let W_t be a Brownian motion, $b > 0$, $T = \inf\{t : W_t \geq b\}$, and $x < b$. Then*

$$\mathbb{P}(\sup_{s \leq t} W_s \geq b, W_t < x) = \mathbb{P}(W_t > 2b - x). \tag{4.5}$$

Proof Let T_n be defined by (3.2). We first show that

$$\mathbb{P}(T_n \leq t, W_t - W_{T_n} < x - b) = \mathbb{P}(T_n \leq t, W_t - W_{T_n} > b - x). \tag{4.6}$$

Writing $[x]$ for the integer part of x, the left-hand side of (4.6) is equal to

$$\sum_{k=0}^{[2^n t]} \mathbb{P}(T_n = k/2^n, W_t - W_{T_n} < x - b)$$

$$= \sum_{k=0}^{[2^n t]} \mathbb{P}(T_n = k/2^n, W_t - W_{k/2^n} < x - b)$$

$$= \sum_{k=0}^{[2^n t]} \mathbb{P}(T_n = k/2^n)\mathbb{P}(W_t - W_{T_n} < x - b),$$

using the independent increments property of Brownian motion and the fact that we have $(T_n = k/2^n) \in \mathcal{F}_{k/2^n}$. Using the symmetry of the normal distribution, that is, that $W_t - W_s$ and $W_s - W_t$ have the same law, this is the same as

$$\sum_{k=0}^{[2^n t]} \mathbb{P}(T_n = k/2^n)\mathbb{P}(W_t - W_{T_n} > b - x),$$

and reversing the steps above, this equals the right-hand side of (4.6).

Since W has continuous paths, $W_T = b$, so $(T = t) \subset (W_t = b)$. Because W_t is a normal random variable, then $\mathbb{P}(T = t) = 0$. Also, $\mathbb{P}(W_t - W_T = b - x)$ and $\mathbb{P}(W_t - W_T = x - b)$ are both zero. If we now let $n \to \infty$ in (4.6), we obtain

$$\mathbb{P}(T \leq t, W_t - W_T < x - b) = \mathbb{P}(T \leq t, W_t - W_T > b - x).$$

Since $W_T = b$, this is the same as

$$\mathbb{P}(T \leq t, W_t < x) = \mathbb{P}(T \leq t, W_t > 2b - x). \tag{4.7}$$

By the definition of T and the continuity of the paths of W, the left-hand side is equal to the left-hand side of (4.5). If $W_t > 2b - x$, then automatically $T \leq t$, so the right-hand side of (4.7) is equal to the right-hand side of (4.5). □

Our second application will be useful when studying local time in Chapter 14.

Proposition 4.5 *Let W_t be a Brownian motion with respect to a filtration $\{\mathcal{F}_t\}$ satisfying the usual conditions. Let T be a finite stopping time and $s > 0$. If $a < b$, then*

$$\mathbb{P}(W_{T+s} \in [a, b] \mid \mathcal{F}_T) \leq \frac{|b - a|}{\sqrt{2\pi s}}.$$

Proof If $A \in \mathcal{F}_T$, let $k > 0$ and write

$$\mathbb{P}(W_{T+s} \in [a, b], A)$$

$$= \sum_{j=-\infty}^{\infty} \mathbb{P}(W_{T+s} \in [a, b], A, j/k \leq W_T < (j+1)/k)$$

$$\leq \sum_{j=-\infty}^{\infty} \mathbb{P}(W_{T+s} - W_T \in [a - (j+1)/k, b - j/k],$$

$$A, j/k \leq W_T \leq (j+1)/k).$$

Using the fact that $W_{T+s} - W_T$ is a Brownian motion independent of $\mathcal{F}_T$, this is less than or equal to

$$\sum_{j=-\infty}^{\infty} \mathbb{P}(W_s \in [a - (j+1)/k, b - j/k]) \, \mathbb{P}(A, j/k \leq W_T \leq (j+1)/k)$$

$$\leq \sum_{j=-\infty}^{\infty} \frac{1}{\sqrt{2\pi}} \frac{b - a + 1/k}{\sqrt{s}} \mathbb{P}(A, j/k \leq W_T \leq (j+1)/k)$$

$$\leq \frac{1}{\sqrt{2\pi}} \frac{b - a + 1/k}{\sqrt{s}} \mathbb{P}(A).$$

We used here the formula for the density of a normal random variable with mean zero and variance s. This is true for all k, so letting $k \to \infty$ yields our result. $\qquad \square$

Exercises

4.1 If W is a Brownian motion, let $S_t = \sup_{s \leq t} W_s$. Find the density for S_t.

4.2 With W and S as in Exercise 4.1, find the joint density of (S_t, W_t).

4.3 Let W be a Brownian motion started at $a > 0$ and let T_0 be the first time W hits 0. Find the law of $\sup_{t \leq T_0} W_t$.

4.4 Use the reflection principle to prove that if W is a Brownian motion and $T = \inf\{t > 0 : W_t \in (0, \infty)\}$, then

$$\mathbb{P}(T = 0) = 1.$$

In other words, Brownian motion enters the interval $(0, \infty)$ immediately. By symmetry it enters the interval $(-\infty, 0)$ immediately. Conclude that Brownian motion hits 0 infinitely often in every time interval $[0, t]$.

4.5 Let W_t be a Brownian motion and $\{\mathcal{F}_t\}$ be the minimal augmented filtration generated by W. Let

$$T = \inf\{t > 0 : W_t = \sup_{0 \leq s \leq 1} W_s\}.$$

Show that T is *not* a stopping time with respect to $\{\mathcal{F}_t\}$.

4.6 Let W and S be as in Exercise 4.1.

(1) Let $0 < s < t < u$ and let $a < b$ with $b - a \leq 1$. Show that there exists a constant c, depending on s, t, and u, but not a or b, such that

$$\mathbb{P}(S_s \in [a, b], \sup_{t \leq r \leq u} W_r \in [a, b]) \leq c(b - a)^2.$$

(2) Show that the path of a Brownian motion does not take on the same value as a local maximum twice. That is, if S and T are times when W has a local maximum, then $W_S \neq W_T$, a.s.

4.7 Let V_t be the number of upcrossings of $[0, 1]$ by a Brownian motion W up to time t. This means we let $S_1 = 0$, $T_i = \inf\{t > S_i : W_t \geq 1\}$, and $S_{i+1} = \inf\{T > T_i : W_t \leq 0\}$ for $i = 1, 2, \ldots$, and we set $V_t = \sup\{k : T_k \leq t\}$. Show that $V_t \to \infty$, a.s., as $t \to \infty$.

4.8   Let W be a Brownian motion. The *zero set* of Brownian motion is the random set

$$Z(\omega) = \{t \in [0, 1] : W_t(\omega) = 0\}.$$

(1) Show that $Z(\omega)$ is a closed set for each ω.

(2) Show that with probability one, every point of $Z(\omega)$ is a limit point of $Z(\omega)$. Conclude that $Z(\omega)$ is an uncountable set.

4.9   Let W be a one-dimensional Brownian motion and $\delta > 0$.

(1) Prove that there exists γ such that if $t \leq \gamma$, then

$$\mathbb{P}(0 \leq W_t \leq \delta/2) \geq 1/4 \qquad \text{and} \qquad \mathbb{P}(-\delta/2 \leq W_t \leq 0) \geq 1/4.$$

(2) Prove there exists γ such that

$$\mathbb{P}(\sup_{s \leq \gamma} |W_s| > \delta/2) \leq 1/8.$$

(3) Prove that if $m \geq 1$, then

$$\mathbb{P}(\sup_{m\gamma \leq s \leq (m+1)\gamma} |W_s - W_{m\gamma}| \leq \delta/2, W_{m\gamma} \in [0, \delta/2], |W_{(m+1)\gamma}| \leq \delta/2 \mid \mathcal{F}_{m\gamma})$$

$$\geq \tfrac{1}{8} \mathbb{P}(\sup_{m\gamma \leq s \leq (m+1)\gamma} |W_s - W_{m\gamma}| \leq \delta/2, W_{m\gamma} \in [0, \delta/2])$$

and the same with $W_{m\gamma} \in [-\delta/2, 0]$ in place of $W_{m\gamma} \in [0, \delta/2]$. Conclude that

$$\mathbb{P}(\sup_{m\gamma \leq s \leq (m+1)\gamma} |W_s - W_{m\gamma}| \leq \delta/2, |W_{m\gamma}| \leq \delta/2, |W_{(m+1)\gamma}| \leq \delta/2 \mid \mathcal{F}_{m\gamma})$$

$$\geq \tfrac{1}{8} \mathbb{P}(\sup_{m\gamma \leq s \leq (m+1)\gamma} |W_s - W_{m\gamma}| \leq \delta/2, |W_{m\gamma}| \leq \delta/2).$$

(4) Use induction to prove that if $t_0 > 0$, there exists $c_1 > 0$ such that

$$\mathbb{P}(\sup_{s \leq t_0} |W_s| \leq \delta) > c_1.$$

(5) Prove that if W is a d-dimensional Brownian motion, $t_0 > 0$, and $\delta > 0$, there exists c_2 such that

$$\mathbb{P}(\sup_{s \leq t_0} |W_s| \leq \delta) > c_2.$$

4.10   The *p-variation* of a function f on the interval $[0, 1]$ is defined by

$$V^p(f) = \sup\left\{ \sum_{i=0}^{n-1} |f(t_{i+1}) - f(t_i)|^p : n \geq 1, 0 = t_0 < t_1, \cdots < t_n = 1 \right\};$$

the supremum is over all partitions P of $[0, 1]$. In this exercise we will prove that if $p < 2$ and W is a Brownian motion, then $V^p(W) = \infty$, a.s.

(1) Let X_i be an i.i.d. sequence of random variables with finite mean. Use the strong law of large numbers to prove that if $K > \mathbb{E}X_1$, then

$$\mathbb{P}\left(\sum_{i=1}^{n} X_i > Kn \right) \to 0$$

as $n \to \infty$.

(2) If $p < 2$, take $r \in (p, 2)$, and let $\varepsilon_n = n^{-1/r}$. Let $S_0 = 0$ and for $i \geq 0$, set $S_{i+1} = \inf\{t > S_i : |W_t - W_{S_i}| > \varepsilon_n\}$. Set $X_i = \varepsilon_n^{-2}(S_i - S_{i-1})$. Prove that the X_i are i.i.d. with finite mean.

(3) Use (1) to show that

$$\mathbb{P}(S_n > 1) = \mathbb{P}\left(\sum_{i=1}^n X_i > \varepsilon_n^{-2}\right) \to 0$$

as $n \to \infty$.

(4) Using the partition $\{S_0, S_1, \ldots, S_n\}$, show that $V^p(W) \geq n\varepsilon_n^p$ on the event $(S_n \leq 1)$.

(5) Conclude $V^p(W) = \infty$, a.s.

5

The Poisson process

At the opposite extreme from Brownian motion is the Poisson process. This is a process that only changes value by means of jumps, and even then, the jumps are nicely spaced. The Poisson process is the prototype of a pure jump process, and later we will see that it is the building block for an important class of stochastic processes known as Lévy processes.

Definition 5.1 Let $\{\mathcal{F}_t\}$ be a filtration, not necessarily satisfying the usual conditions. A *Poisson process* with parameter $\lambda > 0$ is a stochastic process X satisfying the following properties:
(1) $X_0 = 0$, a.s.
(2) The paths of X_t are right continuous with left limits.
(3) If $s < t$, then $X_t - X_s$ is a Poisson random variable with parameter $\lambda(t - s)$.
(4) If $s < t$, then $X_t - X_s$ is independent of $\mathcal{F}_s$.

Define $X_{t-} = \lim_{s \to t, s < t} X_s$, the left-hand limit at time t, and $\Delta X_t = X_t - X_{t-}$, the size of the jump at time t. We say a function f is increasing if $s < t$ implies $f(s) \le f(t)$. We use "strictly increasing" when $s < t$ implies $f(s) < f(t)$. We have the following proposition.

Proposition 5.2 *Let X be a Poisson process. With probability one, the paths of X_t are increasing and are constant except for jumps of size 1. There are only finitely many jumps in each finite time interval.*

Proof For any fixed $s < t$, we have that $X_t - X_s$ has the distribution of a Poisson random variable with parameter $\lambda(t - s)$, hence is non-negative, a.s.; let $N_{s,t}$ be the null set of ω's where $X_t(\omega) < X_s(\omega)$. The set of pairs (s, t) with s and t rational is countable, and so $N = \cup_{s,t \in \mathbb{Q}_+} N_{s,t}$ is also a null set, where we write $\mathbb{Q}_+$ for the non-negative rationals. For $\omega \notin N$, $X_t \ge X_s$ whenever $s < t$ are rational. In view of the right continuity of the paths of X, this shows the paths of X are increasing with probability one.

Similarly, since Poisson random variables only take values in the non-negative integers, X_t is a non-negative integer, a.s. Using this fact for every t rational shows that with probability one, X_t takes values only in the non-negative integers when t is rational, and the right continuity of the paths implies this is also the case for all t. Since the paths have left limits, there can only be finitely many jumps in finite time.

It remains to prove that ΔX_t is either 0 or 1 for all t. Let $t_0 > 0$. If there were a jump of size 2 or larger at some time t strictly less than t_0, then for each n sufficiently large there exists $0 \leq k_n \leq 2^n$ such that $X_{(k_n+1)t_0/2^n} - X_{k_n t_0/2^n} \geq 2$. Therefore

$$\mathbb{P}(\exists s < t_0 : \Delta X_s \geq 2) \leq \mathbb{P}(\exists k \leq 2^n : X_{(k+1)t_0/2^n} - X_{kt_0/2^n} \geq 2) \qquad (5.1)$$
$$\leq 2^n \sup_{k \leq 2^n} \mathbb{P}(X_{(k+1)t_0/2^n} - X_{kt_0/2^n} \geq 2)$$
$$= 2^n \mathbb{P}(X_{t_0/2^n} \geq 2^n)$$
$$\leq 2^n (1 - \mathbb{P}(X_{t_0/2^n} = 0) - \mathbb{P}(X_{t_0/2^n} = 1))$$
$$= 2^n \left(1 - e^{-\lambda t_0/2^n} - (\lambda t_0/2^n)e^{-\lambda t_0/2^n}\right).$$

We used property 5.1(3) for the two equalities. By l'Hôpital's rule, $(1 - e^{-x} - xe^{-x})/x \to 0$ as $x \to 0$. We apply this with $x = \lambda t_0/2^n$, and see that the last line of (5.1) tends to 0 as $n \to \infty$. Since the left-hand side of (5.1) does not depend on n, it must be 0. This holds for each t_0. □

Another characterization of the Poisson process is as follows. Let $T_1 = \inf\{t : \Delta X_t = 1\}$, the time of the first jump. Define $T_{i+1} = \inf\{t > T_i : \Delta X_t = 1\}$, so that T_i is the time of the ith jump.

Proposition 5.3 *The random variables $T_1, T_2 - T_1, \ldots, T_{i+1} - T_i, \ldots$ are independent exponential random variables with parameter λ.*

Proof In view of Corollary 4.3 it suffices to show that T_1 is an exponential random variable with parameter λ. If $T_1 > t$, then the first jump has not occurred by time t, so X_t is still zero. Hence

$$\mathbb{P}(T_1 > t) = \mathbb{P}(X_t = 0) = e^{-\lambda t},$$

using the fact that X_t is a Poisson random variable with parameter λt. □

We can reverse the characterization in Proposition 5.3 to construct a Poisson process. We do one step of the construction, leaving the rest as Exercise 5.4.

Let $U_1, U_2, \ldots$ be independent exponential random variables with parameter λ and let $T_j = \sum_{i=1}^{j} U_i$. Define

$$X_t(\omega) = k \qquad \text{if } T_k(\omega) \leq t < T_{k+1}(\omega). \qquad (5.2)$$

An examination of the densities shows that an exponential random variable has a gamma distribution with parameters λ and $r = 1$, so by Proposition A.49, T_j is a gamma random variable with parameters λ and j. Thus

$$\mathbb{P}(X_t < k) = \mathbb{P}(T_k > t) = \int_t^\infty \frac{\lambda e^{-\lambda x}(\lambda x)^{k-1}}{\Gamma(k)} \, dx.$$

Performing the integration by parts repeatedly shows that

$$\mathbb{P}(X_t < k) = \sum_{i=0}^{k-1} e^{-\lambda t} \frac{(\lambda t)^i}{i!},$$

and so X_t is a Poisson random variable with parameter λt.

We will use the following proposition later.

Proposition 5.4 *Let $\{\mathcal{F}_t\}$ be a filtration satisfying the usual conditions. Suppose $X_0 = 0$, a.s., X has paths that are right continuous with left limits, $X_t - X_s$ is independent of $\mathcal{F}_s$ if $s < t$, and $X_t - X_s$ has the same law as X_{t-s} whenever $s < t$. If the paths of X are piecewise constant, increasing, all the jumps of X are of size 1, and X is not identically 0, then X is a Poisson process.*

Proof Let $T_0 = 0$ and $T_{i+1} = \inf\{t > T_i : \Delta X_t = 1\}$, $i = 1, 2, \ldots$ We will show that if we set $U_i = T_i - T_{i-1}$, then the U_i are i.i.d. exponential random variables and then appeal to Exercise 5.4.

By Corollary 4.3, the U_i are independent and have the same law. Hence it suffices to show U_1 is an exponential random variable. We observe

$$
\begin{aligned}
\mathbb{P}(U_1 > s + t) &= \mathbb{P}(X_{s+t} = 0) = \mathbb{P}(X_{s+t} - X_s = 0, X_s = 0) \\
&= \mathbb{P}(X_{t+s} - X_s = 0)\mathbb{P}(X_s = 0) = \mathbb{P}(X_t = 0)\mathbb{P}(X_s = 0) \\
&= \mathbb{P}(U_1 > t)\mathbb{P}(U_1 > s).
\end{aligned}
$$

Setting $f(t) = \mathbb{P}(U_1 > t)$, we thus have $f(t + s) = f(t)f(s)$. Since $f(t)$ is decreasing and $0 < f(t) < 1$, we conclude $\mathbb{P}(U_1 > t) = f(t) = e^{-\lambda t}$ for some $\lambda > 0$, or U_1 is an exponential random variable. $\qquad\square$

Exercises

5.1 Suppose P_t is a Poisson process and we write $X_t = P_{t-}$. Is $P_1 - X_{1-t}$ a Poisson process on $[0, 1]$? Why or why not?

5.2 Let P be a Poisson process with parameter λ. Show that

$$
\lim_{n \to \infty} \sup_{t \leq 1} \left| \frac{P_{nt}}{n} - \lambda t \right| = 0, \qquad \text{a.s.}
$$

5.3 Show that if $P^{(1)}$ and $P^{(2)}$ are independent Poisson processes with parameters λ_1 and λ_2, respectively, then $P_t^{(1)} + P_t^{(2)}$ is a Poisson process with parameter $\lambda_1 + \lambda_2$.

5.4 If X is defined by (5.2), show that X is a Poisson process.

5.5 Let X_t be a stochastic process and let $\{\mathcal{F}_t^{00}\}$ be the filtration generated by X. Suppose X is a Poisson process with respect to the filtration $\{\mathcal{F}_t^{00}\}$. Show that X is a Poisson process with respect to the minimal augmented filtration generated by X.
   Hint: Imitate the proof of Proposition 2.5.

5.6 Suppose P_t is a Poisson process and f and g are non-negative bounded deterministic functions with compact support. Find necessary and sufficient conditions on f and g so that $\int_0^\infty f(s)\, dP_s$ and $\int_0^\infty g(s)\, dP_s$ are independent.
   Hint: First show that the characteristic function of $F = \int_0^\infty f(s)\, dP_s$ is

$$
\mathbb{E}\, e^{iuF} = \exp\left(\int_0^\infty (e^{iuf(s)} - 1)\, ds \right).
$$

5.7 We will talk about weak convergence in general metric spaces in Chapters 30–35. This exercise is concerned with the weak convergence of real-valued random variables as defined in Section A.12.

Suppose for each n, P_n is a Poisson random variable with parameter λ_n and $\lambda_n \to \infty$ as $n \to \infty$. Prove that

$$\frac{P_n - \lambda_n}{\sqrt{\lambda_n}}$$

converges weakly to a normal random variable with mean zero and variance one.

Hint: Imitate the proof of Theorem A.51.

6

Construction of Brownian motion

There are several ways of constructing Brownian motion, none of them easy. Here we give two constructions. The first is the one that Wiener used, which is based on Fourier series. The second uses martingale techniques. A method due to Lévy can be found in Bass (1995); see also Exercises 6.4 and 6.5. We will see several other constructions in later chapters.

6.1 Wiener's construction

For any of the constructions of Brownian motion, the main step is to construct W_t for $t \in [0, 1]$. Once we have done this, we get Brownian motion for all t rather easily. More specifically, suppose we have a Brownian motion $Y^{(0)}$ started at 0 on the time interval $[0, 1]$. Take independent copies $Y^{(1)}, Y^{(2)}, \ldots$, each on $[0, 1]$. We have $Y_0^{(i)} = 0$ for each i, and now to get Brownian motion started at 0, define W_t to be equal to $Y_t^{(0)}$ if $t \leq 1$, equal to $Y_1^{(0)} + Y_{t-1}^{(1)}$ if $1 < t \leq 2$, and more generally

$$W_t = \left(\sum_{i=0}^{[t]-1} Y_1^{(i)} \right) + Y_{t-[t]}^{[t]}$$

if $t \geq 1$, where $[t]$ is the largest integer less than or equal to t. This will give Brownian motion started at 0 on the time interval $[0, \infty)$.

Therefore the crux of the problem is to construct Brownian motion on $[0, 1]$. Because we are working with Fourier series, it is more convenient to look at Brownian motion on $[0, \pi]$; we can just disregard times between 1 and π when we are done.

Throughout this chapter we make the supposition that we can find a countable sequence $Z_1, Z_2, \ldots$ of independent and identically distributed mean zero normal random variables with variance one that are $\mathcal{F}$ measurable, where $(\Omega, \mathcal{F}, \mathbb{P})$ is our probability space. This is an extremely mild condition.

Theorem 6.1 *There exists a process $\{W_t; 0 \leq t \leq 1\}$ that is Brownian motion.*

Proof If we fix $t \in [0, \pi]$ and compute the Fourier series for the function $f(s) = s \wedge t$, it is an exercise in calculus to get the Fourier coefficients. We end up with

$$s \wedge t = \frac{st}{\pi} + \frac{2}{\pi} \sum_{k=1}^{\infty} \frac{\sin ks \sin kt}{k^2}. \tag{6.1}$$

This suggests letting $Z_0, Z_1, \ldots$ be i.i.d. normal random variables with mean 0 and variance 1 and setting

$$W_t = \frac{t}{\sqrt{\pi}} Z_0 + \sum_{k=1}^{\infty} \left(\sqrt{\frac{2}{\pi}} \frac{\sin kt}{k} \right) Z_k. \tag{6.2}$$

Assuming there is no problem with convergence, we see that W_t has mean zero, since each of the Z_i does, and that

$$\mathbb{E}[W_s W_t] = \frac{st}{\pi} + \sum_{k=1}^{\infty} \frac{2}{\pi} \frac{\sin ks \sin kt}{k^2} = s \wedge t \tag{6.3}$$

as required. We used the independence of the Z_i here to show that $\mathbb{E}[Z_i Z_j] = 0$ if $i \neq j$.

We argue that there is in fact no difficulty with the convergence. Note that $\sum_{k=1}^{m} \frac{\sin^2 kt}{k^2}$ increases as m increases to a finite limit. Therefore

$$\mathbb{E}\left[\left(\sum_{k=m}^{n} Z_k \frac{\sin kt}{k} \right)^2 \right] = \sum_{k=m}^{n} \frac{\sin^2 kt}{k^2} \to 0$$

in L^2 as $m, n \to \infty$. This means that the sum on the right of (6.2) is a Cauchy sequence in L^2. By the completeness of L^2, the sum on the right of (6.2) converges in L^2. A use of the Cauchy–Schwarz inequality allows us to justify the formula for the expectation of $W_s W_t$.

If we let

$$W_t^j = \frac{t}{\sqrt{\pi}} Z_0 + \sum_{k=1}^{j} \left(\sqrt{\frac{2}{\pi}} \frac{\sin kt}{k} \right) Z_k,$$

then $(W_{t_1}^j, \ldots, W_{t_m}^j)$ is a jointly normal collection of random variables for each j whenever $t_1, \ldots, t_n \in [0, \pi]$. By Remark A.56, it follows that $(W_{t_1}, \ldots, W_{t_m})$ is a jointly normal collection of random variables. Therefore W_t is a Gaussian process. Since each W_t has mean zero and $\mathrm{Cov}(W_s, W_t) = s \wedge t$, then W_t has the correct finite-dimensional distributions to be a Brownian motion.

The only part remaining to the construction is to show that W_t as constructed above has continuous paths, for we can then use Theorem 2.4. In what follows, pay attention to where the absolute values are placed. If one is cavalier about placing them, one will very likely run into trouble.

Define

$$S_m(t) = \sum_{k=m}^{2m-1} \frac{\sin kt}{k} Z_k$$

and let $T_m = \sup_{0 \le t \le \pi} |S_m(t)|$. We write

$$W_t = \frac{t}{\sqrt{\pi}} Z_0 + \sqrt{\frac{2}{\pi}} \sum_{n=0}^{\infty} S_{2^n}(t).$$

We will show

$$\mathbb{E}\, T_m^2 \le \frac{c}{m^{1/2}}. \tag{6.4}$$

Once we have this, then by the Fubini theorem and then Jensen's inequality,

$$\mathbb{E}\sum_{n=0}^{\infty} T_{2^n} = \sum_{n=0}^{\infty}\mathbb{E}\, T_{2^n} \le \sum_{n=0}^{\infty}\left(\mathbb{E}\,[T_{2^n}^2]\right)^{1/2} < \infty.$$

Therefore $\sum_{n=0}^{\infty} T_{2^n} < \infty$, a.s., and by the Weierstrass M-test (see, e.g., Rudin, 1976), we have that with probability 1, $\sum_{n=0}^{\infty} S_{2^n}(t)$ converges uniformly in t. Since each $S_{2^n}(t)$ is a continuous function of t, we see that the uniform limit is also continuous and we are done.

We therefore have to prove (6.4). Using $|\sum_k a_k|^2 = \sum_{j,k} a_k \bar{a}_j$ for a_k complex valued, we have

$$T_m^2 \le \sup_{0\le t\le\pi}\left|\sum_{k=m}^{2m-1}\frac{e^{ikt}}{k}Z_k\right|^2$$

$$\le \sup_{0\le t\le\pi}\left|\sum_{j,k=m}^{2m-1}\frac{e^{ikt}e^{-ijt}}{jk}Z_j Z_k\right|$$

$$\le \sum_{k=m}^{2m-1}\frac{1}{k^2}Z_k^2 + 2\sup_{0\le t\le\pi}\left|\sum_{\ell=1}^{m-1}\sum_{j=m}^{2m-\ell-1}\frac{e^{i\ell t}}{j(j+\ell)}Z_j Z_{j+\ell}\right|$$

$$\le \sum_{k=m}^{2m-1}\frac{1}{k^2}Z_k^2 + 2\sum_{\ell=1}^{m-1}\left|\sum_{j=m}^{2m-\ell-1}\frac{1}{j(j+\ell)}Z_j Z_{j+\ell}\right|. \tag{6.5}$$

In the third inequality we wrote

$$\sum_{j,k=m}^{2m-1} = \sum_{m\le j=k\le 2m-1} + 2\sum_{m\le j<k\le 2m-1},$$

and then set $\ell = k - j$. Write I for the first sum on the last line of (6.5) and J_ℓ for $\sum_{j=m}^{2m-\ell-1}\frac{1}{j(j+\ell)}Z_j Z_{j+\ell}$. The expectation of I is equal to

$$\sum_{k=m}^{2m-1}\frac{1}{k^2} \le \frac{c}{m}.$$

We next look at the expectation of the J_ℓ. Since the Z_i are mean zero and independent, $\mathbb{E}[Z_{i_1}Z_{i_2}Z_{i_3}Z_{i_4}]$ is zero unless either all four subscripts are equal or else two subscripts are equal and the other two subscripts are also equal. By Jensen's inequality,

$$\mathbb{E}\,|J_\ell| \le \left(\mathbb{E}\left[\sum_{j=m}^{2m-\ell-1}\frac{1}{j(j+\ell)}Z_j Z_{j+\ell}\right]^2\right)^{1/2}$$

$$= \left(\sum_{j=m}^{2m-\ell-1}\frac{1}{j^2(j+\ell)^2}\right)^{1/2}. \tag{6.6}$$

The last equality follows by multiplying out

$$\left(\sum\frac{Z_j Z_{j+\ell}}{j(j+\ell)}\right)^2$$

and noting that expectations of the cross-product terms are zero. Since $j \geq m$ in the last line of (6.6) and there are at most m terms in the sum, the last line of (6.6) is bounded by $(cm/m^4)^{1/2} = cm^{-3/2}$. Therefore

$$\mathbb{E} \sum_{\ell=1}^{m-1} |J_\ell| \leq c/m^{1/2}.$$

Substituting in (6.5) completes the proof of (6.4). □

By Proposition 2.5, the Brownian motion that we constructed is a Brownian motion with respect to the minimal augmented filtration.

6.2 Martingale methods

Here, we use martingale methods to take care of the continuity of the paths. We proceed as in the previous section to construct $\{W_t; 0 \leq t \leq \pi\}$, where W_t is a Gaussian process with $\mathbb{E} W_t = 0$ and $\text{Cov}(W_s, W_t) = s \wedge t$, and we need to show that W has a version with continuous paths. We show that W is a martingale, and so has a version with paths that are right continuous with left limits. We use Doob's inequalities to control the oscillation of W over short time intervals, and then use the Borel–Cantelli lemma to show continuity.

Theorem 6.2 *If $\{W_t; t \leq 1\}$ is a Gaussian process with $\mathbb{E} W_t = 0$ for all $t \leq 1$ and $\text{Cov}(W_s, W_t) = s \wedge t$ for all $s, t \leq 1$, then there is a version of W that is a Brownian motion on $[0, 1]$.*

Proof As in the proof of Theorem 6.1, we need to show that W has a version with continuous paths. Since $\text{Cov}(W_t - W_s, W_r) = r - r = 0$ if $r \leq s < t$, we see by Proposition A.55 that $W_t - W_s$ is independent of $\mathcal{F}_s^{00} = \sigma(W_r; r \leq s)$. Then

$$\mathbb{E}[W_t - W_s \mid \mathcal{F}_s^{00}] = \mathbb{E}[W_t - W_s] = 0,$$

so W_t is a martingale. By Theorem 3.12, with probability one, W has left and right limits along $\mathcal{D}$, the dyadic rationals. Let $W_t' = \lim_{u>t, u\in\mathcal{D}, u\to t} W_u$. Since $\mathbb{E}(W_u - W_t)^2 = u - t \to 0$ as $u \to t$, then $W_t' = W_t$, a.s., or W' is a version of W with paths that are right continuous with left limits. We now drop the primes. Set $W_t = W_1$ if $t \geq 1$.

For any $t_0 \in [0, 1]$, $W_{t+t_0} - W_{t_0}$ is also a martingale, and by Jensen's inequality for conditional expectations (Proposition A.21), $|W_{t+t_0} - W_{t_0}|^4$ is a submartingale. Using Doob's inequalities (Theorem 3.6), if $\lambda > 0$ and $t_0, \delta \in [0, 1]$,

$$\mathbb{P}(\sup_{t_0 \leq t \leq t_0 + \delta} |W_t - W_{t_0}| \geq \lambda) = \mathbb{P}(\sup_{t_0 \leq t \leq t_0 + \delta} |W_t - W_{t_0}|^4 \geq \lambda^4)$$

$$\leq c \frac{\mathbb{E}|W_{t_0+\delta} - W_{t_0}|^4}{\lambda^4}.$$

Since $W_{t_0+\delta} - W_{t_0}$ is a mean zero normal random variable with variance δ if $t_0 + \delta \leq 1$, we have

$$\mathbb{P}(\sup_{t_0 \leq t \leq t_0 + \delta} |W_t - W_{t_0}| \geq \lambda) \leq c \frac{\delta^2}{\lambda^4}. \tag{6.7}$$

Let

$$A_n = \{\exists\, k \leq 2^n : \sup_{k/2^n \leq t \leq (k+2)/2^n} |W_t - W_{k/2^n}| > 2^{-n/8}\}.$$

From (6.7) with $\delta = 2^{-n+1}$ and $\lambda = 2^{-n/8}$,

$$\mathbb{P}(A_n) \leq 2^n \max_{k \leq 2^n} \mathbb{P}(\sup_{k/2^n \leq t \leq (k+2)/2^n} |W_t - W_{k/2^n}| > 2^{-n/8})$$

$$\leq \frac{c 2^n 2^{-2n}}{2^{-n/2}} = c 2^{-n/2},$$

which is summable. By the Borel–Cantelli lemma, $\mathbb{P}(A_n \text{ i.o.}) = 0$. (The event $(A_n \text{ i.o.})$ is the event where ω is in infinitely many of the A_n.)

Except for a set of ω's in a null set, there exists a positive integer N (which will depend on ω) such that if $n \geq N$, then $\omega \notin A_n$. Given $\varepsilon > 0$, take $n \geq N$ such that $2^{-n/8} < \varepsilon/2$. If $|t - s| \leq 2^{-n}$ with $s, t \in [0, 1]$, then $s, t \in [k/2^n, (k+2)/2^n]$ for some $k \leq 2^n$. Since $\omega \notin A_n$,

$$|W_t - W_s| \leq |W_t - W_{k/2^n}| + |W_s - W_{k/2^n}| \leq 2 \cdot 2^{-n/8} < \varepsilon.$$

This proves the continuity of W_t. $\qquad\qquad\square$

There is nothing special about the trigonometric polynomials in this second construction. Let $\langle f, g \rangle = \int_0^1 f(r)g(r)\, dr$ be the inner product for the Hilbert space $L^2[0, 1]$; we consider only real-valued functions for simplicity. Let $\{\varphi_n\}$ be a complete orthonormal system for $L^2[0, 1]$: we have $\langle \varphi_m, \varphi_n \rangle = 0$ if $m \neq n$, $\langle \varphi_n, \varphi_n \rangle = 1$ for each n, and $f = 0$, a.e., if $\langle f, \varphi_n \rangle = 0$ for all n. One property of a complete orthonormal system is Parseval's identity, which says that

$$\langle f, f \rangle = \sum_{n=1}^{\infty} |\langle f, \varphi_n \rangle|^2;$$

see Folland (1999). If we replace f by g and then by $f + g$ and use

$$\langle f, g \rangle = \tfrac{1}{2}[\langle f + g, f + g \rangle - \langle f, f \rangle - \langle g, g \rangle],$$

we obtain

$$\langle f, g \rangle = \sum_{n=1}^{\infty} \langle f, \varphi_n \rangle \langle g, \varphi_n \rangle.$$

Now let

$$a_n(t) = \langle 1_{[0,t]}, \varphi_n \rangle = \int_0^t \varphi_n(r)\, dr.$$

If $Z_1, Z_2, \ldots$ are independent mean zero normal random variables with variance one, let

$$W_t = \sum_{n=1}^{\infty} a_n(t) Z_k. \tag{6.8}$$

Assuming there is no difficulty with the convergence, we have

$$\text{Cov}\,(W_s, W_t) = \sum_{n=1}^{\infty} a_n(s) a_n(t) = \sum_{n=1}^{\infty} \langle 1_{[0,s]}, \varphi_n \rangle \langle 1_{[0,t]}, \varphi_n \rangle$$

$$= \langle 1_{[0,s]}, 1_{[0,t]} \rangle = s \wedge t.$$

Exercise 6.2 asks you to verify that the process W defined by (6.8) is a mean zero Gaussian process on $[0, 1]$ with the same covariances as a Brownian motion.

Exercises

6.1   Let $Z_0, Z_1, Z_2, \ldots$ be a sequence of independent identically distributed mean zero normal random variables with variance one. Define

$$X_t = \frac{t^2}{2\sqrt{\pi}} Z_0 + \sum_{k=1}^{\infty} \left(\sqrt{\frac{2}{\pi}} \frac{\cos kt}{k^2} \right) Z_k. \tag{6.9}$$

(1) Show that the convergence in (6.9) is absolute and uniform over $t \in [0, 1]$.

(2) Show that X_t is a Gaussian process.

(3) If W_t is a Brownian motion and

$$Y_t = \int_0^t W_r \, dr, \qquad t \in [0, 1],$$

show that X and Y have the same finite-dimensional distributions. Show that X and Y have the same law when viewed as random variables taking values in $C[0, 1]$. (The process X is sometimes known as *integrated Brownian motion*.)

(4) Find $\text{Cov}\,(X_s, X_t)$.

6.2   Let $\{\varphi_n\}$ be a complete orthonormal system for $L^2[0, 1]$. Show that the sum (6.8) converges in L^2 and give the details of the proof that the resulting process W is a mean zero Gaussian process with $\text{Cov}\,(W_s, W_t) = s \wedge t$ if $s, t \in [0, 1]$.

6.3   Let $\mathcal{D} = \{k/2^n : n \geq 1, k = 0, 1, \ldots, 2^n\}$ be the dyadic rationals. Suppose the collection of random variables $\{V_t : t \in \mathcal{D}\}$ is jointly normal, each V_t has mean zero, and $\text{Cov}\,(V_s, V_t) = s \wedge t$.

(1) Prove that the paths of V are uniformly continuous over $t \in \mathcal{D}$.

(2) If we define $W_t = \lim_{s \in \mathcal{D}, s \to t} V_s$, prove that W is a Brownian motion.

6.4   In this and the next exercise we give the Haar function construction of Brownian motion. Let $\varphi_{00} = 1$ on $[0, 1]$ and for $i = 1, 2, \ldots$, and $1 \leq j \leq 2^{i-1}$, set

$$\varphi_{ij}(x) = \begin{cases} 2^{(i-1)/2}, & (2j-2)/2^i \leq x < (2j-1)/2^i, \\ -2^{(i-1)/2}, & (2j-1)/2^i \leq x < 2j/2^i, \\ 0, & \text{otherwise.} \end{cases}$$

It is a well-known and easily proved result from analysis (see, e.g., Bass (1995), Section I.2) that the collection $\{\varphi_{ij}\}$ is a complete orthonormal system for $L^2[0, 1]$.

For each i, j, define

$$\psi_{ij}(t) = \int_0^t \varphi_{ij}(s) \, ds,$$

for each i and j, let Y_{ij} be independent mean zero normal random variables with variance one, and let

$$V_i(t) = \sum_{j=1}^{2^{i-1}} Y_{ij}\varphi_{ij}(t)$$

for $i \geq 1$. Set $V_0 = Y_{00}\varphi_{00}$.

(1) Fix $i \geq 1$. Prove that each ψ_{ij} is bounded by $2^{(-i-1)/2}$. Prove that the sets $\{t : \psi_{ij}(t) > 0\}$, $j = 1, \ldots, 2^{i-1}$, are disjoint.

(2) Fix $i \geq 1$. Write

$$\mathbb{P}(\exists t \in [0,1] : |V_i(t)| > i^{-2}) \leq \mathbb{P}(\exists j \leq 2^{i-1} : |Y_{ij}|2^{(-i-1)/2} > i^{-2}),$$

use Proposition A.52 to estimate this, and conclude that

$$\sum_{i=1}^{\infty} \mathbb{P}(\sup_{0 \leq t \leq 1} |V_i(t)| > i^{-2}) < \infty. \tag{6.10}$$

6.5 This is a continuation of Exercise 6.4. With φ_{ij}, ψ_{ij}, Y_{ij}, and V_i as in that problem, let

$$W_t = \sum_{i=0}^{\infty} V_i(t).$$

(1) Prove that W is a jointly normal Gaussian process with mean zero and Cov $(W_s, W_t) = s \wedge t$.

(2) Use (6.10) and the Borel–Cantelli lemma to show that $\sum_{i=1}^{n} |V_i(t)|$ converges uniformly over $[0, 1]$. Conclude that W is a Brownian motion.

7

Path properties of Brownian motion

The paths of Brownian motion are continuous, but we will see that they are not differentiable. How continuous are they? We will see that the paths satisfy what is known as a Hölder continuity condition. A precise description of the oscillatory behavior of Brownian motion will be given by the law of the iterated logarithm.

A function $f \colon [0, 1] \to \mathbb{R}$ is said to be *Hölder continuous of order* α if there exists a constant M such that

$$|f(t) - f(s)| \le M|t - s|^{\alpha}, \qquad s, t \in [0, 1]. \tag{7.1}$$

We show that the paths of Brownian motion are Hölder continuous of order α if $\alpha < \frac{1}{2}$. (They are also not Hölder continuous of order α if $\alpha \ge \frac{1}{2}$; we will see this from the law of the iterated logarithm.)

Theorem 7.1 *If $\alpha < \frac{1}{2}$, the paths of Brownian motion are Hölder continuous of order α on $[0, 1]$.*

Proof *Step 1.* First we apply the Borel–Cantelli lemma to a certain sequence of sets. Let W be a Brownian motion and set

$$A_n = \{\exists k \le 2^n - 1 : \sup_{k/2^n \le t \le (k+1)/2^n} |W_t - W_{k/2^n}| > 2^{-n\alpha}\}.$$

Since $W_{t+k/2^n} - W_{k/2^n}$ is a Brownian motion,

$$\begin{aligned}
\mathbb{P}(A_n) &\le 2^n \sup_{k \le 2^n} \mathbb{P}(\sup_{t \le 1/2^n} |W_{t+k/2^n} - W_{k/2^n}| > 2^{-n\alpha}) \\
&\le 2^n \mathbb{P}(\sup_{t \le 1/2^n} |W_t| > 2^{-n\alpha}) \\
&\le 2 \cdot 2^n \exp(-2^{-2n\alpha}/2(2^{-n})).
\end{aligned} \tag{7.2}$$

Here we used Proposition 3.15. Since $\alpha < \frac{1}{2}$, then $2^{n(1-2\alpha)} > 2n$ for n large, and the last line of (7.2) is less than

$$2^{n+1} \exp(-2^{n(1-2\alpha)}/2) \le 2^{n+1} e^{-n}$$

if n is large. Hence $\sum \mathbb{P}(A_n) < \infty$, and $\mathbb{P}(A_n \text{ i.o.}) = 0$ by the Borel–Cantelli lemma.

Step 2. Next we show that this implies the Hölder continuity. For almost every ω there exists N (depending on ω) such that if $n \ge N$, then $\omega \notin A_n$. Let $s \le t$ be two points in $[0, 1]$. If $2^{-(n+2)} \le t - s \le 2^{-(n+1)}$ for some $n \ge N$ and k is the largest integer such that

43

$k/2^{n+2} \leq s$, then

$$
\begin{aligned}
|W_t - W_s| &\leq |W_t - W_{t \wedge ((k+1)/2^{n+2})}| + |W_{t \wedge ((k+1)/2^{n+2})} - W_{k/2^{n+2}}| \\
&\quad + |W_s - W_{k/2^{n+2}}| \\
&\leq 3 \cdot 2^{-n\alpha} \leq 3 \cdot 4^\alpha |t - s|^\alpha.
\end{aligned}
$$

We know $|W_t(\omega)|$ is bounded on $[0, 1]$ since the paths are continuous; let K (depending on ω) be the bound. If $|t - s| \geq 2^{-(N+1)}$, then

$$
|W_t - W_s| \leq 2K \leq (2K)(2^{N+1})|t - s| \leq (2K)(2^{N+1})|t - s|^\alpha.
$$

Thus, no matter whether $|t - s|$ is small or large, there exists L (depending on ω) such that $|W_t(\omega) - W_s(\omega)| \leq L|t - s|^\alpha$ for all $s, t \in [0, 1]$. $\qquad\square$

One of the most beautiful theorems in probability theory is the law of the iterated logarithm (LIL). It describes precisely how Brownian motion oscillates.

Theorem 7.2 *Let W be a Brownian motion. We have*

$$
\limsup_{t \to \infty} \frac{|W_t|}{\sqrt{2t \log \log t}} = 1, \qquad \text{a.s.}
$$

and

$$
\limsup_{t \to 0} \frac{|W_t|}{\sqrt{2t \log \log(1/t)}} = 1, \qquad \text{a.s.}
$$

Proof   The second assertion follows from the first by time inversion; see Exercise 2.5. Thus we only need to prove the first assertion.

Proof of upper bound: We use the Borel–Cantelli lemma. Let $\varepsilon > 0$ and then choose q larger than 1 but close enough to 1 so that $(1 + \varepsilon)^2/q > 1$. Let

$$
A_n = (\sup_{s \leq q^n} |W_s| > (1 + \varepsilon)\sqrt{2q^{n-1} \log \log q^{n-1}}).
$$

By Proposition 3.15,

$$
\begin{aligned}
\mathbb{P}(A_n) &\leq 2 \exp\left(-\frac{(1 + \varepsilon)^2 2q^{n-1} \log \log q^{n-1}}{2q^n} \right) \\
&= 2 \exp\left(-\frac{(1 + \varepsilon)^2}{q} (\log(n - 1) + \log \log q) \right) = \frac{c}{(n - 1)^{(1+\varepsilon)^2/q}},
\end{aligned}
$$

where we are using our convention that the letter c denotes a constant whose exact value is unimportant. This is summable in n, so $\sum \mathbb{P}(A_n) < \infty$.

By the Borel–Cantelli lemma, $\mathbb{P}(A_n \text{ i.o.}) = 0$. Hence, except for a null set, there exists $N = N(\omega)$ such that $\omega \notin A_n$ if $n \geq N(\omega)$. If $t \geq q^N$, then for some $n \geq N + 1$ we have $q^{n-1} \leq t \leq q^n$, and

$$
|W_t| \leq \sup_{s \leq q^n} |W_s| \leq (1 + \varepsilon)\sqrt{2q^{n-1} \log \log q^{n-1}} \leq (1 + \varepsilon)\sqrt{2t \log \log t}.
$$

Therefore

$$\limsup_{t \to \infty} \frac{|W_t|}{\sqrt{2t \log \log t}} \leq 1 + \varepsilon, \qquad \text{a.s.} \tag{7.3}$$

Since $\varepsilon > 0$ is arbitrary, the upper bound is proved.

Proof of lower bound: We start with the second half of the Borel–Cantelli lemma. Let $\varepsilon > 0$ and then take $q > 1$ very large so that

$$\frac{(1 - \varepsilon)^2(1 + \varepsilon)}{1 - q^{-1}} < 1$$

and $2/\sqrt{q} < \varepsilon/2$. This is possible because $(1 - \varepsilon)^2(1 + \varepsilon) = (1 - \varepsilon^2)(1 - \varepsilon) < 1$. Let

$$B_n = (W_{q^{n+1}} - W_{q^n} > (1 - \varepsilon)\sqrt{2q^{n+1} \log \log q^{n+1}}).$$

Since Brownian motion has independent increments, the events B_n are independent. Let

$$Z = \frac{W_{q^{n+1}} - W_{q^n}}{\sqrt{q^{n+1} - q^n}}.$$

Then Z is a mean zero normal random variable with variance one. By Proposition A.52, we see that

$$\begin{aligned}
\mathbb{P}(B_n) &= \mathbb{P}(Z > (1 - \varepsilon)\sqrt{2q^{n+1} \log \log q^{n+1}}/\sqrt{q^{n+1} - q^n}) \\
&\geq \exp\left(-\frac{(1 - \varepsilon)^2(1 + \varepsilon)2q^{n+1} \log \log q^{n+1}}{2(q^{n+1} - q^n)}\right) \\
&= c \exp\left(-(1 - \varepsilon)^2(1 + \varepsilon)\frac{\log(n + 1) + \log \log q}{1 - q^{-1}}\right)
\end{aligned}$$

for n large. Hence

$$\sum_n \mathbb{P}(B_n) \geq c \sum_n \frac{1}{(n + 1)^{(1-\varepsilon)^2(1+\varepsilon)/(1-q^{-1})}} = \infty.$$

By the Borel–Cantelli lemma, with probability one, ω is in infinitely many B_n. Consequently, with probability one, infinitely often

$$W_{q^{n+1}} - W_{q^n} > (1 - \varepsilon)\sqrt{2q^{n+1} \log \log q^{n+1}}. \tag{7.4}$$

The inequality (7.4) is not exactly what we want, as we want a lower bound for $W_{q^{n+1}}$, but we can derive the desired lower bound by using the upper bound we proved in Step 1. We know from (7.3) that for n large enough,

$$|W_{q^n}| \leq 2\sqrt{2q^n \log \log q^n} \leq \frac{2}{\sqrt{q}}\sqrt{2q^{n+1} \log \log q^{n+1}} < \frac{\varepsilon}{2}\sqrt{2q^{n+1} \log \log q^{n+1}}.$$

Thus infinitely often

$$W_{q^{n+1}} > (1 - 3\varepsilon/2)\sqrt{2q^{n+1} \log \log q^{n+1}}.$$

This proves

$$\limsup_{n \to \infty} \frac{W_{q^{n+1}}}{\sqrt{2q^{n+1} \log \log q^{n+1}}} \geq 1 - \frac{3\varepsilon}{2}, \qquad \text{a.s.}$$

Since ε is arbitrary, the lower bound follows. $\square$

The law of the iterated logarithm show that the paths of W_t are not differentiable at time 0, a.s. Applying this to $W_{s+t} - W_t$, we see that for each t, W is not differentiable at time t, a.s. But the null set N_t might depend on t, and it is even conceivable that $\cup_{t \in [0,1]} N_t$ is not a null set. We have the following stronger result, which says that except for a set of ω's that form a null set, $t \to W_t(\omega)$ is a function that does not have a derivative at any time $t \in [0, 1]$.

Theorem 7.3 *With probability one, the paths of Brownian motion are nowhere differentiable.*

Proof   Note that if Z is a normal random variable with mean 0 and variance 1, then

$$\mathbb{P}(|Z| \leq r) = \frac{1}{\sqrt{2\pi}} \int_{-r}^{r} e^{-x^2/2} \, dx \leq 2r. \qquad (7.5)$$

Let $M, h > 0$ and let

$$A_{M,h} = \{\exists s \in [0, 1] : |W_t - W_s| \leq M|t - s| \text{ if } |t - s| \leq h\},$$
$$B_n = \{\exists k \leq 2n : |W_{k/n} - W_{(k-1)/n}| \leq 4M/n,$$
$$|W_{(k+1)/n} - W_{k/n}| \leq 4M/n, |W_{(k+2)/n} - W_{(k+1)/n}| \leq 4M/n\}.$$

We check that $A_{M,h} \subset B_n$ if $n \geq 2/h$. To see this, if $\omega \in A_{M,h}$, there exists an s such that $|W_t - W_s| \leq M|t - s|$ if $|t - s| \leq 2/n$; let k/n be the largest multiple of $1/n$ less than or equal to s. Then

$$|(k + 2)/n - s| \leq 2/n \qquad \text{and} \qquad |(k + 1)/n - s| \leq 2/n,$$

and therefore

$$|W_{(k+2)/n} - W_{(k+1)/n}| \leq |W_{(k+2)/n} - W_s| + |W_s - W_{(k+1)/n}|$$
$$\leq 2M/n + 2M/n < 4M/n.$$

Similarly $|W_{(k+1)/n} - W_{k/n}|$ and $|W_{k/n} - W_{(k-1)/n}|$ are less than $4M/n$.

Using the independent increments property, the stationary increments property, and (7.5),

$$\mathbb{P}(B_n) \leq 2n \sup_{k \leq 2n} \mathbb{P}(|W_{k/n} - W_{(k-1)/n}| < 4M/n, |W_{(k+1)/n} - W_{k/n}| < 4M/n,$$
$$|W_{(k+2)/n} - W_{(k+1)/n}| < 4M/n)$$
$$\leq 2n\mathbb{P}(|W_{1/n}| < 4M/n, |W_{2/n} - W_{1/n}| < 4M/n,$$
$$|W_{3/n} - W_{2/n}| < 4M/n)$$
$$= 2n\mathbb{P}(|W_{1/n}| < 4M/n)\mathbb{P}(|W_{2/n} - W_{1/n}| < 4M/n)$$
$$\times \mathbb{P}(|W_{3/n} - W_{2/n}| < 4M/n)$$
$$= 2n(\mathbb{P}(|W_{1/n}| < 4M/n))^3$$
$$\leq cn\left(\frac{4M}{\sqrt{n}}\right)^3,$$

which tends to 0 as $n \to \infty$. Hence for each M and h,

$$\mathbb{P}(A_{M,h}) \leq \limsup_{n \to \infty} \mathbb{P}(B_n) = 0.$$

This implies that the probability that there exists $s \leq 1$ such that

$$\limsup_{h \to 0} \frac{|W_{s+h} - W_s|}{|h|} \leq M$$

is zero. Since M is arbitrary, this proves the theorem. $\qquad\square$

Exercises

7.1 Here you are asked to find a more precise description of the modulus of continuity of Brownian paths. Prove that

$$\lim_{\delta \to 0} \sup_{s,t \in [0,1], 0 < |t-s| < \delta} \frac{|W_t - W_s|}{\sqrt{\delta \log(1/\delta)}} < \infty, \qquad \text{a.s.}$$

Hint: Imitate the proof of Theorem 7.1.

7.2 The following is part of what is known as *Chung's law of the iterated logarithm*. We will see in Section 40.3 that there exists c_1 such that

$$\mathbb{P}(\sup_{s \leq t} |W_s| \leq \lambda) \leq c_1 e^{-\pi^2 t / 8\lambda^2}$$

for t/λ^2 sufficiently large. Prove that

$$\liminf_{t \to \infty} \frac{\sup_{s \leq t} |W_s|}{\sqrt{t/\log \log t}} < \infty, \qquad \text{a.s.}$$

7.3 Let W_t be a one-dimensional Brownian motion. We will see in Section 40.3 that there exists c_2 such that

$$\mathbb{P}(\sup_{s \leq t} |W_s| \leq \lambda) \geq c_2 e^{-\pi^2 t / 8\lambda^2}$$

if t/λ^2 is sufficiently large. Prove that

$$\liminf_{t \to \infty} \frac{\sup_{s \leq t} |W_s|}{\sqrt{t/\log \log t}} > 0, \qquad \text{a.s.}$$

This is the other half of Chung's law of the iterated logarithm. In fact,

$$\liminf_{t \to \infty} \frac{\sup_{s \leq t} |W_s|}{\sqrt{t/\log \log t}} = c, \qquad \text{a.s.} \qquad (7.6)$$

Identify c and prove (7.6).

7.4 A function f is Hölder continuous of order α at a point t if there exists c such that $|f(u) - f(t)| \leq c|u - t|^\alpha$ for all u. Suppose $\alpha > 1/2$ and W_t is a Brownian motion. Show that the event

$$A = \{\exists t \in [0, 1] : W \text{ is Hölder continuous of order } \alpha \text{ at } t\}$$

has probability 0.

Hint: Imitate the proof of nowhere differentiability, but use more than three time intervals.

7.5   Let W be a one-dimensional Brownian motion and let $M_t = \sup_{s \leq t} W_s$ (with no absolute value signs). Prove that if $\varepsilon > 0$, then

$$\liminf_{t \to \infty} \frac{M_t}{\sqrt{t}/(\log t)^{1+\varepsilon}} > 0, \qquad \text{a.s.}$$

7.6   This is a complement to Exercise 4.10. Prove that if $p > 2$ and W is a Brownian motion, then the p-variation of W, defined in Exercise 4.10, is finite, a.s.

   Hint: Use the fact that the paths of Brownian motion are Hölder continuous of order α if $\alpha < 1/2$.

7.7   Let W be a Brownian motion and let Z be the zero set: $Z = \{t \in [0, 1] : W_t = 0\}$.
   (1) Show there exists a constant c not depending on x or δ such that

$$\mathbb{P}(\exists s \leq \delta : W_s = -x) \leq \mathbb{P}(\sup_{s \leq \delta} |W_s| \geq |x|) \leq ce^{-x^2/2\delta}.$$

   (2) Use the Markov property of Brownian motion to show that there exists a constant c not depending on s or t such that

$$\mathbb{P}(Z \cap [s, t] \neq \emptyset) \leq c\left(1 \wedge \sqrt{\frac{t-s}{s}}\right).$$

7.8   Given a Borel measurable subset A of $[0, 1]$, define

$$H_\gamma(A) = \limsup_{\delta \to 0} \left[\inf\left\{\sum_{i=1}^{\infty}[b_i - a_i]^\gamma : A \subset \cup_{i=1}^{\infty}[a_i, b_i], \sup_i |b_i - a_i| \leq \delta\right\}\right].$$

In other words, cover A by the union of intervals $[a_i, b_i]$ and define the analog of Lebesgue measure. The differences are that we look at $|b_i - a_i|^\gamma$ but do not require that γ be one, and we require that none of the intervals be longer than δ. The quantity $H_\gamma(A)$ is called the *Hausdorff measure* of A with respect to the function x^γ. The *Hausdorff dimension* of a set A is defined to be

$$\inf\{\gamma : H_\gamma(A) > 0\} = \sup\{\gamma : H_\gamma(A) = \infty\}.$$

(For subsets of $\mathbb{R}^d$, we replace the intervals $[a_i, b_i]$ by balls of radius r_i.) As a warm-up to this exercise, prove that the Hausdorff dimension of the standard Cantor set in $[0, 1]$ is $\log 2/\log 3$.

   The purpose of this exercise is to show that if W is a Brownian motion and $Z = \{t \in [0, 1] : W_t = 0\}$ is the zero set, then the Hausdorff dimension of Z is no more than $1/2$.

   (1) For each n, let $\mathcal{C}_n$ be the collection of intervals $[i/2^n, (i+1)/2^n]$ contained in $[0, 1]$ that intersect Z. ($\mathcal{C}_n$ is random.) If $\#\mathcal{C}_n$ is the cardinality of $\mathcal{C}_n$, use Exercise 7.7 to show

$$\mathbb{E}[\#\mathcal{C}_n] \leq \sum_{i=0}^{2^n-1} \mathbb{P}(Z \cap [i/2^n, (i+1)/2^n] \neq \emptyset) \leq c2^{n/2}.$$

   (2) Write

$$\sum_{[i/2^n, (i+1)/2^n] \in \mathcal{C}_n} |2^{-n}|^\gamma = 2^{-n\gamma} \#\mathcal{C}_n.$$

Use the Chebyshev inequality and (1) to conclude that the Hausdorff dimension of Z is less than or equal to $1/2$, a.s. (We will show that it is at least $1/2$ in Exercise 14.10.)

8

The continuity of paths

It is often important to know whether a stochastic path has continuous paths. An important sufficient condition is the Kolmogorov continuity criterion. This criterion is also useful in showing the continuity of a family of random variables X^a in the variable a, where a is a parameter other than time. Kolmogorov's continuity criterion is part (2) of Theorem 8.1.

Let $\mathcal{D}_n = \{k/2^n : k \leq 2^n\}$ and let $\mathcal{D} = \cup_n \mathcal{D}_n$. The set $\mathcal{D}$ is known as the set of *dyadic rationals* in $[0, 1]$. We will use

$$\sum_{i=1}^{\infty} i^{-2} \leq 1 + \int_1^{\infty} x^{-2} \, dx = 2.$$

(In fact by a standard exercise using Parseval's identity in the theory of Fourier series, $\sum_{i=1}^{\infty} i^{-2}$ is actually equal to $\pi^2/6$.)

We will be considering at first a real-valued process $\{X_t : t \in \mathcal{D}\}$. To show continuity by considering $X_t - X_s$ for all pairs (s, t) doesn't work – there are too many pairs. Kolmogorov's proof circumvents this problem by considering only a restricted collection of pairs. To bound $X_{15/32} - X_{11/32}$, for example, we compare $X_{15/32}$ to $X_{7/16}$, compare $X_{7/16}$ to $X_{3/8}$, and compare $X_{3/8}$ to $X_{1/4}$, and we also compare $X_{11/32}$ to $X_{5/16}$ and compare $X_{5/16}$ to $X_{1/4}$. The advantage of this complicated way of matching pairs is that each comparison, say, for example $X_{3/8}$ to $X_{1/4}$, is used for a great many of the possible pairs (s, t).

The proof of Theorem 8.1 has three main steps. Step 1 is to reduce the problem to proving the bound (8.3). The second step is to set up the comparisons that we need, and the third is to obtain estimates on all the comparisons.

Theorem 8.1 *Suppose $\{X_t : t \in \mathcal{D}\}$ is a real-valued process and there exist c_1, ε, and $p > 0$ such that*

$$\mathbb{E}\left[|X_t - X_s|^p\right] \leq c_1 |t - s|^{1+\varepsilon}, \qquad s, t \in \mathcal{D}. \tag{8.1}$$

Then the following hold.

(1) There exists c_2 depending only on c_1, p, and ε such that for $M > 0$,

$$\mathbb{P}\left(\sup_{s,t \in \mathcal{D}, s \neq t} \frac{|X_t - X_s|}{|t - s|^{\varepsilon/4p}} \geq M \right) \leq c_1/M^p. \tag{8.2}$$

(2) With probability one, X_t is uniformly continuous on $\mathcal{D}$.

Proof *Step 1.* Let $\lambda_n = M 2^{-(n+1)\varepsilon/4p}$ and

$$A_n = \left\{ |X_t - X_s| \geq \lambda_n \text{ for some } s, t \in \mathcal{D} \text{ with } |t - s| \leq 2^{-n} \right\}.$$

49

Recall our convention that the letter c denotes unimportant constants which can change from line to line. We will show

$$\mathbb{P}(A_n) \le c2^{-n\varepsilon/4}M^{-p}. \tag{8.3}$$

This implies (1) and (2) as follows. If $|X_t - X_t| \ge M|t - s|^{\varepsilon/4p}$ for some $s, t \in \mathcal{D}$ with $s \neq t$, choose n such that $2^{-(n+1)} < |t - s| \le 2^{-n}$, and then A_n holds. The event on the left-hand side of (8.2) is contained in $\cup_n A_n$, and using (8.3) shows that

$$\mathbb{P}(\cup_n A_n) \le cM^{-p} \sum_{n=1}^{\infty} 2^{-n\varepsilon/4} = cM^{-p},$$

which implies (1). Let

$$B_M = \{ \sup_{s,t \in \mathcal{D}, s \neq t} |X_t - X_t|/|t - s|^{\varepsilon/4p} \ge M\}.$$

Note B_M decreases as M increases and from (1) we have $\mathbb{P}(\cap_{M=1}^{\infty} B_M) = 0$. Thus except for an event of probability zero, each ω is in B_M^c for some M (where M depends on ω), and this implies (2). Thus we must show (8.3).

Step 2. Define $a(j, t)$ to be the integer multiple of 2^{-j} that is closest to t (if there are two different multiples that are equally close, we use some convention to break the tie). If $t \in \mathcal{D}_m$, then $a(m, t) = t$. If $|t - s| \le 2^{-n}$, then $|a(n, t) - a(n, s)| \le 2^{-n+2}$.

Now if $s, t \in \mathcal{D}_m$ and $m \ge n$, we use the triangle inequality to write

$$\begin{aligned}
|X_t - X_s| &= |X_{a(m,t)} - X_{a(m,s)}| \tag{8.4}\\
&\le |X_{a(n,t)} - X_{a(n,s)}|\\
&\quad + |X_{a(n+1,t)} - X_{a(n,t)}| + \cdots + |X_{a(m,t)} - X_{a(m-1,t)}|\\
&\quad + |X_{a(n+1,s)} - X_{a(n,s)}| + \cdots + |X_{a(m,s)} - X_{a(m-1,s)}|.
\end{aligned}$$

If $|X_{a(n,t)} - X_{a(n,s)}| < \lambda_n/2$ and for each i

$$|X_{a(n+i+1,t)} - X_{a(n+i,t)}| < \frac{\lambda_n}{8(i+1)^2}$$

and the same with t replaced by s, then by (8.4)

$$|X_t - X_s| < \frac{\lambda_n}{2} + 2\sum_{i=0}^{\infty} \frac{\lambda_n}{8(i+1)^2} \le \lambda_n.$$

Hence if $|X_t - X_s| \ge \lambda_n$ for some $s, t \in \mathcal{D}_m$, then at least one of the events E, F_i, or G_i, $i \ge 0$, must hold, where

$$\begin{aligned}
E &= \{|X_{a(n,t)} - X_{a(n,s)}| \ge \lambda_n/2 \text{ for some } s, t \in \mathcal{D}_n \text{ with } |s - t| \le 2^{-n}\},\\
F_i &= \{|X_{a(n+i+1,t)} - X_{a(n+i,t)}| \ge \lambda_n/8(i+1)^2 \text{ for some } t\},\\
G_i &= \{|X_{a(n+i+1,s)} - X_{a(n+i,s)}| \ge \lambda_n/8(i+1)^2 \text{ for some } s\}.
\end{aligned}$$

Step 3. For the event E to hold, we must have $|X_r - X_q| \geq \lambda_n/2$ for some $q, r \in \mathcal{D}_n$ with $|q - r| \leq 2^{-n+2}$. There are at most $c2^n$ such pairs (q, r), so the probability of E is bounded, using Chebyshev's inequality and (8.1), by

$$(c2^n) \sup_{q \in \mathcal{D}_n, r \in \mathcal{D}_{n+1}, |r-q| \leq 2^{-n+2}} \mathbb{P}(|X_r - X_q| \geq \lambda_n/2)$$

$$\leq c2^n \frac{\sup_{q \in \mathcal{D}_n, r \in \mathcal{D}_{n+1}, |r-q| \leq 2^{-n+2}} \mathbb{E}[|X_r - X_q|^p]}{(\lambda_n/2)^p}$$

$$\leq \frac{c2^n}{\lambda_n^p}(2^{-n+2})^{1+\varepsilon}$$

$$\leq \frac{c2^{-n\varepsilon}}{\lambda_n^p}.$$

For F_i to hold, that is, for $|X_{a(n+i+1,t)} - X_{a(n+i,t)}|$ to be greater than $\lambda_n/8(i+1)^2$ for some t, we must have $|X_r - X_q| \geq \lambda_n/8(i+1)^2$ for some $r \in \mathcal{D}_{n+i}, q \in \mathcal{D}_{n+i+1}$ with $|r-q| \leq 2^{-n-i+2}$. There are at most $c2^{n+i}$ such pairs, and so the probability of F_i is bounded by

$$(c2^{n+i}) \sup_{r \in \mathcal{D}_{n+i}, q \in \mathcal{D}_{n+i+1}, |r-q| \leq 2^{-n-i+2}} \mathbb{P}\left(|X_r - X_q| \geq \frac{\lambda_n}{8(i+1)^2}\right)$$

$$\leq c \frac{2^{n+i}2^{(-n-i+2)(1+\varepsilon)}(8(i+1)^2)^p}{\lambda_n^p}$$

$$\leq \frac{c2^{-n\varepsilon}2^{-i\varepsilon/2}}{\lambda_n^p}.$$

Here we used the fact that $2^{-i\varepsilon}(i+1)^{2p} \leq c2^{-i\varepsilon/2}$ for some constant c depending on p and ε but not i. We have the same bound for G_i. Therefore

$$\mathbb{P}(\cup_i(F_i \cup G_i) \cup E) \leq \sum_{i=0}^{\infty} \frac{c2^{-n\varepsilon/2}2^{-i\varepsilon/2}}{\lambda_n^p} + \frac{c2^{-n\varepsilon/2}}{\lambda_n^p} \leq c2^{-n\varepsilon/2}\lambda_n^{-p}.$$

Letting $m \to \infty$ we have

$$\mathbb{P}(A_n) \leq c2^{-n\varepsilon/2}\lambda_n^{-p} = c2^{-n\varepsilon/4}M^{-p}$$

as required. □

The proof of Theorem 8.1 is an example of what is known as a *metric entropy* or *chaining* argument.

In the above, the only place we relied on the fact that we were using real-valued processes was in using the triangle inequality. Therefore with only slight changes in notation, we have the following theorem.

Theorem 8.2 *Suppose X takes values in some metric space $\mathcal{S}$ with metric $d_{\mathcal{S}}$ and there exist c_1, ε, and $p > 0$ such that*

$$\mathbb{E}\left[d_{\mathcal{S}}(X_s, X_t)^p\right] \leq c_1 |t - s|^{1+\varepsilon}, \qquad s, t \in \mathcal{D}. \tag{8.5}$$

Then the following hold.

(1) There exists c_2 depending only on c_1, p, and ε such that for $M > 0$,

$$\mathbb{P}\left(\sup_{s,t \in \mathcal{D}, s \neq t} \frac{d_{\mathcal{S}}(X_s, X_t)}{|t - s|^{\varepsilon/2p}} \geq M \right) \leq c_1/M^p.$$

(2) With probability one, X_t is uniformly continuous on $\mathcal{D}$.

Remark 8.3 Theorem 8.2 holds for random variables indexed by time, but the analogous result holds for the continuity in a of random variables X^a indexed by some parameter a running through $\mathcal{D}$. We may also let the parameter a run instead through the dyadic rationals in $[b_1, b_2]$ for any $b_1 < b_2$.

The proof of the following corollary is an adaptation of the proof of Theorem 8.1 and is left as Exercise 8.1.

Corollary 8.4 *Suppose there exist c_1, ε, N, and $p > 0$ such that if $n \leq N$,*

$$\mathbb{E}\left[d_{\mathcal{S}}(X_s, X_t)^p\right] \leq c|t - s|^{1+\varepsilon}, \qquad s, t \in \mathcal{D}_n.$$

Then there exists c_2 depending on c_1, ε, and p but not N such that for $M > 0$ and $n \leq N$ we have

$$\mathbb{P}\left(\sup_{s,t \in \mathcal{D}_n, s \neq t} \frac{d_{\mathcal{S}}(X_s, X_t)}{|t - s|^{\varepsilon/2p}} \geq M \right) < c_2 M^{-p}.$$

Recall the definition of Hölder continuity from (7.1).

Proposition 8.5 *If $\alpha < 1/2$, then the paths of a one-dimensional Brownian motion $\{W_t; 0 \leq t \leq 1\}$ are Hölder continuous of order α with probability one.*

Proof By the stationary increments property and scaling,

$$\mathbb{E}\,|W_t - W_s|^p = \mathbb{E}\,|W_{t-s}|^p = |t - s|^{p/2} \mathbb{E}\,|W_1|^p.$$

If $\alpha < 1/2$, choose p large enough so that $((p/2) - 1)/p > \alpha$ and then take $\varepsilon = (p/2) - 1$. (Here ε is large!) Take γ sufficiently small that $(\varepsilon/p) - \gamma > \alpha$. Then by Exercise 8.2 the paths of W_t are Hölder continuous of order α, with probability one, provided we restrict t to $\mathcal{D}$. But the paths of Brownian motion are continuous, so we see that we have Hölder continuity of order α when $t \in [0, 1]$. $\square$

Exercises

8.1   Prove Corollary 8.4.

8.2   If the hypothesis of Theorem 8.1 holds and $\gamma < \varepsilon/p$, show that there exists c_2 depending only on c_1, ε, γ, and p such that for $M > 0$

$$\mathbb{P}\left(\sup_{s,t \in \mathcal{D}, s \neq t} \frac{d_{\mathcal{S}}(X_s, X_t)}{|t - s|^{(\varepsilon/p)-\gamma}} \geq M \right) \leq cM^{-p}.$$

8.3 Suppose X is a real-valued process and there exist constants c_1, c_2 such that

$$\mathbb{P}(|X_t - X_s| > \lambda) \leq c_1 e^{-c_2 \lambda \log^4(1/|t-s|)}, \qquad s, t \in [0, 1].$$

Prove that with probability one, X has a version which is uniformly continuous on the dyadic rationals in $[0, 1]$.

8.4 Suppose $(X_t, t \in [0, 1])$ is a mean zero Gaussian process and there exist c and ε such that

$$\text{Var}\,(X_t - X_s) \leq c|t - s|^\varepsilon, \qquad s, t \in [0, 1].$$

Prove that there is a version of X that has continuous paths on $[0, 1]$.

8.5 Let X be as in Exercise 8.4. For what values α will X have paths that are Hölder continuous of order α? (α will depend on ε.)

8.6 Let $\{X_{s,t} : s, t \in [0, 1]\}$ be a collection of random variables. Suppose there exist c, p, and $\varepsilon > 0$ such that

$$\mathbb{E}\,|X_{s',t'} - X_{s,t}|^p \leq c(|t' - t| + |s' - s|)^{2+\varepsilon}.$$

Prove that with probability one, the map $(s, t) \to X_{s,t}(\omega)$ is uniformly continuous on $\mathcal{D} \times \mathcal{D} = \{(s, t); s, t \in \mathcal{D}\}$.

9

Continuous semimartingales

Roughly speaking, a semimartingale is the sum of a martingale and a process whose paths are of bounded variation. In this chapter we consider semimartingales whose paths are continuous. We will give definitions, and then investigate in more detail the class of martingales that are square integrable. Finally we present a proof of the Doob–Meyer decomposition for continuous supermartingales. The Doob–Meyer decomposition used to be considered a very hard theorem, but at least in the continuous case, an elementary proof is possible. For a proof for the general case, see Chapter 16.

9.1 Definitions

Let $\{\mathcal{F}_t\}$ be a filtration satisfying the usual conditions and let

$$\mathcal{F}_\infty = \bigvee_{t \geq 0} \mathcal{F}_t = \sigma\left(\bigcup_{t \geq 0} \mathcal{F}_t\right).$$

We say a process X has *increasing paths* or that X is an *increasing process* if the functions $t \to X_t(\omega)$ are increasing with probability one. Throughout this book saying f is "increasing" means that $s < t$ implies $f(s) \leq f(t)$, while saying f is "strictly increasing" means that $s < t$ implies $f(s) < f(t)$. A process X with *paths of bounded variation* is just what one would expect: with probability one, the functions $t \to X_t(\omega)$ are of bounded variation. We say X has *paths locally of bounded variation* if there exist stopping times $R_n \to \infty$ such that the process $X_{t \wedge R_n}$ has paths of bounded variation for each n.

We turn to martingales. A martingale M is a *uniformly integrable martingale* if the family of random variables $\{M_t\}$ is uniformly integrable. A process X is a *local martingale* if there exist stopping times $R_n \to \infty$ such that $M_t^n = X_{t \wedge R_n}$ is a uniformly integrable martingale for each n. A martingale whose paths are continuous is called a *continuous martingale* and we similarly define a *right-continuous martingale*.

A *semimartingale* is a process X of the form $X_t = M_t + A_t$, where M_t is a local martingale and A_t is a process whose paths are locally of bounded variation. As a consequence of the Doob–Meyer decomposition we will see that submartingales and supermartingales are semimartingales.

As an example, a Brownian motion W_t is a martingale and is a local martingale (let R_n be identically equal to n), but is not a uniformly integrable martingale. We will define what it means to be a square integrable martingale in the next section; Brownian motion is not a square integrable martingale.

9.2 Square integrable martingales

Definition 9.1 A martingale is a *square integrable martingale* if there exists a $\mathcal{F}_\infty$ measurable random variable M_∞ such that $\mathbb{E}\, M_\infty^2 < \infty$ and $M_t = \mathbb{E}\,[M_\infty \mid \mathcal{F}_t]$ for all t.

An example of a square integrable martingale would be $M_t = W_{t \wedge t_0}$, where W_t is a Brownian motion and t_0 is a fixed time; in this case $M_\infty = W_{t_0}$.

Proposition 9.2 *Let $\{\mathcal{F}_t\}$ be a filtration satisfying the usual conditions and M a right continuous process. The following are equivalent:*
(1) M_t is a square integrable martingale.
(2) M is a martingale with $\sup_{t \geq 0} \mathbb{E}\, M_t^2 < \infty$.
(3) M is a martingale with $\mathbb{E}\,[\sup_{t \geq 0} M_t^2] < \infty$.

Proof To show (1) implies (2), suppose M is a square integrable martingale. Then by Jensen's inequality for conditional expectations (Proposition A.21),

$$\mathbb{E}\, M_t^2 = \mathbb{E}\,[(\mathbb{E}\,[M_\infty \mid \mathcal{F}_t])^2] \leq \mathbb{E}\,[\mathbb{E}\,[M_\infty^2 \mid \mathcal{F}_t]] = \mathbb{E}\, M_\infty^2.$$

To show (2) implies (3), for each N,

$$\mathbb{E}\,[\sup_{0 \leq t \leq N} M_t^2] \leq 4\mathbb{E}\, M_N^2$$

by Doob's inequalities. That (2) implies (3) follows by letting $N \to \infty$ and using Fatou's lemma.

Now suppose (3) holds, and we will show (1) holds. Since $\mathbb{E}\, M_n^2$ is uniformly bounded in n, the martingale convergence theorem (Theorem A.35) implies that M_n converges almost surely and in L^2. Let us call the limit M_∞; we have $\mathbb{E}\, M_\infty^2 < \infty$ by the L^2 convergence. Since $\mathbb{E}\, M_n^2$ is uniformly bounded, then M_n is a uniformly integrable martingale, and by Proposition A.37, $M_n = \mathbb{E}\,[M_\infty \mid \mathcal{F}_n]$. If $n - 1 \leq t \leq n$, we have

$$M_t = \mathbb{E}\,[M_n \mid \mathcal{F}_t] = \mathbb{E}\,[\mathbb{E}\,[M_\infty \mid \mathcal{F}_n] \mid \mathcal{F}_t] = \mathbb{E}\,[M_\infty \mid \mathcal{F}_t],$$

as required. $\square$

For the remainder of this section all our martingales will have paths that are right continuous with left limits.

Proposition 9.3 *If M is a square integrable martingale and $S \leq T$ are finite stopping times, then $\mathbb{E}\,[M_T \mid \mathcal{F}_S] = M_S$.*

Proof Let $A \in \mathcal{F}_S$ and define $U(\omega) = S(\omega)1_A(\omega) + T(\omega)1_{A^c}(\omega)$. Thus U is equal to S if $\omega \in A$ and otherwise is equal to T. Since $A \in \mathcal{F}_S \subset \mathcal{F}_T$, then we have $(U \leq t) = [(S \leq t) \cap A] \cup [(T \leq t) \cap A^c]$ is in $\mathcal{F}_t$, and therefore U is a stopping time. By Proposition 3.11,

$$\mathbb{E}\, M_0 = \mathbb{E}\, M_U = \mathbb{E}\,[M_S; A] + \mathbb{E}\,[M_T; A^c]$$

and

$$\mathbb{E}\, M_0 = \mathbb{E}\, M_T = \mathbb{E}\,[M_T; A] + \mathbb{E}\,[M_T; A^c].$$

These two equations imply that $\mathbb{E}[M_S; A] = \mathbb{E}[M_T; A]$, which is what we needed to prove. $\square$

By Exercise 3.11, the conclusion is valid if M is a uniformly integrable martingale. As an immediate corollary we have

Corollary 9.4 *Suppose M is a square integrable martingale and T is a stopping time. Then $X_t = M_{t \wedge T}$ is a martingale with respect to $\{\mathcal{F}_{t \wedge T}\}$.*

The proof of the following proposition is similar to that of Proposition 9.3. It may be viewed as a converse of the optional stopping theorem.

Proposition 9.5 *Suppose $\{\mathcal{F}_t\}$ is a filtration satisfying the usual conditions and M is a process that is adapted to $\{\mathcal{F}_t\}$ such that M_t is integrable for each t. If $\mathbb{E} M_T = 0$ for every bounded stopping time T, then M_t is a martingale.*

Proof Suppose $s < t$ and $A \in \mathcal{F}_s$. Define T to be equal to s if $\omega \in A$ and equal to t if $\omega \notin A$. As in the proof of Proposition 9.3, but even more simply, T is a stopping time, so

$$0 = \mathbb{E} M_T = \mathbb{E}[M_s; A] + \mathbb{E}[M_t; A^c].$$

The fixed time t is a stopping time, hence

$$0 = \mathbb{E} M_t = E[M_t; A] + \mathbb{E}[M_t; A^c].$$

Comparing, $\mathbb{E}[M_t; A] = \mathbb{E}[M_s; A]$, which proves M is a martingale. $\square$

Proposition 9.6 *Suppose M_t is a square integrable martingale. Then*

$$\mathbb{E}[(M_T - M_S)^2 \mid \mathcal{F}_S] = \mathbb{E}[M_T^2 - M_S^2 \mid \mathcal{F}_S]. \tag{9.1}$$

Proof By Proposition 9.3

$$\begin{aligned}
\mathbb{E}[(M_T - M_S)^2 \mid \mathcal{F}_S] &= \mathbb{E}[M_T^2 \mid \mathcal{F}_S] - 2M_S \mathbb{E}[M_T \mid \mathcal{F}_S] + M_S^2 \\
&= \mathbb{E}[M_T^2 \mid \mathcal{F}_S] - M_S^2 \\
&= \mathbb{E}[M_T^2 - M_S^2 \mid \mathcal{F}_S]
\end{aligned}$$

and we are done. $\square$

If we take expectations in (9.1), we obtain

$$\mathbb{E}[(M_T - M_S)^2] = \mathbb{E} M_T^2 - \mathbb{E} M_S^2. \tag{9.2}$$

Theorem 9.7 *Suppose $M_0 = 0$, M_t is a continuous local martingale, and the paths of M_t are locally of bounded variation. Then M is identically 0, a.s., that is, $\mathbb{P}(M_t = 0 \text{ for all } t) = 1$.*

Proof Using the definition of local martingale, it suffices to suppose M is a continuous uniformly integrable martingale. Let t_0 be fixed and let A_t denote the total variation of the paths of M up to time t. If $T_N = \inf\{t : A_t \geq N\}$, we look at $M_t^N = M_{T_N \wedge t \wedge t_0}$. Using Proposition 9.3 and the remark following it, we see that M^N is also a continuous martingale with paths of bounded variation, and if M^N is identically zero, then letting $N \to \infty$ and $t_0 \to \infty$, we obtain our result. Therefore it suffices to suppose the total variation of M_t is bounded by N, a.s. In particular, M_t is bounded by N.

Let $n \geq 1$ and set

$$V_n = \sup_{k \leq 2^n - 1} |M_{(k+1)t_0/2^n} - M_{kt_0/2^n}|.$$

Note $V_n \leq 2N$, a.s., and $V_n \to 0$, a.s., as $n \to \infty$ by the uniform continuity of the paths of M on $[0, t_0]$. By dominated convergence, $\mathbb{E} V_n \to 0$ as $n \to \infty$. We write

$$\mathbb{E} M_{t_0}^2 = \mathbb{E} \left[\sum_{k=0}^{2^n - 1} (M_{(k+1)t_0/2^n}^2 - M_{kt_0/2^n}^2) \right]$$

$$= \mathbb{E} \left[\sum_{k=0}^{2^n - 1} (M_{(k+1)t_0/2^n} - M_{kt_0/2^n})^2 \right]$$

$$\leq \mathbb{E} \left[V_n \sum_{k=0}^{2^n - 1} |M_{(k+1)t_0/2^n} - M_{kt_0/2^n}| \right]$$

$$\leq N \mathbb{E} V_n.$$

The second equality follows by (9.2). Since n is arbitrary and $\mathbb{E} V_n \to 0$, then $\mathbb{E} M_{t_0}^2 = 0$. By Doob's inequalities, $\mathbb{E}[\sup_{s \leq t_0} M_s^2] = 0$. Hence M is identically 0 up to time t_0. $\qquad \square$

9.3 Quadratic variation

Definition 9.8 A continuous square integrable martingale M_t has *quadratic variation* $\langle M \rangle_t$ (sometimes written $\langle M, M \rangle_t$) if $M_t^2 - \langle M \rangle_t$ is a martingale, where $\langle M \rangle_t$ is a continuous adapted increasing process with $\langle M \rangle_0 = 0$.

In the case where W is a Brownian motion, t_0 is fixed, and $M_t = W_{t \wedge t_0}$ the quadratic variation of M is just $\langle M \rangle_t = t \wedge t_0$ by Example 3.3. Brownian motion itself does not fit perfectly into the framework of stochastic integration because it is not a square integrable martingale, although it is a martingale; we will be dealing with this point several times in what follows.

We will show existence and uniqueness of $\langle M \rangle_t$ by means of the Doob–Meyer decomposition, Theorem 9.12, below. However we defer the proof of the Doob–Meyer decomposition until the next section. A process Z is of *class D* if $\{Z_T : T \text{ a finite stopping time}\}$ is a uniformly integrable family of random variables.

Theorem 9.9 *Let M_t be a continuous square integrable martingale. There exists a continuous adapted increasing process $\langle M \rangle_t$ with $\langle M \rangle_0 = 0$ and with increasing paths such that $M_t^2 - \langle M \rangle_t$ is a martingale.*

If A_t is a continuous adapted increasing process such that $M_t^2 - A_t$ is a martingale, then $\mathbb{P}(A_t \neq \langle M \rangle_t \text{ for some } t) = 0$.

Proof By Jensen's inequality for conditional expectations,

$$\mathbb{E}[M_t^2 \mid \mathcal{F}_s] \geq (\mathbb{E}[M_t \mid \mathcal{F}_s])^2 = M_s^2$$

if $s < t$, and so M_t^2 is a submartingale. Since M_∞ is square integrable, given ε there exists δ such that $\mathbb{E}[M_\infty^2; A] < \varepsilon$ if $\mathbb{P}(A) < \delta$. Since M_t^2 is a submartingale, if $K > \mathbb{E} M_\infty^2 / \delta$, then

$$\mathbb{P}(M_t^2 > K) \leq \mathbb{E} M_t^2 / K \leq \mathbb{E} M_\infty^2 / K < \delta,$$

and consequently

$$\mathbb{E}\,[M_t^2; M_t^2 > K] \le \mathbb{E}\,[M_\infty^2; M_t^2 > K] < \varepsilon.$$

By Exercise 3.11, M_t^2 is of class D. Applying the Doob–Meyer decomposition (Theorem 9.12) to $-M_t^2$, we write $-M_t^2 = N_t - B_t$, where N_t is a martingale and B_t has increasing paths. We then set $\langle M \rangle_t = B_t$. The uniqueness follows from the uniqueness part of the Doob–Meyer decomposition. $\square$

In view of Proposition 9.3 and the definition of $\langle M \rangle$, we have

$$\mathbb{E}\,[(M_T - M_S)^2 - (\langle M \rangle_T - \langle M \rangle_S) \mid \mathcal{F}_S] \tag{9.3}$$
$$= \mathbb{E}\,[M_T^2 - M_S^2 - (\langle M \rangle_T - \langle M \rangle_S) \mid \mathcal{F}_S] = 0$$

if S and T are finite stopping times and M is a continuous square integrable martingale.

If M and N are two square integrable martingales, we define $\langle M, N \rangle_t$ by

$$\langle M, N \rangle_t = \tfrac{1}{2}[\langle M + N \rangle_t - \langle M \rangle_t - \langle N \rangle_t]. \tag{9.4}$$

This is sometimes called the *covariation* of M and N.

An alternative representation of $\langle M \rangle_t$ is the following. A proof could be given now, but it is a bit messy. After we have Itô's formula this will be easier.

Theorem 9.10 *Let M be a square integrable martingale and let $t_0 > 0$. Then $\langle M \rangle_t$ is the limit in probability of*

$$\sum_{k=0}^{[2^n t_0]} (M_{(k+1)/2^n} - M_{k/2^n})^2,$$

where $[2^n t_0]$ is the largest integer less than or equal to $2^n t_0$.

9.4 The Doob–Meyer decomposition

In this section we give a proof of the Doob–Meyer decomposition for continuous supermartingales. First we need the following inequality, which has many other uses as well.

Proposition 9.11 *Suppose A^1 and A^2 are two increasing adapted continuous processes starting at zero with $A_\infty^i = \lim_{t \to \infty} A_t^i < \infty$, a.s., $i = 1, 2$, and suppose there exists a positive real K such that for all t,*

$$\mathbb{E}\,[A_\infty^i - A_t^i \mid \mathcal{F}_t] \le K, \qquad \text{a.s.}, \qquad i = 1, 2. \tag{9.5}$$

Let $B_t = A_t^1 - A_t^2$. Suppose there exists a non-negative random variable V with $\mathbb{E}\,V^2 < \infty$ such that for all t,

$$|\mathbb{E}\,[B_\infty - B_t \mid \mathcal{F}_t]| \le \mathbb{E}\,[V \mid \mathcal{F}_t], \qquad \text{a.s.} \tag{9.6}$$

Then

$$\mathbb{E}\,\sup_{t \ge 0} B_t^2 \le 8\mathbb{E}\,V^2 + 8\sqrt{2}K(\mathbb{E}\,V^2)^{1/2}. \tag{9.7}$$

Proof We start by showing

$$\mathbb{E}\,(A_\infty^i)^2 \le 2K^2, \qquad i = 1, 2. \tag{9.8}$$

First suppose A^i_∞ is bounded by a positive real number L. Note that we have $\mathbb{E} A^i_\infty = \mathbb{E}\left[\mathbb{E}\left[A^i_\infty - A^i_0 \mid \mathcal{F}_0\right]\right] \leq K$. A simple calculation shows that

$$(A^i_\infty)^2 = 2\int_0^\infty (A^i_\infty - A^i_t)\, dA^i_t.$$

We then have, using Proposition 3.14,

$$\begin{aligned}
\mathbb{E}\,(A^i_\infty)^2 &= 2\mathbb{E}\int_0^\infty (A^i_\infty - A^i_t)\, dA^i_t \\
&= 2\mathbb{E}\int_0^\infty (\mathbb{E}\left[A^i_\infty \mid \mathcal{F}_t\right] - A^i_t)\, dA^i_t \\
&= 2\mathbb{E}\int_0^\infty \mathbb{E}\left[A^i_\infty - A^i_t \mid \mathcal{F}_t\right] dA^i_t \\
&\leq 2K\mathbb{E}\int_0^\infty dA^i_t = 2K\mathbb{E} A^i_\infty \leq 2K^2.
\end{aligned}$$

If we let $T_L = \inf\{t : A^1_t + A^2_t \geq L\}$ and $A^{i,L}_t = A^i_{t \wedge T_L}$, then (9.5) still holds if we replace A^i_t by $A^{i,L}_t$. We obtain $\mathbb{E}\,(A^{i,L}_\infty)^2 \leq 2K^2$, and then letting $L \to \infty$ and using Fatou's lemma proves (9.8).

We next write

$$B^2_\infty = 2\int_0^\infty (B_\infty - B_t)\, dB_t,$$

and hence

$$\begin{aligned}
\mathbb{E}\, B^2_\infty &= 2\mathbb{E}\int_0^\infty \mathbb{E}\left[B_\infty - B_t \mid \mathcal{F}_t\right] dB_t \\
&\leq \mathbb{E}\int_0^\infty \mathbb{E}\left[V \mid \mathcal{F}_t\right] d(A^1_t + A^2_t) \\
&= \mathbb{E}\int_0^\infty V\, d(A^1_t + A^2_t) \\
&= \mathbb{E}\left[V(A^1_\infty + A^2_\infty)\right].
\end{aligned}$$

The bound (9.8) takes care of the integrability concerns. By the Cauchy–Schwarz inequality we obtain

$$\mathbb{E}\, B^2_\infty \leq (\mathbb{E}\left[(A^1_\infty + A^2_\infty)^2\right])^{1/2}(\mathbb{E}\, V^2)^{1/2} \leq 2\sqrt{2}K(\mathbb{E}\, V^2)^{1/2}.$$

Now let $M_t = \mathbb{E}\left[B_\infty \mid \mathcal{F}_t\right]$, $N_t = \mathbb{E}\left[V \mid \mathcal{F}_t\right]$, where we take the right–continuous versions (see Corollary 3.13), and let $X_t = M_t - B_t$. We have

$$|X_t| = |\mathbb{E}\left[B_\infty - B_t \mid \mathcal{F}_t\right]| \leq N_t,$$

and using Doob's inequalities,

$$\mathbb{E}\,\sup_{t\geq 0} X^2_t \leq \mathbb{E}\,\sup_{t\geq 0} N^2_t \leq 4\mathbb{E}\, N^2_\infty = 4\mathbb{E}\, V^2.$$

Also by Doob's inequalities,

$$\mathbb{E}\,\sup_{t\geq 0} M^2_t \leq 4\mathbb{E}\, M^2_\infty = 4\mathbb{E}\, B^2_\infty.$$

Since $\sup_{t\geq 0} |B_t| \leq \sup_{t\geq 0} |X_t| + \sup_{t\geq 0} |M_t|$, our result follows. $\square$

We now prove the Doob–Meyer decomposition for continuous supermartingales. In view of the proof of Proposition A.30, we would like to let

$$A_t = \int_0^t \mathbb{E}\left[\frac{dZ_s}{ds} \mid \mathcal{F}_s\right] ds,$$

but this doesn't make sense. We instead define an approximation A_t^h by (9.9) and show that A_t^h converges to what we want as $h \to 0$.

Theorem 9.12 *Suppose Z_t is a continuous adapted supermartingale of class D. Then there exists an increasing adapted continuous process A_t with paths locally of bounded variation started at 0 and a continuous local martingale M_t such that*

$$Z_t = M_t - A_t.$$

If M' and A' are two other such processes with $Z_t = M_t' - A_t'$, then $M_t = M_t'$ and $A_t = A_t'$ for all t, a.s.

Proof Let us prove the second assertion first. Let S_N be the first time that $|M_t| + |M_t'|$ exceeds N. If

$$Z_t = M_t - A_t = M_t' - A_t',$$

then $M_{t \wedge S_N} - M_{t \wedge S_N}' = A_{t \wedge S_N} - A_{t \wedge S_N}'$ is a martingale whose paths are locally of bounded variation. By Theorem 9.7, $M_{t \wedge S_N} = M_{t \wedge S_N}'$, a.s. Since this is true for all N, then $M_t = M_t'$.

Now let us prove the existence of M and A. Let $T_N = \inf\{t : |Z_t| \geq N\} \wedge N$ and $Z_t^N = Z_{t \wedge T_N}$. By Exercise 9.2, Z^N is a supermartingale. If we prove the decomposition $Z_t^N = M_t^N - A_t^N$ for each N, then by the uniqueness assertion, if $N_1 < N_2$, we have $A_t^{N_1}$ and $M_t^{N_1}$ agreeing with $A_t^{N_2}$ and $M_t^{N_2}$, respectively, for $t \leq T_{N_1}$. Hence given t, we can choose N large enough so that $t \leq T_N$ and then define $M_t = M_t^N$, $A_t = A_t^N$. Clearly this gives the desired decomposition. Thus we may suppose that Z_t is bounded by some N and that Z_t is constant for $t \geq N$.

Let $V_\delta = \sup_{|t-s| \leq \delta} |Z_t - Z_s|$. Since Z has continuous paths,

$$V_\delta = \sup_{s,t \in \mathbb{Q}_+, |t-s| \leq \delta} |Z_t - Z_s|,$$

and therefore V_δ is measurable with respect to $\mathcal{F}_\infty$. Since the paths of Z are uniformly continuous, $V_\delta \to 0$, a.s., as $\delta \to 0$, and since $|V_\delta| \leq 2N$, we have by dominated convergence that $\mathbb{E} V_\delta^2 \to 0$ as $\delta \to 0$.

We define

$$A_t^h = \frac{1}{h} \int_0^t (Z_s - \mathbb{E}[Z_{s+h} \mid \mathcal{F}_s]) \, ds. \tag{9.9}$$

At this point we do not know even that $\mathbb{E}[Z_{s+h} \mid \mathcal{F}_s]$ has any nice measurability properties (it is not a martingale, for example); let us assume that it has a version that has continuous paths, is adapted, and is jointly measurable in t and ω, and prove this fact a bit later on. Because Z is a supermartingale, A^h is increasing. We have (note Exercise 9.6)

$$
\begin{aligned}
E[A^h_\infty - A^h_t \mid \mathcal{F}_t] &= \frac{1}{h} \mathbb{E}\Big[\int_t^\infty \mathbb{E}[Z_s - Z_{s+h} \mid \mathcal{F}_s] \, ds \mid \mathcal{F}_t \Big] \\
&= \frac{1}{h} \int_t^\infty \mathbb{E}[Z_s - Z_{s+h} \mid \mathcal{F}_t] \, ds \\
&= \frac{1}{h} \mathbb{E}\Big[\int_t^\infty Z_s \, ds - \int_{t+h}^\infty Z_s \, ds \mid \mathcal{F}_t \Big] \\
&= \frac{1}{h} \mathbb{E}\Big[\int_t^{t+h} Z_s \, ds \mid \mathcal{F}_t \Big] \\
&= \mathbb{E}\Big[\int_0^1 Z_{t+uh} \, du \mid \mathcal{F}_t \Big].
\end{aligned}
$$

Since Z is bounded by N, it follows that A^h satisfies (9.5). If $k < h$, then

$$
\begin{aligned}
|\mathbb{E}[(A^h_\infty - A^h_t) - (A^k_\infty - A^k_t) \mid \mathcal{F}_t]| &= \Big| \mathbb{E}\Big[\int_0^1 (Z_{t+uh} - Z_{t+uk}) \, du \mid \mathcal{F}_t \Big] \Big| \\
&\leq \mathbb{E}[V_h \mid \mathcal{F}_t].
\end{aligned}
$$

Now apply Proposition 9.11 to see that $\mathbb{E}\sup_{t\geq 0}(A^h_t - A^k_t)^2 \to 0$ as $k, h \to 0$. This shows that whenever h_n decreases to 0, then A^{h_n} is a Cauchy sequence in a normed linear space, where the norm is given by

$$
\|X\| = (\mathbb{E}\sup_{t\geq 0} |X_t|^2)^{1/2}, \tag{9.10}
$$

which is complete by Exercise 9.5. Therefore there exists a limit A. Since

$$
\mathbb{E}\sup_{t\geq 0}(A^h_t - A_t)^2 \to 0
$$

as $h \to 0$, there exists a subsequence $h_n \to 0$ such that $\sup_{t\geq 0}(A^{h_n}_t - A_t)^2 \to 0$, a.s., which proves that A_t is continuous and increasing.

We calculate

$$
\begin{aligned}
\mathbb{E}[A_\infty - A_t \mid \mathcal{F}_t] &= \lim_{h\to 0} \mathbb{E}[A^h_\infty - A^h_t \mid \mathcal{F}_t] \\
&= \lim_{h\to 0} \mathbb{E}\Big[\int_0^1 Z_{t+uh} \, du \mid \mathcal{F}_t \Big] \\
&= \mathbb{E}\Big[\int_0^1 Z_t \, du \mid \mathcal{F}_t \Big] \\
&= Z_t.
\end{aligned}
$$

Therefore

$$
Z_t = \mathbb{E}[A_\infty \mid \mathcal{F}_t] - A_t,
$$

which is the decomposition of Z into a martingale minus an increasing process.

Fix h. It remains to show that there is a version of $\mathbb{E}[Z_{s+h} \mid \mathcal{F}_s]$ that is a continuous jointly measurable adapted process. Define $Y_t = Z_{t+h}$ and define Y_t^n to be equal to $Y_{k/2^n}$ if $k/2^n \le t < (k+1)/2^n$. Take the right-continuous version $\widetilde{Y}_t^{k,n}$ of the martingale $\mathbb{E}[Y_{k/2^n} \mid \mathcal{F}_t]$ (see Corollary 3.13) and let

$$\widetilde{Y}_t^n(\omega) = \sum_{k=0}^{\infty} 1_{[k/2^n, (k+1)/2^n)}(t)\widetilde{Y}_t^{k,n}(\omega).$$

Note that $\widetilde{Y}_t^n = \mathbb{E}[Y_t^n \mid \mathcal{F}_t]$, a.s., for all t. Moreover, $\widetilde{Y}_t^n$ is right continuous, so we see that it is jointly measurable in t and ω. Now for $n > m$,

$$\sup_{t \ge 0} |\widetilde{Y}_t^n - \widetilde{Y}_t^m| \le \sup_{t \ge 0} \mathbb{E}[V_{2^{-m}} \mid \mathcal{F}_t]. \tag{9.11}$$

We have already seen that there exists a subsequence such that the right-hand side of (9.11) converges to 0 almost surely. Hence along the appropriate subsequence, $\widetilde{Y}_t^n$ converges uniformly. If we call the limit $\widetilde{Y}$, we see that $\widetilde{Y}_t$ is right continuous, adapted, and jointly measurable. If $k/2^n \le t \le (k+1)/2^n$, then $|Y_t^n - Y_{k/2^n}^n| \le V_{2^{-n}}$, so

$$|\widetilde{Y}_t^n - \widetilde{Y}_{k/2^n}^n| = |\mathbb{E}[Y_t^n - Y_{k/2^n}^n \mid \mathcal{F}_t]| \le \mathbb{E}[V_{2^{-n}} \mid \mathcal{F}_t].$$

By the triangle inequality,

$$|\widetilde{Y}_t^n - \widetilde{Y}_s^n| \le 2 \sup_{t \ge 0} \mathbb{E}[V_{2^{-n}} \mid \mathcal{F}_t]$$

if $k/2^n \le s, t \le (k+1)/2^n$. Therefore the largest jump of $\widetilde{Y}_t^n$ is bounded by $2 \sup_{t \ge 0} \mathbb{E}[V_{2^{-n}} \mid \mathcal{F}_t]$, and we conclude the limit $\widetilde{Y}$ has continuous paths. Finally, Y_t^n differs from Y_t by at most $V_{2^{-n}}$, so we see by passing to the limit that $\widetilde{Y}_t$ is a version of $\mathbb{E}[Z_{t+h} \mid \mathcal{F}_t]$. $\qquad\square$

Exercises

9.1   Let W_t be a Brownian motion started at 1 and $T_0 = \inf\{t > 0 : W_t = 0\}$. Is $M_t = W_{t \wedge T_0}$ a square integrable martingale? A locally square integrable martingale? A uniformly integrable martingale? A martingale? A local martingale? A semimartingale?

9.2   Prove that if M is a submartingale such that the paths of M are continuous, $\sup_t |M_t|$ is integrable, and $S \le T$ are finite stopping times, then $\mathbb{E}[M_T \mid \mathcal{F}_S] \ge M_S$. Note that the last part of the proof of Proposition 9.3 breaks down here.

9.3   Suppose M_t is a local martingale with continuous paths. Show that if $N > 0$, $T_N = \inf\{t : |M_t| \ge N\}$, and $M_t^N = M_{t \wedge T_N}$, then M^N is a uniformly integrable martingale.

9.4   Suppose W_t^1 and W_t^2 are two independent Brownian motions, $t_0 > 0$, and $M_t^i = W_{t \wedge t_0}^i$, $i = 1, 2$. Show $\langle M^1, M^2 \rangle_t = 0$.

9.5   Show that the norm defined in (9.10) is complete.

9.6   Let Z_t be a bounded supermartingale with continuous paths that is constant from some time t_0 on. Show that for each t

$$\mathbb{E}\left[\int_t^{\infty} \mathbb{E}[Z_s - Z_{s+h} \mid \mathcal{F}_s]\, ds \mid \mathcal{F}_t\right] = \int_t^{\infty} \mathbb{E}[Z_s - Z_{s+h} \mid \mathcal{F}_t]\, ds, \qquad \text{a.s.}$$

9.7   We mentioned that one can prove the existence of $\langle M \rangle$ without using the Doob–Meyer theorem. Here is how that argument starts. Let M be a bounded continuous martingale and for each n, define

$$I_n(t) = \sum_{i=0}^{[t2^n]} (M_{(i+1)/2^n} - M_{i/2^n})^2.$$

Here $[x]$ is the integer part of x. Prove that for each $t > 0$, $\mathbb{E}\, |I_n(t) - I_m(t)|^2 \to 0$ as $n, m \to \infty$. One can then define $\langle M \rangle_t$ as the L^2 limit of $I_n(t)$.

Hint: If $n > m$, note that

$$M_{(i+1)/2^m} - M_{i/2^m} = \sum_{j=2^{n-m}i}^{2^{n-m}(i+1)-1} (M_{(j+1)/2^n} - M_{j/2^n}).$$

Notes

The first proof of the Doob–Meyer decomposition was by Meyer in the early 1960s and was a major breakthrough. There are now a number of alternate proofs. The proof we give here for continuous supermartingales is new.

10

Stochastic integrals

This chapter is devoted to the construction of stochastic integrals, primarily with respect to continuous square integrable martingales. The motivating example is $\int_0^t H_s \, dW_s$, where W is a Brownian motion and H is an adapted process satisfying certain conditions. We cannot define this integral as a Lebesgue–Stieltjes integral because the paths of Brownian motion are nowhere differentiable (Theorem 7.3).

One way to visualize a stochastic integral is to think of dW_s as "white noise," on a radio and H_s as the volume control which increases or decreases the white noise by a factor. For another model, if W_s is supposed to represent a stock price at time s (of course, stock prices can't be negative, while Brownian motion can!) and H_s is the number of shares held at time s, then the stochastic integral represents the net profit.

10.1 Construction

Let M_t be a continuous square integrable martingale with respect to a filtration $\{\mathcal{F}_t\}$ satisfying the usual conditions, and suppose H_t is an adapted process. Under appropriate additional assumptions on H, we want to define

$$N_t = \int_0^t H_s \, dM_s, \tag{10.1}$$

the stochastic integral of H with respect to M.

We impose two conditions on the integrand H_t, a measurability one and an integrability one. First we define the *predictable σ-field* $\mathcal{P}$ on $[0, \infty) \times \Omega$. This is the smallest σ-field of subsets of $[0, \infty) \times \Omega$ with respect to which all left continuous, bounded, and adapted processes are measurable. In symbols,

$$\mathcal{P} = \sigma(X : X \text{ is left continuous, bounded, and adapted to } \{\mathcal{F}_t\}).$$

This can be rephrased by saying $\mathcal{P}$ is the σ-field on $[0, \infty) \times \Omega$ generated by the collection of all sets of the form

$$\{(t, \omega \in [0, \infty) \times \Omega : X_t(\omega) > a\},$$

where $a \in \mathbb{R}$ and X is a bounded, adapted, left continuous process. We require $H : [0, \infty) \times \Omega \to \mathbb{R}$ to be measurable with respect to $\mathcal{P}$. When this happens, we say H is *predictable*. The integrability is easier to state: we require

$$\mathbb{E} \int_0^\infty H_s^2 \, d\langle M \rangle_s < \infty. \tag{10.2}$$

Observe that H will meet both requirements if H is bounded, adapted, and has continuous paths.

We define $\int_0^t H_s \, dM_s$ in three steps:

Step 1. When $H_s(\omega) = K(\omega) 1_{(a,b]}(s)$, where K is bounded and $\mathcal{F}_a$ measurable.

Step 2. When H_s is a sum of processes of the form in Step 1.

Step 3. When H is predictable and satisfies (10.2).

If $M_t = W_{t \wedge t_0}$, where W is a Brownian motion and t_0 is a fixed time, then $\langle M \rangle_t = t \wedge t_0$, and it might help the reader to work through the proofs in this special case. Even in this situation, all the elements of the general construction are present.

We will need the following easy lemma.

Lemma 10.1 *The predictable σ-field $\mathcal{P}$ is generated by the collection $\mathcal{C}$ of processes of the form $X_t(\omega) = \sum_{i=1}^n K_i(\omega) 1_{(a_i, b_i]}(t)$, where for each i, K_i is a bounded $\mathcal{F}_{a_i}$ measurable random variable.*

Proof If $X \in \mathcal{C}$, then X is bounded, adapted, and left continuous, hence X is a predictable process. Thus $\mathcal{C} \subset \mathcal{P}$.

On the other hand, if Y is a bounded, adapted, left-continuous process, we can approximate Y by the processes

$$Y_t^n(\omega) = \sum_{i=0}^{n2^n} Y_{i/2^n}(\omega) 1_{(i/2^n, (i+1)/2^n]}(t).$$

Each such Y^n is in $\mathcal{C}$. Therefore the σ-field generated by $\mathcal{C}$ contains $\mathcal{P}$. $\square$

Proposition 10.2 *Suppose H is as in Step 1 above. Then*

$$N_t = K(M_{t \wedge b} - M_{t \wedge a})$$

is a continuous martingale,

$$\mathbb{E} N_\infty^2 = \mathbb{E} \int_0^\infty K^2 1_{(a,b]}(s) \, d\langle M \rangle_s = \mathbb{E}[K^2 (\langle M \rangle_b - \langle M \rangle_a)],$$

and

$$\langle N \rangle_t = \int_0^t K^2 1_{(a,b]}(s) \, d\langle M \rangle_s.$$

Proof The continuity of the paths of N is clear. Set $N_\infty = K(M_b - M_a)$. Since K is bounded and M is square integrable, $\mathbb{E} N_\infty^2 < \infty$. We will show $N_t = \mathbb{E}[N_\infty \mid \mathcal{F}_t]$, which will prove that N_t is a martingale.

If $t \geq b$, then since K, M_b, and M_a are $\mathcal{F}_t$ measurable,

$$\mathbb{E}[N_\infty \mid \mathcal{F}_t] = K(M_b - M_a) = N_t.$$

If $a \leq t \leq b$, K is $\mathcal{F}_t$ measurable, and

$$\mathbb{E}[K(M_b - M_a) \mid \mathcal{F}_t] = K\mathbb{E}[M_b - M_a \mid \mathcal{F}_t] = K(M_t - M_a) = N_t.$$

In particular, $N_a = \mathbb{E}[N_\infty \mid \mathcal{F}_a] = 0$. Finally, if $t \leq a$,

$$\mathbb{E}[N_\infty \mid \mathcal{F}_t] = \mathbb{E}[\mathbb{E}[N_\infty \mid \mathcal{F}_a] \mid \mathcal{F}_t] = 0 = N_t.$$

For $\mathbb{E}\,N_\infty^2$, we have by (9.2) with $S = a$ and $T = b$,

$$
\begin{aligned}
\mathbb{E}\,N_\infty^2 = \mathbb{E}\,[K^2(M_b - M_a)^2] &= \mathbb{E}\,[K^2\mathbb{E}\,[(M_b - M_a)^2 \mid \mathcal{F}_a]] \\
&= \mathbb{E}\,[K^2\mathbb{E}\,[\langle M \rangle_b - \langle M \rangle_a \mid \mathcal{F}_a]] = \mathbb{E}\,[K^2(\langle M \rangle_b - \langle M \rangle_a)].
\end{aligned}
$$

To verify the formula for $\langle N \rangle_t$, let

$$
\begin{aligned}
L_\infty &= K^2(M_b - M_a)^2 - K^2(\langle M \rangle_b - \langle M \rangle_a), \\
L_t &= K^2(M_{b\wedge t} - M_{a\wedge t})^2 - K^2(\langle M \rangle_{b\wedge t} - \langle M \rangle_{a\wedge t}).
\end{aligned}
$$

Then

$$
L_t = N_t^2 - \int_0^t K^2 1_{(a,b]}(s)\,d\langle M \rangle_s,
$$

and we must show that L_t is a martingale. To do this, it suffices to show $L_t = \mathbb{E}\,[L_\infty \mid \mathcal{F}_t]$.
If $t \geq b$, then L_∞ is $\mathcal{F}_t$ measurable, so $\mathbb{E}\,[L_\infty \mid \mathcal{F}_t] = L_\infty = L_t$. If $a \leq t \leq b$, then

$$
\begin{aligned}
\mathbb{E}\,[L_\infty \mid \mathcal{F}_t] &= K^2\mathbb{E}\,[(M_b - M_a)^2 - (\langle M \rangle_b - \langle M \rangle_a) \mid \mathcal{F}_t] \\
&= K^2\mathbb{E}\,[M_b^2 - M_a^2 - (\langle M \rangle_b - \langle M \rangle_a) \mid \mathcal{F}_t] \\
&= K^2\mathbb{E}\,[M_t^2 - M_a^2 - (\langle M \rangle_t - \langle M \rangle_a) \mid \mathcal{F}_t] \\
&= K^2\mathbb{E}\,[(M_t - M_a)^2 - (\langle M \rangle_t - \langle M \rangle_a) \mid \mathcal{F}_t] \\
&= L_t,
\end{aligned}
$$

using (9.1) and (9.3) with the stopping times there being fixed positive real numbers. In particular, $\mathbb{E}\,[L_\infty \mid \mathcal{F}_a] = L_a = 0$. Finally, if $t \leq a$,

$$
\mathbb{E}\,[L_\infty \mid \mathcal{F}_t] = \mathbb{E}\,[\mathbb{E}\,[L_\infty \mid \mathcal{F}_a] \mid \mathcal{F}_t] = 0 = L_a
$$

as required. □

Next suppose

$$
H_s(\omega) = \sum_{j=1}^{J} K_j 1_{(a_j, b_j]}(s), \tag{10.3}
$$

where each K_j is $\mathcal{F}_{a_j}$ measurable and bounded. We may rewrite H so that the intervals $(a_j, b_j]$ satisfy $a_1 < b_1 \leq a_2 < b_2 \leq \cdots \leq a_J < b_J$. For example, if $H_s = K_1 1_{(a_1, b_1]} + K_2 1_{(a_2, b_2]}$ with $a_1 < a_2 < b_1 < b_2$, we may rewrite H_s as

$$
K_1 1_{(a_1, a_2]} + (K_1 + K_2) 1_{(a_2, b_1]} + K_2 1_{(b_1, b_2]}.
$$

Define

$$
N_t = \sum_{j=1}^{J} K_j (M_{t\wedge b_j} - M_{t\wedge a_j}). \tag{10.4}
$$

We need to check that rewriting H_s so that $a_1 < b_1 \leq a_2 < \cdots < b_J$ does not affect the value of N_t, but this is routine.

Proposition 10.3 *With H as in* (10.3) *and N defined by* (10.4), N_t *is a continuous martingale,*

$$\mathbb{E}\,N_\infty^2 = \mathbb{E}\int_0^\infty H_s^2\,d\langle M\rangle_s,$$

and

$$\langle N\rangle_t = \int_0^t H_s^2\,d\langle M\rangle_s. \tag{10.5}$$

Proof By linearity, N_t is a continuous martingale. We have

$$\mathbb{E}\,N_\infty^2 = \mathbb{E}\left[\sum_j H_j^2(M_{b_j} - M_{a_j})^2\right] \tag{10.6}$$

$$+ 2\mathbb{E}\left[\sum_{i<j} H_i H_j(M_{b_i} - M_{a_i})(M_{b_j} - M_{a_j})\right].$$

The cross terms vanish, because when $i < j$ and we condition on $\mathcal{F}_{a_j}$, we have

$$\mathbb{E}\left[H_i H_j(M_{b_i} - M_{a_i})\mathbb{E}\left[(M_{b_j} - M_{a_j}) \mid \mathcal{F}_{a_j}\right]\right] = 0.$$

For the terms in the first sum in (10.6), by (9.3)

$$\mathbb{E}\left[H_j^2(M_{b_j} - M_{a_j})^2\right] = \mathbb{E}\left[H_j^2\mathbb{E}\left[(M_{b_j} - M_{a_j})^2 \mid \mathcal{F}_{a_j}\right]\right]$$
$$= \mathbb{E}\left[H_j^2\mathbb{E}\left[\langle M\rangle_{b_j} - \langle M\rangle_{a_j} \mid \mathcal{F}_{a_j}\right]\right]$$
$$= \mathbb{E}\left[H_j^2\left(\left[\langle M\rangle_{b_j} - \langle M\rangle_{a_j}\right]\right)\right].$$

Therefore

$$\mathbb{E}\,N_\infty^2 = \mathbb{E}\int_0^\infty H_s^2\,d\langle M\rangle_s. \tag{10.7}$$

The argument for $\langle N\rangle_t$ is similar. $\qquad\square$

Now suppose H_s is predictable and (10.2) holds. Choose H_s^n of the form given in (10.3) above such that

$$\mathbb{E}\int_0^\infty (H_s^n - H_s)^2\,d\langle M\rangle_s \to 0.$$

To see that this can be done, define

$$\|Y\|_2 = \left(\mathbb{E}\int_0^\infty Y_t^2\,d\langle M\rangle_t\right)^{1/2}$$

for Y predictable. Then $\|Y\|_2$ is an L^2 norm on functions on $[0,\infty) \times \Omega$, so by Lemma 10.1 we can approximate H in this norm by processes of the form given in (10.3). (When H is bounded, adapted, and has continuous paths, taking H_s^n equal to $H_{k/2^n}$ if $k/2^n < s \le (k+1)/2^n$ for $s < n$ and $H_s^n = 0$ if $s \ge n$ will work.)

By Doob's inequalities we have

$$\mathbb{E}\left[\sup_{t\ge 0}\left(\int_0^t (H_s^n - H_s^m)\,dM_s\right)^2\right] \le 4\mathbb{E}\left(\int_0^\infty (H_s^n - H_s^m)\,dM_s\right)^2$$

$$= 4\mathbb{E}\int_0^\infty (H_s^n - H_s^m)^2\,d\langle M\rangle_s \to 0.$$

The norm

$$\|Y\|_\infty = (\mathbb{E}\,[\sup_t |Y_t|^2])^{1/2} \tag{10.8}$$

is complete; this was shown in Exercise 9.5. Thus there exists a process N_t such that $\sup_{t \geq 0} |N_t - \int_0^t H_s^n\, dM_s| \to 0$ in L^2.

If H_s^n and $\overline{H_s^n}$ are two sequences converging in the $\|\cdot\|_2$ norm to H, then

$$\mathbb{E}\left(\int_0^t (H_s^n - \overline{H_s^n})\, dM_s\right)^2 = \mathbb{E}\int_0^t (H_s^n - \overline{H_s^n})^2\, d\langle M\rangle_s \to 0,$$

or the limit is independent of which sequence H^n we choose.

It is easy to see, because of the L^2 convergence, that N_t is a martingale: if $A \in \mathcal{F}_s$, then

$$\mathbb{E}\left[\int_0^t H_r^n\, dM_r; A\right] = \mathbb{E}\left[\int_0^s H_r^n\, dM_r; A\right]$$

by Proposition 10.3. Now use that

$$\left|\mathbb{E}\left[\int_0^t H_r^n\, dM_r - N_t; A\right]\right| \leq \mathbb{E}\left|\int_0^t H_r^n\, dM_r - N_t\right|$$

$$\leq \left(E\left(\int_0^t H_r^n\, dM_r - N_t\right)^2\right)^{1/2} \to 0$$

and similarly with t replaced by s.

Similar arguments using the L^2 convergence show that

$$\mathbb{E}\,N_t^2 = \mathbb{E}\int_0^t H_s^2\, d\langle M\rangle_s, \tag{10.9}$$

and

$$\langle N\rangle_t = \int_0^t H_s^2\, d\langle M\rangle_s. \tag{10.10}$$

Because $\sup_{t \geq 0} |N_t - \int_0^t H_s^n\, dM_s| \to 0$ in L^2, there exists a subsequence $\{n_k\}$ such that the convergence takes place almost surely, that is

$$\sup_{t \geq 0}\left|\int_0^t H_s^{n_k}\, dM_s - N_t\right| \to 0, \qquad \text{a.s.}$$

Since each $\int_0^t H_s^n\, dM_s$ has continuous paths, with probability one, N_t has continuous paths. We write $N_t = \int_0^t H_s\, dM_s$ and call N_t the stochastic integral of H with respect to M.

We summarize our construction as follows.

Theorem 10.4 *Suppose the filtration $\{\mathcal{F}_t\}$ satisfies the usual conditions and M_t is a square integrable martingale with continuous paths. Suppose H is of the form*

$$\sum_{i=1}^J K_j(\omega) 1_{(a_j, b_j]}(s), \tag{10.11}$$

where each K_j is bounded and $\mathcal{F}_{a_j}$ measurable. In this case define

$$\int_0^t H_s \, dM_s = \sum_{j=1}^J K_j (M_{t \wedge b_j} - M_{t \wedge a_j}).$$

If H is predictable and

$$\mathbb{E} \int_0^\infty H_s^2 \, d\langle M \rangle_s < \infty,$$

choose H^n of the form given in (10.11) with $\mathbb{E} \int_0^\infty (H_s^n - H_s)^2 \, d\langle M \rangle_s \to 0$, and define

$$N_t = \int_0^t H_s \, dM_s$$

to be the limit with respect to the norm (10.8) of $\int_0^t H_s^n \, dM_s$. Then N_t is a continuous martingale,

$$\mathbb{E} N_\infty^2 = \mathbb{E} \int_0^\infty H_s^2 \, d\langle M \rangle_s,$$

and

$$\langle N \rangle_t = \int_0^t H_s^2 \, d\langle M \rangle_s.$$

Moreover the definition of N_t is independent of the particular choice of the H^n.

10.2 Extensions

There are some extensions of the definition that are fairly routine.

Extension 1. If

$$\int_0^\infty H_s^2 \, d\langle M \rangle_s < \infty, \qquad \text{a.s.,}$$

but without the expectation being finite, let

$$T_N = \inf \left\{ t : \int_0^t H_s^2 \, d\langle M \rangle_s > N \right\}.$$

$M_t' = M_{t \wedge T_N}$ is a square integrable martingale with $\langle M' \rangle_t = \langle M \rangle_{t \wedge T_N}$, so $\int_0^t H_s^2 \, d\langle M' \rangle_t \leq N$. Define $\int_0^t H_s \, dM_s$ to be the quantity $\int_0^t H_s \, dM_{s \wedge T_N}$ if $t \leq T_N$. If $t \leq T_K \leq T_N$, we need to check that $\int_0^t H_s \, d\langle M \rangle_{t \wedge T_K} = \int_0^t H_s \, d\langle M \rangle_{t \wedge T_N}$, so that our definition is consistent. This is part of Exercise 10.2.

Extension 2. If M_t is a continuous local martingale (see Section 9.1 for the definition), let $S_n = \inf\{t : |M_t| \geq n\}$. By Exercise 9.3, $M_{t \wedge S_n}$ will be a uniformly integrable martingale, and in fact, since $M_{t \wedge S_n}$ is bounded, it is square integrable. For $t \leq S_n$ we set

$$\int_0^t H_s \, dM_s = \int_0^t H_s \, dM_{s \wedge S_n}$$

and $\langle M \rangle_t = \langle M \rangle_{t \wedge S_n}$. Again there is consistency to check, which is also part of Exercise 10.2.

Extension 3. Suppose that $X_t = M_t + A_t$ is a semimartingale with continuous paths, so that M is a local martingale and A is a process with paths locally of bounded variation. If $\int_0^\infty H_s^2 \, d\langle M \rangle_s + \int_0^\infty |H_s| \, |dA_s| < \infty$, we define

$$\int_0^t H_s \, dX_s = \int_0^t H_s \, dM_s + \int_0^t H_s \, dA_s,$$

where the first integral on the right is a stochastic integral and the second is a Lebesgue–Stieltjes integral.

For a semimartingale, we define

$$\langle X \rangle_t = \langle M \rangle_t. \tag{10.12}$$

Given two semimartingales X and Y we define $\langle X, Y \rangle_t$ by:

$$\langle X, Y \rangle_t = \tfrac{1}{2}[\langle X + Y \rangle_t - \langle X \rangle_t - \langle Y \rangle_t].$$

Exercises

10.1   Prove (10.5) in Proposition 10.3.

10.2   Check the consistency of the first two extensions of the definition of stochastic integrals.

10.3   Show that if M is a continuous square integrable martingale, and T a finite stopping time, then

$$\int_0^\infty 1_{[0,T]} \, dM_s = M_T.$$

10.4   Show that if $N_t = \int_0^t H_s \, dM_s$ where M is a continuous square integrable martingale, H is predictable, and $\mathbb{E} \int_0^\infty H_s^2 \, d\langle M \rangle_s < \infty$, and $L_t = \int_0^t K_s \, dN_s$, where K is predictable and $\mathbb{E} \int_0^\infty K_s^2 \, d\langle N \rangle_s < \infty$, then

$$L_t = \int_0^t H_s K_s \, dM_s.$$

10.5   Show that if M, H, and N are as in Exercise 10.4, then $\langle M, N \rangle_t = \int_0^t H_s \, d\langle M \rangle_s$.
       Hint: Derive a formula for $\langle N + M \rangle_t$ from the fact that

$$N_t + M_t = \int_0^t (1 + H_s) \, dM_s.$$

10.6   Suppose that M and L are square integrable martingales, H is predictable and satisfies (10.2), and $N_t = \int_0^t H_s \, dM_s$. Show that

$$\langle N, L \rangle_t = \int_0^t H_s \, d\langle M, L \rangle_s. \tag{10.13}$$

Sometimes the stochastic integral of H with respect to M is defined to be the square integrable martingale N for which (10.13) holds for all square integrable martingales L.

10.7   Show that if M and N are square integrable martingales with continuous paths, then

$$\langle M, N \rangle_t \le (\langle M \rangle_t)^{1/2} (\langle N \rangle_t)^{1/2}.$$

Hint: Imitate an appropriate proof of the Cauchy–Schwarz inequality. This result is a special case of the inequality of Kunita–Watanabe.

11

Itô's formula

The most important result in the theory of stochastic integration is Itô's formula. This is also known as the *change of variables formula.*

Let C^k be the functions that are k times continuously differentiable and C_b^k those functions C^k such that the function and its ith-order derivatives are bounded for $i \leq k$.

Theorem 11.1 *Let X_t be a semimartingale with continuous paths and suppose $f \in C^2$. Then for almost every ω*

$$f(X_t) = f(X_0) + \int_0^t f'(X_s)\, dX_s + \frac{1}{2} \int_0^t f''(X_s)\, d\langle X \rangle_s, \qquad t \geq 0. \qquad (11.1)$$

Step 1 will be to reduce to the case when $f \in C_b^3$ and X has appropriate boundedness conditions. Step 2 is a use of Taylor's formula; see (11.2). Step 3 shows that each term converges to the appropriate quantity, and Step 4 removes the restriction that f be in C_b^3.

Proof *Step 1.* If $X_t = M_t + A_t$ is the decomposition of X into a local martingale M and a process A that has paths locally of bounded variation, let V_t be the total variation of A up to time t: $V_t = \int_0^t |dA_s|$. Let

$$T_N = \inf\{t : |M_t| > N \text{ or } \langle M \rangle_t > N \text{ or } V_t > N\}.$$

By the continuity of paths, $T_N \to \infty$, a.s., as $N \to \infty$, so for almost every ω and for each t, $t \wedge T_N = t$ for N large enough. Since Itô's formula is a path-by-path result, it suffices to prove Itô's formula for $X_{t \wedge T_N}$ for each N, or what amounts to the same thing, we may take N arbitrary and assume M_t, $\langle M \rangle_t$, A_t, and V_t are all bounded by N. In this case, X_t is bounded by $2N$.

Since X is bounded, we may modify f, f', and f'' outside of $[-2N, 2N]$ without affecting the validity of Itô's formula. Therefore we will also assume $f \in C^2$ with compact support. Let us temporarily assume in addition that f''' exists and is continuous; we will remove this last assumption later on.

Let $t_0 > 0$, $\varepsilon > 0$, $S_0 = 0$, and define

$$S_{i+1} = S_{i+1}(\varepsilon) = \inf\{t > S_i : |M_t - M_{S_i}| > \varepsilon \text{ or } \langle M \rangle_t - \langle M \rangle_{S_i} > \varepsilon$$
$$\text{or } V_t - V_{S_i} > \varepsilon\} \wedge t_0.$$

Note $S_i = t_0$ for i sufficiently large (how large depends on ω) by the continuity of the paths.

Step 2. The key idea to proving Itô's formula is Taylor's theorem. We write

$$f(X_{t_0}) - f(X_0) = \sum_{i=0}^{\infty} [f(X_{S_{i+1}}) - f(X_{S_i})] \tag{11.2}$$

$$= \sum_{i=0}^{\infty} f'(X_{S_i})(X_{S_{i+1}} - X_{S_i}) + \tfrac{1}{2} \sum_{i=0}^{\infty} f''(X_{S_i})(X_{S_{i+1}} - X_{S_i})^2$$

$$+ \sum_{i=0}^{\infty} R_i,$$

where R_i is the remainder term. We have $|R_i| \leq c \|f'''\|_\infty |X_{S_{i+1}} - X_{S_i}|^3$.

Step 3. Let us first look at the terms with f' in them. Let $H_s^\varepsilon = f'(X_{S_i})$ if $S_i \leq s < S_{i+1}$. By the continuity of f' and X_s, we see that H_s^ε converges boundedly and pointwise to $f'(X_s)$. In particular, $\int_0^{t_0} |H_s^\varepsilon - f'(X_s)| \, dV_s \to 0$ boundedly, hence

$$\mathbb{E} \int_0^{t_0} |H_s^\varepsilon - f'(X_s)| \, dV_s \to 0.$$

Also,

$$\mathbb{E} \left(\int_0^{t_0} (H_s^\varepsilon - f'(X_s)) \, dM_s \right)^2 = \mathbb{E} \int_0^{t_0} |H_s^\varepsilon - f'(X_s)|^2 \, d\langle M \rangle_s \to 0$$

as $\varepsilon \to 0$. We then have

$$\sum_i f'(X_{S_i})(X_{S_{i+1}} - X_{S_i}) = \int_0^{t_0} H_s^\varepsilon \, (dM_s + dA_s) \to \int_0^{t_0} f'(X_s) \, (dM_s + dA_s),$$

which leads to the f' term in Itô's formula.

Next let us look at the f'' terms. We can write

$$(X_{S_{i+1}} - X_{S_i})^2 = (M_{S_{i+1}} - M_{S_i})^2 + 2(M_{S_{i+1}} - M_{S_i})(A_{S_{i+1}} - A_{S_i}) + (A_{S_{i+1}} - A_{S_i})^2.$$

Note $\sum_i f''(X_{S_i})(M_{S_{i+1}} - M_{S_i})(A_{S_{i+1}} - A_{S_i})$ is bounded in absolute value by

$$\sum_i \varepsilon \|f''\|_\infty |A_{S_{i+1}} - A_{S_i}| \leq \varepsilon \|f''\|_\infty \int_0^{t_0} dV_s \leq \varepsilon \|f''\|_\infty N,$$

which goes to 0 as $\varepsilon \to 0$; this follows from the definition of S_i. Similarly the expression $\sum f''(X_{S_i})(A_{S_{i+1}} - A_{S_i})^2$ also goes to 0. Therefore we need to show

$$\sum_i f''(X_{S_i})(M_{S_{i+1}} - M_{S_i})^2 \to \int_0^{t_0} f''(X_s) \, d\langle X \rangle_s.$$

By an argument very similar to the one for the f' terms,

$$\tfrac{1}{2} \sum_i f''(X_{S_i})(\langle M \rangle_{S_{i+1}} - \langle M \rangle_{S_i}) \to \tfrac{1}{2} \int_0^{t_0} f''(X_s) \, d\langle M \rangle_s, \tag{11.3}$$

and since $\langle X \rangle_t = \langle M \rangle_t$ for semimartingales (see (10.12)), the right-hand side of (11.3) is the correct f'' term. We thus need to show that

$$\sum_i f''(X_{S_i})[(M_{S_{i+1}} - M_{S_i})^2 - (\langle M \rangle_{S_{i+1}} - \langle M \rangle_{S_i})] \to 0 \tag{11.4}$$

as $\varepsilon \to 0$.

We will show

$$\mathbb{E}\left(\sum_{i=0}^{\infty} B_i\right)^2 \to 0, \tag{11.5}$$

where

$$B_i = f''(X_{S_i})[(M_{S_{i+1}} - M_{S_i})^2 - (\langle M \rangle_{S_{i+1}} - \langle M \rangle_{S_i})].$$

We have

$$\mathbb{E}\left(\sum_i B_i\right)^2 = \mathbb{E}\sum_i B_i^2 + 2\sum_{i<j} B_i B_j.$$

If $i < j$, then

$$\mathbb{E}[B_i B_j] = \mathbb{E}[B_i \mathbb{E}[B_j \mid \mathcal{F}_{S_{i+1}}]].$$

By (9.2) and the fact that $S_{i+1} \le S_j$, we see that

$$\mathbb{E}[B_j \mid \mathcal{F}_{S_{i+1}}] = f''(X_{S_j})\mathbb{E}[(M_{S_{j+1}} - M_{S_j})^2 - (\langle M \rangle_{S_{j+1}} - \langle M \rangle_{S_j}) \mid \mathcal{F}_{S_{i+1}}] = 0,$$

so the cross-products vanish.

Therefore to prove (11.5) it remains to show $\mathbb{E}\sum_i B_i^2 \to 0$ as $\varepsilon \to 0$. We use the easy inequality $(x+y)^2 \le 2x^2 + 2y^2$. Since f'' is bounded,

$$\mathbb{E}\sum_i B_i^2 \le 2\|f''\|_\infty^2 \sum_i \mathbb{E}[(M_{S_{i+1}} - M_{S_i})^4] \tag{11.6}$$

$$+ 2\|f''\|_\infty^2 \sum_i \mathbb{E}[(\langle M \rangle_{S_{i+1}} - \langle M \rangle_{S_i})^2].$$

The first sum on the right-hand side of (11.6) is bounded by

$$2\varepsilon^2 \|f''\|_\infty^2 \sum_i \mathbb{E}[(M_{S_{i+1}} - M_{S_i})^2] = 2\varepsilon^2 \|f''\|_\infty^2 \mathbb{E}[M_{t_0}^2 - M_0^2]$$

$$\le 8\varepsilon^2 \|f''\|_\infty^2 N^2.$$

The second sum on the right-hand side of (11.6) is bounded by

$$2\varepsilon \|f''\|_\infty^2 \sum_i \mathbb{E}[(\langle M \rangle_{S_{i+1}} - \langle M \rangle_{S_i}] \le 2\varepsilon \|f''\|_\infty^2 \mathbb{E}\langle M \rangle_{t_0} \le 2\varepsilon \|f''\|_\infty^2 N.$$

Both of these tend to 0 as $\varepsilon \to 0$. Therefore $\mathbb{E}\sum_i B_i^2 \to 0$, and the proof of the convergence for the f'' term is complete.

The final terms to examine are the remainder terms. We have shown that $\mathbb{E} \sum_i (X_{S_{i+1}} - X_{S_i})^2$ remains bounded as $\varepsilon \to 0$. Since

$$|R_i| \leq c\varepsilon \|f'''\|_\infty (X_{S_{i+1}} - X_{S_i})^2,$$

we see $\mathbb{E} \sum_i |R_i| \to 0$ as $\varepsilon \to 0$.

Step 4. To finish up, we remove the assumption that $f \in C^3$. (We still assume that $f \in C^2$ with compact support.) Take a sequence $\{f_m\}$ of C^3 functions such that f_m, f'_m, and f''_m converge uniformly to f, f', and f'', respectively. Apply Itô's formula with f_m and then let $m \to \infty$. The terms $f_m(X_t)$ and $f_m(X_0)$ clearly converge to $f(X_t)$ and $f(X_0)$. The f'_m terms converge because

$$\mathbb{E} \left(\int_0^{t_0} (f'_m(X_s) - f'(X_s)) \, dM_s \right)^2 = \mathbb{E} \int_0^{t_0} |f'_m(X_s) - f'(X_s)|^2 \, d\langle M \rangle_s \to 0$$

and

$$\mathbb{E} \left| \int_0^{t_0} (f'_m(X_s) - f'(X_s)) \, dA_s \right| \leq \mathbb{E} \int_0^{t_0} |f'_m(X_s) - f'(X_s)| \, dV_s \to 0$$

as $m \to \infty$. The f''_m terms converge by dominated convergence. This shows that (11.1) holds for each t_0, except for a null set N_{t_0} depending on t_0. Let $N = \cup_{t \in \mathbb{Q}_+} N_t$, where $\mathbb{Q}_+$ denotes the non-negative rationals. If $\omega \notin N$, then (11.1) holds for every t_0 rational. Each term in (11.1) is continuous, a.s. (with a null set N' independent of t_0). Therefore if $\omega \notin N \cup N'$, (11.1) holds for all t_0. $\qquad\square$

There is a multivariate version of Itô's formula, which is proved in a very similar way:

Theorem 11.2 *Suppose $X_t^1, \ldots, X_t^d$ are continuous semimartingales, $X_t = (X_t^1, \ldots, X_t^d)$, and f is a C^2 function on $\mathbb{R}^d$. Then with probability one,*

$$f(X_t) = f(X_0) + \int_0^t \sum_{i=1}^d \frac{\partial f}{\partial x_i}(X_s) \, dX_s^i \qquad (11.7)$$

$$+ \frac{1}{2} \int_0^t \sum_{i,j=1}^d \frac{\partial^2 f}{\partial x_i \partial x_j}(X_s) \, d\langle X^i, X^j \rangle_s$$

for all $t \geq 0$.

The following is known as the *integration by parts formula* or *Itô's product formula*, and is very useful.

Corollary 11.3 *If X and Y are semimartingales with continuous paths, then*

$$X_t Y_t = X_0 Y_0 + \int_0^t X_s \, dY_s + \int_0^t Y_s \, dX_s + \langle X, Y \rangle_t.$$

Proof By Itô's formula,

$$X_t^2 = X_0^2 + 2 \int_0^t X_s \, dX_s + \langle X \rangle_t.$$

The analogous formula holds when X is replaced by Y and when X is replaced by $X + Y$. We then use

$$X_t Y_t = \tfrac{1}{2}[(X_t + Y_t)^2 - X_t^2 - Y_t^2];$$

substituting the formulas for X_t^2, Y_t^2, and $(X_t + Y_t)^2$ that we obtained from Itô's formula and doing some algebra yields our result. □

Exercises

11.1 Suppose W_t is a Brownian motion and $a \in \mathbb{R}$. Show that the amount of time Brownian motion spends at the point a is zero, i.e., that

$$\int_0^t 1_{\{a\}}(W_s)\, ds = 0, \qquad \text{a.s.}$$

for all $t > 0$.

11.2 Let $a < b$ and let $f_{a,b}$ be the C^1 function such that $f_{a,b}(0) = f'_{a,b}(0) = 0$ and

$$f'_{a,b}(x) = \int_0^x 1_{[a,b]}(y)\, dy, \qquad x \in \mathbb{R}.$$

In other words, $f_{a,b}$ is the function whose second derivative is $1_{[a,b]}$, except that the second derivative is not defined at a and b. Show Itô's formula holds for $f_{a,b}$:

$$f_{a,b}(W_t) = \int_0^t f'_{a,b}(W_s)\, dW_s + \tfrac{1}{2} \int_0^t 1_{[a,b]}(W_s)\, ds.$$

11.3 If W_t is a Brownian motion, $a > 0$, and $T = \inf\{t > 0 : |W_t| = a\}$, calculate $\mathbb{E} \int_0^T (W_s)^k\, ds$ for each non-negative integer k. Also calculate

$$\mathbb{E} \int_0^T 1_{[b_1, b_2]}(W_s)\, ds$$

if $[b_1, b_2] \subset [-a, a]$.

11.4 Let W be a Brownian motion, let $t_0 < t_1 < \cdots < t_n = 1$, and let

$$B_i = (W_{t_i} - W_{t_{i-1}})^2 - (t_i - t_{i-1}).$$

Show there exists a constant c_1 not depending on $\{t_0, \ldots, t_n\}$ such that

$$\mathbb{E}\left(\sum_{i=1}^n B_i \right)^2 \le c_1 \max_{1 \le i \le n} |t_i - t_{i-1}|.$$

11.5 Use Exercise 11.4 and the Borel–Cantelli lemma to prove that if W is a Brownian motion, then

$$\lim_{n \to \infty} \sum_{k=1}^{2^n} (W_{k/2^n} - W_{(k-1)/2^n})^2 = 1, \qquad \text{a.s.}$$

11.6 In our proof of Itô's formula, the use of stopping times simplifies the proof considerably. This exercise considers a proof of Itô's formula using fixed times. Suppose M is a bounded continuous martingale, A is a continuous process whose paths have total variation bounded by $N > 0$, a.s., and $X_t = M_t + A_t$.

(1) Writing $[x]$ for the integer part of x, prove that for each t,

$$\sum_{i=1}^{[2^n t]+1} (X_{(i+1)/2^n} - X_{i/2^n})^2$$

converges in probability to $\langle X \rangle_t$.

(2) Prove that if f is a C^2 function whose second derivative is bounded, then

$$\sum_{i=1}^{[2^n t]+1} f''(X_{i/2^n})(X_{(i+1)/2^n} - X_{i/2^n})^2$$

converges in probability to

$$\int_0^t f''(X_s) \, d\langle X_s \rangle.$$

Since the increments of M and A are not uniformly bounded by something small, this is much harder than the proof of Theorem 11.1 given in this chapter.

11.7 Here is an alternate way to prove Itô's formula.

(1) Suppose $X = M + A$, where M and A are as in Exercise 11.6. Write

$$X_t^2 - X_0^2 = \sum_{i=0}^{[t2^n]-1} (X_{(i+1)/2^n}^2 - X_{i/2^n}^2)$$

$$= \sum_{i=0}^{[t2^n]-1} 2X_{i/2^n}(X_{(i+1)/2^n} - X_{i/2^n}) + \sum_{i=0}^{[t2^n]-1} (X_{(i+1)/2^n} - X_{i/2^n})^2.$$

Use Exercise 11.6 to show that Itô's formula holds when $f(x) = x^2$.

(2) Derive the Itô product formula. Then use induction to show that Itô's formula holds when $f(x) = x^n$, n a positive integer.

(3) Given $f \in C^2$, find polynomials P_m such that P_m, P_m', P_m'' converge uniformly to f, f', f'', respectively, on a compact interval as $m \to \infty$. Apply Itô's formula for P_m and show that one can take limits to derive Itô's formula for f.

12

Some applications of Itô's formula

We will be using Itô's formula throughout the book. In this chapter we give some applications, each of which will turn out to be quite useful.

12.1 Lévy's theorem

The following is known as Lévy's theorem. Recall that if M is a local martingale with continuous paths and $T_N = \inf\{t : |M_t| \geq N\}$, we defined $\langle M \rangle_t$ to be equal to $\langle M \rangle_{t \wedge T_N}$ if $t \leq T_N$; see Section 10.2. Moreover, by Exercise 9.3, $M_{t \wedge N}$ is a square integrable martingale for each N.

Theorem 12.1 *Let M_t be a continuous local martingale with respect to a filtration $\{\mathcal{F}_t\}$ satisfying the usual conditions such that $M_0 = 0$ and $\langle M \rangle_t = t$. Then M_t is a Brownian motion with respect to $\{\mathcal{F}_t\}$.*

Proof Fix t_0 and let $N_t = M_{t+t_0} - M_{t_0}$, $\mathcal{F}'_t = \mathcal{F}_{t+t_0}$. It is routine to check that N_t is a martingale with respect to $\mathcal{F}'_t$ and that $\langle N \rangle_t = t$. Note $\mathcal{F}'_0$ will not be the trivial σ-field in general. We see that

$$\mathbb{E}\, N_t^2 = \mathbb{E}\, M_{t+t_0}^2 - \mathbb{E}\, M_{t_0}^2 = t < \infty.$$

If f is a function mapping the reals to the complex numbers, we may still use Itô's formula: just apply Itô's formula to the real and imaginary parts of f. Doing this for $f(x) = e^{iux}$, where u and x are real, we have

$$e^{iuN_t} = 1 + iu \int_0^t e^{iuN_s}\, dN_s - \frac{u^2}{2} \int_0^t e^{iuN_s}\, ds. \tag{12.1}$$

If we take $T_K = \inf\{t : |N_t| \geq K\}$, then

$$e^{iuN_{t \wedge T_K}} = 1 + iu \int_0^{t \wedge T_K} e^{iuN_s}\, dN_s - \frac{u^2}{2} \int_0^{t \wedge T_K} e^{iuN_s}\, ds. \tag{12.2}$$

Take $A \in \mathcal{F}'_0$, multiply (12.2) by 1_A, and take expectations. The stochastic integral is a martingale, so this term will have 0 expectation. Then let $K \to \infty$, and we are left with

$$\mathbb{E}\, [e^{iuN_t}; A] = \mathbb{P}(A) - \frac{u^2}{2} \int_0^t \mathbb{E}\, [e^{iuN_s}; A]\, ds. \tag{12.3}$$

We used the Fubini theorem here. (The reason we introduced the stopping time T_K is that $N_{t \wedge T_K}$ is a square integrable martingale, and hence the stochastic integral is a martingale. We might run into integrability problems if we worked with (12.1) instead of (12.2).)

Write $J(t) = \mathbb{E}\left[e^{iuN_t}; A\right]$, so we have

$$J(t) = \mathbb{P}(A) - \frac{u^2}{2}\int_0^t J(s)\,ds. \qquad (12.4)$$

Since J is bounded, (12.4) shows that J is continuous. Since J is continuous, using (12.4) again shows that J is differentiable. Hence $J'(t) = -\frac{u^2}{2}J(t)$ with $J(0) = \mathbb{P}(A)$. The only solution to this ordinary differential equation is

$$J(t) = \mathbb{P}(A)e^{-u^2t/2}. \qquad (12.5)$$

If we set $A = \Omega$, this tells us that $\mathbb{E}\,e^{iuN_t} = e^{-u^2t/2}$, and by the uniqueness theorem for characteristic functions (Theorem A.48), $M_{t+t_0} - M_{t_0}$ is a mean zero normal random variable with variance t. Equation (12.5) also tells us that

$$\mathbb{E}\left[e^{iuN_t}; A\right] = \mathbb{E}\left[e^{iuN_t}\right]\mathbb{P}(A) \qquad (12.6)$$

when $A \in \mathcal{F}_0'$. Let f be a C^∞ function with compact support. The Fourier transform $\widehat{f}(u)$ will be in the Schwartz class; see Section B.2. Replacing u by $-u$ in (12.6), multiplying the resulting equation by $\widehat{f}(u)$, and integrating over $u \in \mathbb{R}$, we have

$$\int \widehat{f}(u)\mathbb{E}\left[e^{-iuN_t}; A\right]du = \int \widehat{f}(u)\mathbb{E}\left[e^{-iuN_t}\right]du\,\mathbb{P}(A).$$

Using the Fubini theorem and the Fourier inversion theorem, and dividing by a constant, we conclude

$$\mathbb{E}\left[f(N_t); A\right] = \mathbb{E}\left[f(N_t)\right]\mathbb{P}(A).$$

Since $\widehat{f}$ is in the Schwartz class, integrability is not a problem when applying the Fubini theorem. A limit argument shows that this equation holds with f equal to 1_B, where B is a Borel subset of $\mathbb{R}$, hence

$$\mathbb{P}(M_{t+t_0} - M_{t_0} \in B, A) = \mathbb{P}(M_{t+t_0} - M_{t_0} \in B)\,\mathbb{P}(A).$$

This shows the independence of $M_{t+t_0} - M_{t_0}$ and $\mathcal{F}_{t_0}$. We thus see that M_t is a continuous process starting at 0 with $M_{t+t_0} - M_{t_0}$ being a mean zero normal random variable with variance t independent of $\mathcal{F}_{t_0}$, and therefore M is a Brownian motion. $\square$

12.2 Time changes of martingales

The next theorem says that most continuous martingales arise from Brownian motion via a time change. That is, the paths are the same, but the rate at which one moves along the paths varies. In fact, it is possible to show that all continuous martingales arise from a time change of a Brownian motion that is possibly stopped at a random time.

Theorem 12.2 *Suppose M_t is a continuous local martingale, $M_0 = 0$, $\langle M\rangle_t$ is strictly increasing, and $\lim_{t\to\infty}\langle M\rangle_t = \infty$, a.s. Let*

$$\tau(t) = \inf\{u : \langle M\rangle_u \geq t\}.$$

Then $W_t = M_{\tau(t)}$ is a Brownian motion with respect to $\mathcal{F}_t' = \mathcal{F}_{\tau(t)}$.

Proof Let us first suppose that W_t^2 is integrable. We have by Proposition 9.3 that

$$\mathbb{E}\left[W_t \mid \mathcal{F}_s'\right] = \mathbb{E}\left[M_{\tau(t)} \mid \mathcal{F}_{\tau(s)}\right] = M_{\tau(s)} = W_s,$$

or W_t is a continuous martingale. Similarly, $W_t^2 - t$ is a martingale. Now apply Lévy's theorem, Theorem 12.1. Removing the assumption that W_t^2 is integrable is left as Exercise 12.1. □

12.3 Quadratic variation

Itô's formula allows us to prove Theorem 9.10 fairly simply.

Proof of Theorem 9.10 If $T_K = \inf\{t : |M_t| \geq K\}$, we will show that

$$\sum_{k=0}^{[t_0 2^n]} (M_{T_K \wedge (k+1)/2^n} - M_{T_K \wedge k/2^n})^2$$

converges to $\langle M \rangle_{t_0 \wedge T_K}$. Since $T_K \to \infty$ as $K \to \infty$, this will prove the proposition. Thus we may assume M is bounded by K.

If $s > 0$ and we let $N_t = M_{s+t} - M_s$, then N_t is a martingale with respect to the filtration $\mathcal{F}_t' = \mathcal{F}_{s+t}$ and we can check that $\langle N \rangle_t = \langle M \rangle_{t+s} - \langle M \rangle_s$. By Itô's formula applied to the process N, we obtain

$$(M_{t+s} - M_s)^2 = 2 \int_0^t (M_{r+s} - M_s)\, dM_r + (\langle M \rangle_{t+s} - \langle M \rangle_s).$$

Applying this with $t = 1/2^n$ and $s = k/2^n$ and summing, we see that

$$\sum_{k=0}^{[t_0 2^n]} (M_{(k+1)/2^n} - M_{k/2^n})^2 - \langle M \rangle_t = 2 \int_0^{t_0} L_r^n\, dM_r + R, \tag{12.7}$$

where $L_r^n = M_r - M_{k/2^n}$ for $k/2^n \leq r < (k+1)/2^n$ and

$$R = \langle M \rangle_{([t_0 2^n]+1)/2^n} - \langle M \rangle_{t_0}.$$

Note

$$\mathbb{E}\left(2 \int_0^{t_0} L_r^n\, dM_r\right)^2 = 4\mathbb{E} \int_0^{t_0} (L_r^n)^2\, d\langle M \rangle_r. \tag{12.8}$$

The integrand $(L_r^n)^2$ is bounded by $4K^2$, $\mathbb{E}\langle M \rangle_t = \mathbb{E} M_t^2 \leq K^2$ is finite, and L_r^n tends to 0 as $n \to \infty$. By dominated convergence, the right-hand side of (12.8) tends to 0 as $n \to \infty$. As for the remainder term, R goes to 0 by the continuity of the paths of $\langle M \rangle_t$. The reason we only have convergence in probability rather than in L^2 is due to the stopping time argument involving T_K. □

12.4 Martingale representation

The next theorem says that every martingale adapted to the filtration of a Brownian motion can be expressed as a stochastic integral with respect to the Brownian motion. This used to be a rather arcane result that was of interest only to probabilists specializing in martingales. But then it turned out that this theorem is the basis for showing the completeness of the market in the theory of financial mathematics; see Chapter 28. The martingale representation theorem is also key to the innovations approach to stochastic filtering; see Chapter 29.

Theorem 12.3 *Let $\mathcal{F}_t$ be the minimal augmented filtration generated by a one-dimensional Brownian motion W_t, let $t_0 > 0$, and let Y be $\mathcal{F}_{t_0}$ measurable with $\mathbb{E}\, Y^2 < \infty$. There exists a predictable process H_s with $\mathbb{E} \int_0^{t_0} H_s^2\, ds < \infty$ such that*

$$Y = \mathbb{E}\, Y + \int_0^{t_0} H_s\, dW_s, \qquad \text{a.s.} \tag{12.9}$$

The proof consists of showing (12.9) holds for successively larger classes of random variables. Step 1 of the proof shows that the equation holds for random variables of the form $e^{iu(W_t - W_s)}$ and Step 2 shows that (12.9) holds for products of such random variables. In Step 3, it is shown that if the equation holds for a set of random variables, it holds for the closure of that set with respect to the L^2 norm.

Proof Step 1. Let $X_t = iuW_t + u^2 t/2$. Note $\langle X \rangle_t = (iu)^2 \langle W \rangle_t$. By Itô's formula applied with $f(x) = e^x$,

$$e^{iuW_t + u^2 t/2} = 1 + \int_0^t e^{X_r}\, d(iuW_r - u^2 r/2) + \tfrac{1}{2} \int_0^t (-u^2) e^{X_r}\, dr$$

$$= 1 + \int_0^t iu\, e^{iuW_r + u^2 r/2}\, dW_r.$$

Therefore

$$e^{iuW_t} = e^{-u^2 t/2} + \int_0^t iu\, e^{iuW_r + u^2 r/2 - u^2 t/2}\, dW_r. \tag{12.10}$$

The integrand in the stochastic integral in (12.10) is e^{iuW_r} times a deterministic function, hence is predictable. Therefore (12.9) holds when $Y = e^{iuW_t}$ and moreover, the support of H in this case is contained in $[0, t]$, that is, $H_r = 0$ if $r \notin [0, t]$. Similarly, (12.9) holds when $Y = e^{iu(W_t - W_s)}$, and in this case the support of the corresponding H is $[s, t]$.

Step 2. Suppose now that Y_1 and Y_2 are two random variables for which (12.9) holds with the supports of the corresponding H_1 and H_2 overlapping by at most finitely many points. To be more precise, if $Y_i = \mathbb{E}\, Y_i + \int_0^{t_0} H_i(s)\, dW_s$, $i = 1, 2$, then we suppose that, with probability one, $H_1(s) H_2(s) = 0$ except for finitely many points s. This implies

$$\int_0^{t_0} H_1(s) H_2(s)\, ds = 0.$$

Let $Z_i(t) = \mathbb{E}\, Y_i + \int_0^t H_i(s)\, dW_s$, $i = 1, 2$. Note $Z_i(0) = \mathbb{E}\, Y_i$ and $Z_i(t_0) = Y_i$. Then by the product formula (Corollary 11.3),

$$
\begin{aligned}
Y_1 Y_2 &= (\mathbb{E}\, Y_1)(\mathbb{E}\, Y_2) + \int_0^{t_0} Z_1(s)\, dZ_2(s) + \int_0^{t_0} Z_2(s)\, dZ_1(s) + \langle Z_1, Z_2 \rangle_{t_0} \\
&= (\mathbb{E}\, Y_1)(\mathbb{E}\, Y_2) + \int_0^{t_0} Z_1(s) H_2(s)\, dW_s + \int_0^{t_0} Z_2(s) H_1(s)\, dW_s \\
&\quad + \int_0^{t_0} H_1(s) H_2(s)\, ds \\
&= (\mathbb{E}\, Y_1)(\mathbb{E}\, Y_2) + \int_0^{t_0} K_s\, dW_s,
\end{aligned}
\tag{12.11}
$$

where $K_s = Z_1(s) H_2(s) + Z_s(s) H_1(s)$, and so the support of K_s is contained in the union of the supports of $H_1(s)$ and $H_2(s)$. Taking an expectation in (12.11), $\mathbb{E}\,[Y_1 Y_2] = (\mathbb{E}\, Y_1)(\mathbb{E}\, Y_2)$. Thus (12.9) holds for $Y_1 Y_2$. Using induction, (12.9) will hold for the product of n random variables Y_i, $i = 1, \ldots, n$, provided the supports of any two of the corresponding H_i overlap by at most finitely many values of s. Combining this with Step 1, we see that if $s_1 < s_2 < \cdots < s_{n+1} \leq t_0$, then the random variables of the form

$$
Y = \exp\left(i \sum_{j=1}^n u_j (W_{s_{j+1}} - W_{s_j}) \right)
\tag{12.12}
$$

satisfy (12.9).

Step 3. We claim that random variables of the form (12.12) generate $\sigma(W_s; s \leq t_0)$. To see this, we proceed as in the last paragraph of the proof of Theorem 12.1, namely, we replace each u_j by $-u_j$, multiply by $\widehat{f}(u_1, \ldots, u_n)$, the Fourier transform of a C^∞ function f with compact support, integrate over $(u_1, \ldots, u_n) \in \mathbb{R}^n$, use the Fubini theorem and the Fourier inversion theorem, and we obtain random variables of the form

$$
f(W_{s_2} - W_{s_1}, \ldots, W_{s_{n+1}} - W_{s_n})
$$

for f in C^∞ with compact support. By a limit argument, such random variables generate $\sigma(W_s; s \leq t_0)$. We will prove that whenever Y_n satisfies (12.9) and $Y_n \to Y$ in L^2, then Y satisfies (12.9). By Exercise 2.7 and Proposition 2.5, this will prove our theorem.

Suppose each Y_n satisfies (12.9) with integrand $H_n(s)$ and suppose $Y_n \to Y$ in L^2. Then $\mathbb{E}\, Y_n \to \mathbb{E}\, Y$, and $Y_n - \mathbb{E}\, Y_n$ converges in L^2 to $Y - \mathbb{E}\, Y$. Since

$$
\mathbb{E} \int_0^{t_0} (H_n(s) - H_m(s))^2\, ds = \mathbb{E}\, ((Y_n - \mathbb{E}\, Y_n) - (Y_m - \mathbb{E}\, Y_m))^2 \to 0,
$$

the sequence H_n is a Cauchy sequence with respect to the norm $\|X\| = (\mathbb{E} \int_0^{t_0} X_s^2\, ds)^{1/2}$, which is an L^2 norm and hence complete. Therefore there exists H_s (which is predictable because each $H_n(s)$ is predictable) such that $\mathbb{E} \int_0^{t_0} H_s^2\, ds < \infty$ and $\mathbb{E} \int_0^{t_0} (H_n(s) - H_s)^2\, ds \to 0$. Hence

$$
\mathbb{E}\left((Y_n - \mathbb{E}\, Y_n) - \int_0^{t_0} H_s\, dW_s \right)^2 = \mathbb{E} \int_0^{t_0} (H_n(s) - H_s)^2\, ds \to 0.
$$

Since $Y_n - \mathbb{E}\, Y_n$ converges in L^2 to $Y - \mathbb{E}\, Y$, it follows that $Y - \mathbb{E}\, Y = \int_0^{t_0} H_s\, dW_s$, a.s. $\qquad\square$

Corollary 12.4 *Suppose M_t is a right-continuous square integrable martingale with respect to the minimal augmented filtration $\{\mathcal{F}_t\}$ generated by a one-dimensional Brownian motion and suppose $M_0 = 0$. Let $t_0 > 0$. Then there exists a predictable process H_s with $\mathbb{E} \int_0^{t_0} H_s^2 \, ds < \infty$ such that with probability one*

$$M_t = \int_0^t H_s \, dW_s$$

for all $t \leq t_0$.

Proof Since M_t is a martingale, $\mathbb{E}[M_{t_0} \mid \mathcal{F}_0] = M_0$, and taking expectations, $\mathbb{E} M_{t_0} = \mathbb{E} M_0 = 0$. By Theorem 12.3, there exists a predictable process H with $\mathbb{E} \int_0^{t_0} H_s^2 \, ds < \infty$ such that $M_{t_0} = \int_0^t H_s \, dW_s$.

Taking conditional expectations with respect to $\mathcal{F}_t$, we obtain $M_t = \int_0^t H_s \, dW_s$. This holds almost surely for each t. Thus except for a null set of ω's, it holds for all t rational. Since M_t is right continuous, it holds for all t.   □

Corollary 12.5 *If M_t is a square integrable martingale with respect to the minimal augmented filtration of a one-dimensional Brownian motion W, then M_t has a version with continuous paths.*

Proof By Corollary 3.13, M has a version with right continuous paths. By Corollary 12.4, M can be written as a stochastic integral with respect to W. But such stochastic integrals have continuous paths by Theorem 10.4.   □

It is important for the martingale representation theorem that M_t be a martingale with respect to the minimal augmented filtration of W and not a larger filtration. For example, let (X, Y) be a two-dimensional Brownian motion and let $\{\mathcal{F}_t\}$ be the minimal augmented filtration generated by (X, Y). We show that we cannot write Y_1 as a stochastic integral with respect to X_t. If it were possible to do so, since Y_1 has mean zero, we would have

$$Y_1 = \int_0^1 H_s \, dX_s.$$

Taking conditional expectations, $Y_t = \int_0^t H_s \, dX_s$. Then $\langle X, Y \rangle_t = \int_0^t H_s \, ds$ by Exercise 10.5. But if (X, Y) is two-dimensional Brownian motion, then X and Y are independent, and so $\langle X, Y \rangle_t = 0$ by Exercise 9.4, a contradiction. (However, it is true, by a proof similar to that of Theorem 12.3, if $\{\mathcal{F}_t\}$ is the minimal augmented filtration of a d-dimensional Brownian motion $(W^1, \ldots, W^d)$ and Y is square integrable and $\mathcal{F}_{t_0}$ measurable, then there exist suitable processes H_s^i such that $Y = \mathbb{E} Y + \sum_{i=1}^d \int_0^{t_0} H_s^i \, dW_s^i$.)

12.5 The Burkholder–Davis–Gundy inequalities

Next we turn to a pair of basic inequalities, those of Burkholder, Davis, and Gundy. In both of the following theorems, the constant depends on p, the exponent. As stated and proved below, we require $p \geq 2$ for Theorems 12.6 and 12.7; in fact, the two theorems are true (with a different proof) as long as $p > 0$; see Bass (1995), pp. 62–4, or Exercise 12.12. The proof we present here is a nice application of Itô's formula.

Define

$$M_t^* = \sup_{s \le t} |M_s|.$$

Theorem 12.6 *Let M_t be a continuous local martingale with $M_0 = 0$, a.s., and suppose $2 \le p < \infty$. There exists a constant c_1 depending on p such that for any finite stopping time T,*

$$\mathbb{E}\, (M_T^*)^p \le c_1 \mathbb{E}\, \langle M \rangle_T^{p/2}.$$

Proof There is nothing to prove if the left-hand side is zero, so we may assume it is positive. First suppose M_T^* is bounded by a positive constant K. Note for $p \ge 2$ the function $x \to |x|^p$ is C^2. By Doob's inequalities and then Itô's formula (and the fact that $|M_s| \ge 0$), we have

$$
\begin{aligned}
\mathbb{E}\, |M_T^*|^p &\le c \mathbb{E}\, |M_T|^p \\
&= c \mathbb{E} \int_0^T p|M_s|^{p-1}\, dM_s + \tfrac{1}{2} c \mathbb{E} \int_0^T p(p-1)|M_s|^{p-2}\, d\langle M \rangle_s \\
&\le c \mathbb{E} \int_0^T (M_T^*)^{p-2}\, d\langle M \rangle_s \\
&= c \mathbb{E}\, [(M_T^*)^{p-2} \langle M \rangle_T].
\end{aligned}
$$

(Recall our convention about constants and the letter c.) Using Hölder's inequality with exponents $p/(p-2)$ and $p/2$, we obtain

$$\mathbb{E}\, (M_T^*)^p \le c(\mathbb{E}\, (M_T^*)^p)^{\frac{p-2}{p}} (\mathbb{E}\, (\langle M \rangle_T^{\frac{p}{2}})^{\frac{2}{p}}.$$

Dividing both sides by $(\mathbb{E}\, (M_T^*)^p)^{(p-2)/p}$ and then taking both sides to the power $p/2$ gives our result.

We then apply the above to $T \wedge U_K$, where $U_K = \inf\{t : |M_t| \ge K\}$, let $K \to \infty$, and use Fatou's lemma. $\qquad \square$

Theorem 12.7 *Let M_t be a continuous local martingale with $M_0 = 0$, a.s., and suppose $2 \le p < \infty$. There exists a constant c_2 depending on p such that for any finite stopping time T,*

$$\mathbb{E}\, \langle M \rangle_T^{p/2} \le c_2 \mathbb{E}\, (M_T^*)^p.$$

Proof As in the previous theorem, we may assume the left-hand side is positive. Set $r = p/2$. Let us first suppose $\langle M \rangle_T$ and M_T^* are bounded by a positive constant K. Let $N_t = M_{t \wedge T}$, so that $\langle N \rangle_\infty = \langle M \rangle_T$, and let $A_t = \langle M \rangle_{t \wedge T}^{r-1}$. Using integration by parts,

$$
\begin{aligned}
\int_0^\infty \langle N \rangle_s\, dA_s &= \langle N \rangle_\infty A_\infty - \int_0^\infty A_s\, d\langle N \rangle_s \\
&= \langle N \rangle_\infty^r - \frac{1}{r} \langle N \rangle_\infty^r.
\end{aligned}
$$

Since

$$\int_0^\infty \langle N \rangle_\infty\, dA_s = \langle N \rangle_\infty^r,$$

we then have

$$\langle N \rangle_\infty^r = r \int_0^\infty (\langle N \rangle_\infty - \langle N \rangle_s)\, dA_s.$$

Using Propositions 3.14 and 9.6,

$$\begin{aligned}
\mathbb{E}\, \langle M \rangle_T^r = \mathbb{E}\, \langle N \rangle_\infty^r &= r\mathbb{E} \int_0^\infty (\langle N \rangle_\infty - \langle N \rangle_s)\, dA_s \\
&= r\mathbb{E} \int_0^\infty (\mathbb{E}\,[\langle N \rangle_\infty \mid \mathcal{F}_s] - \langle N \rangle_s)\, dA_s \\
&= r\mathbb{E} \int_0^\infty \mathbb{E}\,[\langle N \rangle_\infty - \langle N \rangle_s \mid \mathcal{F}_s]\, dA_s \\
&= r\mathbb{E} \int_0^\infty \mathbb{E}\,[N_\infty^2 - N_s^2 \mid \mathcal{F}_s]\, dA_s \\
&\leq c\mathbb{E} \int_0^\infty \mathbb{E}\,[(N_\infty^*)^2 \mid \mathcal{F}_s]\, dA_s \\
&= c\mathbb{E}\,[(N_\infty^*)^2 A_\infty] \\
&= c\mathbb{E}\,[(M_T^*)^2 \langle M \rangle_T^{r-1}].
\end{aligned}$$

We use Hölder's inequality with exponents r and $r/(r-1)$, divide both sides by the quantity $(\mathbb{E}\, \langle M \rangle_T^r)^{(r-1)/r}$, and then take both sides to the rth power. We then get

$$\mathbb{E}\, \langle M \rangle_T^r \leq c\mathbb{E}\,(M_T^*)^{2r},$$

which is what we wanted.

To remove the restriction that $\langle M \rangle$ and M^* are bounded, we apply the above to $T \wedge V_K$ in place of T, where $V_K = \inf\{t : \langle M \rangle_t + M_t^* \geq K\}$, let $K \to \infty$, and use Fatou's lemma. $\square$

12.6 Stratonovich integrals

For stochastic differential geometry and also many other purposes, the Stratonovich integral is more convenient than the Itô integral. If X and Y are continuous semimartingales, the *Stratonovich integral*, denoted $\int_0^t X_s \circ dY_s$, is defined by

$$\int_0^t X_s \circ dY_s = \int_0^t X_s\, dY_s + \tfrac{1}{2}\langle X, Y \rangle_t.$$

Both the beauty and the difficulty of Itô's formula are due to the quadratic variation term. The change of variables formula for the Stratonovich integral avoids this.

Theorem 12.8 *Suppose $f \in C^3$ and X is a continuous semimartingale. Then*

$$f(X_t) = f(X_0) + \int_0^t f'(X_s) \circ dX_s.$$

Proof By Itô's formula applied to the function f and the definition of the Stratonovich integral, it suffices to show that

$$\langle f'(X), X \rangle_t = \int_0^t f''(X_s)\, d\langle X \rangle_s. \tag{12.13}$$

Applying Itô's formula to the function f', which is in C^2,

$$f'(X_t) = f'(X_0) + \int_0^t f''(X_s)\,dX_s + \tfrac{1}{2}\int_0^t f'''(X_s)\,d\langle X\rangle_s,$$

from which (12.13) follows. □

If X and Y are continuous semimartingales and we apply the change of variables formula with $f(x) = x^2$ to $X + Y$ and $X - Y$, we obtain

$$(X_t + Y_t)^2 = (X_0 + Y_0)^2 + 2\int_0^t (X_s + Y_s) \circ d(X_s + Y_s)$$

and

$$(X_t - Y_t)^2 = (X_0 - Y_0)^2 + 2\int_0^t (X_s - Y_s) \circ d(X_s - Y_s).$$

Taking the difference and then dividing by 4, we have the *product formula for Stratonovich integrals*

$$X_t Y_t = X_0 Y_0 + \int_0^t X_s \circ dY_s + \int_0^t Y_s \circ dX_s. \qquad (12.14)$$

The Stratonovich integral $\int H_s \circ dX_s$ can be represented as a limit of Riemann sums.

Proposition 12.9 *Suppose H and X are continuous semimartingales and $t_0 > 0$. Then $\int_0^t H_s \circ dX_s$ is the limit in probability as $n \to \infty$ of*

$$\sum_{k=0}^{2^n-1} \frac{H_{kt_0/2^n} + H_{(k+1)t_0/2^n}}{2}(X_{(k+1)t_0/2^n} - X_{kt_0/2^n}).$$

Proof We write the sum as

$$\sum H_{kt_0/2^n}(X_{(k+1)t_0/2^n} - X_{kt_0/2^n})$$
$$+ \tfrac{1}{2}\sum (H_{(k+1)t_0/2^n} - H_{kt_0/2^n})(X_{(k+1)t_0/2^n} - X_{kt_0/2^n}).$$

The first sum tends to $\int_0^t H_s\,dX_s$ while by Exercise 12.10 the second sum tends to $\tfrac{1}{2}\langle H, X\rangle_t$. This proves the proposition. □

Exercises

12.1 Show that W_t and $W_t^2 - t$ are local martingales, where W is defined in the statement of Theorem 12.2.

12.2 Suppose $\{\mathcal{F}_t\}$ is a filtration satisfying the usual conditions, X is a Brownian motion with respect to $\{\mathcal{F}_t\}$, and T is a finite stopping time with respect to this same filtration. Let Y be another Brownian motion that is independent of $\{\mathcal{F}_t\}$ and define

$$Z_t = \begin{cases} X_t, & t < T \\ X_T + Y_{t-T}, & t \geq T. \end{cases}$$

Show that Z is a Brownian motion (although not necessarily with respect to $\{\mathcal{F}_t\}$).

12.3   Suppose M_t is a continuous local martingale with respect to a filtration $\{\mathcal{F}_t\}$ satisfying the usual conditions, T is a stopping time with respect to $\{\mathcal{F}_t\}$, and $\langle M \rangle_t = t \wedge T$. Prove that $M_{t \wedge T}$ has the same law as a Brownian motion stopped at time T.

12.4   Here is a multidimensional version of Lévy's theorem. Let $\{\mathcal{F}_t\}$ be a filtration satisfying the usual conditions. Suppose $(M_t^1, \ldots, M_t^d)$ is a d-dimensional process such that each component M_t^i is a continuous martingale with respect to $\{\mathcal{F}_t\}$ with $\langle M^i \rangle_t = t$. Suppose that $\langle M^i, M^j \rangle_t = 0$ if $i \neq j$. Prove that $(M_t^1, \ldots, M_t^d)$ is a d-dimensional Brownian motion.

12.5   Let $\{\mathcal{F}_t\}$ be a filtration satisfying the usual conditions. Let A_t be a strictly increasing continuous process adapted to $\{\mathcal{F}_t\}$ with $\lim_{t \to \infty} A_t = \infty$, a.s. Suppose $(M_t^1, \ldots, M_t^d)$ is a d-dimensional process such that each component M_t^i is a continuous martingale with respect to $\{\mathcal{F}_t\}$ and $\langle M^i \rangle_t = A_t$. Suppose that $\langle M^i, M^j \rangle_t = 0$ if $i \neq j$. Prove that $(M_t^1, \ldots, M_t^d)$ is a time change of d-dimensional Brownian motion.

12.6   Suppose M is a continuous local martingale such that $\langle M \rangle_t$ is deterministic. Prove that M is a Gaussian process.

12.7   Suppose M is a continuous local martingale with $M_0 = 0$, a.s. Show that there exists a Brownian motion W, an increasing process τ_t, and a stopping time T such that $M_t = W_{\tau_t \wedge T}$ for all t.

12.8   Let M_t be a continuous local martingale. Show that the events $(M_\infty^* < \infty)$ and $(\langle M \rangle_\infty < \infty)$ differ by at most a null set.

12.9   Let M_t be a continuous local martingale. Prove that

$$\mathbb{P}(\sup_{t \geq 0} |M_t| > x, \langle M \rangle_\infty < y) \leq 2e^{-x^2/2y}.$$

12.10 Suppose X and Y are continuous semimartingales and $t_0 > 0$. Prove that

$$\sum_{k=0}^{2^n - 1} (X_{(k+1)t_0/2^n} - X_{kt_0/2^n})(Y_{(k+1)t_0/2^n} - Y_{kt_0/2^n})$$

converges to $\langle X, Y \rangle_{t_0}$ in probability.

12.11 Let $p > 0$. Suppose X and Y are non-negative random variables, $\beta > 1$, $\delta \in (0, 1)$, and $\varepsilon \in (0, \beta^{-p}/2)$ such that

$$\mathbb{P}(X > \beta\lambda, Y < \delta\lambda) \leq \varepsilon\mathbb{P}(X \geq \lambda)$$

for all $\lambda > 0$. This inequality is known as a *good-λ inequality*. Prove that there exists a constant c (depending on β, δ, ε, and p but not X or Y) such that

$$\mathbb{E}\, X^p \leq c\,\mathbb{E}\, Y^p.$$

Hint: First assume X is bounded. Write

$$\mathbb{P}(X/\beta > \lambda) = \mathbb{P}(X > \beta\lambda, Y < \delta\lambda) + \mathbb{P}(Y \geq \delta\lambda)$$
$$\leq \varepsilon\mathbb{P}(X \geq \lambda) + \mathbb{P}(Y/\delta \geq \lambda).$$

Multiply by $p\lambda^{p-1}$, integrate over λ, and use the fact that $\varepsilon < \beta^{-p}/2$.

12.12 Use Exercise 12.11 to prove that the Burkholder–Davis–Gundy inequalities hold for all $p > 0$.

Hint: Use time change to reduce to the case of a Brownian motion W. If T is a stopping time and $U = \inf\{t : W_T^* > \lambda\}$, write

$$\mathbb{P}(W_T^* > \beta\lambda, T^{1/2} < \delta\lambda) = \mathbb{P}(W_T^* > \beta\lambda, T < \delta^2\lambda^2, U < \infty)$$
$$\leq \mathbb{P}(\sup_{U \leq t \leq U + \delta^2\lambda^2} |W_t - W_U| > (\beta - 1)\lambda, U < \infty).$$

Condition on $\mathcal{F}_U$, use Theorem 4.2, and notice that $\mathbb{P}(U < \infty) = \mathbb{P}(W_T^* > \lambda)$.

12.13 Define the H^1 norm of a martingale by

$$\|M\|_{H^1} = \mathbb{E}\left[\sup_{t \geq 0} |M_t|\right].$$

Prove that this is a norm. Does there exist a uniformly integrable continuous martingale that is not in H^1?

12.14 Let W be a Brownian motion and let T be a stopping time. Prove that if $\mathbb{E}\, T^{1/2} < \infty$, then $\mathbb{E}\, W_T = 0$.

12.15 Suppose $W = (W^1, \ldots, W^d)$ is a d-dimensional Brownian motion started at 0, and let $\{\mathcal{F}_t\}$ be the minimal augmented filtration of W. Suppose Y is a $\mathcal{F}_1$ measurable random variable with mean zero and finite variance. Prove there exist predictable processes $H^1, \ldots, H^d$ such that $\mathbb{E}\int_0^1 (H_s^i)^2\, ds < \infty$ for each i and

$$Y = \sum_{i=1}^d \int_0^1 H_s^i\, dW_s^i.$$

12.16 Suppose W is a Brownian motion and H is adapted, bounded, and right continuous. Let $t \geq 0$. Show

$$\frac{1}{W_{t+h} - W_t} \int_t^{t+h} H_s\, dW_s$$

converges in probability to H_t.

12.17 Let W be a Brownian motion and $\alpha > 0$. Show that

$$\int_0^t \frac{1}{|W_s|^\alpha}\, ds$$

is infinite almost surely if $\alpha \geq 1$ but finite almost surely if $\alpha < 1$.

12.18 Here is a useful inequality. Suppose A is an increasing process with $A_0 = 0$, a.s., and suppose there exists a non-negative random variable B such that for each t,

$$\mathbb{E}[A_\infty - A_t \mid \mathcal{F}_t] \leq \mathbb{E}[B \mid \mathcal{F}_t], \qquad \text{a.s.}$$

Prove that for each integer $p \geq 1$, there exists a constant c_p depending only on p such that

$$\mathbb{E}\, A_\infty^p \leq c_p \mathbb{E}\, B^p.$$

Hint: Write

$$A_\infty = p! \int_0^\infty (A_\infty - A_t)\, dA_t,$$

take expectations, and use Proposition 3.14.

12.19 Let W be a one-dimensional Brownian motion with filtration $\{\mathcal{F}_t\}$ and let $f(r, s)$ be a deterministic function. Define the *multiple stochastic integral* by

$$\int_0^t \int_0^s f(r, s) \, dW_r \, dW_s = \int_0^t \left(\int_0^s f(r, s) \, dW_r \right) dW_s,$$

provided

$$\int_0^t \int_0^s f(r, s)^2 \, dr \, ds < \infty,$$

and similarly for higher-order multiple stochastic integrals.

(1) If $f : \mathbb{R}^m \to \mathbb{R}$ and $g : \mathbb{R}^n \to \mathbb{R}$ are bounded and deterministic, $n \neq m$,

$$M_t^f = \int_0^t \cdots \int_0^{r_{m-1}} f \, dW_{r_1} \cdots dW_{r_m},$$

and M_t^g is defined similarly, show that $\mathbb{E}\,[M_t^f M_t^g] = 0$ for all t.

(2) Show that the collection of random variables

$$\{M_1^f : f \text{ has domain } \mathbb{R}^m \text{ for some } m \text{ and is bounded and deterministic}\}$$

is dense in the set of mean zero $\mathcal{F}_1$ measurable random variables with respect to the $L^2(\mathbb{P})$ norm.

13

The Girsanov theorem

We look at what happens to a Brownian motion when we change $\mathbb{P}$ to another probability measure $\mathbb{Q}$. This may seem strange, but there are many applications of this, including to financial mathematics and to filtering; see Chapters 28 and 29. Another application we will give (at the end of this chapter in Section 13.2) is to determine the probability a Brownian motion W_s crosses a line $a + bs$ before time t.

13.1 The Brownian motion case

We start with an observation. Suppose Y_t is a continuous local martingale with $Y_0 = 0$ and let $Z_t = e^{Y_t - \langle Y \rangle_t / 2}$. Applying Itô's formula to $X_t = Y_t - \frac{1}{2}\langle Y \rangle_t$ with the function e^x yields

$$Z_t = e^{Y_t - \langle Y \rangle_t / 2} = 1 + \int_0^t e^{X_s} d\left(Y_s - \tfrac{1}{2}\langle Y \rangle_s\right) + \tfrac{1}{2}\int_0^t e^{X_s} d\langle Y \rangle_s$$

$$= 1 + \int_0^t Z_s \, dY_s. \tag{13.1}$$

This can be abbreviated by $dZ_t = Z_t \, dY_t$. Z_t is called the *exponential of the martingale Y*, and since Z is the stochastic integral with respect to a local martingale, it is itself a local martingale.

Before stating the Girsanov theorem, we need two technical lemmas.

Lemma 13.1 *Suppose Y is a continuous local martingale with $Y_0 = 0$ and $Z_t = e^{Y_t - \langle Y \rangle_t / 2}$. If $\langle Y \rangle_t$ is a bounded random variable for each t, then $\mathbb{E}\,|Z_t|^p < \infty$ for each $p > 1$ and each t.*

Proof Let us first suppose Y is bounded in absolute value by N. Since $Z_t \geq 0$, we have by the Cauchy–Schwarz inequality

$$\mathbb{E}\,Z_t^p = \mathbb{E}\,e^{pY_t - p\langle Y \rangle_t / 2} \tag{13.2}$$

$$= \mathbb{E}\left[e^{pY_t - p^2 \langle Y \rangle_t} e^{(p^2 - (p/2))\langle Y \rangle_t}\right]$$

$$\leq \left(\mathbb{E}\,e^{2pY_t - 2p^2 \langle Y \rangle_t}\right)^{1/2} \left(\mathbb{E}\,e^{(2p^2 - p)\langle Y \rangle_t}\right)^{1/2}.$$

By the exact same calculation as in (13.1) but with Y replaced by $2pY$, we see $e^{2pY_t - 2p^2 \langle Y \rangle_t}$ is a stochastic integral of a bounded integrand with respect to a bounded martingale, and hence is a martingale. This shows that the first factor on the last line of (13.2) is 1. By our assumption that $\langle Y \rangle_t$ is bounded, the second factor on this line is finite and does not depend on N.

If Y is not bounded, let $T_N = \inf\{s : |Y_s| \geq N\}$, apply the above argument to $Y_{t \wedge T_N}$, and let $N \to \infty$. □

The second lemma is the following.

Lemma 13.2 *Suppose A_t is a continuous increasing process adapted to a filtration $\{\mathcal{F}_t\}$ satisfying the usual conditions. Let X be a bounded random variable, H a bounded adapted process, $s < t$, and $B \in \mathcal{F}_s$. Then*

$$\mathbb{E}\Big[\int_s^t X H_r \, dA_r; B\Big] = \mathbb{E}\Big[\int_s^t \mathbb{E}[X \mid \mathcal{F}_r] H_r \, dA_r; B\Big].$$

Proof By linearity, it suffices to suppose X and H are non-negative. Let $A'_r = A_{r+s}$, $H'_r = H_{r+s}$, and $\mathcal{F}'_r = \mathcal{F}_{r+s}$. Let $C_r = \int_0^r H'_s 1_B \, dA'_s$, and so we must show

$$\mathbb{E}\int_0^{t-s} X \, dC_r = \mathbb{E}\int_0^{t-s} \mathbb{E}[X \mid \mathcal{F}'_r] \, dC_r.$$

This follows by Proposition 3.14. □

Let M_t be a non-negative continuous martingale with $M_0 = 1$, a.s. Define a new probability measure $\mathbb{Q}$ by $\mathbb{Q}(A) = \mathbb{E}[M_t; A]$ if $A \in \mathcal{F}_t$. Note $\mathbb{Q}$ is a probability measure because $\mathbb{Q}(\Omega) = \mathbb{E} M_t = \mathbb{E} M_0 = 1$. $\mathbb{Q}$ is well-defined because if $A \in \mathcal{F}_s \subset \mathcal{F}_t$, then since M is a martingale, we have $\mathbb{E}[M_t; A] = \mathbb{E}[M_s; A]$.

A more general version of the Girsanov theorem is possible (see Exercise 13.5), but the Girsanov theorem is most frequently used with Brownian motion.

Theorem 13.3 *Suppose W_t is a Brownian motion with respect to $\mathbb{P}$, H is bounded and predictable,*

$$M_t = \exp\Big(\int_0^t H_r \, dW_r - \tfrac{1}{2}\int_0^t H_r^2 \, dr\Big), \tag{13.3}$$

and

$$\mathbb{Q}(B) = \mathbb{E}_\mathbb{P}[M_t; B] \qquad \text{if } B \in \mathcal{F}_t. \tag{13.4}$$

Then $W_t - \int_0^t H_r \, dr$ is a Brownian motion with respect to $\mathbb{Q}$.

Proof We prove the theorem by showing $W_t - \int_0^t H_r \, dr$ satisfies the hypotheses of Lévy's theorem (Theorem 12.1). We first show $W_t - \int_0^t H_r \, dr$ is a martingale with respect to $\mathbb{Q}$. By (13.1) with $Y_t = \int_0^t H_r \, dW_r$ and $Z_t = M_t$,

$$M_t = 1 + \int_0^t M_r H_r \, dW_r.$$

By Exercise 10.5,

$$\langle M, W \rangle_t = \int_0^t M_r H_r \, dr. \tag{13.5}$$

We want to show that if $B \in \mathcal{F}_s$, then

$$\mathbb{E}_\mathbb{Q}\Big[W_t - \int_0^t H_r \, dr; B\Big] = \mathbb{E}_\mathbb{Q}\Big[W_s - \int_0^s H_r \, dr; B\Big]. \tag{13.6}$$

If $B \in \mathcal{F}_s$, then using the definition of $\mathbb{Q}$ and the product formula (Corollary 11.3),

$$\mathbb{E}_\mathbb{Q}[W_t; B] = \mathbb{E}_\mathbb{P}[M_t W_t; B] \tag{13.7}$$
$$= \mathbb{E}_\mathbb{P}\left[\int_0^t M_r \, dW_r; B\right] + E_\mathbb{P}\left[\int_0^t W_r \, dM_r; B\right]$$
$$+ \mathbb{E}_\mathbb{P}[\langle M, W\rangle_t; B]$$

and

$$\mathbb{E}_\mathbb{Q}[W_s; B] = \mathbb{E}_\mathbb{P}[M_s W_s; B] \tag{13.8}$$
$$= \mathbb{E}_\mathbb{P}\left[\int_0^s M_r \, dW_r; B\right] + E_\mathbb{P}\left[\int_0^s W_r \, dM_r; B\right]$$
$$+ \mathbb{E}_\mathbb{P}[\langle M, W\rangle_s; B].$$

Since H is bounded, $\langle \int_0^{\cdot} H_r \, dW_r\rangle_t \le ct$. By Lemma 13.1, M_t is a martingale and $\mathbb{E}\,|M_t|^p < \infty$ for each t and each $p \ge 1$. Since stochastic integrals with respect to martingales are martingales,

$$\mathbb{E}_\mathbb{P}\left[\int_0^t M_r \, dW_r; B\right] = \mathbb{E}_\mathbb{P}\left[\int_0^s M_r \, dW_r; B\right] \tag{13.9}$$

and

$$\mathbb{E}_\mathbb{P}\left[\int_0^t W_r \, dM_r; B\right] = \mathbb{E}_\mathbb{P}\left[\int_0^s W_r \, dM_r; B\right]. \tag{13.10}$$

Combining (13.7), (13.8), (13.9), and (13.10), we see that (13.6) will follow if we show

$$\mathbb{E}_\mathbb{P}[\langle M, W\rangle_t - \langle M, W\rangle_s; B] = \mathbb{E}_\mathbb{Q}\left[\int_s^t H_r \, dr; B\right]. \tag{13.11}$$

Using Lemma 13.2 and (13.5), we have

$$\mathbb{E}_\mathbb{Q}\left[\int_s^t H_r \, dr; B\right] = \mathbb{E}_\mathbb{P}\left[M_t \int_s^t H_r \, dr; B\right] = \mathbb{E}_\mathbb{P}\left[\int_s^t M_t H_r \, dr; B\right]$$
$$= \mathbb{E}_\mathbb{P}\left[\int_0^t \mathbb{E}\,[M_t \mid \mathcal{F}_r] H_r \, dr; B\right] = \mathbb{E}_\mathbb{P}\left[\int_s^t M_r H_r \, dr; B\right]$$
$$= \mathbb{E}_\mathbb{P}[\langle M, W\rangle_t - \langle M, W\rangle_s; B],$$

which proves (13.11).

A similar proof shows that $(W_t - \int_0^t H_r \, dr)^2 - t$ is a martingale with respect to $\mathbb{Q}$, and hence the quadratic variation of $W_t - \int_0^t H_r \, dr$ under $\mathbb{Q}$ is still t (or see Exercise 13.2). Since the process $W_t - \int_0^t H_r \, dr$ has continuous paths, by Lévy's theorem, $W_t - \int_0^t H_r \, dr$ is a Brownian motion under $\mathbb{Q}$. $\qquad \square$

The assumption that H be bounded can be weakened, but in practice it is more common to use a stopping time argument; for an example, see the proof of Theorem 29.3.

13.2 An example

Let us give an example of the use of the Girsanov theorem, namely, to compute the probability that Brownian motion crosses a line $a + bt$ by time t_0, $a > 0$. We want to find an exact expression for $\mathbb{P}(\exists t \leq t_0 : W_t = a + bt)$, where W is a Brownian motion.

Let W_t be a Brownian motion under $\mathbb{P}$. Define $\mathbb{Q}$ on $\mathcal{F}_{t_0}$ by

$$d\mathbb{Q}/d\mathbb{P} = M_t = e^{-bW_t - b^2 t/2}.$$

By the Girsanov theorem, under $\mathbb{Q}$, $\widetilde{W}_t = W_t + bt$ is a Brownian motion, and $W_t = \widetilde{W}_t - bt$.

Let $A = (\sup_{s \leq t_0} W_s \geq a)$. If we set $S = \inf\{t > 0 : W_t = a\}$, then $A = (S \leq t_0)$ and $A \in \mathcal{F}_{S \wedge t_0}$. We write

$$\mathbb{P}(\exists t \leq t_0 : W_t = a + bt) = \mathbb{P}(\exists t \leq t_0 : W_t - bt = a) \qquad (13.12)$$
$$= \mathbb{P}(\sup_{s \leq t_0}(W_s - bs) \geq a).$$

W_t is a Brownian motion under $\mathbb{P}$ while $\widetilde{W}_t$ is a Brownian motion under $\mathbb{Q}$. Therefore the last line of (13.12) is equal to

$$\mathbb{Q}(\sup_{s \leq t_0}(\widetilde{W}_s - bs) \geq a).$$

This in turn is equal to

$$\mathbb{Q}(\sup_{s \leq t_0} W_s \geq a) = \mathbb{Q}(A).$$

To evaluate $\mathbb{Q}(A)$, note $M_S = e^{-ab - b^2 S/2}$ and by (3.19) with b replaced by a,

$$\mathbb{P}(S \in ds) = \frac{a}{\sqrt{2\pi s^3}} e^{-a^2/2s}.$$

Now we use optional stopping to obtain

$$\mathbb{P}(\exists t \leq t_0 : W_t = a + bt) = \mathbb{Q}(A) = \mathbb{E}_{\mathbb{P}}[M_{t_0}; A] \qquad (13.13)$$
$$= \mathbb{E}_{\mathbb{P}}[M_{S \wedge t_0}; S \leq t_0]$$
$$= \mathbb{E}_{\mathbb{P}}[M_S; S \leq t_0]$$
$$= \int_0^{t_0} e^{-ab - b^2 s/2} \frac{a}{\sqrt{2\pi s^3}} e^{-a^2/2s} \, ds.$$

Exercises

13.1 Whether a filtration satisfies the usual conditions depends on the class of null sets and hence the probability measure involved matters. Suppose $\{\mathcal{F}_t\}$ satisfies the usual conditions with respect to $\mathbb{P}$, H is a bounded predictable process, W a Brownian motion with respect to $\mathbb{P}$, M defined by (13.3), and $\mathbb{Q}$ defined by (13.4). If $t_0 > 0$ and $A \in \sigma(W_s; s \leq t_0)$, show $\mathbb{P}(A) = 0$ if and only if $\mathbb{Q}(A) = 0$.

13.2 Theorem 9.10 allows us to avoid some calculations in the last paragraph of the proof of Theorem 13.3. Suppose X is a continuous semimartingale under $\mathbb{P}$ and $\mathbb{Q}$ is a probability measure equivalent to $\mathbb{P}$. That is, a set is a null set for $\mathbb{P}$ if and only if it is a null set for $\mathbb{Q}$. Show X is a semimartingale under $\mathbb{Q}$ and the quadratic variation of X under $\mathbb{P}$ equals the quadratic variation of X under $\mathbb{Q}$.

13.3 Let $W = (W^1, \ldots, W^d)$ be a d-dimensional Brownian motion with minimal augmented filtration $\{\mathcal{F}_t\}$ and let $H_1, \ldots, H_d$ be bounded predictable processes. Let

$$M_t = \exp\left(\sum_{i=1}^{d} \int_0^t H_i(s)\, dW_s^i - \tfrac{1}{2} \sum_{i=1}^{d} \int_0^t |H_i(s)|^2\, ds \right).$$

Define a probability measure $\mathbb{Q}$ by setting $\mathbb{Q}(A) = \mathbb{E}_{\mathbb{P}}[M_t; A]$ if $A \in \mathcal{F}_t$. Let $\widetilde{W}_t^i = W_t^i - \int_0^t H_i(s)\, ds$ for each i. Prove that $\widetilde{W} = (\widetilde{W}^1, \ldots, \widetilde{W}^d)$ is a d-dimensional Brownian motion under $\mathbb{Q}$.

13.4 Let W_t be a d-dimensional Brownian motion and let $\delta, t_0 > 0$. Let $f : [0, t_0] \to \mathbb{R}^d$ be a continuous function. Prove that there exists a constant c such that

$$\mathbb{P}(\sup_{s \leq t_0} |W_s - f(s)| < \delta) > c.$$

This is known as the *support theorem* for Brownian motion.

Hint: First assume that f has a bounded derivative. Use Exercise 4.9 and the Girsanov theorem.

13.5 Here is a more general form of the Girsanov theorem. Suppose L_t is a bounded continuous martingale under $\mathbb{P}$, $M_t = e^{L_t - \langle L \rangle_t / 2}$, and $\mathbb{Q}$ is a probability measure defined by $\mathbb{Q}(A) = \mathbb{E}_{\mathbb{P}}[M_{t_0}; A]$ if $A \in \mathcal{F}_{t_0}$. Suppose $\{\mathcal{F}_t\}$ is a filtration satisfying the usual conditions with respect to both $\mathbb{P}$ and $\mathbb{Q}$. Show that if X is a martingale under $\mathbb{P}$, then $X_t - \langle X, L \rangle_t$ is a martingale under $\mathbb{Q}$.

Local times

Let W_t be a one-dimensional Brownian motion. Although the Lebesgue measure of the random set $\{t : W_t = 0\}$ is 0, a.s., nevertheless there is an increasing continuous process which grows only when the Brownian motion is at 0. This increasing process is known as local time at 0. We want to derive some of its properties.

14.1 Basic properties

Let W be a Brownian motion. By Jensen's inequality for conditional expectations (Proposition A.21), $|W_t|$ is a submartingale, and by the Doob–Meyer decomposition (Theorem 9.12), it can be written as a martingale plus an increasing process. Since W_t is itself a martingale, the increasing process grows only at times when the Brownian motion is at 0.

Rather than appealing to the Doob–Meyer decomposition, we give the explicit decomposition of $|W_t|$. We define

$$\text{sgn}\,(x) = \begin{cases} 1, & x > 0; \\ 0, & x = 0; \\ -1, & x < 0. \end{cases}$$

Theorem 14.1 *Let W_t be a one-dimensional Brownian motion.*

(1) There exists a non-negative increasing continuous adapted process L_t^0 such that

$$|W_t| = \int_0^t \text{sgn}\,(W_s)\,dW_s + L_t^0. \tag{14.1}$$

(2) L_t^0 increases only when W is at 0. More precisely, if $W_s(\omega) \neq 0$ for $r \leq s \leq t$, then $L_r^0(\omega) = L_t^0(\omega)$.

L_t^0 is called the *local time at 0*. The equation (14.1) is called the *Tanaka formula*.

Proof Define

$$f_\varepsilon(x) = \begin{cases} x^2/2\varepsilon, & |x| < \varepsilon; \\ |x| - (\varepsilon/2), & |x| \geq \varepsilon. \end{cases}$$

The function f_ε is an approximation to the function $|\cdot|$, and note that $f_\varepsilon(0) = f_\varepsilon'(0)$, while $f_\varepsilon''(x) = \varepsilon^{-1} 1_{[-\varepsilon,\varepsilon]}(x)$, except at $x = \pm\varepsilon$.

We apply the extension of Itô's formula given in Exercise 11.2 to $f_\varepsilon(W_t)$ and obtain

$$f_\varepsilon(W_t) = \int_0^t f_\varepsilon'(W_s)\,dW_s + \frac{1}{2}\int_0^t f_\varepsilon''(W_s)\,ds.$$

As we let $\varepsilon \to 0$, we see that $f_\varepsilon(x) \to |x|$ uniformly, and $f_\varepsilon'(x) \to \mathrm{sgn}\,(x)$ boundedly. By Doob's inequalities, if $t_0 > 0$,

$$\mathbb{E}\sup_{t \le t_0}\left|\int_0^t f_\varepsilon'(W_s)\,dW_s - \int_0^t \mathrm{sgn}\,(W_s)\,dW_s\right|^2 \to 0, \qquad (14.2)$$

while $\sup_{t \le t_0}\big||f_\varepsilon(W_t)| - |W_t|\big| \to 0$, a.s. Therefore there exists an increasing process L_t^0 and a subsequence $\varepsilon_n \to 0$ such that

$$\sup_{t \le t_0}\left|\frac{1}{2\varepsilon_n}\int_0^t 1_{[-\varepsilon_n,\varepsilon_n]}(W_s)\,ds - L_t^0\right| \to 0, \qquad \text{a.s.} \qquad (14.3)$$

Hence for almost every ω there is convergence uniformly over t in finite intervals, so L_t^0 is continuous in t. Since $\frac{1}{2\varepsilon_n}\int_0^t 1_{[-\varepsilon_n,\varepsilon_n]}(W_s)\,ds$ increases only for those times t where $|W_t| \le \varepsilon_n$, then L_t^0 increases only on the set of times when $W_t = 0$. $\qquad \square$

In the Tanaka formula, the stochastic integral term is a martingale, say N_t. Note $\langle N \rangle_t = t$, since $\mathrm{sgn}\,(x)^2 = 1$ unless $x = 0$, and we have seen that Brownian motion spends 0 time at 0 (Exercise 11.1). Hence we have exhibited reflecting Brownian motion, namely $|W_t|$, as the sum of another Brownian motion, N_t, and a continuous process that increases only when W is at zero.

Let M_t denote $\sup_{s \le t} W_s$. Note we do not have an absolute value here. The following, due to Lévy, is often useful.

Theorem 14.2 *The two-dimensional processes* $(|W|, L^0)$ *and* $(M - W, M)$ *have the same law.*

Proof Let $V_t = -N_t$ in the Tanaka formula, so that

$$|W_t| = -V_t + L_t^0. \qquad (14.4)$$

Let $S_t = \sup_{s \le t} V_s$. We will show $S_t = L_t^0$. This will prove the result, since V is a Brownian motion, and hence $(M - W, M)$ is equal in law to $(S - V, S) = (|W|, L^0)$.

From (14.4), $V_t = L_t^0 - |W_t|$, or $V_t \le L_t^0$ for all t, hence $S_t \le L_t^0$, since L^0 is increasing. L_t^0 increases only when $W_t = 0$ and at those times

$$L_t^0 = V_t + |W_t| = V_t \le S_t.$$

Given two increasing functions with $f \le g$, if $f(t) = g(t)$ at those times when f increases, a little thought shows that f and g are equal for all t. Hence $L_t^0 = S_t$ for all t. $\qquad \square$

Just as we defined L_t^0 via the Tanaka formula, we can construct local time at the level a by the formula

$$|W_t - a| - |W_0 - a| = \int_0^t \mathrm{sgn}\,(W_s - a)\,dW_s + L_t^a, \qquad (14.5)$$

and the same proof as above shows that L_t^a is the limit in L^2 of

$$\frac{1}{2\varepsilon} \int_0^t 1_{[a-\varepsilon,a+\varepsilon]}(W_s) \, ds.$$

14.2 Joint continuity of local times

Next we will prove that L_t^a can be taken to be jointly continuous in both t and a.

Theorem 14.3 *Let W be a one-dimensional Brownian motion and let L_t^a be the local time of W at level a. For each $a \in \mathbb{R}$ there exists a version $\widetilde{L}_t^a$ of L_t^a so that with probability one, $\widetilde{L}_t^a$ is jointly continuous in t and a.*

Recall that two processes X and Y are versions of each other if for each t, $X_t = Y_t$, a.s. We will use the Kolmogorov continuity criterion, Corollary 8.2, together with Remark 8.3. We will obtain an estimate on $\widetilde{N}_t^a - \widetilde{N}_t^b$, where $\widetilde{N}_t^a = \int_0^t \text{sgn}\,(W_s - a)\,dW_s$, by means of the Burkholder–Davis–Gundy inequalities.

Proof Let $M > 0$ be arbitrary. It suffices to show the joint continuity for times less than or equal to M and for $|a| \le M$. Let

$$N_t^a = \int_0^{M \wedge t} \text{sgn}\,(W_s - a)\,dW_s.$$

Since $|W_t - a|$ is uniformly continuous in t and a for $|t| \le M$, $|a| \le M$, by the Tanaka formula (14.5) it suffices to establish the same fact for N_t^a.

Let T be a stopping time bounded by M and $a < b$. Since $(N_t^a - N_t^b)^2 - \langle N^a - N^b \rangle_t$ is a martingale,

$$\mathbb{E}\left[((N_M^a - N_M^b) - (N_T^a - N_T^b))^2 | \mathcal{F}_T\right]$$

$$= \mathbb{E}\left[\int_T^M (\text{sgn}\,(W_s - a) - \text{sgn}\,(W_s - b))^2 \, ds | \mathcal{F}_T\right]$$

$$= 4\mathbb{E}\left[\int_T^M 1_{[a,b]}(W_s) \, ds | \mathcal{F}_T\right]$$

$$\le 4\mathbb{E}\left[\int_T^{M+T} 1_{[a,b]}(W_s) \, ds | \mathcal{F}_T\right]$$

$$= 4\mathbb{E}\left[\int_0^M 1_{[a,b]}(W_{s+T}) \, ds | \mathcal{F}_T\right];$$

recall Exercise 11.1. From Proposition 4.5 we deduce

$$\mathbb{E}\left[\int_0^M 1_{[a,b]}(W_{s+T}) \, ds | \mathcal{F}_T\right] \le \int_0^M \frac{c(b-a)}{\sqrt{s}} \, ds \le c(b-a).$$

Thus

$$\mathbb{E}\left[((N_M^a - N_M^b) - (N_T^a - N_T^b))^2 | \mathcal{F}_T\right] \le c|b-a|,$$

and so by (9.3)

$$\mathbb{E}\left[\langle N^a - N^b \rangle_M - \langle N^a - N^b \rangle_T \mid \mathcal{F}_T\right] \le c|b-a|.$$

If we write $A_t = \langle N^a - N^b \rangle_t$, then we have by Proposition 3.14

$$\mathbb{E} A_M^2 = 2\mathbb{E} \int_0^M (A_M - A_t)\, dA_t$$

$$= 2\mathbb{E} \left[\int_0^M (\mathbb{E}[A_M \mid \mathcal{F}_t] - A_t)\, dA_t \right]$$

$$= 2\mathbb{E} \left[\int_0^M \mathbb{E}[A_M - A_t \mid \mathcal{F}_t]\, dA_t \right]$$

$$\leq c|b - a|\mathbb{E} \int_0^M dA_t \leq c|b - a|^2.$$

Applying the Burkholder–Davis–Gundy inequalities,

$$\mathbb{E} \left[\sup_{t \leq M} |N_t^a - N_t^b|^4 \right] \leq c|b - a|^2. \tag{14.6}$$

By the Kolmogorov continuity criterion applied on the Banach space of continuous functions with the metric $d(f, g) = \sup_{t \leq M} |f(t) - g(t)|$, we see N_t^a is continuous as a function of a for a in the dyadic rationals in $[-M, M]$, uniformly over $t \leq M$. Therefore L_a^t is continuous over a in the dyadic rationals in $[-M, M]$, uniformly for $t \leq M$. Also, (14.5) and (14.6) imply

$$\mathbb{E} \left[\sup_{t \leq M} |L_t^a - L_t^b|^4 \right] \leq c\big(|a - b| \wedge 1\big)^2. \tag{14.7}$$

Note that if we define $\widetilde{L}_t^a = \lim L_t^{b_n}$ where the limit is as $b_n \to a$ and b_n is in the dyadic rationals, then (14.7) implies that $\widetilde{L}_t^a = L_t^a$, a.s. The uniform continuity of L_t^a over a in the dyadic rationals and $t \leq M$ implies the joint continuity of $\widetilde{L}_t^a$. $\qquad \square$

14.3 Occupation times

If we integrate local times over a set, we obtain occupation times. More precisely, we have the following.

Theorem 14.4 *Let W_t be a Brownian motion and L_t^y the local time at the level y, where we take L_t^y to be jointly continuous in t and y. If f is non-negative and Borel measurable,*

$$\int f(y) L_t^y\, dy = \int_0^t f(W_s)\, ds, \qquad \text{a.s.} \tag{14.8}$$

with the null set independent of f and t.

Proof Suppose we prove the above equality for each C^2 function f with compact support and denote the null set by N_f. Taking a countable collection $\{f_i\}$ of non-negative C^2 functions with compact support that are dense in the set of non-negative continuous functions on $\mathbb{R}$ with compact support and letting $N = \cup_i N_{f_i}$, then if $\omega \notin N$ we have the above equality for all f_i. By taking limits, we have (14.8) for all bounded and continuous f. A further limiting procedure implies our result.

Suppose f is bounded and C^2 with compact support. Notice that the process $\int f(y)L_t^y\,dy$ is increasing and continuous. Define

$$g(x) = \int f(y)|x - y|\,dy. \qquad (14.9)$$

By Exercise 14.1, g is C^2 with $\frac{1}{2}g'' = f$. If we take the Tanaka formula (14.5), replace a by y, multiply by $f(y)$, and integrate over $\mathbb{R}$ with respect to y, we see that

$$g(W_t) - g(W_0) = \text{martingale} + \int_0^t f(y)L_t^y\,dy.$$

Using Itô's formula,

$$g(W_t) - g(W_0) = \text{martingale} + \frac{1}{2}\int_0^t g''(W_s)\,ds$$

$$= \text{martingale} + \int_0^t f(W_s)\,ds.$$

Thus

$$\int_0^t f(y)L_t^y\,dy - \int_0^t f(W_s)\,ds$$

is a continuous martingale with paths locally of bounded variation, hence by Theorem 9.7 it is identically 0.   □

Exercises

14.1 Suppose f is C^2 with compact support and

$$g(x) = \int f(y)|x - y|\,dy.$$

Show that g is C^2 and $g'' = 2f$.

14.2 Let L_t^y be the jointly continuous local times of a Brownian motion W. Show

$$\frac{1}{2\varepsilon}\int_0^t 1_{[y-\varepsilon,y+\varepsilon]}(W_s)\,ds \to L_t^y, \qquad \text{a.s.}$$

Show the null set can be taken to be independent of y. Thus there is no need to take a subsequence ε_n to get almost sure convergence to L_t^y.

14.3 Let W be a Brownian motion and fix t. Show that the function $x \to \int_0^t 1_{(-\infty,x]}(W_s)\,ds$ is continuous, a.s., but that the function $x \to 1_{(-\infty,x]}(W_t)$ is not continuous.

14.4 Let $\{\mathcal{F}_t\}$ be a filtration satisfying the usual conditions. Suppose W_t is a Brownian motion and $X_t = W_t + A_t$, where $X_t \geq 0$ for all t, a.s., and A_t is an increasing continuous adapted process such that A increases only at those times when $X_t = 0$. Suppose also that $X_t' = W_t + A_t'$, where $X_t' \geq 0$ for all t, a.s., and A_t' is an increasing continuous adapted process that increases only when $X_t' = 0$. Show that $X_t' = X_t$ and $A_t = A_t'$, a.s., for all $t \geq 0$.

14.5 Let W be a Brownian motion and L_t^0 the local time at 0. Since L_t^0 is increasing, for each ω there is a Lebesgue–Stieltjes measure dL_t^0. Show that the support of dL_t^0 is equal to $\{t : W_t = 0\}$.

Since Theorem 14.1(2) states that L_t^0 does not increase when W_t is not equal to 0, what you need to show is that with probability one, if $W_u(\omega) = 0$ and $t < u < v$, then $L_v^0(\omega) > L_t^0(\omega)$.

14.6 Use Tanaka's formula to show that if L_t^y is the local time of Brownian motion at level y, $a \leq x \leq y \leq b$, and $T = \inf\{t > 0 : W_t \notin [a, b]\}$, then

$$\mathbb{E}^x L_T^y = \frac{2(x - a)(b - y)}{b - a}.$$

14.7 If L_t^0 is the local time of a Brownian motion at 0, show that L_{at}^0 has the same law as $\sqrt{a} L_t^0$.

14.8 Let W be a Brownian motion with local times L_t^y. Set $L_t^* = \sup_y L_t^y$. Let $p > 0$. Prove that there exist constants c_1, c_2 such that if T is any finite stopping time,

$$c_1 \mathbb{E}\, T^{p/2} \leq \mathbb{E} L_T^* \leq c_2 \mathbb{E}\, T^{p/2}.$$

The constants c_1, c_2 can depend on p, but not on T.

Hint: Use Exercise 12.11.

14.9 This exercise defines the local time of a continuous martingale. If M is a continuous martingale, then M_t^2 is a submartingale and so equals a martingale plus an increasing process. The increasing process L_t^0 is called the local time of M at 0.

(1) Prove the analog of Tanaka's formula.

(2) Define the local time L_t^a of M at a. Prove that L_t^a is jointly continuous in t and a.

(3) Prove that

$$\int_0^t f(M_s)\, d\langle M \rangle_s = \int_{\mathbb{R}} L_t^a f(a)\, da, \qquad \text{a.s.}$$

if f is non-negative and measurable.

14.10 This exercise is a complement to Exercise 7.8. Let W be a Brownian motion and let us define $Z = \{t \in [0, 1] : W_t = 0\}$, the zero set. Let $\varepsilon \in (0, 1/2)$ and let $\delta > 0$. Fix ω and let $\{B_i\}$ be any countable covering of $Z(\omega)$ by closed intervals such that the interiors of the B_i's are pairwise disjoint and the length of each B_i is less than or equal to δ. We write $B_i = [a_i, b_i]$.

Let $\varepsilon > 0$. Since L^0 has the same law of the maximum of Brownian motion, there exists a c (depending on ω) such that

$$L_t^0 - L_s^0 \leq c(t - s)^{\frac{1}{2} - \frac{\varepsilon}{2}}$$

for each $0 \leq s \leq t \leq 0$. Write

$$\sum_i |b_i - a_i|^{\frac{1}{2} - \varepsilon} \geq \frac{\delta^{-\varepsilon/2}}{c} c \sum_i |b_i - a_i|^{\frac{1}{2} - \frac{\varepsilon}{2}}$$

$$\geq \frac{\delta^{-\varepsilon/2}}{c} \sum_i (L_{b_i}^0 - L_{a_i}^0)$$

$$= \frac{\delta^{-\varepsilon/2}}{c} [L_1^0 - L_0^0].$$

Show that this implies that the Hausdorff dimension of Z is at least $1/2$.

Skorokhod embedding

Suppose Y is a random variable with mean zero and finite variance. Skorokhod proved the remarkable fact that if W is a Brownian motion, there exists a stopping time T such that W_T has the same law as Y. Without any restrictions on T, there is a trivial solution (see Exercise 15.1), so one wants to require that $\mathbb{E}\, T < \infty$. Skorokhod's construction required an additional random variable that is independent of the Brownian motion, but since that time there have been 15 or 20 other constructions, most of which don't require the extra randomization, that is, T is a stopping time for the minimal augmented filtration generated by W.

Although conceptually some constructions are easier than others, none is easy from the point of view of technical details. We will give a construction that doesn't have any optimality properties, but is a nice example of stochastic calculus. Then we will use this to prove an embedding for random walks.

15.1 Preliminaries

A function $f : \mathbb{R} \to \mathbb{R}$ is a *Lipschitz function* if there exists a constant k such that

$$|f(y) - f(x)| \leq k|y - x|, \qquad x, y \in \mathbb{R}. \tag{15.1}$$

By the mean value theorem, if f has a bounded derivative, then f is a Lipschitz function.

We will need the following well-known theorem from the theory of ordinary differential equations.

Theorem 15.1 *Suppose $F : [0, \infty) \times \mathbb{R} \to \mathbb{R}$ is a bounded function and there exists a positive real k such that*

$$|F(t, x) - F(t, y)| \leq k|x - y|$$

for all $t \geq 0$ and all $x, y \in \mathbb{R}$. Let $y_0 \in \mathbb{R}$, define the function y^0 by $y^0(t) = y_0$ for all $t \geq 0$, and define the function y^i inductively by

$$y^{i+1}(t) = y_0 + \int_0^t F(s, y^i(s))\, ds, \qquad t \geq 0. \tag{15.2}$$

Then the functions y^i converge uniformly on bounded intervals to a function y that satisfies

$$y(t) = y_0 + \int_0^t F(s, y(s))\, ds. \tag{15.3}$$

For any s such that $F(s, y(s))$ is continuous at s, y satisfies

$$\frac{dy}{ds} = F(s, y(s)). \tag{15.4}$$

The solution to (15.3) is unique.

This inductive procedure for obtaining the solution to (15.4) is known as *Picard iteration*.

Proof Note each $y^i(t)$ is bounded in absolute value by $|y_0| + t \sup |F|$. Let $g_i(t) = \sup_{s \le t} |y^{i+1}(s) - y^i(s)|$. If $s \le t$, then

$$|y^{i+1}(s) - y^i(s)| = \left| \int_0^s [F(r, y^i(r)) - F(r, y^{i-1}(r))] \, dr \right|$$

$$\le \int_0^t |F(r, y^i(r)) - F(r, y^{i-1}(r))| \, dr$$

$$\le k \int_0^t |y^i(r) - y^{i-1}(r)| \, dr$$

$$\le k \int_0^t g_{i-1}(r) \, dr.$$

Taking the supremum over $s \le t$, we have

$$g_i(t) \le k \int_0^t g_{i-1}(r) \, dr.$$

Fix t_0. Now $g_1(t)$ is bounded for $t \le t_0$, say by L. Then $g_2(t) \le k \int_0^t L \, dr = kLt$ for each $t \le t_0$, and then $g_3(t) \le k \int_0^t (kLr) \, dr = k^2 L t^2 / 2$ and $g_4(t) \le k \int_0^t (k^2 L r^2 / 2) \, dr = k^3 L t^3 / 3!$ By induction $g_i(t) \le k^{i-1} L t^{i-1} / (i-1)!$ We conclude $\sum_{i=1}^{\infty} g_i(t_0) < \infty$.

Then

$$\sup_{s \le t_0} |y^n(s) - y^m(s)| \le \sum_{i=m}^{n-1} g_i(t_0),$$

which tends to zero as m and n tend to infinity. By the completeness of the space $C[0, t_0]$, there exists a continuous function y such that $\sup_{s \le t_0} |y^n(s) - y(s)| \to 0$ as $n \to \infty$.

F is continuous in the x variable, so taking the limit in (15.2) shows that y solves (15.3). If F is continuous at a particular value of s, then (15.4) holds by the fundamental theorem of calculus.

To prove uniqueness, suppose x and y are solutions to (15.4) and let us set $g(t) = \sup_{s \le t} |x(s) - y(s)|$. If $s \le t$, then

$$|x(s) - y(s)| \le \int_0^s |F(r, x(r)) - F(r, y(r))| \, dr$$

$$\le k \int_0^t |x(r) - y(r)| \, dr$$

$$\le k \int_0^t g(r) \, dr.$$

Taking the supremum over $s \leq t$, we obtain

$$g(t) \leq k \int_0^t g(r)\, dr.$$

For $t \leq t_0$, we have $|x(t)|$ and $|y(t)|$ bounded by a constant, say L, so $g(t)$ is bounded for $t \leq t_0$. We then have $g(t) \leq k \int_0^t L\, dr = kLt$ for each $t \leq t_0$ and then $g(t) \leq k \int_0^t kLr\, dr = k^2 Lt^2/2$. Iterating, we have $g(t) \leq k^i t^i L/i!$ for each i, and hence $g(t) = 0$. This is true for each t, hence $x(s) = y(s)$ for all $s \leq t_0$. $\qquad\square$

If the random variable Y that we are considering is equal to 0, a.s., we can just let our stopping time T equal 0, a.s., and then $W_T = 0 = Y$ if W is a Brownian motion. In the remainder of this section and the next we assume $\mathbb{E}\, Y = 0$, $\mathbb{E}\, Y^2 < \infty$, but that Y is not identically zero.

Define

$$p_s(y) = \frac{1}{\sqrt{2\pi s}} e^{-y^2/2s},$$

the density of a mean zero normal random variable with variance s. Use $p_s'(x)$ to denote the derivative of p_s with respect to x.

Lemma 15.2 *Suppose W is a Brownian motion and $g : \mathbb{R} \to \mathbb{R}$ such that $\mathbb{E}\,[g(W_1)^2] < \infty$. For $0 < s < 1$, let*

$$a(s, x) = -\int p_{1-s}'(z - x)g(z)\, dz \tag{15.5}$$

and

$$b(s, x) = \int p_{1-s}(z - x)g(z)\, dz. \tag{15.6}$$

We have

$$g(W_1) = \mathbb{E}\, g(W_1) + \int_0^1 a(s, W_s)\, dW_s, \qquad \text{a.s.} \tag{15.7}$$

and

$$\mathbb{E}\,[g(W_1) \mid \mathcal{F}_s] = b(s, W_s), \qquad \text{a.s.} \tag{15.8}$$

Proof   We will first prove (15.7), and we will first look at the case when $g(x) = e^{iux}$.

By Itô's formula with the function $f(x) = e^x$ applied to the semimartingale $X_t = iuW_t + u^2 t/2$

$$e^{iuW_t + u^2 t/2} = 1 + \int_0^t e^{X_s}\, d(iuW_s + u^2 s/2) + \frac{1}{2} \int_0^t (-u^2) e^{X_s}\, ds$$

$$= 1 + iu \int_0^t e^{iuW_s + u^2 s/2}\, dW_s,$$

so

$$e^{iuW_1} = e^{-u^2/2} + \int_0^1 iu e^{iuW_s} e^{u^2(s-1)/2}\, dW_s.$$

We need to check that

$$iu e^{iux} e^{u^2(s-1)/2} = a(s,x).$$

Using integration by parts,

$$a(s,x) = -\int p'_{1-s}(z-x)g(z)\,dz = \int p_{1-s}(z-x)g'(z)\,dz$$

$$= iu \int \frac{1}{\sqrt{2\pi(1-s)}} e^{-(z-x)^2/2(1-s)} e^{iuz}\,dz.$$

This is iu times the characteristic function of a normal random variable with mean x and variance $1-s$, and so by (A.25) equals

$$iu e^{iux} e^{-u^2(1-s)/2},$$

as desired. We therefore have

$$e^{iuW_1} = \mathbb{E}\,e^{iuW_1} - \int_0^1 \int p'_{1-s}(z-W_s) e^{iuz}\,dz\,dW_s. \tag{15.9}$$

Now suppose g is in the Schwartz class (see Section B.2), replace u by $-u$ in (15.9), multiply by the Fourier transform of g, and integrate over $u \in \mathbb{R}$. We then obtain

$$(2\pi)^{-1} g(W_1) = (2\pi)^{-1} \mathbb{E}\,g(W_1) \tag{15.10}$$

$$- \int \int_0^1 \int p'_{1-s}(z-W_s) e^{-iuz} \widehat{g}(u)\,dz\,dW_s\,du,$$

where $\widehat{g}$ is the Fourier transform of g. Using the Fubini theorem (check that there is no trouble with the stochastic integral; see Exercise 15.2) and the inversion formula for Fourier transforms, the triple integral on the right-hand side of (15.10) is equal to

$$- (2\pi)^{-1} \int_0^1 \int p'_{1-s}(z-W_s) g(z)\,dz\,dW_s, \tag{15.11}$$

which gives us (15.7) when g is the Schwartz class. A limit argument gives us (15.7) for all g that we are interested in.

To prove (15.8) we again start with the case $g(x) = e^{iux}$. We have

$$\mathbb{E}\,[e^{iuW_1} \mid \mathcal{F}_s] = e^{iuW_s} \mathbb{E}\,[e^{iu(W_1-W_s)} \mid \mathcal{F}_s] = e^{iuW_s} \mathbb{E}\,[e^{iu(W_1-W_s)}]$$

$$= e^{iuW_s} e^{-u^2(1-s)/2},$$

using the independent increments property of Brownian motion and (A.25). On the other hand, the definition of $b(s,x)$ shows that when $g(x) = e^{iux}$, $b(s,x)$ is the characteristic function of a normal random variable with mean x and variance $1-s$, so

$$b(s,x) = e^{iux} e^{-u^2(1-s)/2}.$$

Replacing x by W_s proves (15.8) in the case $g(x) = e^{iux}$. We extend this to general g in the same way as in the proof of (15.7). $\qquad\square$

Next, we want to find a reasonable function g such that $g(W_1)$ is equal in law to Y, where again W is a Brownian motion. Let $F_Y(x) = \mathbb{P}(Y \leq x)$, the distribution function of Y and let $\Phi(x) = \mathbb{P}(W_1 \leq x)$. Then

$$\mathbb{P}(\Phi(W_1) \leq x) = \mathbb{P}(W_1 \leq \Phi^{-1}(x)) = \Phi(\Phi^{-1}(x)) = x$$

for $x \in [0, 1]$, so $\Phi(W_1)$ is a uniform random variable on $[0, 1]$. Define

$$g(x) = F_Y^{-1}(\Phi(x)). \tag{15.12}$$

We use the right-continuous version of F_Y^{-1} if F_Y^{-1} is not continuous. Then

$$\mathbb{P}(g(W_1) \leq x) = \mathbb{P}(\Phi(W_1) \leq F_Y(x)) = F_Y(x),$$

or Y is equal in law to $g(W_1)$ as desired. Note g is an increasing function.

We will need the following estimates.

Proposition 15.3 *Let g be defined by* (15.12) *and define a and b by* (15.5) *and* (15.6).

(1) For each $L > 0$ and $s_0 < 1$, a is continuously differentiable on $[0, s_0] \times [-L, L]$. Also, for each $L > 0$ and $s_0 < 1$, a is bounded below by a positive constant on $[0, s_0] \times [-L, L]$.

(2) For each $L > 0$ and $s_0 < 1$, b is continuously differentiable on $[0, s_0] \times [-L, L]$.

(3) For each $s \in [0, s_0]$, the function $x \to b(s, x)$ is strictly increasing. For each fixed s, let $B(s, x)$ be the inverse of $b(s, x)$ (so that $B(s, b(s, x)) = x$ and $b(s, B(s, x)) = x$). For each $L > 0$ and $s_0 < 1$, B is continuously differentiable on $[0, s_0] \times [-L, L]$.

Proof   To start, we observe that for every $r > 0$,

$$\mathbb{E} \, e^{r|W_1|} \leq \mathbb{E} \, e^{rW_1} + \mathbb{E} \, e^{-rW_1} < \infty.$$

Since $|z|^m \leq m! e^{|z|}$ if m is a non-negative integer, then by the Cauchy–Schwarz inequality and the fact that $\mathbb{E} \, Y^2 < \infty$,

$$\int |z|^m e^{r|z|} e^{-z^2/2} |g(z)| \, dz \leq m! \int e^{(r+1)|z|} e^{-z^2/2} |g(z)| \, dz \tag{15.13}$$

$$= m! \mathbb{E} \left[e^{(r+1)|W_1|} |g(W_1)| \right]$$

$$\leq m! \left(\mathbb{E} \, e^{2(r+1)|W_1|} \right)^{1/2} (\mathbb{E} \, |g(W_1)|^2)^{1/2}$$

$$\leq m! \left(\mathbb{E} \, e^{2(r+1)|W_1|} \right)^{1/2} (\mathbb{E} \, Y^2)^{1/2} < \infty.$$

We now turn to (1).

$$|p'_{1-s}(z - x)| \leq c \frac{|z - x|}{(1 - s)^{3/2}} e^{-(z-x)^2/2(1-s)}$$

$$\leq c|z - x| e^{-x^2/2(1-s)} e^{zx/2(1-s)} e^{-z^2/2(1-s)}$$

$$\leq c(|z| + L) e^{|z|L/2(1-s_0)} e^{-z^2/2}$$

$$\leq c|z| e^{c'|z|} e^{-z^2/2} + c e^{c'|z|} e^{-z^2/2}.$$

Therefore

$$|a(s, x)| \leq \int c|z| e^{c'|z|} e^{-z^2/2} |g(z)| \, dz + \int c e^{c'|z|} e^{-z^2/2} |g(z)| \, dz,$$

which is bounded by (15.13). This gives an upper bound for a.

By the mean value theorem,

$$|p'_{1-s}(z-x) - p'_{1-s}(z-(x+h))| \le c|h|(1+|z|^2+L^2)e^{-(z-x)^2/2(1-s)}$$

if $s \le s_0$, $|x| \le L$, and $|h| \le 1$, so

$$\left|\frac{1}{h}(p'_{1-s}(z-x) - p'_{1-s}(z-(x+h)))\right| \le c(1+|z|^2)e^{c'|z|}e^{-z^2/2}.$$

In view of (15.13), we can use dominated convergence to conclude that

$$\frac{\partial a}{\partial x}(s,x) = \int p''_{1-s}(z-x)g(z)\,dz$$

and that $|\partial a(s,x)/\partial x|$ is bounded above on $[0, s_0] \times [-L, L]$.

By a similar argument we obtain that $|\partial a(s,x)/\partial s|$ is also bounded above on $[0, s_0] \times [-L, L]$. The same argument shows that the second partial derivatives of a are bounded, and hence the first partial derivatives are continuous.

Using integration by parts,

$$a(s,x) = \int p_{1-s}(z-x)\,dg(z),$$

where the integral is a Lebesgue–Stieltjes integral; recall that g is an increasing function. Since we are working under the assumption that Y is not identically zero, then g is not identically zero, which implies that a is bounded below for $s \le s_0$ and $|x| \le L$.

The proof of (2) is quite similar. To prove (3), as above, we can use a dominated convergence argument to prove

$$\frac{\partial b(s,x)}{\partial x} = a(s,x).$$

Since $a(s,x) > 0$ for each x and for each $s < s_0$, we conclude that $x \to b(s,x)$ is strictly increasing. The estimates for B follow from the implicit function theorem applied to $f(s,x,y) = 0$, where $f(s,x,y) = b(s,x) - y$. $\qquad\square$

15.2 Construction of the embedding

Theorem 15.4 *Suppose Y is a random variable with $\mathbb{E}\,Y = 0$ and $\mathbb{E}\,Y^2 < \infty$. There exists a Brownian motion N and a stopping time T with respect to the minimal augmented filtration of N such that N_T is equal in law to Y. Moreover $\mathbb{E}\,T = \mathbb{E}\,Y^2$.*

Proof The idea is to define M by (15.14) below and do a time change so that $N_T = M_1 = g(W_1)$. To show that T is a stopping time relative to the minimal augmented filtration for N, we set up an ordinary differential equation that the time change solves and use Picard iteration to show that the solution can be obtained in a constructive way.

The case where Y is identically zero is trivial for we take $T = 0$, so we suppose Y is not identically zero. Let W_t be a Brownian motion and let $\{\mathcal{F}_t\}$ be its minimal augmented filtration. Define the function g by (15.12) and define a and b for $s < 1$ by (15.5) and (15.6). Define $a(s,x) = 1$ and $b(s,x) = x$ if $s \ge 1$.

Now let

$$M_t = \int_0^t a(s, W_s)\, dW_s, \tag{15.14}$$

and hence

$$\langle M \rangle_t = \int_0^t a(s, W_s)^2\, ds.$$

Note $\langle M \rangle_t \to \infty$, a.s., as $t \to \infty$. Since $\mathbb{E}\, Y = 0$, then $\mathbb{E}\, g(W_1) = 0$, so $M_1 = g(W_1)$ by (15.7). Let

$$\tau_t = \inf\{s : \langle M \rangle_s \geq t\},$$

the inverse of $\langle M \rangle$. By Theorem 12.2, if we set $N_t = M_{\tau_t}$, then N is a Brownian motion. Let $\{\mathcal{G}_t\}$ be the minimal augmented filtration generated by N.

We let $T = \langle M \rangle_1$. Then

$$N_T = N_{\langle M \rangle_1} = M_{\tau_{\langle M \rangle_1}} = M_1 = g(W_1),$$

and N_T has the same law as Y.

For the integrability of T we have

$$\mathbb{E}\, T = \mathbb{E}\, \langle M \rangle_1 = \mathbb{E}\, M_1^2 = \mathbb{E}\, [g(W_1)^2] = \mathbb{E}\, Y^2 = \operatorname{Var} Y < \infty. \tag{15.15}$$

It remains to show that T is a stopping time with respect to $\{\mathcal{G}_t\}$. Since $T = \lim_{s \uparrow 1} \langle M \rangle_s$, it suffices to show that $\langle M \rangle_s$ is a stopping time with respect to $\{\mathcal{G}_t\}$ for each $s < 1$. Fix K. We will show

$$(\tau_t \leq s,\ \sup_{s \leq t} |N_s| \leq K) \in \mathcal{G}_t, \qquad s < 1. \tag{15.16}$$

Letting $K \to \infty$ will then show $(\langle M \rangle_s \geq t) = (\tau_t \leq s) \in \mathcal{G}_t$ for $s < 1$.

Since τ is the inverse of $\langle M \rangle$, then

$$\frac{d\tau_t}{dt} = \frac{1}{d\langle M \rangle_{\tau_t} / d\tau_t} = \frac{1}{a(\tau_t, W_{\tau_t})^2}$$

with $\tau_0 = 0$, a.s. With $B(s, x)$ being the inverse of $b(s, x)$ in the x variable,

$$M_s = \mathbb{E}\, [M_1 \mid \mathcal{F}_s] = \mathbb{E}\, [g(W_1) \mid \mathcal{F}_s] = b(s, W_s),$$

or

$$W_s = B(s, M_s), \qquad s < 1.$$

Therefore

$$W_{\tau_t} = B(\tau_t, M_{\tau_t}) = B(\tau_t, N_t)$$

on the event $(\tau_t \leq s)$ if $s < 1$. Thus τ_t solves the equation

$$\frac{d\tau_t}{dt} = \frac{1}{a(\tau_t, B(\tau_t, N_t))^2}, \qquad \tau_0 = 0,$$

or

$$\tau_t = \int_0^t \frac{1}{a(\tau_u, B(\tau_u, N_u))^2}\, du.$$

Fix s and t and choose $s_0 \in (s, 1)$. Let $S_K = \inf\{t : |N_t| \ge K\}$ and let $N_t^K = N_{t \wedge S_K}$. Define

$$\Phi(q, r) = \frac{1}{(a(r, B(r, N_q^K(\omega))))^2}$$

if $r \le s_0$. Observe that Φ depends on ω. Define $\Phi(q, r) = 1$ for $r \ge 1$ and define $\Phi(q, r)$ by linear interpolation for $r \in (s_0, 1)$. Note that by Proposition 15.3, Φ is continuous, bounded, and there exists $k > 0$ such that

$$|\Phi(q, r) - \Phi(q, r')| \le k|r - r'|, \qquad r \in \mathbb{R}, \ q \in [0, \infty).$$

τ_t solves the equation

$$\tau_t = \int_0^t \Phi(u, \tau_u) \, du.$$

We solve the differential equation

$$y(t) = \int_0^t \Phi(u, y(u)) \, du \tag{15.17}$$

using Theorem 15.1. The function $y^0(t)$ in the statement of Theorem 15.1 is identically zero, and the function $y^1(t) = \int_0^t \Phi(u, y^0(u)) \, du$ (which depends on ω because Φ does) will be $\mathcal{G}_t$ measurable, and by induction, the functions $y^i(t)$ will be $\mathcal{G}_t$ measurable. Therefore the limit, $y(t)$, will be $\mathcal{G}_t$ measurable. Since $|N_q^K(\omega)| \le K$ for all q and we are only interested in the solution to (15.17) for $y(t) \le s$, then $\tau_t = y(t)$ as long as $\tau_t \le s$; therefore (15.16) holds and the proof is complete. $\qquad \square$

In the above theorem, we started with a Brownian motion W, constructed a new Brownian motion N, and then defined our stopping time T in terms of N. We can actually start with a Brownian motion W and define a stopping time that is a stopping time with respect to the minimal augmented filtration of W.

Corollary 15.5 *Let W be a Brownian motion and let $\{\mathcal{F}_t\}$ be the minimal augmented filtration for W. Let Y be a random variable with $\mathbb{E}\, Y = 0$ and $\operatorname{Var} Y < \infty$. There exists a stopping time V with respect to $\{\mathcal{F}_t\}$ such that W_V has the same law as Y.*

Proof We sketch the proof and ask you to give the details in Exercise 15.3. Define

$$\overline{\Phi}(q, r) = \frac{1}{(a(r, B(r, W_q(\omega))))^2}$$

and solve the equation

$$\frac{d\overline{\tau}_t}{dt} = \Phi(t, \overline{\tau}_t), \qquad \tau_0 = 0$$

by Picard iteration. The proof of Theorem 15.4 shows that the solution $\overline{\tau}_t$ will satisfy $(\overline{\tau}_t \le s) \in \mathcal{F}_t$ for every t as long as $s < 1$. Let A be the inverse of $\overline{\tau}$, and define $V = \lim_{s \uparrow 1} A_s$. Then V will be the desired stopping time. $\qquad \square$

15.3 Embedding random walks

Let us give an application of Skorokhod embedding to show that we can find a Brownian motion that is relatively close to a random walk. Suppose $Y_1, Y_2, \ldots$ is an i.i.d. sequence of real-valued random variables with mean zero and variance one. Given a Brownian motion W_t we can find a stopping time T_1 such that W_{T_1} has the same distribution as Y_1. We use the strong Markov property at time T_1 and find a stopping time T_2 for $W_{T_1+t} - W_{T_1}$ so that $W_{T_1+T_2} - W_{T_1}$ has the same distribution as Y_2 and is independent of $\mathcal{F}_{T_1}$. We continue. We see that the T_i are i.i.d. and by Theorem 15.4, $\mathbb{E}\, T_i = \mathbb{E}\, Y_i^2 = 1$. Let $U_k = \sum_{i=1}^{k} T_i$. Then for each n, $S_n = \sum_{i=1}^{n} Y_i$ has the same distribution as W_{U_n}.

Theorem 15.6

$$\sup_{i \leq n} |W_{U_i} - W_i| / \sqrt{n}$$

tends to 0 in probability as $n \to \infty$.

Proof   We will show that for each $\varepsilon > 0$

$$\limsup_{n \to \infty} \mathbb{P}(\sup_{k \leq n} |W_{U_k} - W_k| > \varepsilon \sqrt{n}) \leq \varepsilon. \tag{15.18}$$

Since the paths of Brownian motion are continuous, we can find $\delta \leq 1$ small such that

$$\mathbb{P}(\sup_{s,t \leq 2, |t-s| \leq \delta} |W_t - W_s| > \varepsilon) < \varepsilon/2.$$

By scaling,

$$\mathbb{P}(\sup_{s,t \leq 2n, |t-s| \leq \delta n} |W_t - W_s| > \varepsilon \sqrt{n}) < \varepsilon/2. \tag{15.19}$$

The strong law of large numbers (Theorem A.38) says that $U_n/n \to \mathbb{E}\, T_1 = 1$, a.s., and in fact, by Proposition A.39, we even have

$$\frac{\max_{k \leq n} |U_k - k|}{n} \to 0, \qquad \text{a.s.} \tag{15.20}$$

Therefore

$$\mathbb{P}(\max_{k \leq n} |W_{U_k} - W_k| > \varepsilon \sqrt{n})$$

$$\leq \mathbb{P}(\max_{k \leq n} |U_k - k| > \delta n) + \mathbb{P}(\sup_{s,t \leq 2n, |t-s| \leq \delta n} |W_t - W_s| > \varepsilon \sqrt{n})$$

$$\leq \mathbb{P}\left(\max_{k \leq n} \frac{|U_k - k|}{n} > \delta \right) + \frac{\varepsilon}{2}.$$

By (15.20) this will be less than ε if we take n sufficiently large. $\square$

Exercises

15.1   Without some supplemental conditions on T, the problem of Skorokhod embedding is trivial. Suppose W is a Brownian motion with respect to a filtration $\{\mathcal{F}_t\}$ satisfying the usual conditions. Suppose Y is a finite random variable and suppose h is a real-valued function such that $h(W_1)$ has the same law as Y.

(1) Show that if $T = \inf\{t > 1 : W_t = h(W_1)\}$, then W_T and Y have the same law.

(2) Give an example of a mean zero random variable Y with finite variance such that if T is defined as in (1), then $\mathbb{E}\, T = \infty$.

15.2 Show that the triple integral on the right-hand side of (15.10) is equal to the expression in (15.11).

15.3 A sketch was given for the proof of Corollary 15.5. Provide a detailed proof.

15.4 Here is another approach to proving Corollary 15.5. Let Y, N, T, and $\{\mathcal{G}_t\}$ be as in the proof of Theorem 15.4.

(1) Show that there is a random variable U that is measurable with respect to $\sigma(N_s : 0 \le s < \infty)$ such that $U = T$, a.s.

(2) Show there is a Borel measurable map $H : C[0, \infty) \to [0, \infty)$ such that $U = H(N)$.

(3) If W is a Brownian motion, define $V = H(W)$. Show V is a stopping time with respect to the minimal augmented filtration generated by W such that W_V has the same law as Y.

15.5 Suppose $p \in (0, 1/2)$ and Y is a random variable such that $\mathbb{P}(Y = 1) = \mathbb{P}(Y = -1) = p$ and $\mathbb{P}(Y = 0) = 1 - 2p$. Let W be a Brownian motion. Let $S_x = \inf\{t > 0 : W_t = x\}$ and let $T = \inf\{t > S_x \wedge S_{-x} : W_t \in \{-1, 0, 1\}\}$. Determine x such that W_T and Y have the same law.

15.6 Suppose Y is a mean zero random variable and there exists a real number $K > 0$ such that $|Y| \le K$, a.s. Let W be a Brownian motion and let T be a stopping time with $\mathbb{E}\, T < \infty$ such that W_T and Y have the same law. (We do not necessarily assume that T was constructed by the method of Section 15.2.) Let $S_K = \inf\{t : |W_t| \ge K\}$. Prove that $T \le S_K$, a.s.

15.7 Let Y_i be a sequence of i.i.d. random variables with $\mathbb{P}(Y_i = 1) = \mathbb{P}(Y_i = -1) = \frac{1}{2}$, and let $S_n = \sum_{i=1}^{n} Y_i$. S_n is called a *simple symmetric random walk*. Let $T_1, T_2, \ldots$ and $U_1, U_2, \ldots$ be as in Section 15.3.

(1) Prove that $\mathbb{E}\, T_1^p < \infty$ for all $p \ge 1$.

(2) Prove that if $\varepsilon > 0$,

$$\lim_{n \to \infty} \frac{\sup_{k \le n} |U_k - k|}{n^{(1/2)+\varepsilon}} = 0, \qquad \text{a.s.}$$

Hint: Use Doob's inequalities to estimate

$$\mathbb{P}(\sup_{k \le n} |U_k - k| \ge \delta n^{(1/2)+\varepsilon}).$$

(3) Show that

$$\sup_{i \le n} |W_{U_i} - W_i|/n^{(1/4)+(\varepsilon/2)}$$

tends to zero in probability as $n \to \infty$.

15.8 Let S_n, T_i, and U_i be as in Exercise 15.7. Prove that

$$\lim_{n \to \infty} \frac{\sup_{i \le n} |W_{U_i} - W_i|}{\sqrt{n}} = 0, \qquad \text{a.s.}$$

15.9 Let S_n be a simple symmetric random walk; see Exercise 15.7. Let Y be a bounded symmetric random variable that takes values only in $\mathbb{Z}$. (Y being symmetric means that Y and $-Y$ have the same law.) Does there necessarily exist a stopping time N such that S_N and Y have the same law? Why or why not?

Notes

The survey article Obłój (2004) summarizes many different methods of Skorokhod embedding. The embedding presented here is from Bass (1983); see also Stroock (2003), pp. 213–17.

16

The general theory of processes

The name "general theory of processes" refers to the foundations of stochastic processes. Specific topics include measurability issues and classifications of stopping times. This chapter is fairly technical and abstract and should only be skimmed on the first reading of this book: read the definitions and statements of theorems, propositions, and lemmas, but not the proofs.

The two main results we discuss are the measurability of hitting times, and the Doob–Meyer decomposition of submartingales, Theorem 16.29.

16.1 Predictable and optional processes

Suppose $(\Omega, \mathcal{F}, \mathbb{P})$ is a probability space. The *outer probability* $\mathbb{P}^*$ associated with $\mathbb{P}$ is given by

$$\mathbb{P}^*(A) = \inf\{\mathbb{P}(B) : A \subset B, B \in \mathcal{F}\}. \tag{16.1}$$

A set A is a $\mathbb{P}$-*null set* if $\mathbb{P}^*(A) = 0$. We suppose throughout this chapter that $\{\mathcal{F}_t\}$ is a filtration satisfying the usual conditions; recall from Chapter 1 that this means that each $\mathcal{F}_t$ contains all the $\mathbb{P}$-null sets and that $\cap_{\varepsilon>0}\mathcal{F}_{t+\varepsilon} = \mathcal{F}_t$ for each t. Let $\pi : [0, \infty) \times \Omega \to \Omega$ be defined by

$$\pi(t, \omega) = \omega. \tag{16.2}$$

We define the *predictable σ-field* $\mathcal{P}$ to be the σ-field on $[0, \infty) \times \Omega$ generated by the collection of all bounded left continuous processes adapted to $\mathcal{F}_t$. That is, $\mathcal{P}$ is the σ-field on $[0, \infty) \times \Omega$ generated by the collection of all sets of the form

$$\{(t, \omega) \in [0, \infty) \times \Omega : X_t(\omega) > a\},$$

where $a \in \mathbb{R}$ and X is a bounded, adapted, left-continuous process. The *optional σ-field* $\mathcal{O}$ is the σ-field on $[0, \infty) \times \Omega$ generated by the collection of all bounded right-continuous processes adapted to $\mathcal{F}_t$. The word for predictable in French is "prévisible." The older literature uses "well measurable" in place of the word "optional."

If S and T are random variables taking values in $[0, \infty]$, let $[S, T) = \{(t, \omega) \in [0, \infty) \times \Omega : S(\omega) \leq t < T(\omega)\}$, and define $(S, T]$, (S, T), etc. similarly. With this notation, $[T, T]$, the *graph* of T, is equal to $\{(t, \omega) \in [0, \infty) \times \Omega : T(\omega) = t < \infty\}$. Note that $[T, T]$ is a subset of $[0, \infty) \times \Omega$, so $\pi([T, T]) = (T < \infty)$.

111

Recall that a stopping time can take the value ∞. A stopping time T is *predictable* if there exists a sequence of stopping times T_n such that for all ω

(1) $T_1(\omega) \leq T_2(\omega) \leq \cdots$,

(2) $\lim_{n \to \infty} T_n(\omega) = T(\omega)$, and

(3) if $T(\omega) > 0$, then $T_n(\omega) < T(\omega)$ for each n.

In this case, the stopping times T_n *predict* T or *announce* T. If T is a stopping time satisfying (1)–(3) above and $S = T$, a.s., then we call S a predictable stopping time as well. A stopping time T is *totally inaccessible* if $\mathbb{P}(T = S < \infty) = 0$ for every predictable stopping time S.

For an example of a predictable stopping time, let W_t be a Brownian motion started at 0 and let $T = \inf\{t > 0 : W_t = 1\}$. The stopping time T is predicted by the stopping times $T_n = \inf\{t > 0 : W_t = 1 - (1/n)\}$.

For an example of a totally inaccessible stopping time, let P_t be a Poisson process with parameter 1 and let $T = \inf\{t : P_t = 1\}$, the first time the Poisson process jumps. Since P_t has independent increments, $P_t - t$ is a martingale, just as in Example 3.2. By (A.8), $\mathbb{E}[(P_t - t)^2] < \infty$. If S is a bounded predictable stopping time, by the optional stopping theorem, $\mathbb{E}\,P_S = \mathbb{E}\,S$. If S_n are stopping times predicting S, then by monotone convergence

$$\mathbb{E}\,P_{S-} = \lim_{n \to \infty} \mathbb{E}\,P_{S_n} = \lim_{n \to \infty} \mathbb{E}\,S_n = \mathbb{E}\,S.$$

Therefore $\mathbb{E}\,[P_S - P_{S-}] = 0$, and since P_t is an increasing process, this says that P does not jump at time S. Applying this to $S \wedge M$ and letting $M \to \infty$, we see that P does not jump at any predictable time S, whether or not S is bounded. Therefore $\mathbb{P}(T = S < \infty) = 0$, so T is totally inaccessible.

The proof of the following proposition is reminiscent of that of the Vitali covering theorem from measure theory.

Proposition 16.1 *Let T be a stopping time. There exist predictable stopping times $S_1, S_2, \ldots$ and a totally inaccessible stopping time U such that $[T, T] = [U, U] \cup (\cup_{i=1}^{\infty}[S_i, S_i])$.*

Proof Let

$$a_1 = \sup\{\mathbb{P}(S = T < \infty) : S \text{ is a predictable stopping time}\}$$

and choose S_1 to be a predictable stopping time such that $\mathbb{P}(S_1 = T < \infty) \geq \frac{1}{2}a_1$. Given $S_1, \ldots, S_n$, let

$$a_{n+1} = \sup\{\mathbb{P}(S = T < \infty, S \neq S_1, \ldots, S \neq S_n)) :$$
$$S \text{ is a predictable stopping time}\}$$

and choose S_{n+1} such that $\mathbb{P}(S_{n+1} = T < \infty, S_{n+1} \neq S_1, \ldots, S_{n+1} \neq S_n) \geq \frac{1}{2}a_{n+1}$.

If this procedure stops after n steps, set $U(\omega)$ equal to $T(\omega)$ if $T(\omega)$ is not equal to any of $S_1(\omega), \ldots, S_n(\omega)$ and equal to infinity otherwise. It is easy to check that U is a stopping time that is totally inaccessible.

The other alternative is that this procedure continues indefinitely. In this case define

$$U(\omega) = \begin{cases} T(\omega), & T(\omega) \neq S_1(\omega), S_2(\omega), \ldots, \\ \infty, & \text{otherwise.} \end{cases}$$

There is no problem checking that U is a stopping time, but we need to show that U is totally inaccessible. Since probabilities are bounded by one, we have $a_n \to 0$. If there exists a predictable stopping time S such that $b = \mathbb{P}(S = U < \infty) > 0$, then $b > 2a_n$ for some n, and in our construction we would have then chosen S in place of the S_n we did choose. Therefore such a stopping time S cannot exist. □

Proposition 16.2 *(1) The optional σ-field $\mathcal{O}$ is generated by the collection of sets*

$$\{[S, T) : S, T \text{ stopping times}\}.$$

(2) $\mathcal{O}$ is generated by the collection of sets of the form $[a, b) \times C$, where $a < b$ and $C \in \mathcal{F}_a$.
(3) The predictable σ-field $\mathcal{P}$ is generated by the collection of sets

$$\{(S, T] : S, T \text{ stopping times}\}.$$

(4) $\mathcal{P}$ is generated by the collection of sets

$$\{[S, T) : S, T \text{ predictable stopping times}\}.$$

(5) $\mathcal{P}$ is generated by the collection of sets of the form $[b, c) \times C$, where $a < b < c$ and $C \in \mathcal{F}_a$.

Proof (1) Since $1_{[S,T)}$ is a bounded right-continuous process that is adapted to $\{\mathcal{F}_t\}$, sets of the form $[S, T)$ are optional. Now suppose X is a bounded adapted process with right-continuous paths. Let $\varepsilon > 0$, let $U_0 = 0$, a.s., and let

$$U_{i+1} = \inf\{t > U_i : |X_t - X_{U_i}| > \varepsilon\}, \qquad i \geq 0. \tag{16.3}$$

Since X has right-continuous paths,

$$(U_1 < t) = \cap_{q \in \mathbb{Q}_+, q < t}\{|X_q - X_0| > \varepsilon\},$$

where $\mathbb{Q}_+$ denotes the positive rationals, and it follows that U_1 is a stopping time. Similarly U_i is a stopping time for each i; Exercise 16.4 asks you to prove this. If we set

$$X_t^\varepsilon(\omega) = \sum_{i=0}^\infty X_{U_i}(\omega) 1_{[U_i(\omega), U_{i+1}(\omega))}(t),$$

then $\sup_t |X_t - X_t^\varepsilon| \leq \varepsilon$. Therefore it suffices to show that each process X^ε is measurable with respect to the σ-field $\widehat{\mathcal{O}}$ generated by the collection of sets of the form $[S, T)$.

To do that, it suffices to show that processes of the form

$$Y_t(\omega) = 1_A(\omega) 1_{[U_i(\omega), U_{i+1}(\omega))}(t),$$

where $A \in \mathcal{F}_{U_i}$, are measurable with respect to $\widehat{\mathcal{O}}$. If we set $S(\omega)$ equal to $U_i(\omega)$ if $\omega \in A$ and equal to ∞ otherwise and we set $T(\omega)$ equal to $U_{i+1}(\omega)$ if $\omega \in A$ and ∞ otherwise, then $Y_t(\omega) = 1_{[S(\omega), T(\omega))}$.

(2) If $C \in \mathcal{F}_a$, then $1_C(\omega) 1_{[a,b)}(t)$ is a bounded right-continuous adapted process, so it is optional. By (1), every bounded right-continuous adapted process can be approximated by linear combinations of processes of the form $1_{[S,T)}$. Now $1_{[S,T)} = 1_{[S,\infty)} - 1_{[T,\infty)}$, and $1_{[S,\infty)}$

is the limit of $1_{[S_n,\infty)}$, where $S_n = k/2^n$ if $(k-1)/2^n \leq S < k/2^n$, and we can similarly approximate $1_{[T,\infty)}$. Note

$$1_{[S_n(\omega),\infty)}(t) = \sum_{k=1}^{\infty} 1_{((k-1)/2^n \leq S(\omega) < k/2^n)} 1_{[k/2^n,\infty)}(t).$$

Since $((k-1)/2^n \leq S(\omega) < k/2^n) \in \mathcal{F}_{k/2^n}$ and $1_{[k/2^n,\infty)}(t)$ is the limit of $1_{[k/2^n,m)}(t)$ as $m \to \infty$, we see that every bounded right-continuous adapted process is measurable with respect to the σ-field generated by processes of the form $1_A(\omega)1_{[a,b)}(t)$, where A is $\mathcal{F}_a$ measurable.

For (3), $1_{(S,T]}$ is left continuous, bounded, and adapted, hence predictable. Any left-continuous adapted bounded process can be approximated by processes of the form

$$\sum_{k=0}^{n2^n-1} X_{k/2^n}(\omega)1_{(k/2^n,(k+1)/2^n]}(t),$$

which in turn can be approximated by linear combinations of processes of the form $Y = 1_A(\omega)1_{(a,b]}(t)$, where A is $\mathcal{F}_a$ measurable. Such a process Y is of the form $1_{(S,T]}$ if we define S and T by

$$S(\omega) = \begin{cases} a, & \omega \in A, \\ \infty, & \omega \notin A, \end{cases} \qquad T(\omega) = \begin{cases} b, & \omega \in A, \\ \infty, & \omega \notin A. \end{cases}$$

To prove (4), note that $S + \frac{1}{k}$ is always a predictable stopping time (predicted by the stopping times $S_n = S + \frac{1}{k} - \frac{1}{n}$ for $n > k$). We have

$$(S, T] = \cup_k \{\cap_m [S + \tfrac{1}{k}, T + \tfrac{1}{m})\}.$$

On the other hand, if S and T are predictable and are predicted by sequences S_n and T_m, respectively, then

$$[S, T) = \cap_n \{\cup_m (S_n, T_m]\}.$$

(4) now follows by using (3).

(5) As long as $a + (1/n) < b$, the processes $1_C(\omega)1_{(b-(1/n),c-(1/n)]}(t)$ are left continuous, bounded, and adapted, hence predictable. The process $1_C(\omega)1_{[b,c)}(t)$ is the limit of these processes as $n \to \infty$, so is predictable. On the other hand, if X_t is a bounded adapted left-continuous process, it can be approximated by

$$\sum_{k=1}^{n2^n-1} X_{(k-1)/2^n}(\omega)1_{(k/2^n,(k+1)/2^n]}(t).$$

Each summand can be approximated by linear combinations of processes of the form $1_C(\omega)1_{(b,c]}(t)$, where $C \in \mathcal{F}_a$ and $a < b < c$. Finally, $1_C(\omega)1_{(b,c]}(t)$ is the limit of $1_C(\omega)1_{[b+(1/n),c+(1/n))}(t)$ as $n \to \infty$. $\qquad\square$

A consequence of Proposition 16.2(1) and (4) is that $\mathcal{P} \subset \mathcal{O}$.

16.2 Hitting times

Let S be a separable metric space. Suppose $\{\mathcal{F}_t\}$ is a filtration satisfying the usual conditions and X is a stochastic process taking values S whose paths are right continuous and such that the jump times are totally inaccessible. Saying the jump times are totally inaccessible means that if T is a predictable stopping time, then $X_{T-} = X_T$, a.s., where $X_{T-} = \lim_{s < T, s \to T} X_s$.

If B is a Borel subset of a metric space S and X is a S-valued process, let

$$U_B = \inf\{t \geq 0 : X_t \in B\}$$

and

$$T_B = \inf\{t > 0 : X_t \in B\}.$$

T_B is known as the first *hitting time* of B and U_B as the first *entry time* of B.

Proposition 16.3 *(1) If A is an open set, then T_A and U_A are stopping times.*
(2) If A is a compact set, then T_A and U_A are stopping times.

Proof (1) Since the paths of X_t are right continuous and A is open, for each t,

$$(T_A < t) = \cup_{q \in \mathbb{Q}_+, q < t}(X_t \in A) \in \mathcal{F}_t,$$

where $\mathbb{Q}_+$ denotes the non-negative rationals. Thus T_A is a stopping time. Since

$$(U_A < t) = (T_A < t) \cup (X_0 \in A) \in \mathcal{F}_t, \tag{16.4}$$

then U_A is also a stopping time.

(2) Now suppose A is compact and let $A_n = \{x \in S : d(x, A) < 1/n\}$. Each set A_n is open, hence T_{A_n} is a stopping time for each n. The T_{A_n} increase; let T be the limit. If we show $T = T_A$, a.s., this will prove T_A is a stopping time.

Since $A \subset A_n$, then $T_{A_n} \leq T_A$ for each n. Therefore $T \leq T_A$. On the other hand, if $n > m$, then $X_{T_{A_n}} \in \overline{A}_n \subset \overline{A}_m$, the closure of A_m. Either $T_{A_n}(\omega) = T(\omega)$ for all n sufficiently large, in which case $X_T(\omega) \in \overline{A}_m$, or else $T_{A_n}(\omega) < T(\omega)$ for all n. In the latter case, $X_T(\omega) = \lim_{n \to \infty} X_{T_{A_n}}(\omega) \in \overline{A}_m$ except for ω's in a null set since the jump times of X are totally inaccessible. In either case, $X_T \in \overline{A}_m$. This is true for all m, so $X_T \in \cap_m \overline{A}_m = A$, and therefore $T_A \leq T$.

We conclude T_A is a stopping time. To prove U_A is a stopping time, we argue using (16.4) as above. $\square$

For the proof of the following, which uses Choquet's capacity theorem, we refer the reader to Blumenthal and Getoor (1968), Section I.10. Fix t and define

$$R_t(A) = \{\omega : X_s(\omega) \in A \text{ for some } s \in [0, t]\} = (U_A \leq t). \tag{16.5}$$

Theorem 16.4 *If A is a Borel subset of S, then $R_t(A) \in \mathcal{F}_t$ and there exists an increasing sequence of compact sets K_n contained in A such that $\mathbb{P}(R_t(K_n)) \uparrow \mathbb{P}(R_t(A))$.*

Since $(U_A \leq t) = R_t(A)$, we have the following as an immediate corollary.

Theorem 16.5 *For all Borel sets A, U_A is a stopping time.*

Here is the main theorem of this section.

Theorem 16.6 *Suppose $\{\mathcal{F}_t\}$ is a filtration satisfying the usual conditions and X is a right continuous process whose jump times are totally inaccessible. If B is a Borel subset of S, then T_B is a stopping time.*

Proof If we let $Y_t^\delta = X_{t+\delta}$ and $U_B^\delta = \inf\{t \geq 0 : Y_t^\delta \in B\}$, then by the above, U_B^δ is a stopping time with respect to the filtration $\{\mathcal{F}_t^\delta\}$, where $\mathcal{F}_t^\delta = \mathcal{F}_{t+\delta}$. It follows that $\delta + U_B^\delta$ is a stopping time with respect to the filtration $\{\mathcal{F}_t\}$. Since $(1/m) + U_B^{1/m} \downarrow T_B$, then T_B is a stopping time with respect to $\{\mathcal{F}_t\}$. $\qquad\square$

We now show that the hitting times of Borel sets can be approximated by the hitting times of compact sets.

Proposition 16.7 *There exists an increasing sequence of compact sets K_n contained in B such that $U_{K_n} \downarrow U_B$ on $(U_B < \infty)$, $\mathbb{P}$-a.s.*

Proof For each t we can find an increasing sequence of compact sets L_n^t contained in B with $\mathbb{P}(R_t(L_n^t)) \uparrow \mathbb{P}(R_t(B))$. Let q_j be an enumeration of the non-negative rationals. Let $K_n = L_n^{q_1} \cup \cdots \cup L_n^{q_n}$. Then the K_n are compact, form an increasing sequence, and are all contained in B. Thus U_{K_n} decreases, say to S, and since $U_{K_n} \geq U_B$ for all n, then $S \geq U_B$. If we prove $S \leq U_B$, $\mathbb{P}$-a.s., then $S = U_B$, and we have our result.

If $U_B < S$, there exists a rational q_j with $U_B < q_j < S$. Hence it suffices to prove $\mathbb{P}(U_B < q_j < S) = 0$ for all j. If $U_B < q_j$, then $\omega \in R_{q_j}(B)$. Since $R_{q_j}(L_n^{q_j}) \uparrow R_{q_j}(B)$, a.s., then except for a null set, ω will be in $R_{q_j}(L_n^{q_j})$ for all n large enough, hence in $R_{q_j}(K_n)$ if n is large enough. Then $U_{K_n}(\omega) \leq q_j < U_B$ or $S \leq q_j$. Therefore $\mathbb{P}(U_b < q_j < S) = 0$. $\qquad\square$

Theorem 16.8 *There exists an increasing sequence of compacts K_n contained in B such that $T_{K_n} \downarrow T_B$.*

Proof Let $Y_t^\delta = X_{t+\delta}$ and $U_B^\delta = \inf\{t \geq 0 : Y_t \in B\}$. Applying the above proposition to $Y_t^{1/m}$, for each m there exist compact sets L_n^m, increasing in n and contained in B, such that $U_{L_n^m}^{1/m} \downarrow U_B^{1/m}$. Let $K_n = L_n^1 \cup \cdots \cup L_n^n$. Then K_n is an increasing sequence of compact sets contained in B, and $U_{K_n}^{1/m} \downarrow U_B^{1/m}$. Also, for each n, $1/m + U_{K_n}^{1/m} \downarrow T_{K_n}$ and $1/m + U_B^{1/m} \downarrow T_B$. We write

$$T_B = \lim_m (1/m + U_B^{1/m}) = \lim_m \lim_n (1/m + U_{K_n}^{1/m})$$
$$= \lim_n \lim_m (1/m + U_{K_n}^{1/m}) = \lim_n T_{K_n}.$$

Since $1/m + U_{K_n}^{1/m}$ is decreasing in both m and n, the change in the order of taking limits is justified. Since T_{K_n} is decreasing, this completes the proof. $\qquad\square$

16.3 The debut and section theorems

If $E \subset [0, \infty) \times \Omega$, let $D_E = \inf\{t \geq 0 : (t, \omega) \in E\}$, the *debut* of E. An important generalization of Theorem 16.6 is the following, known as the *debut theorem*.

Theorem 16.9 *If $E \in \mathcal{O}$, then D_E is a stopping time.*

The proof of this theorem is beyond the scope of this book, and we refer the reader to Dellacherie and Meyer (1978) for a proof.

Using Theorem 16.9, we can weaken the assumptions on X in Theorem 16.6.

Theorem 16.10 *If X is an optional process taking values in $\mathcal{S}$ and B is a Borel subset of $\mathcal{S}$, then U_B and T_B are stopping times.*

Proof Since B is a Borel subset of $\mathcal{S}$ and X is an optional process, then $1_B(X_t)$ is also an optional process. U_B is then the debut of the set $E = \{(s, \omega) : 1_B(X_s(\omega)) = 1\}$, and therefore is a stopping time.

To prove that T_B is a stopping time, we argue exactly as in the proof of Theorem 16.6. $\square$

Remark 16.11 In the theory of Markov processes, the notion of completion of a σ-field is a bit different. However it is still the case that the hitting times of Borel sets by right continuous processes are stopping times. See Remark 20.4.

The *optional section theorem* is the following.

Theorem 16.12 *If E is an optional set and $\varepsilon > 0$, there exists a stopping time T such that $[T, T] \subset E$ and $\mathbb{P}(\pi(E)) \leq \mathbb{P}(T < \infty) + \varepsilon$.*

The statement of the *predictable section theorem* is very similar.

Theorem 16.13 *If E is a predictable set and $\varepsilon > 0$, there exists a predictable stopping time T such that $[T, T] \subset E$ and $\mathbb{P}(\pi(E)) \leq \mathbb{P}(T < \infty) + \varepsilon$.*

Again we refer to Dellacherie and Meyer (1978) for proofs. We note that Proposition 16.7 is a precursor of the optional section theorem. To see this, let A be a Borel set and let $E = \{(t, \omega) : X_t \in A\}$. Then $D_E = U_A$. If the process is right continuous, then $X_{U_{K_n}} \in K_n \subset A$, where the K_n are as in Proposition 16.7, and the graphs of the U_{K_n} are contained in E.

Here is a corollary of Theorems 16.12 and 16.13.

Corollary 16.14 *(1) If X and Y are optional processes such that $\mathbb{P}(X_T = Y_T) = 1$ for every finite stopping time T, then X and Y are indistinguishable: $\mathbb{P}(X_t = Y_t$ for all $t) = 1$.*

(2) If X and Y are predictable processes with $\mathbb{P}(X_T = Y_T) = 1$ for every finite predictable stopping time T, then X and Y are indistinguishable.

Proof We prove (1), the proof of (2) being similar. Let $F = \{(t, \omega) : X_t(\omega) \neq Y_t(\omega)\}$. Then F is an optional set, and if $\mathbb{P}(\pi(F)) > 0$, there exists a stopping time U with $[U, U] \subset F$

and $\mathbb{P}(U < \infty) > 0$. By looking at $T = U \wedge N$ for sufficiently large N, we obtain a contradiction. $\qquad\square$

Another application of the section theorems is the following.

Proposition 16.15 *Suppose* $[T, T]$ *is a predictable set. Then* T *is a predictable stopping time.*

Proof Since T is the debut of $[T, T]$, then T is a stopping time. By the predictable section theorem, Theorem 16.13, for each n there exists a predictable stopping time S_n such that $[S_n, S_n] \subset [T, T]$ and

$$\mathbb{P}(\pi([S_n, S_n])) \geq \mathbb{P}(\pi([T, T])) - 2^{-n}.$$

Saying $[S_n, S_n] \subset [T, T]$ implies that for each ω, either $S_n(\omega) = T(\omega)$ or else $S_n(\omega) = \infty$. The set of ω's for which $T(\omega) < \infty$ but $S_n(\omega) = \infty$ has probability at most 2^{-n}.

Let $Q_n = S_1 \wedge \cdots \wedge S_n$. Then the Q_n's are predictable stopping times by Exercise 16.1, they decrease, $[Q_n, Q_n] \subset [T, T]$, and $\mathbb{P}(\pi([Q_n, Q_n])) \geq \mathbb{P}(\pi([T, T])) - 2^{-n+1}$. Let $Q = \lim_n Q_n$. If $Q(\omega) < \infty$, then $Q_n(\omega) < \infty$ for all n sufficiently large (how large depends on ω); since $Q_n(\omega)$ is either equal to $T(\omega)$ or to ∞, $Q_n(\omega) = Q(\omega)$ for all n sufficiently large, and hence $Q(\omega) = T(\omega)$. If $T(\omega) < \infty$, then except for a set of ω's of probability zero, $Q_n(\omega) = T(\omega)$ for n sufficiently large. Therefore $Q = T$, a.s.

Choose R_{nm} predicting Q_n as $m \to \infty$. Choose m_n large enough such that

$$\mathbb{P}(R_{nm_n} + 2^{-n} < Q_n < \infty) < 2^{-n} \quad \text{and} \quad \mathbb{P}(R_{nm_n} < n, Q_n = \infty) < 2^{-n}.$$

Let $U_n = n \wedge R_{nm_n} \wedge R_{n+1,m_{n+1}} \wedge \cdots$. Fix n for the moment. If $0 < Q(\omega) < \infty$, then $R_{jm_j}(\omega) < Q_j(\omega) = Q(\omega)$ for all j sufficiently large. Choosing $j > n$ sufficiently large, $U_n(\omega) \leq R_{jm_j}(\omega) < Q(\omega)$.

The U_n increase; let T be the limit. By the Borel–Cantelli lemma, if $Q(\omega) < \infty$, then $R_{nm_n}(\omega) \geq Q_n(\omega) - 2^{-n} = Q(\omega) - 2^{-n}$ for all n sufficiently large, except for a set of ω's of probability zero. Therefore $U_n(\omega) \geq Q(\omega) - 2^{-n+1}$ for n sufficiently large, and we conclude that $U_n(\omega) \uparrow Q(\omega)$, except for a set of ω's of probability zero.

If $Q(\omega) = \infty$, then $Q_n(\omega) = \infty$ for all n. By the Borel–Cantelli lemma, except for a set of probability zero, $R_{nm_n} \geq n$ for n sufficiently large. Hence $U_n(\omega) = n$ for n sufficiently large, so $U_n(\omega) < Q(\omega)$ and $U_n(\omega) \uparrow Q(\omega)$. Thus Q is predictable and $T = Q$, a.s. (We leave consideration of those ω for which $Q(\omega) = 0$ to the reader.) $\qquad\square$

Proposition 16.16 *Let* X_t *be a predictable process with paths that are right continuous with left limits. If* $a \in \mathbb{R}$ *and* $T = \inf\{t > 0 : X_t \geq a\}$*, then* T *is a predictable stopping time.*

Proof The set $A = \{(t, \omega) : X_t(\omega) \geq a\}$ is a predictable set. Since X_t is right continuous, $[T, \infty) = A \cup (T, \infty) \in \mathcal{P}$ by Proposition 16.2, and so $[T, T] = [T, \infty) \setminus (T, \infty) \in \mathcal{P}$. Now apply Proposition 16.15. $\qquad\square$

16.4 Projection theorems

Let $\mathcal{B}[0, \infty)$ be the Borel σ-field on $[0, \infty)$, let $\mathcal{F}_\infty = \vee_{t \geq 0} \mathcal{F}_t$, and let $\mathcal{H}$ be the product σ-field

$$\mathcal{H} = \mathcal{B}[0, \infty) \times \mathcal{F}_\infty. \tag{16.6}$$

The following is the optional projection theorem.

Theorem 16.17 *Let X be a bounded process that is $\mathcal{H}$ measurable. There exists a unique optional process ^oX such that*

$$^oX_T 1_{(T < \infty)} = \mathbb{E}\left[X_T 1_{(T < \infty)} \mid \mathcal{F}_T\right] \tag{16.7}$$

for all stopping times T, including those taking infinite values. If $X \geq 0$, then $^oX \geq 0$.

^oX is called the *optional projection* of X. If X is already optional, then by the uniqueness result, Corollary 16.14, $^oX = X$.

If we take our stopping time T in (16.7) equal to a fixed time t, we have

$$^oX_t = \mathbb{E}\left[X_t \mid \mathcal{F}_t\right], \qquad \text{a.s.} \tag{16.8}$$

This observation is sometimes useful when X is not an adapted process and one wants a version of $\mathbb{E}\left[X_t \mid \mathcal{F}_t\right]$ that is jointly measurable in t and ω.

If (16.7) holds, then taking expectations shows that

$$\mathbb{E}\left[^oX_T; T < \infty\right] = \mathbb{E}\left[X_T; T < \infty\right] \tag{16.9}$$

for all stopping times T. Conversely, suppose (16.9) holds for all stopping times T. If S is a stopping time and $A \in \mathcal{F}_S$, let S_A be defined by

$$S_A(\omega) = \begin{cases} S(\omega) & \omega \in A; \\ \infty & \omega \notin A. \end{cases} \tag{16.10}$$

Then (16.9) with T replaced by S_A implies that

$$\mathbb{E}\left[^oX_S 1_{(S < \infty)}; A\right] = \mathbb{E}\left[X_S 1_{(S < \infty)}; A\right].$$

Since $^oX_S 1_{(S < \infty)}$ is $\mathcal{F}_S$ measurable, this implies (16.7) holds for the stopping time S. Consequently (16.7) holding for all stopping times T is equivalent to (16.9) holding for all stopping times T.

Proof of Theorem 16.17 The uniqueness is immediate from Corollary 16.14. We look at existence. If $X_t(\omega) = 1_F(\omega) 1_{[a,b)}(t)$ where $F \in \mathcal{F}_\infty$, we set ^oX_t equal to $\mathbb{E}\left[1_F \mid \mathcal{F}_t\right] 1_{[a,b)}(t)$, where we use Corollary 3.13 to take the right continuous version of the martingale $\mathbb{E}\left[1_F \mid \mathcal{F}_t\right]$. We check:

$$\begin{aligned} \mathbb{E}\left[^oX_T; T < \infty\right] &= \mathbb{E}\left[\mathbb{E}\left[1_F \mid \mathcal{F}_T\right] 1_{[a,b)}(T); T < \infty\right] \\ &= \mathbb{E}\left[1_F 1_{[a,b)}(T); T < \infty\right] \\ &= \mathbb{E}\left[X_T; T < \infty\right] \end{aligned}$$

since $(T < \infty)$ and $1_{[a,b)}(T)$ are both $\mathcal{F}_T$ measurable. We then use linearity and limits to define ^oX for bounded measurable X. The positivity of ^oX when $X \geq 0$ is clear from the construction. $\qquad\square$

Almost the same proof gives

Theorem 16.18 *Let X be a bounded measurable process. There exists a unique predictable process ^pX, called the predictable projection of X, such that*

$$\mathbb{E}\left[^pX_T; T < \infty\right] = \mathbb{E}\left[X_T; T < \infty\right]$$

for every predictable stopping time T. If $X \geq 0$, then $^pX \geq 0$.

Proof Uniqueness is as before. If $X_t = 1_F(\omega)1_{(a,b]}(t)$, we let $^pX_t = 1_{(a,b]}(t)Z_{t-}(\omega)$, where Z_{t-} denotes the left-hand limit of Z_t at time t and Z_t is the right-continuous version of the martingale $\mathbb{E}[1_F \mid \mathcal{F}_t]$. We use linearity and limits to define ^pX for bounded measurable X. The positivity of ^pX when $X \geq 0$ is clear. $\qquad\square$

16.5 More on predictability

If U is a random time, i.e., a $\mathcal{F}_\infty$ measurable map from Ω to $[0, \infty]$, define

$$\mathcal{F}_{U-} = \sigma\{X_U : X \text{ is bounded and predictable}\}.$$

Lemma 16.19 *Suppose T is a predictable stopping time predicted by stopping times T_n. Then $\mathcal{F}_{T-} = \bigvee_{n=1}^\infty \mathcal{F}_{T_n}$.*

Proof If X is left continuous, adapted, and bounded, then $X_T = \lim X_{T_m}$ and $X_{T_m} \in \mathcal{F}_{T_m} \subset \bigvee_n \mathcal{F}_{T_n}$, so $X_T \in \bigvee_n \mathcal{F}_{T_n}$. An argument using the monotone class theorem shows $\mathcal{F}_{T-} \subset \bigvee_n \mathcal{F}_{T_n}$.

On the other hand, suppose $A \in \mathcal{F}_{T_n}$ for some n. Define $X = 1_{(U_n,\infty)}$, where $U_n = T_n$ if $\omega \in A$ and ∞ otherwise. Since $T_n < T$ on $(T > 0)$, then $X_T = 1_A$. (We leave consideration of what happens on the event $(T = 0)$ to the reader.) X is predictable since it is left continuous, adapted, and bounded, so A is $\mathcal{F}_{T-}$ measurable. Therefore $\mathcal{F}_{T_n} \subset \mathcal{F}_{T-}$ for all n, and we conclude $\bigvee_n \mathcal{F}_{T_n} \subset \mathcal{F}_{T-}$. $\qquad\square$

Corollary 16.20 *Suppose T is a predictable stopping time. If M is a uniformly integrable martingale with right-continuous paths, then*

$$\mathbb{E}[M_T \mid \mathcal{F}_{T-}] = M_{T-}.$$

Proof If $X_t = M_{t-}$, then X is left continuous, hence predictable, so $M_{T-} = X_T$ is $\mathcal{F}_{T-}$ measurable by the definition of $\mathcal{F}_{T-}$ and a limit argument. Suppose the sequence T_n predicts T. If $A \in \mathcal{F}_{T_m}$ and $n > m$, then $A \in \mathcal{F}_{T_m} \subset \mathcal{F}_{T_n}$, and by optional stopping (see Exercise 3.12),

$$\mathbb{E}[M_T; A] = \mathbb{E}[M_{T_n}; A] \to \mathbb{E}[M_{T-}; A]$$

as $n \to \infty$. Since $\mathcal{F}_{T-} = \bigvee_m \mathcal{F}_{T_m}$, we have $\mathbb{E}[M_T; A] = \mathbb{E}[M_{T-}; A]$ for all $A \in \mathcal{F}_{T-}$. Now use the definition of conditional expectation. $\qquad\square$

Corollary 16.21 *Let S be a predictable stopping time, M a square integrable martingale, and $N_t = \Delta M_S 1_{(t \geq S)}$. Then N_t is a square integrable martingale.*

Proof Since $|N_t| \leq 2\sup_{s\geq 0}|M_s|$, N is square integrable. We will show N is a martingale by showing $\mathbb{E}\,N_T = 0$ for all bounded stopping times T, and then appealing to Proposition 9.5.

If T is a bounded stopping time, then $(T \geq S) \in \mathcal{F}_{S-}$; to see this, if S_m is a sequence of stopping times predicting S, then $(T \geq S) = \cap_m (T \geq S_m) \in \vee_m \mathcal{F}_{S_m}$. Using Corollary 16.20,

$$\mathbb{E}\,N_T = \mathbb{E}\,\Delta M_S 1_{(T \geq S)} = \mathbb{E}\,[M_S;\, T \geq S] - \mathbb{E}\,[M_{S-};\, T \geq S] = 0,$$

and we are done. □

We now show that every stopping time for Brownian motion is predictable.

Proposition 16.22 *Let $\{\mathcal{F}_t\}$ be the minimal augmented filtration of a Brownian motion. If T is a stopping time with respect to $\{\mathcal{F}_t\}$, then T is a predictable stopping time.*

Proof Let T be a stopping time for Brownian motion. Let g be a continuous strictly increasing function from $[0, \infty]$ to $[0, 1]$, e.g., $g(s) = (2/\pi)\arctan s$. Let M_t be the right-continuous modification of the martingale $\mathbb{E}\,[g(T) \mid \mathcal{F}_t]$. The property of Brownian motion that is key here is that every martingale adapted to the filtration of a Brownian motion is continuous; see Corollary 12.5. Hence M_t can be taken to be continuous.

Let $V_t = M_t - g(T \wedge t)$. Then V_t has continuous paths and since $g(T \wedge t)$ increases with t, V is a supermartingale. We have

$$V_t = \mathbb{E}\,[g(T) - g(T \wedge t) \mid \mathcal{F}_t],$$

so V is non-negative. Clearly $V_T = 0$. If S is the first time that V_t is 0, then $S \leq T$. Also,

$$0 = \mathbb{E}\,V_S = \mathbb{E}\,[g(T) - g(T \wedge S)],$$

so $S \geq T$.

We let $T_n = \inf\{t : V_t = 1/n\}$. By the continuity of V, it is clear that each T_n is strictly less than T if $T > 0$ and the T_n increase up to T. Hence T is predictable. □

Now let us suppose that A_t is a right-continuous adapted process whose paths are increasing. We call such a process an *increasing process*. ΔA_t denotes the jump of A at time t, that is, $\Delta A_t = A_t - A_{t-}$.

Proposition 16.23 *Suppose A_t is an increasing process such that*
(1) $\Delta A_T = 0$ whenever T is a totally inaccessible stopping time, and
(2) ΔA_T is $\mathcal{F}_{T-}$ measurable whenever T is a predictable stopping time.
Then A is predictable.

Proof Let U_{mi} be the ith time $|\Delta A_t| \in (2^{-m}, 2^{-m+1}]$. The U_{mi} are predictable stopping times by Exercise 16.5. We decompose each U_{mi} as in Proposition 16.1. Since A does not jump at totally inaccessible times, none of the U_{mi} has a totally inaccessible part.

We do this for each m and i and obtain a countable collection of predictable stopping times, the union of whose graphs contains all the jump times of A. We order them in some way as $R_1, R_2, \ldots$ Define $T_1 = R_1$, define T_2 by setting $T_2(\omega) = R_2(\omega)$ if $R_2(\omega) \neq R_1(\omega)$ and infinity otherwise. Set $T_n(\omega) = R_n(\omega)$ if $R_n(\omega) \neq R_1(\omega), \ldots, R_{n-1}(\omega)$ and $T_n(\omega) = \infty$ otherwise. We thus get a sequence of predictable stopping times T_n with disjoint graphs and

$\cup_n [T_n, T_n]$ includes all the jumps of A, except for the set of ω's of probability zero. The T_n are predictable stopping times by Exercise 16.6.

Since A jumps only at the predictable stopping times T_n, we see that we can write $A_t = A_t^c + \sum_i (\Delta A_{T_n}) 1_{[T_n, \infty)}$, where A^c is a continuous increasing process. By hypothesis, ΔA_{T_n} is $\mathcal{F}_{T_n-}$ measurable. Therefore the proof will be complete once we show $(\Delta A_{T_n}) 1_{[T_n, \infty)}$ is a predictable process.

It therefore suffices to show that the process $Y_t = 1_B(\omega) 1_{[T, \infty)}(t)$ is predictable if T is a predictable stopping time and $B \in \mathcal{F}_{T-}$. Since $Y_t = 1_{[T_B, \infty)}(t)$, where T_B is equal to T if $\omega \in B$ and equal to infinity otherwise, the predictability of Y follows by Exercise 16.3. $\square$

16.6 Dual projection theorems

In this section A_t is a right-continuous increasing process with $A_0 = 0$, a.s. We do not necessarily assume that A_t is adapted, only that A is measurable with respect to $\mathcal{H}$ defined by (16.6). Define μ_A on elements of $\mathcal{H}$ by

$$\mu_A(B) = \mathbb{E} \int_0^\infty 1_B(t, \omega) \, dA_t(\omega).$$

We define $\mu_A(X)$ by $\mathbb{E} \int_0^\infty X_t \, dA_t$ if X is bounded and $\mathcal{H}$ measurable. Note that if $X = 0$, then $\mu_A(X) = 0$.

Theorem 16.24 *Suppose μ is a bounded positive measure on $\mathcal{H}$ such that $\mu(X) = 0$ whenever $X = 0$. Then there exists a unique right-continuous increasing process A with $A_0 = 0$, a.s., such that $\mu = \mu_A$.*

Proof First, uniqueness. If $\mu = \mu_A = \mu_B$, let $t > 0$ and let C be the set of ω's where $A_t(\omega) > B_t(\omega) + \varepsilon$. Then $\mu_A([0, t] \times C) \geq \mu_B([0, t] \times C) + \varepsilon \mathbb{P}(C)$, which implies $\mathbb{P}(C) = 0$. Since ε is arbitrary, then $A_t = B_t$, a.s. Since A and B are right continuous, we conclude $A = B$.

To prove existence, for each rational q, define $\nu_q(C) = \mu([0, q] \times C)$. Clearly ν_q is absolutely continuous with respect to $\mathbb{P}$. Let $\widetilde{A}_q$ be the Radon–Nikodym derivative of ν_q with respect to $\mathbb{P}$. Since μ is positive, $\widetilde{A}$ is increasing in q. Let $A_t = \limsup_{q \to t, q > t} \widetilde{A}_q$. It is easy to check that $\mu_A = \mu$. $\square$

Theorem 16.25 *Suppose A is right continuous, $A_0 = 0$, a.s., and $\mu_A(X) = \mu_A({}^o X)$ for every bounded $\mathcal{H}$ measurable process X. Then A_t is optional.*

Proof Since A_t is right continuous, we need only show that A_t is adapted. Fix t and let Y be a bounded $\mathcal{F}_\infty$ measurable random variable,

$$Z = Y - \mathbb{E}[Y \mid \mathcal{F}_t],$$

and $X_s(\omega) = 1_{[0,t]}(s) Z(\omega)$. If T is a stopping time, then $(T \leq t) \in \mathcal{F}_t$, and so by the definitions of X and Z,

$$\mathbb{E}[{}^o X_T; T < \infty] = \mathbb{E}[X_T; T < \infty] = \mathbb{E}[Z; T \leq t] = 0.$$

This implies ${}^o X = 0$ by the definition of ${}^o X$. Hence

$$\mathbb{E}[A_t Z] = \mathbb{E}\left[\int_0^\infty X_s \, dA_s \right] = \mu_A(X) = \mu_A({}^o X) = 0.$$

Thus $\mathbb{E}[A_t Y] = \mathbb{E}[A_t \mathbb{E}[Y \mid \mathcal{F}_t]]$. We write

$$\begin{aligned}
\mathbb{E}[A_t Y] &= \mathbb{E}[A_t \mathbb{E}[Y \mid \mathcal{F}_t]] = \mathbb{E}[\mathbb{E}[(A_t \mathbb{E}[Y \mid \mathcal{F}_t]) \mid \mathcal{F}_t]] \\
&= \mathbb{E}[\mathbb{E}[A_t \mid \mathcal{F}_t] \mathbb{E}[Y \mid \mathcal{F}_t]] = \mathbb{E}[\mathbb{E}[(Y \mathbb{E}[A_t \mid \mathcal{F}_t]) \mid \mathcal{F}_t]] \\
&= \mathbb{E}[Y \mathbb{E}[A_t \mid \mathcal{F}_t]].
\end{aligned}$$

Hence $\mathbb{E}[A_t Y] = \mathbb{E}[Y \mathbb{E}[A_t \mid \mathcal{F}_t]]$ for all bounded Y, or $A_t = \mathbb{E}[A_t \mid \mathcal{F}_t]$, a.s., which says that A_t is $\mathcal{F}_t$ measurable. $\qquad\square$

Theorem 16.26 *If $\mu_A(X) = \mu_A(^P X)$ for all bounded X, then A is predictable and can be taken to be right continuous.*

Proof By hypothesis, together with Exercise 16.8,

$$\mu_A(^o X) = \mu_A(^P(^o X)) = \mu_A(^P X) = \mu_A(X).$$

By Theorem 16.25, A_t is right continuous and optional. We need to show that A does not jump at totally inaccessible times and that ΔA_T is $\mathcal{F}_{T-}$ measurable at predictable times T; we then use Proposition 16.23.

Let T be a totally inaccessible stopping time and let $B = (\Delta A_T > 0)$. Set T_B equal to T on B and equal to infinity otherwise. It is easy to check that T_B is also totally inaccessible. Let $X = 1_{[T_B, T_B]}$. If U is a predictable stopping time, $\mathbb{E}[X_U; U < \infty] = \mathbb{P}(T_B = U < \infty) = 0$. By the definition of predictable projection, $^P X = 0$. Hence

$$\mathbb{E}[\Delta A_T; \Delta A_T > 0] = \mathbb{E}[\Delta A_{T_B}] = \mu_A(X) = \mu_A(^P X) = 0.$$

Now suppose T is a predictable stopping time. Let Y be a bounded $\mathcal{H}$ measurable random variable, set

$$Z = Y - \mathbb{E}[Y \mid \mathcal{F}_{T-}],$$

and $X = Z 1_{[T,T]}$. Let S be any predictable stopping time. Then if $W = 1_{[S,S]}$, $W = \lim_{n \to \infty} 1_{[S, S+(1/n))}$ is a predictable process by Proposition 16.2(4). By the definition of $\mathcal{F}_{T-}$, W_T is $\mathcal{F}_{T-}$ measurable. This is the same as saying $(S = T < \infty) \in \mathcal{F}_{T-}$. Therefore

$$\mathbb{E}[X_S; S < \infty] = \mathbb{E}[Z; S = T < \infty] = 0.$$

This implies $^P X = 0$, and then

$$0 = \mu_A(^P X) = \mu_A(X) = \mathbb{E}[Z \Delta A_T].$$

Similarly to the proof of Theorem 16.25,

$$\begin{aligned}
\mathbb{E}[\Delta A_T Y] &= \mathbb{E}[\Delta A_T \mathbb{E}[Y \mid \mathcal{F}_{T-}]] \\
&= \mathbb{E}[\mathbb{E}[\Delta A_T \mid \mathcal{F}_{T-}] \mathbb{E}[Y \mid \mathcal{F}_{T-}]] \\
&= \mathbb{E}[Y \mathbb{E}[\Delta A_T \mid \mathcal{F}_{T-}]].
\end{aligned}$$

Since this holds for all Y, then $\Delta A_T = \mathbb{E}[\Delta A_T \mid \mathcal{F}_{T-}]$ is $\mathcal{F}_{T-}$ measurable. $\qquad\square$

We now define the dual optional projection and the dual predictable projection of an increasing process. Given a right-continuous increasing, not necessarily adapted process A_t with $A_0 = 0$, a.s., define μ_o by

$$\mu_o(X) = \mu_A(^o X) \tag{16.11}$$

for bounded $\mathcal{H}$ measurable X. Exercise 16.11 asks you to prove that μ_o is a measure. Clearly $\mu_o(^oX) = \mu_A(^o(^oX)) = \mu_A(^oX) = \mu_o(X)$. By Theorem 16.17, we see that $^oX \geq 0$ if $X \geq 0$, hence μ_o is a positive measure. If $X = 0$, then $^oX = 0$, so $\mu_o(X) = \mu_A(^oX) = 0$. Therefore by Theorems 16.24 and 16.25, μ_o corresponds to an optional increasing process A^o, called the *dual optional projection* of A.

The dual optional projection is used in excursion theory. More commonly used is the *dual predictable projection*, which is defined in a very similar way. Define $\mu_p(X) = \mu_A(^pX)$, and let A^p be the predictable increasing process associated with μ_p. We often denote A^p by $\widetilde{A}$ and call it the *compensator* of A. The reason for this terminology is the following proposition.

Proposition 16.27 *Let A_t be an adapted increasing process with $A_0 = 0$, a.s. Then $A_t - \widetilde{A}_t$ is a martingale.*

Proof Let $s < t$, let $B \in \mathcal{F}_s$, define

$$S(\omega) = \begin{cases} s, & \omega \in B, \\ \infty, & \omega \notin B, \end{cases} \quad \text{and} \quad T(\omega) = \begin{cases} t, & \omega \in B, \\ \infty, & \omega \notin B. \end{cases}$$

Let $X = 1_{(S,T]}$. Then

$$\mathbb{E}[A_t - A_s; B] = \mu_A(X) = \mu_A(^pX) = \mu_{A^p}(X) = \mathbb{E}[A_t^p - A_s^p; B],$$

which does it. $\qquad\square$

16.7 The Doob–Meyer decomposition

Proposition 16.28 *If M is a predictable uniformly integrable martingale with paths that are right continuous with left limits, then M is continuous.*

Proof Let $\varepsilon > 0$ and let $T = \inf\{t : |\Delta M_t| > \varepsilon\}$. T is a predictable stopping time by Exercise 16.2. By Corollary 16.20, $\mathbb{E}[M_T \mid \mathcal{F}_{T-}] = M_{T-}$. By the definition of $\mathcal{F}_{T-}$ and a limit argument, M_T is $\mathcal{F}_{T-}$ measurable, and thus $\mathbb{E}[M_T \mid \mathcal{F}_{T-}] = M_T$. Hence $M_T = M_{T-}$ at all predictable stopping times, and in particular at time T. But ε is arbitrary, so M has no jumps. $\qquad\square$

We say a process X is of *class D* if the family $\{X_T : T \text{ a stopping time}\}$ is uniformly integrable. The *Doob–Meyer decomposition* is the following. If Z_t is a supermartingale, then $-Z_t$ is a submartingale, and it is a matter only of convenience whether we state the Doob–Meyer decomposition in terms of submartingales or supermartingales.

Theorem 16.29 *Suppose Z_t is a submartingale of class D with paths that are right continuous with left limits and such that $Z_0 = 0$, a.s. Then $Z_t = M_t + A_t$, where M_t is a uniformly integrable right-continuous martingale with $M_0 = 0$, a.s., and A_t is a predictable increasing process with $A_0 = 0$, a.s. The decomposition is unique.*

The existence is the hard part. We define a measure μ by $\mu((S, T]) = E[Z_T - Z_S]$ for stopping times $S \leq T$, and then let A be the increasing process such that $\mu_A(X) = \mu(^pX)$.

Proof We start with uniqueness. If $Z_t = M_t + A_t = N_t + B_t$, then $M_t - N_t = B_t - A_t$, and so $M_t - N_t$ is a predictable uniformly integrable martingale. By Proposition 16.28, $M_t - N_t$ is a continuous martingale. Since $M_t - N_t = B_t - A_t$, then $M_t - N_t$ is a continuous martingale whose paths are of bounded variation on each finite time interval, hence $M_t - N_t = 0$ by Theorem 9.7. This proves uniqueness.

We turn to existence. By the martingale convergence theorem (Theorem 3.12), $Z_\infty = \lim_{t \to \infty} Z_t$ exists, a.s. By Fatou's lemma, $\mathbb{E} |Z_\infty| < \infty$.

Let $\mathcal{I}$ denote the collection of finite unions of subsets of $[0, \infty) \times \Omega$ of the form $(S, T]$, where $S \leq T$ are stopping times. Define $\mu((S, T]) = \mathbb{E}[Z_T - Z_S]$. Since Z is a submartingale, then μ is non-negative. We note that $\mathcal{I}$ is an algebra and that μ is finitely additive on $\mathcal{I}$.

If $K = (S_1, T_1] \cup \cdots \cup (S_n, T_n]$ with $S_1 \leq T_1 \leq S_2 \leq \cdots \leq T_n$, set $\overline{K} = [S_1, T_1] \cup \cdots \cup [S_n, T_n]$.

If $H = (S, T]$ and $\varepsilon > 0$, let

$$S_n(\omega) = \begin{cases} S(\omega) + (1/n), & S(\omega) + (1/n) < T(\omega), \\ \infty, & \text{otherwise}, \end{cases}$$

and

$$T_n(\omega) = \begin{cases} T(\omega), & S(\omega) + (1/n) < T(\omega), \\ \infty, & \text{otherwise}. \end{cases}$$

Then $[S_n, T_n] \subset (S, T]$ and $S_n \downarrow S$, $T_n \downarrow T$. Since Z is right continuous and of class D, then $\mu(S_n, T_n] = \mathbb{E}[Z_{T_n} - Z_{S_n}] \to \mathbb{E}[Z_T - Z_S] = \mu(H)$. Thus if n is sufficiently large and we take $K = (S_n, T_n]$, then $\overline{K} \subset H$ and $\mu(K) > \mu(H) - \varepsilon$.

We now prove that μ is countably additive on $\mathcal{I}$. Suppose $H_n \in \mathcal{I}$ with $H_n \downarrow \emptyset$. We need to show that $\mu(H_n) \downarrow 0$.

Let $\varepsilon > 0$ and choose $K_n \in \mathcal{I}$ such that $\overline{K}_n \subset H_n$ with $\mu(K_n) > \mu(H_n) - \varepsilon/2^n$. Let $L_n = \overline{K}_1 \cap \cdots \cap \overline{K}_n$. Then for each n we have $\mu(H_n) \leq \mu(L_n) + \varepsilon$. Since $L_n \subset \overline{K}_n \subset H_n$, we have $L_n \downarrow \emptyset$.

Let D_{L_n} be the debut of L_n. The stopping times D_{L_n} increase; let R be the limit. Let $F_n = F_n(\omega) = \{t : (t, \omega) \in L_n\}$. This is a closed subset of $[0, \infty)$, and $D_{L_n}(\omega) \in F_n \subset F_m$ whenever $n \geq m$ and $D_{L_n}(\omega) < \infty$. If $R(\omega) < \infty$, then $R(\omega) \in F_m$ for each m, which contradicts $\cap_m L_m = \emptyset$. Therefore $R = \infty$. Since Z is of class D, then $Z_{D_{L_n}}$ converges almost surely and in L^1 to Z_∞. Thus $\mu(L_n) \leq \mathbb{E}[Z_\infty - Z_{D_{L_n}}] \to 0$. Hence $\limsup \mu(H_n) < \varepsilon$, and since ε is arbitrary, $\mu(H_n) \to 0$.

This proves that μ is countably additive on $\mathcal{I}$. By the Carathéodory extension theorem, μ may be extended to a measure on $\mathcal{P}$.

Define $\widetilde{\mu}(X) = \mu(^pX)$. Then $\widetilde{\mu}(^pX) = \mu(^p(^pX)) = \mu(^pX) = \widetilde{\mu}(X)$, and so there exists a predictable right-continuous increasing process A_t such that $\widetilde{\mu} = \mu_A$. Since

$$\mathbb{E} A_\infty = \mu_A(1_{(0,\infty)}) = \mu(^p1_{(0,\infty)}) = \mu(1_{(0,\infty)}) = \mathbb{E}[Z_\infty - Z_0] < \infty,$$

A_∞ is integrable, and since A_t is an increasing process, the collection of random variables $\{A_t\}$ is uniformly integrable.

If S is any stopping time, then by Proposition 16.2, (S, ∞) is a predictable set, hence $^p1_{(S,\infty)} = 1_{(S,\infty)}$. We thus have

$$\mathbb{E}[A_\infty - A_S] = \widetilde{\mu}((S, \infty)) = \mu(^p1_{(S,\infty)}) = \mu(1_{(S,\infty)}) = \mathbb{E}[Z_\infty - Z_S].$$

Letting $t > 0$ and $B \in \mathcal{F}_t$, define $S = t$ if $\omega \in B$ and equal to infinity otherwise. Then

$$\mathbb{E}\left[A_\infty - A_t; B\right] = \mathbb{E}\left[A_\infty - A_S\right] = \mathbb{E}\left[Z_\infty - Z_S\right] = \mathbb{E}\left[Z_\infty - Z_t; B\right],$$

or $M_t = Z_t - A_t$ is a martingale. Proposition A.17 tells us that M is a uniformly integrable martingale. $\square$

A process X is of *class DL* if there exist stopping times $V_n \to \infty$ such that $X_{t \wedge V_n}$ is of class D for each n. It is clear that there is a version of the Doob–Meyer decomposition for submartingales of class *DL*.

Proposition 16.30 *The process A is continuous if and only if $\mathbb{E}\, Z_{T_n} \to \mathbb{E}\, Z_T$ whenever $T_n \uparrow T$ and $T_n < T$ on $(T > 0)$.*

Proof Let T be a predictable stopping time predicted by the sequence T_n. Since we know $\mathbb{E}\left[A_\infty - A_{T_n}\right] = \mathbb{E}\left[Z_\infty - Z_{T_n}\right]$, then taking limits,

$$\mathbb{E}\left[A_\infty - A_{T-}; T < \infty\right] = \mathbb{E}\left[Z_\infty - Z_{T-}; T < \infty\right],$$

using the fact that Z is of class D. Also $\mathbb{E}\left[A_\infty - A_T\right] = \mathbb{E}\left[Z_\infty - Z_T\right]$. Thus $\mathbb{E}\left[A_T - A_{T-}\right] = \mathbb{E}\left[Z_T - Z_{T-}\right]$. Then $\mathbb{E}\left[A_T - A_{T-}\right] = 0$ if and only if $\mathbb{E}\, Z_T = \mathbb{E}\, Z_{T-}$. $\square$

Corollary 16.31 *Let S be a totally inaccessible stopping time, Y a non-negative bounded random variable that is $\mathcal{F}_S$ measurable, and $A_t = Y 1_{(t \geq S)}$. Let $\widetilde{A}$ be the compensator of A. Then $\widetilde{A}$ has continuous paths.*

Proof Let T be a stopping time and let T_n be stopping times increasing to T. If we have $\mathbb{P}(T = S) = 0$, then $\lim_{n \to \infty} A_{T_n} = A_T$, a.s., since A jumps only at time S. If $\mathbb{P}(T = S) > 0$, then $[T, T]$ cannot contain the graph of a predictable stopping time since S is totally inaccessible. Therefore we cannot have $T_n < T$ for all n with positive probability, hence $T_n(\omega) = T(\omega)$ for all n sufficiently large (depending on ω). Thus again $\lim_{n \to \infty} A_{T_n} = A_T$, a.s. By Proposition 16.30, $\widetilde{A}$ is continuous. $\square$

16.8 Two inequalities

Proposition 16.32 *Suppose $Z_t = M_t - A_t$, where M_t is a uniformly integrable martingale and A_t is an increasing predictable process with $A_0 = 0$, a.s. Suppose Z is bounded, that is, there exists $K > 0$ such that $\mathbb{P}(|Z_t| > K$ for some $t) = 0$. If p is any positive integer,*

$$\mathbb{E}\, A_\infty^p < \infty.$$

Proof Let $\lambda > 0$ and let $M = 4K$. Let $T = \inf\{t : A_t \geq \lambda\}$. Because $A_{T-} \leq \lambda$,

$$
\begin{aligned}
\mathbb{P}(A_\infty \geq \lambda + M) &= \mathbb{P}(A_\infty \geq \lambda + M, T < \infty) \\
&\leq \mathbb{P}(A_\infty - A_{T-} \geq M, T < \infty) \\
&\leq \mathbb{E}\left[\frac{A_\infty - A_{T-}}{M}; A_\infty - A_{T-} \geq M, T < \infty\right] \\
&\leq \frac{1}{M}\mathbb{E}\left[A_\infty - A_{T-}; T < \infty\right].
\end{aligned}
$$

We will show

$$\frac{1}{M}\mathbb{E}\left[A_\infty - A_{T-}; T < \infty\right] \le \tfrac{1}{2}\mathbb{P}(T < \infty), \tag{16.12}$$

which, since $\mathbb{P}(T < \infty) = \mathbb{P}(A_\infty \ge \lambda)$, implies

$$\mathbb{P}(A_\infty \ge \lambda + M) \le \tfrac{1}{2}\mathbb{P}(A_\infty \ge \lambda). \tag{16.13}$$

Taking $\lambda = kM$ in (16.13) yields

$$\mathbb{P}(A_\infty \ge (k+1)M) \le \tfrac{1}{2}\mathbb{P}(A_\infty \ge kM).$$

Since $\mathbb{P}(A_\infty \ge M) \le 1$, induction tells us

$$\mathbb{P}(A_\infty \ge kM) \le \frac{1}{2^{k-1}},$$

which implies our conclusion.

Therefore we need to prove (16.12). T is a predictable stopping time by Proposition 16.16. Let T_n be stopping times with $T_n \uparrow T$ and $T_n < T$ on $(T > 0)$. Let n be fixed for the moment and let $N > 0$. If $j > n$,

$$\begin{aligned}
\mathbb{E}\left[A_\infty - A_{T_j}; T_n < N\right] &= \mathbb{E}\left[\mathbb{E}\left[A_\infty - A_{T_j} \mid \mathcal{F}_{T_j}\right]; T_n < N\right] \\
&= -\mathbb{E}\left[\mathbb{E}\left[Z_\infty - Z_{T_j} \mid \mathcal{F}_{T_j}\right]; T_n < N\right] \\
&\le 2K\mathbb{P}(T_n < N)
\end{aligned}$$

since $Z_t + A_t$ is a martingale, $(T_n < N) \in \mathcal{F}_{T_n} \subset \mathcal{F}_{T_j}$, and $|Z|$ is bounded by K. Letting $j \to \infty$ and using Fatou's lemma, we get

$$\mathbb{E}\left[A_\infty - A_{T-}; T_n < N\right] \le 2K\mathbb{P}(T_n < N).$$

Letting $n \to \infty$, by Fatou's lemma again,

$$\mathbb{E}\left[A_\infty - A_{T-}; T < N\right] \le 2K\mathbb{P}(T \le N).$$

Finally, letting $N \to \infty$, by monotone convergence,

$$\mathbb{E}\left[A_\infty - A_{T-}; T < \infty\right] \le 2K\mathbb{P}(T < \infty).$$

By our choice of M, this gives (16.12). □

For use in the reduction theorem in Chapter 17, we will need a variation of the preceding proposition.

Proposition 16.33 *Let U be a stopping time, Y a non-negative integrable random variable that is $\mathcal{F}_U$ measurable. Let N_t be the right-continuous version of $\mathbb{E}\left[Y \mid \mathcal{F}_t\right]$. Suppose there exists $K > 0$ such that $N_t \le K$ if $t < U$. Let $Z_t = Y\mathbf{1}_{(t \ge U)}$, which is an increasing process, and let A_t be its compensator. If p is a positive integer, then $\mathbb{E}\,A_\infty^p < \infty$.*

Proof As in the proof of Proposition 16.32, it suffices to show

$$\mathbb{E}\left[A_\infty - A_{T-}; T < \infty\right] \le K\mathbb{P}(T < \infty), \tag{16.14}$$

where $\lambda > 0$ and $T = \inf\{t : A_t \ge \lambda\}$. Since A is a predictable process, then T is a predictable stopping time by Proposition 16.16. Let T_n be stopping times predicting T.

Let $N, n \geq 1$. If $j > n$, then $(T_n < N) \in \mathcal{F}_{T_n} \subset \mathcal{F}_{T_j}$ and

$$\mathbb{E}[A_\infty - A_{T_j}; T_n < N] = \mathbb{E}[Z_\infty - Z_{T_j}; T_n < N]. \tag{16.15}$$

We observe that $Z_\infty - Z_{T_j} = 0$ on the event $(T_j \geq U)$, while $Z_\infty - Z_{T_j} = Y$ on the event $(T_j < U)$. Therefore

$$
\begin{aligned}
\mathbb{E}[Z_\infty - Z_{T_j}; T_n < N] &= \mathbb{E}[Y; T_j < U, T_n < N] \\
&= \mathbb{E}[\mathbb{E}[Y \mid \mathcal{F}_{T_j}]; T_j < U, T_n < N] \\
&= \mathbb{E}[N_{T_j}; T_j < U, T_n < N] \\
&\leq K\mathbb{P}(T_j < U, T_n < N) \\
&\leq K\mathbb{P}(T_n < N).
\end{aligned}
$$

With this and (16.15), we can now proceed as in the proof of Proposition 16.32 to obtain (16.14).   □

Exercises

16.1 Show that if $S_1, \ldots, S_n$ are predictable stopping times, then so are $S_1 \wedge \cdots \wedge S_n$ and $S_1 \vee \cdots \vee S_n$.

16.2 If A_t is a predictable process with paths that are right continuous with left limits and $a > 0$, show $T = \inf\{t > 0 : \Delta A_t > a\}$ is a predictable stopping time.

16.3 Show that if T is a predictable stopping time, $B \in \mathcal{F}_{T-}$, and $T_B(\omega)$ is defined to be equal to $T(\omega)$ if $\omega \in B$ and equal to ∞ otherwise, then T_B is a predictable stopping time.

16.4 Let X be a bounded adapted right-continuous process, let $\varepsilon > 0$, let $U_0 = 0$, a.s., and define U_i by (16.3) for $i \geq 1$. Show each U_i is a stopping time.

16.5 Let A be a predictable increasing process and let S_k be the kth time A jumps more than ε. Thus $S_0 = 0$, a.s., and $S_{k+1} = \inf\{t > S_k : \Delta A_t > \varepsilon\}$. Show each S_k is a predictable stopping time.

16.6 Show that the stopping times T_n defined in the proof of Proposition 16.23 are predictable.

16.7 Show that if P_t is a Poisson process, then $(^pP)_t = P_{t-}$.

16.8 Show that if X is bounded and measurable with respect to the product σ-field $\mathcal{B}[0, \infty) \times \mathcal{F}_\infty$, then $^p(^oX) = {}^pX$.

16.9 Suppose T is a totally inaccessible stopping time. Show that if $X = 1_{[T,T]}$, then $^pX = 0$.

16.10 If P is a Poisson process with parameter λ, determine P_t^o and P_t^p.

16.11 Show that μ_o defined in (16.11) is a measure.

16.12 Let X_t be a continuous process and suppose there exists $K > 0$ such that for all t,

$$\mathbb{E}[|X_\infty - X_t| \mid \mathcal{F}_t] \leq K, \qquad \text{a.s.}$$

Let $X_\infty^* = \sup_{t \geq 0} |X_t|$. Prove that there exists a depending only on K such that

$$\mathbb{E}\, e^{aX_\infty^*} < \infty.$$

This is sometimes called the *John–Nirenberg inequality* after the inequality of the same name in analysis.

Hint: Imitate the proof of Proposition 16.32. This exercise is somewhat easier than the proof of that proposition because X has continuous paths.

16.13 A martingale M is said to be in the space BMO if

$$\sup_{t \geq 0} \mathbb{E}\,[M_\infty^2 - M_t^2 \mid \mathcal{F}_t] < \infty, \qquad \text{a.s.}$$

Let $M_t^* = \sup_{s \leq t} |M_s|$. Show that if M is in BMO, then there exists $a > 0$ such that

$$\mathbb{E}\,e^{aM_\infty^*} < \infty.$$

The name BMO comes from the "bounded mean oscillation" spaces of harmonic analysis.

Hint: Use Exercise 16.12.

Notes

A progressively measurable set is one whose indicator is a progressively measurable process, which is defined in Exercise 1.3. In fact, the debut of a progressively measurable set is a stopping time; see Dellacherie and Meyer (1978).

An elementary proof of the general Doob–Meyer theorem along the lines of the proof given in Chapter 9 can be found in Bass (1996).

See Dellacherie and Meyer (1978) for more on the general theory of processes.

17

Processes with jumps

In this chapter we investigate the stochastic calculus for processes which may have jumps as well as a continuous component. If X is not a continuous process, it is no longer true that $X_{t \wedge T_N}$ is a bounded process when $T_N = \inf\{t : |X_t| \geq N\}$, since there could be a large jump at time T_N. We investigate stochastic integrals with respect to square integrable (not necessarily continuous) martingales, Itô's formula, and the Girsanov transformation. We prove the reduction theorem that allows us to look at semimartingales that are not necessarily bounded.

Since I encouraged you to skim Chapter 16 on the first reading of this book, it is only fair that I tell you the facts that we will need from that chapter. We will need the Doob–Meyer decomposition (Theorem 16.29), Proposition 16.1, Corollaries 16.21 and 16.31, and the two inequalities in Propositions 16.32 and 16.33.

17.1 Decomposition of martingales

We assume throughout this chapter that $\{\mathcal{F}_t\}$ is a filtration satisfying the usual conditions. This means that each $\mathcal{F}_t$ contains every $\mathbb{P}$-null set and $\cap_{\varepsilon > 0} \mathcal{F}_{t+\varepsilon} = \mathcal{F}_t$ for each t.

Let us begin by recalling a few definitions and facts. The *predictable σ-field* is the σ-field of subsets of $[0, \infty) \times \Omega$ generated by the collection of bounded, left-continuous processes that are adapted to $\{\mathcal{F}_t\}$; see Section 10.1. A stopping time T is *predictable* and predicted by the sequence of stopping times T_n if $T_n \uparrow T$, and $T_n < T$ on the event $(T > 0)$. A stopping time T is *totally inaccessible* if $\mathbb{P}(T = S) = 0$ for every predictable stopping time S. The graph of a stopping time T is $[T, T] = \{(t, \omega) : t = T(\omega) < \infty\}$; see Section 16.1. If X_t is a process that is right continuous with left limits, we set $X_{t-} = \lim_{s \to t, s < t} X_s$ and $\Delta X_t = X_t - X_{t-}$. Thus ΔX_t is the size of the jump of X_t at time t.

Suppose A_t is a bounded increasing process whose paths are right continuous with left limits. Recall that a function f is increasing if $s < t$ implies $f(s) \leq f(t)$. Then trivially A_t is a submartingale, and by the Doob–Meyer decomposition, Theorem 16.28, there exists a predictable increasing process $\widetilde{A}_t$ such that $A_t - \widetilde{A}_t$ is a martingale. We call $\widetilde{A}_t$ the *compensator* of A_t.

If $A_t = B_t - C_t$ is the difference of two increasing processes B_t and C_t, then we can use linearity to define $\widetilde{A}_t$ as $\widetilde{B}_t - \widetilde{C}_t$. We can even extend the notion of compensator to the case where A_t is complex valued and has paths that are locally of bounded variation by looking at the real and imaginary parts.

We will use the following lemma.

Lemma 17.1 *If* $A_t = B_t - C_t$, *where* B_t *and* C_t *are increasing right-continuous processes with* $B_0 = C_0 = 0$, *a.s., and in addition* B *and* C *are bounded, then*

$$\mathbb{E} \sup_{t \geq 0} \widetilde{A}_t^2 < \infty.$$

Proof By Proposition 16.32, $\mathbb{E} \widetilde{B}_\infty^2 < \infty$ and $\mathbb{E} \widetilde{C}_\infty^2 < \infty$, and so

$$\mathbb{E} \sup_{t \geq 0} \widetilde{A}_t^2 \leq \mathbb{E} [2 \sup_{t \geq 0} \widetilde{B}_t^2 + 2 \sup_{t \geq 0} \widetilde{C}_t^2] \leq 2\mathbb{E} \widetilde{B}_\infty^2 + 2\mathbb{E} \widetilde{C}_\infty^2 < \infty.$$

We are done. □

A key result is the following *orthogonality lemma*.

Lemma 17.2 *Suppose* A_t *is a bounded increasing right-continuous process with* $A_0 = 0$, *a.s.,* $\widetilde{A}_t$ *is the compensator of* A, *and* $M_t = A_t - \widetilde{A}_t$. *Suppose* N_t *is a right continuous square integrable martingale such that* $(\Delta N_t)(\Delta M_t) = 0$ *for all* t. *Then* $\mathbb{E} M_\infty N_\infty = 0$.

Proof By Lemma 17.1, M is square integrable. Suppose

$$H(s, \omega) = K(\omega) 1_{(a,b]}(s)$$

with K being $\mathcal{F}_a$ measurable. Since M_t is of bounded variation, we have (this is a Lebesgue–Stieltjes integral here)

$$\mathbb{E} \int_0^\infty H_s \, dM_s = \mathbb{E} [K(M_b - M_a)] = \mathbb{E} [K\mathbb{E} [M_b - M_a \mid \mathcal{F}_a]] = 0.$$

We saw in Lemma 10.1 that linear combinations of such H's generate the predictable σ-field. Thus by linearity and taking limits, $\mathbb{E} \int_0^\infty H_s \, dM_s = 0$ if H_s is a predictable process such that $\mathbb{E} \int_0^\infty |H_s| \, |dM_s| < \infty$. In particular, since N_{s-} is left continuous and hence predictable, $\mathbb{E} \int_0^\infty N_{s-} \, dM_s = 0$, provided we check integrability:

$$\mathbb{E} \left| \int_0^\infty |N_{s-}| \, |dM_s| \right| \leq \mathbb{E} \int_0^\infty (\sup_r |N_r|) \, |dM_s|$$
$$= \mathbb{E} [(\sup_r |N_r|) (A_\infty + \widetilde{A}_\infty)] < \infty$$

by the Cauchy–Schwarz inequality.

By hypothesis, $\mathbb{E} \int_0^\infty \Delta N_s \, dM_s = 0$, so $\mathbb{E} \int_0^\infty N_s \, dM_s = 0$. On the other hand, using Proposition 3.14, we see

$$\mathbb{E} M_\infty N_\infty = \mathbb{E} \int_0^\infty N_\infty \, dM_s = \mathbb{E} \int_0^\infty N_s \, dM_s = 0.$$

The proof is complete. □

If we apply the above to $N_{t \wedge T}$, we have $\mathbb{E}\, M_\infty N_T = 0$. If we then condition on $\mathcal{F}_T$,

$$\mathbb{E}\,[M_T N_T] = \mathbb{E}\,[N_T \mathbb{E}\,[M_\infty \mid \mathcal{F}_T]] = \mathbb{E}\,[N_T M_\infty] = 0. \qquad (17.1)$$

The reason for the name "orthogonality lemma" is that by (17.1) and Proposition 9.5, $M_t N_t$ is a martingale. This implies that $\langle M, N \rangle_t$ (which we will define soon, and is defined similarly to the case of continuous martingales) is identically equal to 0.

Let M_t be a square integrable martingale with paths that are right continuous and left limits, so that $\mathbb{E}\, M_\infty^2 < \infty$. For each $i \in \mathbb{Z}$, let $T_{i1} = \inf\{t : |\Delta M_t| \in [2^i, 2^{i+1})\}$, $T_{i2} = \inf\{t > T_{i1} : |\Delta M_t| \in [2^i, 2^{i+1})\}$, and so on; i can be both positive and negative. Since M_t is right continuous with left limits, for each i, $T_{ij} \to \infty$ as $j \to \infty$. We conclude that M_t has at most countably many jumps. Next we decompose each T_{ij} into predictable and totally inaccessible parts by Proposition 16.1. We relabel the jump times as $S_1, S_2, \ldots$ so that each S_k is either predictable or totally inaccessible, the graphs of the S_k are disjoint, M has a jump at each time S_k and only at these times, and $|\Delta M_{S_k}|$ is bounded for each k; of the proof of Proposition 16.23. We do not assume that $S_{k_1} \leq S_{k_2}$ if $k_1 \leq k_2$, and in general it would not be possible to arrange this.

If S_i is a totally inaccessible stopping time, let

$$A_i(t) = \Delta M_{S_i} 1_{(t \geq S_i)} \qquad (17.2)$$

and

$$M_i(t) = A_i(t) - \widetilde{A}_i(t), \qquad (17.3)$$

where $\widetilde{A}_i$ is the compensator of A_i. $A_i(t)$ is the process that is 0 up to time S_i and then jumps an amount ΔM_{S_i}; thereafter it is constant. By Corollary 16.31, $\widetilde{A}$ is continuous. If S_i is a predictable stopping time, let

$$M_i(t) = \Delta M_{S_i} 1_{(t \geq S_i)}. \qquad (17.4)$$

By Corollary 16.21, M_i is a martingale. Note that in either case, $M - M_i$ has no jump at time S_i.

Theorem 17.3 *Suppose M is a square integrable martingale and we define M_i as in (17.3) and (17.4).*

(1) Each M_i is square integrable.

(2) $\sum_{i=1}^\infty M_i(\infty)$ converges in L^2.

(3) If $M_t^c = M_t - \sum_{i=1}^\infty M_i(t)$, then M^c is square integrable and we can find a version that has continuous paths.

(4) For each i and each stopping time T, $\mathbb{E}\,[M_T^c M_i(T)] = 0$.

Proof (1) If S_i is a totally inaccessible stopping time and we let $B_t = (\Delta M_{S_i})^+ 1_{(t \geq S_i)}$ and $C_t = (\Delta M_{S_i})^- 1_{(t \geq S_i)}$, then (1) follows by Lemma 17.1. If S_i is predictable, (1) follows by Corollary 16.21.

(2) Let $V_n(t) = \sum_{i=1}^n M_i(t)$. By the orthogonality lemma (Lemma 17.2), $\mathbb{E}\,[M_i(\infty) M_j(\infty)] = 0$ if $i \neq j$ and $\mathbb{E}\,[M_i(\infty)(M_\infty - V_n(\infty))] = 0$ if $i \leq n$. We thus

have

$$\sum_{i=1}^{n} \mathbb{E}\, M_i(\infty)^2 = \mathbb{E}\, V_n(\infty)^2$$

$$\leq \mathbb{E}\left[M_\infty - V_n(\infty) \right]^2 + \mathbb{E}\, V_n(\infty)^2$$

$$= \mathbb{E}\left[M_\infty - V_n(\infty) + V_n(\infty) \right]^2$$

$$= \mathbb{E}\, M_\infty^2 < \infty.$$

Therefore the series $\mathbb{E}\, \sum_{i=1}^{n} M_i(\infty)^2$ converges. If $n > m$,

$$\mathbb{E}\left[(V_n(\infty) - V_m(\infty)) \right]^2 = \mathbb{E}\left[\sum_{i=m+1}^{n} M_i(\infty) \right]^2 = \sum_{i=m+1}^{n} \mathbb{E}\, M_i(\infty)^2.$$

This tends to 0 as $n, m \to \infty$, so $V_n(\infty)$ is a Cauchy sequence in L^2, and hence converges.

(3) From (2), Doob's inequalities, and the completeness of L^2, the random variables $\sup_{t\geq 0}[M_t - V_n(t)]$ converge in L^2 as $n \to \infty$. Let $M_t^c = \lim_{n\to\infty}[M_t - V_n(t)]$. There is a sequence n_k such that

$$\sup_{t\geq 0} |(M_t - V_{n_k}(t)) - M_t^c| \to 0, \qquad \text{a.s.}$$

We conclude that the paths of M_t^c are right continuous with left limits. By the construction of the M_i, $M - V_{n_k}$ has jumps only at times S_i for $i > n_k$. We therefore see that M^c has no jumps, i.e., it is continuous.

(4) By the orthogonality lemma and (17.1),

$$\mathbb{E}\left[M_i(T)(M_T - V_n(T)) \right] = 0$$

if T is a stopping time and $i \leq n$. Letting n tend to infinity proves (4). $\qquad \square$

17.2 Stochastic integrals

If M_t is a square integrable martingale, then M_t^2 is a submartingale by Jensen's inequality for conditional expectations. Just as in the case of continuous martingales, we can use the Doob–Meyer decomposition (this time, we use Theorem 16.29 instead of Theorem 9.12) to find a predictable increasing process starting at 0, denoted $\langle M \rangle_t$, such that $M_t^2 - \langle M \rangle_t$ is a martingale.

Let us define

$$[M]_t = \langle M^c \rangle_t + \sum_{s\leq t} |\Delta M_s|^2. \tag{17.5}$$

Here M^c is the continuous part of the martingale M as defined in Theorem 17.3. As an example, if $M_t = P_t - t$, where P_t is a Poisson process with parameter 1, then $M_t^c = 0$ and

$$[M]_t = \sum_{s\leq t} \Delta P_s^2 = \sum_{s\leq t} \Delta P_s = P_t,$$

because all the jumps of P_t are of size one. In this case $\langle M \rangle_t = t$; this follows from Proposition 17.4 below.

In defining stochastic integrals, one could work with $\langle M \rangle_t$, but the process $[M]_t$ is the one that shows up naturally in many formulas, such as the product formula.

Proposition 17.4 $M_t^2 - [M]_t$ *is a martingale.*

Proof By the orthogonality lemma and (17.1) it is easy to see that

$$\langle M \rangle_t = \langle M^c \rangle_t + \sum_i \langle M_i \rangle_t.$$

Since $M_t^2 - \langle M \rangle_t$ is a martingale, we need only show $[M]_t - \langle M \rangle_t$ is a martingale. Since

$$[M]_t - \langle M \rangle_t = \left(\langle M^c \rangle_t + \sum_{s \leq t} |\Delta M_s|^2 \right) - \left(\langle M^c \rangle_t + \sum_i \langle M_i \rangle_t \right),$$

it suffices to show that $\sum_i \langle M_i \rangle_t - \sum_i \sum_{s \leq t} |\Delta M_i(s)|^2$ is a martingale.

By Exercise 17.1

$$M_i(t)^2 = 2 \int_0^t M_i(s-) \, dM_i(s) + \sum_{s \leq t} |\Delta M_i(s)|^2, \tag{17.6}$$

where the first term on the right-hand side is a Lebesgue–Stieltjes integral. If we approximate this integral by a Riemann sum and use the fact that M_i is a martingale, we see that the first term on the right in (17.6) is a martingale. Thus $M_i^2(t) - \sum_{s \leq t} |\Delta M_i(s)|^2$ is a martingale. Since $M_i^2(t) - \langle M_i \rangle_t$ is a martingale, summing over i completes the proof. □

If H_s is of the form

$$H_s(\omega) = \sum_{i=1}^n K_i(\omega) 1_{(a_i, b_i]}(s), \tag{17.7}$$

where each K_i is bounded and $\mathcal{F}_{a_i}$ measurable, define the stochastic integral by

$$N_t = \int_0^t H_s \, dM_s = \sum_{i=1}^n K_i [M_{b_i \wedge t} - M_{a_i \wedge t}].$$

Very similar proofs to those in Chapter 10 show that the left-hand side will be a martingale and (with $[\cdot]$ instead of $\langle \cdot \rangle$), $N_t^2 - [N]_t$ is a martingale.

If H is $\mathcal{P}$ measurable and $\mathbb{E} \int_0^\infty H_s^2 \, d[M]_s < \infty$, approximate H by integrands H_s^n of the form (17.7) so that

$$\mathbb{E} \int_0^\infty (H_s - H_s^n)^2 \, d[M]_s \to 0$$

and define N_t^n as the stochastic integral of H^n with respect to M_t. By almost the same proof as that of Theorem 10.4, the martingales N_t^n converge in L^2. We call the limit $N_t = \int_0^t H_s \, dM_s$ the *stochastic integral* of H with respect to M. A subsequence of the N^n converges uniformly over $t \geq 0$, a.s., and therefore the limit has paths that are right continuous with left limits. The same arguments as those of Theorem 10.4 apply to prove that the stochastic integral is a martingale and

$$[N]_t = \int_0^t H_s^2 \, d[M]_s.$$

A consequence of this last equation is that

$$\mathbb{E} \left(\int_0^t H_s \, dM_s \right)^2 = \mathbb{E} \int_0^t H_s^2 \, d[M]_s. \tag{17.8}$$

17.3 Itô's formula

We will first prove Itô's formula for a special case, namely, we suppose $X_t = M_t + A_t$, where M_t is a square integrable martingale and A_t is a process of bounded variation whose total variation is integrable. The extension to semimartingales without the integrability conditions will be done later in the chapter (in Section 17.5) and is easy. Define $\langle X^c \rangle_t$ to be $\langle M^c \rangle_t$.

Theorem 17.5 *Suppose $X_t = M_t + A_t$, where M_t is a square integrable martingale and A_t is a process with paths of bounded variation whose total variation is integrable. Suppose f is C^2 on $\mathbb{R}$ with bounded first and second derivatives. Then*

$$f(X_t) = f(X_0) + \int_0^t f'(X_{s-}) \, dX_s + \tfrac{1}{2} \int_0^t f''(X_{s-}) \, d\langle X^c \rangle_s \tag{17.9}$$

$$+ \sum_{s \leq t} [f(X_s) - f(X_{s-}) - f'(X_{s-}) \Delta X_s].$$

Proof The proof will be given in several steps. Set

$$S(t) = \int_0^t f'(X_{s-}) \, dX_s, \qquad Q(t) = \tfrac{1}{2} \int_0^t f''(X_{s-}) \, d\langle X^c \rangle_s,$$

and

$$J(t) = \sum_{s \leq t} [f(X_s) - f(X_{s-}) - f'(X_{s-}) \Delta X_s].$$

We use these letters as mnemonics for "stochastic integral term," "quadratic variation term," and "jump term," respectively.

Step 1. Suppose X_t has a single jump at time T which is either a predictable stopping time or a totally inaccessible stopping time and there exists $N > 0$ such that $|\Delta M_T| + |\Delta A_T| \leq N$ a.s.

If T is totally inaccessible, let $C_t = \Delta M_T 1_{(t \geq T)}$ and let $\tilde{C}_t$ be the compensator. If we replace M_t by $M_t - C_t + \tilde{C}_t$ and A_t by $A_t + C_t - \tilde{C}_t$, we may assume that M_t is continuous. If T is a predictable stopping time, replace M_t by $M_t - \Delta M_T 1_{(t \geq T)}$ and A_t by $A_t + \Delta M_T 1_{(t \geq T)}$, and again we may assume M is continuous.

Let $B_t = \Delta X_T 1_{(t \geq T)}$. Set $\widehat{X}_t = X_t - B_t$ and $\widehat{A}_t = A_t - B_t$. Then $\widehat{X}_t = M_t + \widehat{A}_t$ and $\widehat{X}_t$ is a continuous process that agrees with X_t up to but not including time T. We have $\widehat{X}_{s-} = \widehat{X}_s$ and $\Delta \widehat{X}_s = 0$ if $s \leq T$. By Theorem 11.1

$$f(\widehat{X}_t) = f(\widehat{X}_0) + \int_0^t f'(\widehat{X}_s) \, d\widehat{X}_s + \tfrac{1}{2} \int_0^t f''(\widehat{X}_s) \, d\langle M \rangle_s$$

$$= f(\widehat{X}_0) + \int_0^t f'(\widehat{X}_{s-}) \, d\widehat{X}_s + \tfrac{1}{2} \int_0^t f''(\widehat{X}_{s-}) \, d\langle \widetilde{X}^c \rangle_s$$

$$+ \sum_{s \leq t} [f(\widehat{X}_s) - f(\widehat{X}_{s-}) - f'(\widehat{X}_{s-}) \Delta \widehat{X}_s],$$

since the sum on the last line is zero. For $t < T$, $\widehat{X}_t$ agrees with X_t. At time T, $f(X_t)$ has a jump of size $f(X_T) - f(X_{T-})$. The integral with respect to $\widehat{X}$, $S(t)$, will jump $f'(X_{T-})\Delta X_T$, $Q(t)$ does not jump at all, and $J(t)$ jumps $f(X_T) - f(X_{T-}) - f'(X_{T-})\Delta X_T$. Therefore both sides of (17.9) jump the same amount at time T, and hence in this case we have (17.9) holding for $t \leq T$.

Step 2. Suppose there exist times $T_1 < T_2 < \cdots$ with $T_n \to \infty$, each T_i is either a totally inaccessible stopping time or a predictable stopping time, for each i, there exists $N_i > 0$ such that $|\Delta M_{T_i}|$ and $|\Delta A_{T_i}|$ are bounded by N_i, and X_t is continuous except at the times $T_1, T_2, \ldots$ Let $T_0 = 0$.

Fix i for the moment. Define $X'_t = X_{(t-T_i)^+}$, define A'_t and M'_t similarly, and apply Step 1 to X' at time $T_i + t$. We have for $T_i \leq t \leq T_{i+1}$

$$f(X_t) = f(X_{T_i}) + \int_{T_i}^t f'(X_{s-})\, dX_s + \tfrac{1}{2} \int_{T_i}^t f''(X_{s-})\, d\langle X^c \rangle_s$$
$$+ \sum_{T_i < s \leq t} [f(X_s) - f(X_{s-}) - f'(X_{s-})\Delta X_s].$$

Thus for any t we have

$$f(X_{T_{i+1}\wedge t}) = f(X_{T_i \wedge t}) + \int_{T_i \wedge t}^{T_{i+1}\wedge t} f'(X_{s-})\, dX_s + \tfrac{1}{2} \int_{T_i \wedge t}^{T_{i+1}\wedge t} f''(X_{s-})\, d\langle X^c \rangle_s$$
$$+ \sum_{T_i \wedge t < s \leq T_{i+1}\wedge t} [f(X_s) - f(X_{s-}) - f'(X_{s-})\Delta X_s].$$

Summing over i, we have (17.9) for each t.

Step 3. We now do the general case. As in the paragraphs preceding Theorem 17.3, we can find stopping times $S_1, S_2, \ldots$ such that each jump of X occurs at one of the times S_i and so that for each i, there exists $N_i > 0$ such that $|\Delta M_{S_i}| + |\Delta A_{S_i}| \leq N_i$. Moreover each S_i is either a predictable stopping time or a totally inaccessible stopping time. Let M be decomposed into M^c and M_i as in Theorem 17.3 and let

$$A^c_t = A_t - \sum_{i=1}^{\infty} \Delta A_{S_i} 1_{(t \geq S_i)}.$$

Since A_t is of bounded variation, then A^c will be finite and continuous. Define

$$M^n_t = M^c_t + \sum_{i=1}^n M_i(t)$$

and

$$A^n_t = A^c_t + \sum_{i=1}^n \Delta A_{S_i} 1_{(t \geq S_i)},$$

and let $X^n_t = M^n_t + A^n_t$. We already know that M^n converges uniformly over $t \geq 0$ to M in L^2. If we let $B^n_t = \sum_{i=1}^n (\Delta A_{S_i})^+ 1_{(t \geq S_i)}$ and $C^n_t = \sum_{i=1}^n (\Delta A_{S_i})^- 1_{(t \geq S_i)}$ and let $B_t = \sup_n B^n_t$, $C_t = \sup_n C^n_t$, then the fact that A has paths of bounded variation implies that with probability one, $B_t^n \to B_t$ and $C_t^n \to C_t$ uniformly over $t \geq 0$ and $A_t = B_t - C_t$. In particular, we have convergence in total variation norm:

$$\mathbb{E} \int_0^\infty |d(A_t^n) - A_t)| \to 0.$$

We define $S^n(t)$, $Q^n(t)$, and $J^n(t)$ analogously to $S(t)$, $Q(t)$, and $J(t)$, respectively. By applying Step 2 to X^n, we have

$$f(X_t^n) = f(X_0^n) + S^n(t) + Q^n(t) + J^n(t),$$

and we need to show convergence of each term. We now examine the various terms.

Uniformly in t, X_t^n converges to X_t in probability, that is,

$$\mathbb{P}(\sup_{t \geq 0} |X_t^n - X_t| > \varepsilon) \to 0$$

as $n \to \infty$ for each $\varepsilon > 0$. Since $\int_0^t d\langle M^c\rangle_s < \infty$, by dominated convergence

$$\int_0^t f''(X_{s-}^n) \, d\langle M^c\rangle_s \to \int_0^t f''(X_{s-}) \, d\langle M^c\rangle_s$$

in probability. Therefore $Q^n(t) \to Q(t)$ in probability. Also, $f(X_t^n) \to f(X_t)$ and $f(X_0^n) \to f(X_0)$, both in probability.

We now show $S^n(t) \to S(t)$. Write

$$\int_0^t f'(X_{s-}^n) \, dA_s^n - \int_0^t f'(X_{s-}) \, dA_s$$

$$= \left[\int_0^t f'(X_{s-}^n) \, dA_s^n - \int_0^t f'(X_{s-}^n) \, dA_s \right]$$

$$+ \left[\int_0^t f'(X_{s-}^n) \, dA_s - \int_0^t f'(X_{s-}) \, dA_s \right]$$

$$= I_1^n + I_2^n.$$

We see that

$$|I_1^n| \leq \|f'\|_\infty \int_0^t |dA_s^n - dA_s| \to 0$$

as $n \to \infty$, while by dominated convergence, $|I_2^n|$ also tends to 0.

We next look at the stochastic integral part of $S^n(t)$.

$$\int_0^t f'(X_{s-}^n) \, dM_s^n - \int_0^t f'(X_{s-}) \, dM_s$$

$$= \left[\int_0^t f'(X_{s-}^n) \, dM_s^n - \int_0^t f'(X_{s-}) \, dM_s^n \right]$$

$$+ \left[\int_0^t f'(X_{s-}) \, dM_s^n - \int_0^t f'(X_{s-}) \, dM_s \right]$$

$$= I_3^n + I_4^n.$$

The L^2 norm of I_3^n is bounded by

$$\mathbb{E} \int_0^t |f'(X_{s-}^n) - f'(X_{s-})|^2 \, d[M^n]_s \le \mathbb{E} \int_0^t |f'(X_{s-}^n) - f'(X_{s-})|^2 \, d[M]_s,$$

which goes to zero by dominated convergence. Also

$$I_4^n = \int_0^t f'(X_{s-}) \sum_{i=n+1}^\infty dM_i(s),$$

so using the orthogonality lemma (Lemma 17.2), the L^2 norm of I_4^n is less than

$$\|f'\|_\infty^2 \sum_{i=n+1}^\infty \mathbb{E} [M_i]_\infty \le \|f'\|_\infty^2 \sum_{i=n+1}^\infty \mathbb{E} M_i(\infty)^2,$$

which goes to zero as $n \to \infty$.

Finally, we look at the convergence of J^n. The idea here is to break both $J(t)$ and $J^n(t)$ into two parts, the jumps that might be relatively large (jumps at times S_i for $i \le N$ where N will be chosen appropriately) and the remaining jumps. Let $N > 1$ be chosen later.

$$
\begin{aligned}
J(t) - J^n(t) &= \sum_{s \le t} [f(X_s) - f(X_{s-}) - f'(X_{s-})\Delta X_s] \\
&\quad - \sum_{s \le t} [f(X_s^n) - f(X_{s-}^n) - f'(X_{s-}^n)\Delta X_s^n] \\
&= \sum_{\{i:S_i \le t\}} [f(X_{S_i}) - f(X_{S_i-}) - f'(X_{S_i-})\Delta X_{S_i}] \\
&\quad - \sum_{\{i:S_i \le t\}} [f(X_{S_i}^n) - f(X_{S_i-}^n) - f'(X_{S_i-}^n)\Delta X_{S_i}^n] \\
&= \sum_{\{i>N:S_i \le t\}} [f(X_{S_i}) - f(X_{S_i-}) - f'(X_{S_i-})\Delta X_{S_i}] \\
&\quad - \sum_{\{i>N:S_i \le t\}} [f(X_{S_i}^n) - f(X_{S_i-}^n) - f'(X_{S_i-}^n)\Delta X_{S_i}^n] \\
&\quad + \sum_{\{i\le N,S_i \le t\}} \Big\{ [f(X_{S_i}) - f(X_{S_i-}) - f'(X_{S_i-})\Delta X_{S_i}] \\
&\qquad\qquad - [f(X_{S_i}^n) - f(X_{S_i-}^n) - f'(X_{S_i-}^n)\Delta X_{S_i}^n] \Big\} \\
&= I_5^N - I_6^{n,N} + I_7^{n,N}.
\end{aligned}
$$

By the fact that M and A are right continuous with left limits, $|\Delta M_{S_i}| \le 1/2$ and $|\Delta A_{S_i}| \le 1/2$ if i is large enough (depending on ω), and then $|\Delta X_{S_i}| \le 1$, and also

$$
\begin{aligned}
|\Delta X_{S_i}|^2 &\le 2|\Delta M_{S_i}|^2 + 2|\Delta A_{S_i}|^2 \\
&\le 2|\Delta M_{S_i}|^2 + |\Delta A_{S_i}|.
\end{aligned}
$$

We have

$$|I_5^N| \leq \|f''\|_\infty \sum_{i > N, S_i \leq t} (\Delta X_{S_i})^2$$

and

$$|I_6^{n,N}| \leq \|f''\|_\infty \sum_{n \geq i > N, S_i \leq t} (\Delta X_{S_i})^2.$$

Since $\sum_{i=1}^\infty |\Delta M_{S_i}|^2 \leq [M]_\infty < \infty$ and $\sum_{i=1}^\infty |\Delta A_{S_i}| < \infty$, then given $\varepsilon > 0$, we can choose N large such that

$$\mathbb{P}(|I_5^N| + |I_6^{n,N}| > \varepsilon) < \varepsilon.$$

Once we choose N, we then see that $I_7^{n,N}$ tends to zero in probability as $n \to \infty$, since X_t^n converges in probability to X_t uniformly over $t \geq 0$. We conclude that $J^n(t)$ converges to $J(t)$ in probability as $n \to \infty$.

This completes the proof. $\qquad\square$

17.4 The reduction theorem

Let M be a process adapted to $\{\mathcal{F}_t\}$. If there exist stopping times T_n increasing to ∞ such that each process $M_{t \wedge T_n}$ is a uniformly integrable martingale, we say M is a *local martingale*. If each $M_{t \wedge T_n}$ is a square integrable martingale, we say M is a *locally square integrable martingale*. We say a stopping time T *reduces* a process M if $M_{t \wedge T}$ is a uniformly integrable martingale.

Lemma 17.6 *(1) The sum of two local martingales is a local martingale.*

(2) If S and T both reduce M, then so does $S \vee T$.

(3) If there exist times $T_n \to \infty$ such that $M_{t \wedge T_n}$ is a local martingale for each n, then M is a local martingale.

Proof (1) If the sequence S_n reduces M and the sequence T_n reduces N, then $S_n \wedge T_n$ will reduce $M + N$.

(2) $M_{t \wedge (S \vee T)}$ is bounded in absolute value by $|M_{t \wedge T}| + |M_{t \wedge S}|$. Both $\{|M_{t \wedge T}|\}$ and $\{|M_{t \wedge S}|\}$ are uniformly integrable families of random variables. Now use Proposition A.17.

(3) Let S_{nm} be a family of stopping times reducing $M_{t \wedge T_n}$ and let $S'_{nm} = S_{nm} \wedge T_n$. Renumber the stopping times into a single sequence $R_1, R_2, \dots$ and let $H_k = R_1 \vee \cdots \vee R_k$. Note $H_k \uparrow \infty$. To show that H_k reduces M, we need to show that R_i reduces M and use (2). But $R_i = S'_{nm}$ for some m, n, so $M_{t \wedge R_i} = M_{t \wedge S_{nm} \wedge T_n}$ is a uniformly integrable martingale. $\qquad\square$

Let M be a local martingale with $M_0 = 0$. We say that a stopping time T *strongly reduces* M if T reduces M and the martingale $\mathbb{E}[|M_T| \mid \mathcal{F}_s]$ is bounded on $[0, T)$, that is, there exists $K > 0$ such that

$$\sup_{0 \leq s < T} \mathbb{E}[|M_T| \mid \mathcal{F}_s] \leq K, \qquad \text{a.s.}$$

Lemma 17.7 *(1) If T strongly reduces M and $S \leq T$, then S strongly reduces M.*

(2) If S and T strongly reduce M, then so does $S \vee T$.

(3) If Y_∞ is integrable, then $\mathbb{E}[\mathbb{E}[Y_\infty \mid \mathcal{F}_T] \mid \mathcal{F}_S] = \mathbb{E}[Y_\infty \mid \mathcal{F}_{S \wedge T}].$

Proof (1) Note $\mathbb{E}[|M_S| \mid \mathcal{F}_s] \leq \mathbb{E}[|M_T| \mid \mathcal{F}_s]$ by Jensen's inequality, hence S strongly reduces M.

(2) It suffices to show that $\mathbb{E}[|M_{S \vee T}| \mid \mathcal{F}_t]$ is bounded for $t < T$, since by symmetry the same will hold for $t < S$. For $t < T$ this expression is bounded by

$$\mathbb{E}[|M_T| \mid \mathcal{F}_t] + \mathbb{E}[|M_S|1_{(S>T)} \mid \mathcal{F}_t].$$

The first term is bounded since T strongly reduces M. For the second term, if $t < T$,

$$\begin{aligned}
1_{(t<T)}\mathbb{E}[|M_S|1_{(S>T)} \mid \mathcal{F}_t] &= \mathbb{E}[|M_S|1_{(S>T)}1_{(t<T)} \mid \mathcal{F}_t] \\
&\leq \mathbb{E}[|M_S|1_{(t<S)} \mid \mathcal{F}_t] \\
&= \mathbb{E}[|M_S| \mid \mathcal{F}_t]1_{(t<S)},
\end{aligned}$$

which in turn is bounded since S strongly reduces M.

(3) This is Exercise 3.10. $\qquad\square$

Lemma 17.8 *If M is a local martingale with $M_0 = 0$, then there exist stopping times $T_n \uparrow \infty$ that strongly reduce M.*

Proof Let $R_n \uparrow \infty$ be a sequence reducing M. Let

$$S_{nm} = R_n \wedge \inf\{t : \mathbb{E}[|M_{R_n}| \mid \mathcal{F}_t] \geq m\}.$$

Arrange the stopping times S_{nm} into a single sequence $\{U_n\}$ and let $T_n = U_1 \vee \cdots \vee U_n$. In view of the preceding lemmas, we need to show U_i strongly reduces M, which will follow if S_{nm} does for each n and m.

Let $Y_t = \mathbb{E}[|M_{R_n}| \mid \mathcal{F}_t]$, where we take a version whose paths are right continuous with left limits. Y is bounded by m on $[0, S_{nm})$. By Jensen's inequality for conditional expectations and Lemma 17.7

$$\begin{aligned}
\mathbb{E}[|M_{S_{nm}}|1_{(t<S_{nm})} \mid \mathcal{F}_t] &\leq \mathbb{E}[\mathbb{E}[|M_{R_n}| \mid \mathcal{F}_{S_{nm}}]1_{(t<S_{nm})} \mid \mathcal{F}_t] \\
&= \mathbb{E}[\mathbb{E}[|M_{R_n}|1_{(t<S_{nm})} \mid \mathcal{F}_{S_{nm}}] \mid \mathcal{F}_t] \\
&= \mathbb{E}[|M_{R_n}|1_{(t<S_{nm})} \mid \mathcal{F}_{S_{nm} \wedge t}] \\
&= Y_{S_{nm} \wedge t}1_{(t<S_{nm})} \\
&= Y_t 1_{(t<S_{nm})} \leq m.
\end{aligned}$$

We are done. $\qquad\square$

Our main theorem of this section is the following.

Theorem 17.9 *Suppose M is a local martingale. Then there exist stopping times $T_n \uparrow \infty$ such that $M_{t \wedge T_n} = U_t^n + V_t^n$, where each U^n is a square integrable martingale and each V^n is a martingale whose paths are of bounded variation and such that the total variation of the paths of V_n is integrable. Moreover, $U_t = U_T$ and $V_t = V_T$ for $t \geq T$.*

The last sentence of the statement of the theorem says that U and V are both constant from time T on.

Proof It suffices to prove that if M is a local martingale with $M_0 = 0$ and T strongly reduces M, then $M_{t \wedge T}$ can be written as $U + V$ with U and V of the described form. Thus we may assume $M_t = M_T$ for $t \geq T$, $|M_T|$ is integrable, and $\mathbb{E}\,[\,|M_T|\,\mid\,\mathcal{F}_t]$ is bounded, say by K, on $[0, T)$.

Let $A_t = M_T 1_{(t \geq T)} = M_t 1_{(t \geq T)}$, let $\widetilde{A}$ be the compensator of A, let $V = A - \widetilde{A}$, and let $U = M - A + \widetilde{A}$. Then V is a martingale of bounded variation. We compute the expectation of the total variation of V. Let $B_t = M_T^+ 1_{(t \geq T)}$ and $C_t = M_T^- 1_{(t \geq T)}$. Then the expectation of the total variation of A is bounded by $\mathbb{E}\,|M_T| < \infty$ and the expectation of the total variation of $\widetilde{A}$ is bounded by

$$\mathbb{E}\,\widetilde{B}_\infty + \mathbb{E}\,\widetilde{C}_\infty = \mathbb{E}\,B_\infty + \mathbb{E}\,C_\infty \leq \mathbb{E}\,|M_T| < \infty.$$

We need to show U is square integrable. Note

$$|M_t - A_t| = |M_t| 1_{(t<T)} = |\mathbb{E}\,[M_\infty \mid \mathcal{F}_t]|\,1_{(t<T)}$$
$$= |\mathbb{E}\,[\mathbb{E}\,[M_\infty \mid \mathcal{F}_{T \vee t}]\,\mid \mathcal{F}_t]|\,1_{(t<T)} = |\mathbb{E}\,[M_{T \vee t} \mid \mathcal{F}_t]|\,1_{(t<T)}$$
$$= |\mathbb{E}\,[M_T \mid \mathcal{F}_t]|\,1_{(t<T)} \leq \mathbb{E}\,[\,|M_T|\,\mid \mathcal{F}_t]1_{(t<T)} \leq K.$$

Therefore it suffices to show $\widetilde{A}$ is square integrable.

Our hypotheses imply that $\mathbb{E}\,[M_T^+ \mid \mathcal{F}_t]$ is bounded by K on $[0, T)$, and by Proposition 16.33, $\mathbb{E}\,\widetilde{B}_\infty^2 < \infty$. Similarly, $\mathbb{E}\,\widetilde{C}_\infty^2 < \infty$. Since $A = B - C$, then $\widetilde{A} = \widetilde{B} - \widetilde{C}$, and it follows that $\sup_{t \geq 0} \widetilde{A}_t$ is square integrable. $\qquad\square$

17.5 Semimartingales

We define a *semimartingale* to be a process of the form $X_t = X_0 + M_t + A_t$, where X_0 is finite, a.s., and is $\mathcal{F}_0$ measurable, M_t is a local martingale, and A_t is a process whose paths have bounded variation on $[0, t]$ for each t.

If M_t is a local martingale, let T_n be a sequence of stopping times as in Theorem 17.9. We set $M_{t \wedge T_n}^c = (U^n)_t^c$ for each n and

$$[M]_{t \wedge T_n} = \langle M^c \rangle_{t \wedge T_n} + \sum_{s \leq t \wedge T_n} \Delta M_s^2.$$

It is easy to see that these definitions are independent of how we decompose M into $U^n + V^n$ and of which sequence of stopping times T_n strongly reducing M we choose. We define $\langle X^c \rangle_t = \langle M^c \rangle_t$ and define

$$[X]_t = \langle X^c \rangle_t + \sum_{s \leq t} \Delta X_s^2.$$

We say an adapted process H is *locally bounded* if there exist stopping times $S_n \uparrow \infty$ and constants K_n such that on $[0, S_n]$ the process H is bounded by K_n. If X_t is a semimartingale and H is a locally bounded predictable process, define $\int_0^t H_s\, dX_s$ as follows. Let $X_t = X_0 + M_t + A_t$. If $R_n = T_n \wedge S_n$, where the T_n are as in Theorem 17.9 and the S_n are the stopping times used in the definition of locally bounded, set $\int_0^{t \wedge R_n} H_s\, dM_s$ to be the stochastic integral as defined in Section 17.2. Define $\int_0^{t \wedge R_n} H_s\, dA_s$ to be the usual Lebesgue–Stieltjes integral. Define the stochastic integral with respect to X as the sum of these two. Since $R_n \uparrow \infty$, this defines $\int_0^t H_s\, dX_s$ for all t. One needs to check that the definition does not depend on the decomposition of X into M and A nor on the choice of stopping times R_n.

We now state the general Itô formula.

Theorem 17.10 *Suppose X is a semimartingale and f is C^2. Then*

$$f(X_t) = f(X_0) + \int_0^t f'(X_{s-})\, dX_s + \frac{1}{2}\int_0^t f''(X_{s-})\, d\langle X^c\rangle_s$$

$$+ \sum_{s\le t}[f(X_s) - f(X_{s-}) - f'(X_{s-})\Delta X_s].$$

Proof First suppose f has bounded first and second derivatives. Let T_n be stopping times strongly reducing M_t, let $S_n = \inf\{t : \int_0^t |dA_s| \ge n\}$, let $R_n = T_n \wedge S_n$, and let $X_t^n = X_{t\wedge R_n} - \Delta A_{R_n}$. Since the total variation of A_t is bounded on $[0, R_n)$, it follows that X^n is a semimartingale which is the sum of a square integrable martingale and a process whose total variation is integrable. We apply Theorem 17.5 to this process. X_t^n agrees with X_t on $[0, R_n)$. As in the proof of Theorem 17.5, by looking at the jump at time R_n, both sides of Itô's formula jump the same amount at time R_n, and so Itô's formula holds for X_t on $[0, R_n]$. If we now only assume that f is C^2, we approximate f by a sequence f_m of functions that are C^2 and whose first and second derivatives are bounded, and then let $m \to \infty$; we leave the details to the reader. Thus Itô's formula holds for t in the interval $[0, R_n]$ and for f without the assumption of bounded derivatives. Finally, we observe that $R_n \to \infty$, so except for a null set, Itô's formula holds for each t. $\square$

The proof of the following corollary is similar to the proof of Itô's formula.

Corollary 17.11 *If $X_t = (X_t^1, \ldots, X_t^d)$ is a process taking values in $\mathbb{R}^d$ such that each component is a semimartingale, and f is a C^2 function on $\mathbb{R}^d$, then*

$$f(X_t) = f(X_0) + \int_0^t \sum_{i=1}^d \frac{\partial f}{\partial x_i}(X_{s-})\, dX_s^i$$

$$+ \frac{1}{2}\int_0^t \sum_{i,j=1}^d \frac{\partial^2 f}{\partial x_i \partial x_j}(X_{s-})\, d\langle (X^i)^c, (X^j)^c\rangle_s$$

$$+ \sum_{s\le t}\left[f(X_s) - f(X_{s-}) - \sum_{i=1}^d \frac{\partial f}{\partial x_i}(X_{s-})\Delta X_s^i\right].$$

If X and Y are real-valued semimartingales, define

$$[X, Y]_t = \frac{1}{2}([X + Y]_t - [X]_t - [Y]_t). \tag{17.10}$$

The following corollary is the product formula for semimartingales with jumps.

Corollary 17.12 *If X and Y are semimartingales of the above form,*

$$X_t Y_t = X_0 Y_0 + \int_0^t X_{s-}\, dY_s + \int_0^t Y_{s-}\, dX_s + [X, Y]_t.$$

Proof Apply Theorem 17.10 with $f(x) = x^2$. Since in this case

$$f(X_s) - f(X_{s-}) - f'(X_{s-})\Delta X_s = \Delta X_s^2,$$

we obtain

$$X_t^2 = X_0^2 + 2 \int_0^t X_{s-} \, dX_s + [X]_t. \tag{17.11}$$

Applying (17.11) with X replaced by Y and by $X + Y$ and using

$$X_t Y_t = \tfrac{1}{2}[(X_t + Y_t)^2 - X_t^2 - Y_t^2]$$

gives our result. $\square$

17.6 Exponential of a semimartingale

A function with finite total variation is *purely discontinuous* if it has no continuous part, i.e., $a(t) = \sum_{s \le t} \Delta a(s)$.

Theorem 17.13 *Let X_t be a semimartingale. Define*

$$Z_t = Z_0 \exp \left(X_t - \tfrac{1}{2} \langle X^c \rangle_t \right) \prod_{0 \le s \le t} (1 + \Delta X_s) e^{-\Delta X_s}. \tag{17.12}$$

Then Z_t is a semimartingale, $\prod_{0 \le s \le t} (1 + \Delta X_s) e^{-\Delta X_s}$ is a process of bounded variation whose paths are purely discontinuous, and Z_t satisfies

$$Z_t = Z_0 + \int_0^t Z_{s-} \, dX_s. \tag{17.13}$$

Proof Since the product of finitely many functions of bounded variation which are purely discontinuous will give a function of the same type and in each finite interval there are only finitely many jumps of X_t of size larger in absolute value than $1/2$, it suffices to consider

$$V_t' = \prod_{0 \le s \le t} (1 + \Delta X_s) e^{-\Delta X_s} 1_{(|\Delta X_s| \le 1/2)}.$$

Note

$$\log V_t' = \sum_{s \le t} (\log(1 + \Delta X_s) - \Delta X_s) 1_{(|\Delta X_s| \le 1/2)},$$

which is bounded in absolute value by a constant times $\sum_{s \le t} \Delta X_s^2 < \infty$. Exercise 17.4 tells us that $V_t' = \exp(\log V_t')$ is a purely discontinuous process, and consequently V is also.

We apply the multivariate version of Itô's formula (Corollary 17.11). Let $f(x, y) = e^x y$ and let $Z_t = f(K_t, V_t)$ where $K_t = X_t - \tfrac{1}{2} \langle X^c \rangle_t$. We obtain

$$Z_t - Z_0 = \int_0^t Z_{s-} dK_s + \int_0^t e^{K_{s-}} dV_s + \tfrac{1}{2} \int_0^t Z_{s-} d\langle K^c \rangle_t$$
$$+ \sum_{s \le t} [Z_s - Z_{s-} - Z_{s-} \Delta K_s - e^{-K_{s-}} \Delta V_s]$$
$$= I_1 + I_2 + I_3 + I_4.$$

We have

$$I_1 = \int_0^t Z_{s-} \, dX_s - \tfrac{1}{2} \int_0^t Z_{s-} \, d\langle X^c \rangle_s.$$

Since V_t is purely discontinuous,

$$I_2 = \sum_{s \leq t} e^{K_{s-}} \Delta V_s.$$

Since $K^c = X^c$,

$$I_3 = \tfrac{1}{2} \int_0^t Z_{s-} \, d\langle X^c \rangle_s.$$

Note that $Z_s = Z_{s-}(1 + \Delta X_s)$ and $Z_{s-} \Delta K_s = Z_{s-} \Delta X_s$, so

$$I_4 = -\sum_{s \leq t} e^{K_{s-}} \Delta V_s.$$

Summing yields (17.13).   □

The solution Z of (17.13) is called the *exponential of the semimartingale* X.

17.7 The Girsanov theorem

Let $\mathbb{P}$ and $\mathbb{Q}$ be two equivalent probability measures, that is, $\mathbb{P}$ and $\mathbb{Q}$ are mutually absolutely continuous. Let M_∞ be the Radon–Nikodym derivative of $\mathbb{Q}$ with respect to $\mathbb{P}$ and let $M_t = \mathbb{E}[M_\infty \mid \mathcal{F}_t]$. The martingale M_t is uniformly integrable since $M_\infty \in L^1(\mathbb{P})$. Once a non-negative martingale hits zero, it is easy to see that it must be zero from then on; this is Exercise 17.5. Since $\mathbb{Q}$ and $\mathbb{P}$ are equivalent, then $M_\infty > 0$, a.s., and so M_t never equals zero, a.s. Observe that M_T is the Radon–Nikodym derivative of $\mathbb{Q}$ with respect to $\mathbb{P}$ on $\mathcal{F}_T$.

Let L_t be the local martingale defined by

$$L_t = \int_0^t \frac{1}{M_{s-}} \, dM_s,$$

so that

$$dM_t = M_{t-} \, dL_t,$$

or M is the exponential of L.

Theorem 17.14 *Suppose X is a local martingale with respect to $\mathbb{P}$. Then $X_t - D_t$ is a local martingale with respect to $\mathbb{Q}$, where*

$$D_t = \int_0^t \frac{1}{M_s} \, d[X, M]_s = \int_0^t \frac{M_{s-}}{M_s} \, d[X, L]_s.$$

Note that in the formula for D, we are using a Lebesgue–Stieltjes integral.

Proof Exercise 17.6 tells us that it suffices to show that $M_t(X_t - D_t)$ is a local martingale with respect to $\mathbb{P}$. By Corollary 17.12,

$$d(M(X - D))_t = (X - D)_{t-} \, dM_t + M_{t-} \, dX_t - M_{t-} \, dD_t$$
$$+ d[M, X - D]_t.$$

The first two terms on the right are local martingales with respect to $\mathbb{P}$. Since D is of bounded variation, the continuous part of D is zero, hence

$$[M, D]_t = \sum_{s \le t} \Delta M_s \Delta D_s = \int_0^t \Delta M_s \, dD_s.$$

Thus

$$M_t(X_t - D_t) = \text{ local martingale } + [M, X]_t - \int_0^t M_s \, dD_s.$$

Using the definition of D shows that $M_t(X_t - D_t)$ is a local martingale. $\qquad \square$

Exercises

17.1 Suppose $a(t)$ is a deterministic right-continuous nondecreasing function of t with $a(0) = 0$. Prove the following formulas:

$$a(t)^2 = \int_0^t [(a(t) - a(s)) + (a(t) - a(s-))] \, da(s), \tag{17.14}$$

$$\text{and} \quad a(t)^2 = \int_0^t (2a(s-) + \Delta a(s)) \, da(s)$$

$$= 2 \int_0^t a(s-) \, da(s) + \sum_s (\Delta a(s))^2. \tag{17.15}$$

Hint: First do the case where a has only finitely many discontinuities.

17.2 If A_t is an increasing process and $\widetilde{A}_t$ is its compensator, show that $\widetilde{A}$ jumps only when A does.

17.3 Let P_t^j, $j \in \mathbb{Z}$, be independent Poisson processes with parameter λ_j. Suppose $\lambda_j = \lambda_{-j}$ for each $j \ne 0$. Suppose λ_j decreases as j increases for $j \ge 1$. Let

$$X_t = \sum_{j \in \mathbb{Z}} P_t^j.$$

Determine reasonable conditions on the sequence λ_j so that X is a semimartingale. A local martingale. A martingale. A locally square integrable martingale.

17.4 Show that if $f(t)$ is a purely discontinuous function, then $e^{f(t)}$ is also.

17.5 Suppose M is a non-negative right-continuous martingale and $T = \inf\{t > 0 : M_t = 0\}$. Show that $M_t = 0$ on $(t > T)$.

17.6 Suppose $\mathbb{P}$ and $\mathbb{Q}$ are two equivalent probability measures, M_∞ is the Radon–Nikodym derivative of $\mathbb{Q}$ with respect to $\mathbb{P}$, and $M_t = \mathbb{E}[M_\infty \mid \mathcal{F}_t]$. Show that Y_t is a local martingale with respect to $\mathbb{Q}$ if and only if $Y_t M_t$ is a local martingale with respect to $\mathbb{P}$.

17.7 Suppose X_t is an increasing process with paths that are right continuous with left limits, $X_0 = 0$, a.s., X is purely discontinuous, and all jumps are of size $+1$ only. Suppose $X_t - t$ is a martingale. Prove that X is a Poisson process.

Hint: Imitate the proof of Theorem 12.1. When using Itô's formula, it is important to use the fact that ΔX_t is always 0 or 1.

17.8 Suppose X_t is an increasing process with paths that are right continuous with left limits, $X_0 = 0$, a.s., X is purely discontinuous, and all jumps are of size $+1$ only. Suppose $\lim_{t \to \infty} X_t = \infty$, a.s. Prove that X is a time change of a Poisson process.

17.9 Suppose P_t is a Poisson process with parameter λ, $\{\mathcal{F}_t\}$ is the minimal augmented filtration for P, and $M_t = P_t - \lambda t$. Suppose Y is a $\mathcal{F}_1$ measurable random variable with finite mean and variance. Prove that there exists a predictable process H such that

$$Y = \mathbb{E}\, Y + \int_0^1 H_s \, dM_s.$$

17.10 Let P_1 and P^2 be two independent Poisson processes with the same parameter. Let $X_t = P_t^1 - P_t^2$ and let $\{\mathcal{F}_t\}$ be the minimal augmented filtration for X. Find a bounded mean zero random variable Y that is $\mathcal{F}_1$ measurable which does not satisfy

$$Y = \int_0^1 H_s \, dX_s$$

for any predictable process H.

18

Poisson point processes

Poisson point processes are random measures that are related to Poisson processes. We will use them when we study Lévy processes in Chapter 42. Poisson point processes are also useful in the study of excursions, even excursions of a continuous process such as Brownian motion (see Chapter 27), and they arise when studying stochastic differential equations with jumps.

Let $\mathcal{S}$ be a metric space, $\mathcal{G}$ the collection of Borel subsets of $\mathcal{S}$, and λ a measure on $(\mathcal{S}, \mathcal{G})$.

Definition 18.1 We say a map

$$N : \Omega \times [0, \infty) \times \mathcal{G} \to \{0, 1, 2, \ldots\}$$

(writing $N_t(A)$ for $N(\omega, t, A)$) is a *Poisson point process* if
(1) for each Borel subset A of $\mathcal{S}$ with $\lambda(A) < \infty$, the process $N_t(A)$ is a Poisson process with parameter $\lambda(A)$, and
(2) for each t and ω, $N(t, \cdot)$ is a measure on $\mathcal{G}$.

A model to keep in mind is where $\mathcal{S} = \mathbb{R}$ and λ is a Lebesgue measure. For each ω there is a collection of points $\{(s, z)\}$ (where the collection depends on ω). The number of points in this collection with $s \leq t$ and z in a subset A is $N_t(A)(\omega)$. Since $\lambda(\mathbb{R}) = \infty$, there are infinitely many points in every time interval.

A consequence of the definition is that since $\lambda(\emptyset) = 0$, then $N_t(\emptyset)$ is a Poisson process with parameter 0; in other words, $N_t(\emptyset)$ is identically zero.

Our main result is that $N_t(A)$ and $N_t(B)$ are independent if A and B are disjoint.

Theorem 18.2 *Let $\{\mathcal{F}_t\}$ be a filtration satisfying the usual conditions. Let $\mathcal{S}$ be a metric space furnished with a positive measure λ. Suppose that $N_t(A)$ is a Poisson point process with respect to the measure λ. If $A_1, \ldots, A_n$ are pairwise disjoint measurable subsets of $\mathcal{S}$ with $\lambda(A_k) < \infty$ for $k = 1, \ldots, n$, then the processes $N_t(A_1), \ldots, N_t(A_n)$ are mutually independent.*

Proof We first make the observation that because $N(t, \cdot)$ is a measure and the $A_1, A_2, \ldots, A_n$ are disjoint, then $\sum_{k=1}^{n} N_t(A_k) = N_t(\cup_{k=1}^{n} A_k)$ is a Poisson process with finite parameter. A Poisson process has jumps of size one only, hence no two of the $N_t(A_k)$ have jumps at the same time.

To prove the theorem, it suffices to let $0 = r_0 < r_1 < \cdots < r_m$ and show that the random variables

$$\{N_{r_j}(A_k) - N_{r_{j-1}}(A_k) : 1 \leq j \leq m, 1 \leq k \leq n\}$$

are independent. Since for each j and each k, $N_{r_j}(A_k) - N_{r_{j-1}}(A_k)$ is independent of $\mathcal{F}_{r_{j-1}}$, it suffices to show that for each $j \leq m$, the random variables

$$\{N_{r_j}(A_k) - N_{r_{j-1}}(A_k) : 1 \leq k \leq n\}$$

are independent. We will do the case $j = m = 1$ and write r for r_j for simplicity; the case when $j, m > 1$ differs only in notation.

We will prove this using induction. We start with the case $n = 2$ and show the independence of $N_r(A_1)$ and $N_r(A_2)$. Each $N_t(A_k)$ is a Poisson process, and so $N_t(A_k)$ has moments of all orders. Let $u_1, u_2 \in \mathbb{R}$ and set

$$\phi_k = \lambda(A_k)(e^{iu_k} - 1), \qquad k = 1, 2.$$

Let

$$M_t^k = e^{iu_k N_t(A_k) - t\phi_k}.$$

We see that M_t^k is a martingale because $\mathbb{E}\, e^{iu_k N_t(A_k)} = e^{t\phi_k}$, and therefore

$$\mathbb{E}\,[M_t^k \mid \mathcal{F}_s] = M_s^k \mathbb{E}\,[e^{iu(N_t(A_k) - N_s(A_k)) - (t-s)\phi_k} \mid \mathcal{F}_s]$$
$$= M_s^k e^{-(t-s)\phi_k} \mathbb{E}\,[e^{iu(N_t(A_k) - N_s(A_k))}] = M_s^k,$$

using the independence and stationarity of the increments of a Poisson process.

Since we have argued that no two of the $N_t(A_k)$ jump at the same time, the same is true for the M_t^k and so $[M^j, M^k]_t = 0$ if $j \neq k$. By the product formula (Corollary 17.12) and Itô's formula (Theorem 17.10)

$$M_t^k = 1 - \phi_k \int_0^t e^{iu_k N_{s-}(A_k) - s\phi_k}\, ds + iu_k \int_0^t e^{iu_k N_{s-}(A_k) - s\phi_k}\, dN_s(A_k)$$
$$+ \sum_{s \leq t} e^{iu_k N_{s-}(A_k) - s\phi_k}[e^{iu_k \Delta N_s(A_k)} - 1 - iu_k \Delta N_s(A_k)]$$
$$= 1 - \phi_k \int_0^t e^{iu_k N_{s-}(A_k) - s\phi_k}\, ds + \sum_{s \leq t} e^{iu_k N_{s-}(A_k) - s\phi_k}[e^{iu_k \Delta N_s(A_k)} - 1]$$
$$= 1 - \widetilde{B}_t^k + B_t^k.$$

We see therefore that $M_t^k - 1$ is of the form $B_t^k - \widetilde{B}_t^k$, where B_t^k is a complex-valued process whose paths are locally of bounded variation, and $\widetilde{B}_t^k$ is the compensator of B_t^k.

Let $\overline{M}_t^k = M_{t \wedge r}^k - 1$. Since the M_t^k do not jump at the same time, by the orthogonality lemma (Lemma 17.2), $\mathbb{E}\,\overline{M}_\infty^1 \overline{M}_\infty^2 = 0$, which translates to

$$\mathbb{E}\, M_r^1 M_r^2 = 1.$$

This implies

$$\mathbb{E}\left[e^{i(u_1 N_r(A_1) + u_2 N_r(A_2))}\right] = e^{r\phi_1} e^{r\phi_2} = \mathbb{E}\left[e^{iu_1 N_r(A_1)}\right] \mathbb{E}\left[e^{iu_2 N_r(A_2)}\right].$$

Since this holds for all u_1, u_2, then $N_r(A_1)$ and $N_r(A_2)$ are independent. We conclude that the processes $N_t(A_1)$ and $N_t(A_2)$ are independent.

To handle the case $n = 3$, we first show that $M_t^1 M_t^2$ is a martingale. We write

$$\mathbb{E}\,[M_t^1 M_t^2 \mid \mathcal{F}_s]$$
$$= M_s^1 M_s^2 e^{-(t-s)(\phi_1 + \phi_2)} \mathbb{E}\,[e^{i(u_1(N_t(A_1) - N_s(A_1)) + u_2(N_t(A_2) - N_s(A_2)))} \mid \mathcal{F}_s]$$
$$= M_s^1 M_s^2 e^{-(t-s)(\phi_1 + \phi_2)} \mathbb{E}\,[e^{i(u_1(N_t(A_1) - N_s(A_1)) + u_2(N_t(A_2) - N_s(A_2)))}]$$
$$= M_s^1 M_s^2,$$

using the fact that $N_t(A_1)$ and $N_t(A_2)$ are independent of each other and each have stationary and independent increments.

Note that $M_t^3 = e^{iu_3 N_t(A_3) - t\phi_3}$ has no jumps in common with M_t^1 or M_t^2. Therefore if $\overline{M}_t^3 = M_{t \wedge r}^3$, then

$$\mathbb{E}\,[\overline{M}_\infty^3 (\overline{M}_\infty^1 \overline{M}_\infty^2)] = 0,$$

and as before this leads to

$$\mathbb{E}\,[M_r^3 (M_r^1 M_r^2)] = 1.$$

As above this implies that $N_r(A_1)$, $N_r(A_2)$, and $N_r(A_3)$ are independent. To prove the general induction step is similar. $\qquad\square$

We will also need the following corollary.

Corollary 18.3 *Let $\mathcal{F}_t$ and $N_t(A_k)$ be as in Theorem 18.2. Suppose Y_t is a process with paths that are right continuous with left limits such that $Y_t - Y_s$ is independent of $\mathcal{F}_s$ whenever $s < t$ and $Y_t - Y_s$ has the same law as Y_{t-s} for each $s < t$. Suppose moreover that Y has no jumps in common with any of the $N_t(A_k)$. Then the processes $N_t(A_1), \ldots, N_t(A_n)$, and Y_t are independent.*

Proof The law of Y_0 is the same as that of $Y_t - Y_t$, so $Y_0 = 0$, a.s. By the fact that Y has stationary and independent increments,

$$\mathbb{E}\,e^{iuY_{s+t}} = \mathbb{E}\,e^{iuY_s}\mathbb{E}\,e^{iu(Y_{s+t} - Y_s)} = \mathbb{E}\,e^{iuY_s}\mathbb{E}\,e^{iuY_t},$$

which implies that the characteristic function of Y is of the form $\mathbb{E}\,e^{iuY_t} = e^{t\psi(u)}$ for some function $\psi(u)$.

We fix $u \in \mathbb{R}$ and define

$$M_t^Y = e^{iuY_t - t\psi(u)}.$$

As in the proof of Theorem 18.2, we see that M_t^Y is a martingale. Since M^Y has no jumps in common with any of the M_t^k, if $\overline{M}_t^Y = M_{t \wedge r}^Y$, we see by Lemma 17.2 that

$$\mathbb{E}\,[\overline{M}_\infty^Y (\overline{M}_\infty^1 \cdots \overline{M}_\infty^n)] = 1,$$

or

$$\mathbb{E}\,[M_r^Y M_r^1 \cdots M_r^n] = 1.$$

This leads as above to the independence of Y from all the $N_t(A_k)$'s. $\qquad\square$

We now turn to stochastic integrals with respect to Poisson point processes. In the same way that a nondecreasing function on the reals gives rise to a measure, so $N_t(A)$ gives rise to a random measure $\mu(dt, dz)$ on the product σ-field $\mathcal{B}[0, \infty) \times \mathcal{G}$, where $\mathcal{B}[0, \infty)$ is the Borel σ-field on $[0, \infty)$; μ is determined by

$$\mu([0, t] \times A)(\omega) = N_t(A)(\omega).$$

Define a nonrandom measure ν on $\mathcal{B}[0, \infty) \times \mathcal{G}$ by $\nu([0, t] \times A) = t\lambda(A)$ for $A \in \mathcal{G}$. If $\lambda(A) < \infty$, then $\mu([0, t] \times A) - \nu([0, t] \times A)$ is the same as a Poisson process minus its mean, hence is locally a square integrable martingale.

We can define a stochastic integral with respect to the random measure $\mu - \nu$ as follows. Suppose $H(\omega, s, z)$ is of the form

$$H(\omega, s, z) = \sum_{i=1}^{n} K_i(\omega) 1_{(a_i, b_i]}(s) 1_{A_i}(z), \tag{18.1}$$

where for each i the random variable K_i is bounded and $\mathcal{F}_{a_i}$ measurable and $A_i \in \mathcal{G}$ with $\lambda(A_i) < \infty$. For such H we define

$$N_t = \int_0^t \int H(\omega, s, z) \, d(\mu - \nu)(ds, dz) \tag{18.2}$$

$$= \sum_{i=1}^{n} K_i(\mu - \nu)(((a_i, b_i] \cap [0, t]) \times A_i).$$

Let us assume without loss of generality that the A_i are disjoint. It is not hard to see (Exercise 18.3) that N_t is a martingale, that $N^c = 0$, and that

$$[N]_t = \int_0^t \int H(\omega, s, z)^2 \, \mu(ds, dz). \tag{18.3}$$

Since $\langle N \rangle_t$ must be predictable and all the jumps of N are totally inaccessible, it follows from Proposition 16.30 that $\langle N \rangle_t$ is continuous. Since $[N]_t - \langle N \rangle_t$ is a martingale, we conclude

$$\langle N \rangle_t = \int_0^t \int H(\omega, s, z)^2 \, \nu(ds, dz). \tag{18.4}$$

Suppose $H(s, z)$ is a predictable process in the following sense: H is measurable with respect to the σ-field generated by all processes of the form (18.1). Suppose also that

$$\mathbb{E} \int_0^\infty \int_S H(s, z)^2 \, \nu(ds, dz) < \infty.$$

Take processes H^n of the form (18.1) converging to H in the space L^2 with norm $(\mathbb{E} \int_0^\infty \int_S H^2 \, d\nu)^{1/2}$. The corresponding $N_t^n = \int_0^t H^n(s, z) \, d(\mu - \nu)$ are easily seen to be a Cauchy sequence in L^2, and the limit N_t we call the *stochastic integral of H with respect to $\mu - \nu$*. As in the continuous case, we may prove that $\mathbb{E} N_t^2 = \mathbb{E} [N]_t = \mathbb{E} \langle N \rangle_t$, and it follows from this, (18.3), and (18.4) that

$$[N]_t = \int_0^t \int_S H(s, z)^2 \, \mu(ds, dz), \qquad \langle N \rangle_t = \int_0^t \int_S H(s, z)^2 \, \nu(ds, dz). \tag{18.5}$$

One may think of the stochastic integral as follows: if μ gives unit mass to a point at time t with value z, then N_t jumps at this time t and the size of the jump is $H(t, z)$.

Exercises

18.1 Suppose $\{\mathcal{F}_t\}$ is a filtration satisfying the usual conditions and P_t^1 and P_t^2 are Poisson processes with respect to $\{\mathcal{F}_t\}$ with parameters λ_1, λ_2, respectively. Suppose $P_t^1 + P_t^2$ is a Poisson process with parameter $\lambda_1 + \lambda_2$. Prove that P^1 and P^2 are independent processes.

18.2 Suppose $\{\mathcal{F}_t\}$ is a filtration satisfying the usual conditions, P_t is a Poisson process with respect to $\{\mathcal{F}_t\}$, and W_t is a Brownian motion with respect to $\{\mathcal{F}_t\}$. Show that if $W_t + P_t$ has stationary and independent increments, then P and W are independent processes.

18.3 If H is as in (18.1) and N is defined by (18.2), show that N is a martingale, $N^c = 0$, and $[N]_t$ is given by (18.3).

18.4 Suppose $\{A_s, 0 < s < \infty\}$ is a collection of subsets of S such that $\lambda(A_s) \to \infty$ as $s \to \infty$. Show that $N_t(A_s)/\lambda(A_s)$ converges to t uniformly over finite intervals, where the convergence is in probability.

18.5 Suppose $\{A_s, 0 < s < \infty\}$ is a collection of subsets of S such that $A_r \subset A_s$ if $r \le s$ and $\lambda(A_s) \to \infty$ as $s \to \infty$. Show that for each t,

$$\sup_{u \le t} \left| \frac{N_u(A_s)}{\lambda(A_s)} - u \right|$$

tends to zero almost surely as $s \to \infty$.

18.6 Let S be a metric space and λ a σ-finite measure on S. Construct a Poisson point process which has λ as the corresponding measure.

18.7 Let $P_t^j, j = 1, 2, \ldots$ be independent Poisson processes with parameter β_j. Let $X_t = \sum_{j=1}^{\infty} a_j P_t^j$, where a_j is a sequence such that X_t is finite, a.s. For $A \subset \mathbb{R} \setminus \{0\}$, define $N_t(A)$ to be the number of times before time t that X has a jump whose size is in A:

$$N_t(A) = \sum_{s \le t} 1_A(X_s - X_{s-}).$$

Prove that N_t is a Poisson point process and determine λ.

Framework for Markov processes

It is not uncommon for a Markov process to be defined as a sextuple $(\Omega, \mathcal{F}, \mathcal{F}_t, X_t, \theta_t, \mathbb{P}^x)$, and for additional notation (e.g., ζ, Δ, $\mathcal{S}$, P_t, R_λ, etc.) to be introduced rather rapidly. This can be intimidating for the beginner. We will explain all this notation in as gentle a manner as possible. We will consider a Markov process to be a pair $(X_t, \mathbb{P}^x)$ (rather than a sextuple), where X_t is a single stochastic process and $\{\mathbb{P}^x\}$ is a family of probability measures, one probability measure $\mathbb{P}^x$ corresponding to each element x of the state space.

19.1 Introduction

The idea that a Markov process consists of one process and many probabilities is one that takes some getting used to. To explain this, let us first look at an example. Suppose $X_1, X_2, \ldots$ is a Markov chain with stationary transition probabilities with K states: $1, 2, \ldots, K$. Everything we want to know about X can be determined if we know $p(i, j) = \mathbb{P}(X_1 = j \mid X_0 = i)$ for each i and j and $\mu(i) = \mathbb{P}(X_0 = i)$ for each i. We sometimes think of having a different Markov chain for every choice of starting distribution $\mu = (\mu(1), \ldots, \mu(K))$. But instead let us define a new probability space by taking Ω' to be the collection of all sequences $\omega = (\omega_0, \omega_1, \ldots)$ such that each ω_n takes one of the values $1, \ldots, K$. Define $X_n(\omega) = \omega_n$. Define $\mathcal{F}_n$ to be the σ-field generated by $X_0, \ldots, X_n$; this is the same as the σ-field generated by sets of the form $\{\omega : \omega_0 = a_0, \ldots, \omega_n = a_n\}$, where $a_0, \ldots, a_n \in \{1, 2, \ldots, K\}$. For each $x = 1, 2, \ldots, K$, define a probability measure $\mathbb{P}^x$ on Ω' by

$$\mathbb{P}^x(X_0 = x_0, X_1 = x_1, \ldots X_n = x_n) \tag{19.1}$$
$$= 1_{\{x\}}(x_0) p(x_0, x_1) \cdots p(x_{n-1}, x_n).$$

We have K different probability measures, one for each of $x = 1, 2, \ldots, K$, and we can start with an arbitrary probability distribution μ if we define $\mathbb{P}^\mu(A) = \sum_{i=1}^K \mathbb{P}^i(A)\mu(i)$. We have lost no information by this redefinition, and it turns out this works much better when doing technical details.

The value of $X_0(\omega) = \omega_0$ can be any of $1, 2, \ldots, K$; the notion of starting at x is captured by $\mathbb{P}^x$, not by X_0. The probability measure $\mathbb{P}^x$ is concentrated on those ω's for which $\omega_0 = x$ and $\mathbb{P}^x$ gives no mass to any other ω.

Let us now look at Brownian motion, and see how this framework plays out there. Let $\mathbb{P}$ be a probability measure and let W_t be a one-dimensional Brownian motion with respect to $\mathbb{P}$ started at 0. Then $W_t^x = x + W_t$ is a one-dimensional Brownian motion started at x. Let $\Omega' = C[0, \infty)$ be the set of continuous functions from $[0, \infty)$ to $\mathbb{R}$, so that each element

ω in Ω' is a continuous function. (We do not require that $\omega(0) = 0$ or that $\omega(0)$ take any particular value of x.) Define

$$X_t(\omega) = \omega(t). \tag{19.2}$$

This will be our process. Let $\mathcal{F}$ be the σ-field on $\Omega' = C[0, \infty)$ generated by the cylindrical subsets of $C[0, \infty)$; see Definition 1.1. Now define $\mathbb{P}^x$ to be the law of W^x. This means that $\mathbb{P}^x$ is the probability measure on $(\Omega', \mathcal{F})$ defined by

$$\mathbb{P}^x(X \in A) = \mathbb{P}(W^x \in A), \qquad x \in \mathbb{R}, A \in \mathcal{F}. \tag{19.3}$$

The probability measure $\mathbb{P}^x$ is determined by the fact that if $n \geq 1$, $t_1 \leq \cdots \leq t_n$, and $B_1, \ldots, B_n$ are Borel subsets of $\mathbb{R}$, then

$$\mathbb{P}(X_{t_1} \in B_1, \ldots, X_{t_n} \in B_n) = \mathbb{P}(W^x_{t_1} \in B_1, \ldots, W^x_{t_n} \in B_n).$$

We call the pair $(X_t, \mathbb{P}^x)$, $x \in \mathbb{R}$, $t \geq 0$, a *Brownian motion*.

19.2 Definition of a Markov process

We want to allow our Markov processes to take values in spaces other than the Euclidean ones. For now, we take our state space $\mathcal{S}$ to be a separable metric space, furnished with the Borel σ-field. For the beginner, just think of $\mathbb{R}$ in place of $\mathcal{S}$.

To define a Markov process, we start with a measurable space $(\Omega, \mathcal{F})$ and suppose we have a filtration $\{\mathcal{F}_t\}$ (not necessarily satisfying the usual conditions).

Definition 19.1 A *Markov process* $(X_t, \mathbb{P}^x)$ is a stochastic process

$$X : [0, \infty) \times \Omega \to \mathcal{S}$$

and a family of probability measures $\{\mathbb{P}^x : x \in \mathcal{S}\}$ on $(\Omega, \mathcal{F})$ satisfying the following.

(1) For each t, X_t is $\mathcal{F}_t$ measurable.
(2) For each t and each Borel subset A of $\mathcal{S}$, the map $x \to \mathbb{P}^x(X_t \in A)$ is Borel measurable.
(3) For each $s, t \geq 0$, each Borel subset A of $\mathcal{S}$, and each $x \in \mathcal{S}$, we have

$$\mathbb{P}^x(X_{s+t} \in A \mid \mathcal{F}_s) = \mathbb{P}^{X_s}(X_t \in A), \qquad \mathbb{P}^x - \text{a.s.} \tag{19.4}$$

Some explanation is definitely in order. Let

$$\varphi(x) = \mathbb{P}^x(X_t \in A), \tag{19.5}$$

so that φ is a function mapping $\mathcal{S}$ to $\mathbb{R}$. Part of the definition of filtration given in Chapter 1 is that each $\mathcal{F}_t \subset \mathcal{F}$. Since we are requiring X_t to be $\mathcal{F}_t$ measurable, that means $(X_t \in A)$ is in $\mathcal{F}$ and it makes sense to talk about $\mathbb{P}^x(X_t \in A)$. Definition 19.1(2) says that the function φ is Borel measurable. This is a very mild assumption, and will be satisfied in the examples we look at.

The expression $\mathbb{P}^{X_s}(X_t \in A)$ on the right-hand side of (19.4) is a random variable and its value at $\omega \in \Omega$ is defined to be $\varphi(X_s(\omega))$, with φ given by (19.5). Note that the randomness in $\mathbb{P}^{X_s}(X_t \in A)$ is thus all due to the X_s term and not the X_t term. Definition 19.1(3) can be rephrased as saying that for each s, t, each A, and each x, there is a set $N_{s,t,x,A} \subset \Omega$ that is a null set with respect to $\mathbb{P}^x$ and for $\omega \notin N_{s,t,x,A}$, the conditional expectation $\mathbb{P}^x(X_{s+t} \in A \mid \mathcal{F}_s)$ is equal to $\varphi(X_s)$.

We have now explained all the terms in the sextuple $(\Omega, \mathcal{F}, \mathcal{F}_t, X_t, \theta_t, \mathbb{P}^x)$ except for θ_t. These are called shift operators and are maps from $\Omega \to \Omega$ such that $X_s \circ \theta_t = X_{s+t}$. We defer the precise meaning of the θ_t and the rationale for them until Section 19.5, where they will appear in a natural way.

In the remainder of the section and in Section 19.3 we define some of the additional notation commonly used for Markov processes. The first one is almost self-explanatory. We use $\mathbb{E}^x$ for expectation with respect to $\mathbb{P}^x$. As with $\mathbb{P}^{X_s}(X_t \in A)$, the notation $\mathbb{E}^{X_s} f(X_t)$, where f is bounded and Borel measurable, is to be taken to mean $\psi(X_s)$ with $\psi(y) = \mathbb{E}^y f(X_t)$.

If we want to talk about our Markov process started with distribution μ, we define

$$\mathbb{P}^{\mu}(B) = \int \mathbb{P}^x(B)\, \mu(dx),$$

and similarly for $\mathbb{E}^{\mu}$; here μ is a probability on $\mathcal{S}$.

19.3 Transition probabilities

If $\mathcal{B}$ is the Borel σ-field on a metric space $\mathcal{S}$, a *kernel* $Q(x, A)$ on $\mathcal{S}$ is a map from $\mathcal{S} \times \mathcal{B} \to \mathbb{R}$ satisfying the following.

(1) For each $x \in \mathcal{S}$, $Q(x, \cdot)$ is a measure on $(\mathcal{S}, \mathcal{B})$.

(2) For each $A \in \mathcal{B}$, the function $x \to Q(x, A)$ is Borel measurable.

The definition of Markov transition probabilities (or simply transition probabilities) is the following.

Definition 19.2 A collection of kernels $\{P_t(x, A); t \geq 0\}$ are *Markov transition probabilities* for a Markov process $(X_t, \mathbb{P}^x)$ if

(1) $P_t(x, \mathcal{S}) = 1$ for each $t \geq 0$ and each $x \in \mathcal{S}$.

(2) For each $x \in \mathcal{S}$, each Borel subset A of $\mathcal{S}$, and each $s, t \geq 0$,

$$P_{t+s}(x, A) = \int_{\mathcal{S}} P_t(y, A) P_s(x, dy). \tag{19.6}$$

(3) For each $x \in \mathcal{S}$, each Borel subset A of $\mathcal{S}$, and each $t \geq 0$,

$$P_t(x, A) = \mathbb{P}^x(X_t \in A). \tag{19.7}$$

Definition 19.2(3) can be rephrased as saying that for each x, the measures $P_t(x, dy)$ and $\mathbb{P}^x(X_t \in dy)$ are the same. We define

$$P_t f(x) = \int f(y) P_t(x, dy) \tag{19.8}$$

when $f : \mathcal{S} \to \mathbb{R}$ is Borel measurable and either bounded or non-negative.

Lemma 19.3 *Suppose P_t are Markov transition probabilities. If f is Borel measurable and either non-negative or bounded, then $P_t f$ is non-negative (respectively, bounded) and Borel measurable and*

$$P_t f(x) = \mathbb{E}^x f(X_t), \qquad x \in \mathcal{S}. \tag{19.9}$$

Proof Using (19.7) and Definition 19.1(2), the Borel measurability and (19.9) hold when f is the indicator of a set A. By linearity they hold for simple functions, and then using monotone convergence they hold for non-negative functions. Using linearity again, we have measurability and (19.9) holding for f bounded and Borel measurable. The non-negativity (respectively, the boundedness) of f follows from (19.9). □

The equations (19.6) are known as the *Chapman–Kolmogorov equations*. They can be rephrased in terms of equality of measures: for each x

$$P_{s+t}(x, dz) = \int_{y \in \mathcal{S}} P_t(y, dz) P_s(x, dy). \tag{19.10}$$

Multiplying (19.10) by a bounded Borel measurable function $f(z)$ and integrating gives

$$P_{s+t} f(x) = \int P_t f(y) P_s(x, dy). \tag{19.11}$$

The right-hand side is the same as $P_s(P_t f)(x)$, so we have

$$P_{s+t} f(x) = P_s P_t f(x), \tag{19.12}$$

i.e., the functions $P_{s+t} f$ and $P_s P_t f$ are the same. The equation (19.12) is known as the *semigroup property*.

By Lemma 19.3, P_t is a linear operator on the space of bounded Borel measurable functions on $\mathcal{S}$. We can then rephrase (19.12) simply as

$$P_{s+t} = P_s P_t. \tag{19.13}$$

Operators satisfying (19.13) are called a *semigroup*, and are much studied in functional analysis. We will show in Chapter 36 how to construct the Markov process corresponding to a given semigroup P_t. More about semigroups can also be found in Chapter 37.

One more observation about semigroups: if we take expectations in (19.4), we obtain

$$\mathbb{P}^x(X_{s+t} \in A) = \mathbb{E}^x\left[\mathbb{P}^{X_s}(X_t \in A)\right].$$

The left-hand side is $P_{s+t} 1_A(x)$ and the right-hand side is

$$\mathbb{E}^x[P_t 1_A(X_s)] = P_s P_t 1_A(x),$$

and so (19.4) encodes the semigroup property.

The *resolvent* or λ-*potential* of a semigroup P_t is defined by

$$R_\lambda f(x) = \int_0^\infty e^{-\lambda t} P_t f(x)\, dt, \qquad \lambda \geq 0, \quad x \in \mathcal{S}.$$

This can be recognized as the Laplace transform of P_t. By Lemma 19.3 and the Fubini theorem, we see that

$$R_\lambda f(x) = \mathbb{E}^x \int_0^\infty e^{-\lambda t} f(X_t)\, dt.$$

Resolvents are useful because they are typically easier to work with than semigroups.

When practitioners of stochastic calculus tire of a martingale, they "stop" it. Markov process theorists are a harsher lot and they "kill" their processes. To be precise, attach an isolated point Δ to $\mathcal{S}$. Thus one looks at $\widehat{\mathcal{S}} = \mathcal{S} \cup \Delta$, and the topology on $\widehat{\mathcal{S}}$ is the one generated by the open sets of $\mathcal{S}$ and $\{\Delta\}$. Δ is called the *cemetery point*. All functions on $\mathcal{S}$ are extended to $\widehat{\mathcal{S}}$ by defining them to be 0 at Δ. At some random time ζ the Markov process is killed, which means that $X_t = \Delta$ for all $t \geq \zeta$. The time ζ is called the *lifetime* of the Markov process.

19.4 An example

Let us give an example, that of Brownian motion, of course. Let X_t and $\mathbb{P}^x$ be defined by (19.2) and (19.3). Define $\mathcal{F}_t = \sigma(X_r; r \leq t)$. Clearly Definition 19.1(1) holds. Observe that since, under $\mathbb{P}$, W_t is a mean zero normal random variable with variance t,

$$\mathbb{P}^x(X_t \in A) = \mathbb{P}(W_t^x \in A) = \mathbb{P}(x + W_t \in A) \tag{19.14}$$

$$= \frac{1}{\sqrt{2\pi t}} \int_A e^{-(y-x)^2/2t} \, dy.$$

By dominated convergence, $x \to \mathbb{P}^x(X_t \in A)$ is continuous, therefore measurable. This proves Definition 19.1(2). It remains to prove Definition 19.1(3), which is the following proposition.

Proposition 19.4 *Let W be a Brownian motion as defined by Definition 2.1, let $W_t^x = x + W_t$, and let $(X_t, \mathbb{P}^x)$ be defined by (19.2) and (19.3). If f is bounded and Borel measurable,*

$$\mathbb{E}^x[f(X_{t+s}) \mid \mathcal{F}_s] = \mathbb{E}^{X_s} f(X_t), \qquad \mathbb{P}^x\text{-a.s.} \tag{19.15}$$

Proof We will first prove

$$\mathbb{E}^x[f(X_{t+s}) \mid \mathcal{F}_s] = \mathbb{E}^{X_s} f(X_t) \tag{19.16}$$

when $f(x) = e^{iux}$. Using independent increments and the fact that $W_{t+s} - W_s$ has the same law as W_t, we see that under each $\mathbb{P}^x$, $X_{t+s} - X_s$ is independent of $\mathcal{F}_s$ and has the same law as a mean zero normal random variable with variance t. We conclude that

$$\mathbb{E}^x e^{iu(X_{t+s}-X_s)} = e^{-u^2 t/2};$$

see (A.25). We then write

$$\mathbb{E}^x\left[e^{iuX_{t+s}} | \mathcal{F}_s\right] = \mathbb{E}^x\left[e^{iu(X_{t+s}-X_s)} | \mathcal{F}_s\right] e^{iuX_s}$$

$$= \mathbb{E}^x\left[e^{iu(X_{t+s}-X_s)}\right] e^{iuX_s}$$

$$= e^{-u^2 t/2} e^{iuX_s}.$$

On the other hand, for any y,

$$\mathbb{E}^y e^{iuX_t} = \mathbb{E} \, e^{iuW_t^y} = \mathbb{E} \, e^{iuW_t} e^{iuy} = e^{-u^2 t/2} e^{iuy}.$$

Replacing y by X_s proves (19.16) for this f.

Now suppose that $f \in C^\infty$ with compact support and let $\widehat{f}$ be the Fourier transform of f. In (19.16) we replace u by $-u$, multiply both sides by $\widehat{f}(u)$, and integrate over $u \in \mathbb{R}$. Using the Fourier inversion formula, we then have

$$\mathbb{E}^x[f(X_{t+s}) \mid \mathcal{F}_s] = (2\pi)^{-1}\mathbb{E}^x\Big[\int e^{-iuX_{t+s}}\widehat{f}(u)\,du \mid \mathcal{F}_s\Big]$$

$$= (2\pi)^{-1}\mathbb{E}^{X_s}\int e^{-iuX_t}\widehat{f}(u)\,du$$

$$= \mathbb{E}^{X_s}f(X_t).$$

We used the Fubini theorem several times to interchange expectation and integration; this is justified because f in C^∞ with compact support implies $\widehat{f}$ is in the Schwartz class; see Section B.2. This proves the proposition for f in C^∞ with compact support, and a limit argument gives it for all bounded and measurable f. □

The same proof works for d-dimensional Brownian motion.

Set

$$P_t(x, A) = \mathbb{P}^x(X_t \in A) = \mathbb{P}(W_t + x \in A) = \frac{1}{\sqrt{2\pi t}}\int_A e^{-(y-x)^2/2t}\,dy. \qquad (19.17)$$

Clearly for each x and t, $P_t(x, \cdot)$ is a measure with total mass 1. As we mentioned earlier, the function $x \to P_t(x, A)$ is continuous, hence Borel measurable. We will show the Chapman–Kolmogorov equations. These follow from the next proposition.

Proposition 19.5 *If $s, t > 0$ and $x, z \in \mathbb{R}$, then*

$$\int_{y\in\mathbb{R}} \frac{1}{\sqrt{2\pi t}}e^{-(y-x)^2/2t}\frac{1}{\sqrt{2\pi s}}e^{-(z-y)^2/2s}\,dy \qquad (19.18)$$

$$= \frac{1}{\sqrt{2\pi(s+t)}}e^{-(z-x)^2/2(s+t)}.$$

Proof This is a well-known property of the Gaussian density, but we can derive (19.18) from Proposition 19.4. Let f be continuous with compact support. Taking expectations in (19.15),

$$\mathbb{E}^x f(X_{t+s}) = \mathbb{E}^x[\mathbb{E}^{X_s}f(X_t)],$$

or

$$P_{t+s}f(x) = P_s P_t f(x).$$

Using Lemma 19.3 and (19.17),

$$\int f(x)\frac{1}{\sqrt{2\pi(s+t)}}e^{-(z-x)^2/2(s+t)}\,dx$$

$$= \int f(x)\int \frac{1}{\sqrt{2\pi t}}e^{-(y-x)^2/2t}\frac{1}{\sqrt{2\pi s}}e^{-(z-y)^2/2s}\,dy\,dx.$$

Since this holds for all continuous f with compact support, (19.18) holds for almost every x. Since both sides of (19.18) are continuous in x, then (19.18) holds for all x. □

19.5 The canonical process and shift operators

Suppose we have a Markov process $(X_t, \mathbb{P}^x)$ where $\mathcal{F}_t = \sigma(X_s; s \leq t)$. Suppose for the moment that X_t has continuous paths. For this to even make sense, it is necessary that the set $\{t \to X_t$ is not continuous$\}$ to be in $\mathcal{F}$, and then we require this event to be $\mathbb{P}^x$-null for each x. Define $\widetilde{\Omega}$ to be the set of continuous functions on $[0, \infty)$. If $\widetilde{\omega} \in \widetilde{\Omega}$, set $\widetilde{X}_t = \widetilde{\omega}(t)$. Define $\widetilde{\mathcal{F}}_t = \sigma(\widetilde{X}_s; s \leq t)$ and $\widetilde{\mathcal{F}}_\infty = \vee_{t \geq 0}\widetilde{\mathcal{F}}_t$. Finally define $\widetilde{\mathbb{P}}^x$ on $(\widetilde{\Omega}, \widetilde{\mathcal{F}}_\infty)$ by $\widetilde{\mathbb{P}}^x(\widetilde{X} \in \cdot) = \mathbb{P}^x(X \in \cdot)$. Thus $\widetilde{\mathbb{P}}^x$ is specified uniquely by

$$\widetilde{\mathbb{P}}^x(\widetilde{X}_{t_1} \in A_1, \ldots, \widetilde{X}_{t_n} \in A_n) = \mathbb{P}^x(X_{t_1} \in A_1, \ldots, X_{t_n} \in A_n)$$

for $n \geq 1, A_1, \ldots, A_n$ Borel subsets of $\mathcal{S}$, and $t_1 < \cdots < t_n$. Clearly there is so far no loss (or gain) by looking at the Markov process $(\widetilde{X}_t, \widetilde{\mathbb{P}}^x)$, which is called the *canonical process*.

Let us now suppose we are working with the canonical process, and we drop the tildes everywhere. We define the *shift operators* $\theta_t : \Omega \to \Omega$ as follows. $\theta_t(\omega)$ will be an element of Ω and therefore is a continuous function from $[0, \infty)$ to $\mathcal{S}$. Define

$$\theta_t(\omega)(s) = \omega(t + s).$$

Then

$$X_s \circ \theta_t(\omega) = X_s(\theta_t(\omega)) = \theta_t(\omega)(s) = \omega(t + s) = X_{t+s}(\omega).$$

The shift operator θ_t takes the path of X and chops off and discards the part of the path before time t.

We will use expressions like $f(X_s) \circ \theta_t$. If we apply this to $\omega \in \Omega$, then

$$(f(X_s) \circ \theta_t)(\omega) = f(X_s(\theta_t(\omega))) = f(X_{s+t}(\omega)),$$

or $f(X_s) \circ \theta_t = f(X_{s+t})$.

If the paths of X are not continuous, but instead only right continuous with left limits, we can follow exactly the above procedure, except we start with $\widetilde{\Omega}$ being the collection of functions from $[0, \infty)$ to $\mathcal{S}$ that are right continuous with left limits.

Even if we are not in this canonical setup, from now on we will suppose there exist shift operators mapping Ω into itself so that

$$X_s \circ \theta_t = X_{s+t}.$$

Exercises

19.1 Suppose $(X_t, \mathbb{P}^x)$ is a Brownian motion and $S_t = \sup_{s \leq t} X_s$. Show that $((X_t, S_t), \mathbb{P}^x)$ is a Markov process and determine the transition probabilities.

19.2 Suppose $(X_t, \mathbb{P}^x)$ is a Brownian motion, f a non-negative, bounded, Borel measurable function, and $A_t = \int_0^t f(X_s)\, ds$. Show that $((X_t, A_t), \mathbb{P}^x)$ is a Markov process.

19.3 Suppose P_t is a Poisson process with parameter λ. Let Ω' be the collection of functions on $[0, \infty)$ which are right continuous and which have left limits, let $\mathcal{F}$ be the σ-field on Ω' generated by the cylindrical subsets of Ω', let $P_t^x = x + P_t$, and let $\mathbb{P}^x$ be the law of $x + P$. Show that $(X_t, \mathbb{P}^x)$ is a Markov process and determine the transition probabilities.

19.4 Suppose m is a measure on the Borel subsets $\mathcal{B}$ of a metric space $\mathcal{S}$. Suppose for each $t > 0$ there exist jointly measurable non-negative functions $p_t : \mathcal{S} \times \mathcal{S} \to \mathbb{R}$ such that $\int p_t(x, y) \, m(dy) = 1$ for each x and t and define

$$P_t(x, A) = \int_A p_t(x, y) \, m(dy).$$

Show that the kernels P_t satisfy the Chapman–Kolmogorov equations if and only if

$$\int p_s(x, y) p_t(y, z) \, m(dy) = p_{s+t}(x, z)$$

for every $s, t \geq 0$, every $x \in \mathcal{S}$, and m-almost every z.

19.5 The Ornstein–Uhlenbeck process Y started at x is a continuous Gaussian process with $\mathbb{E}\, Y_t = e^{-t/2} x$ and covariance

$$\mathrm{Cov}\,(Y_s, Y_t) = e^{-(s+t)/2}(e^{s \wedge t} - 1).$$

If X is the canonical process and $\mathbb{P}^x$ is the law of an Ornstein–Uhlenbeck process started at x, show that $(X_t, \mathbb{P}^x)$ is a Markov process and determine the transition probabilities.

Notes

For more, see Blumenthal and Getoor (1968).

Markov properties

We want to accomplish three things in this chapter. First, we want to talk about what it means in the Markov process context for a filtration to satisfy the usual conditions. This is now more complicated than in Chapter 1 because we have more than one probability measure. Second, we want to extend the Markov property to expressions that are more complicated than $\mathbb{E}^x[f(X_{s+t}) \mid \mathcal{F}_s]$. Third, we want to look at the strong Markov property, which means we look at expressions like $\mathbb{E}^x[f(X_{T+t}) \mid \mathcal{F}_T]$, where T is a stopping time.

Throughout this chapter we assume that X has paths that are right continuous with left limits. To be more precise, if

$$N = \{\omega : \text{ the function } t \to X_t(\omega) \text{ is not right continuous with left limits}\},$$

then we assume $N \in \mathcal{F}$ and N is $\mathbb{P}^x$-null for every $x \in \mathcal{S}$.

20.1 Enlarging the filtration

Let us first introduce some notation. Define

$$\mathcal{F}_t^{00} = \sigma(X_s; s \le t), \qquad t \ge 0. \tag{20.1}$$

This is the smallest σ-field with respect to which each X_s is measurable for $s \le t$. We let $\mathcal{F}_t^0$ be the completion of $\mathcal{F}_t^{00}$, but we need to be careful what we mean by completion here, because we have more than one probability measure present. Let $\mathcal{N}$ be the collection of sets that are $\mathbb{P}^x$-null for every $x \in \mathcal{S}$. Thus $N \in \mathcal{N}$ if $(\mathbb{P}^x)^*(N) = 0$ for each $x \in \mathcal{S}$, where $(\mathbb{P}^x)^*$ is the outer probability corresponding to $\mathbb{P}^x$. The outer probability $(\mathbb{P}^x)^*$ is defined by

$$(\mathbb{P}^x)^*(S) = \inf\{\mathbb{P}^x(B) : A \subset B, B \in \mathcal{F}\}.$$

Let

$$\mathcal{F}_t^0 = \sigma(\mathcal{F}_t^{00} \cup \mathcal{N}). \tag{20.2}$$

Finally, let

$$\mathcal{F}_t = \mathcal{F}_{t+}^0 = \cap_{\varepsilon > 0} \mathcal{F}_{t+\varepsilon}^0. \tag{20.3}$$

We call $\{\mathcal{F}_t\}$ the *minimal augmented filtration* generated by X. Ultimately, we will work only with $\{\mathcal{F}_t\}$, but we need the other two filtrations at intermediate stages. The reason for worrying about which filtrations to use is that $\{\mathcal{F}_t^{00}\}$ is too small to include many interesting sets (such as those arising in the law of the iterated logarithm, for example), while if the filtration is too large, the Markov property will not hold for that filtration.

The filtration matters when defining a Markov process; see Definition 19.1(3). We will assume throughout this section that $(X_t, \mathbb{P}^x)$ is a Markov process with respect to the filtration $\{\mathcal{F}_t^{00}\}$, that is,

$$\mathbb{P}^x(X_{s+t} \in A \mid \mathcal{F}_s^{00}) = \mathbb{P}^{X_s}(X_t \in A), \qquad \mathbb{P}^x\text{-a.s.} \tag{20.4}$$

whenever A is a Borel subset of S and $s, t \geq 0$.

We will also make the following assumption, which will be needed here and also in Section 20.3.

Assumption 20.1 *Suppose $P_t f$ is continuous on S whenever f is bounded and continuous on S.*

Markov processes satisfying Assumption 20.1 are called *Feller processes* or *weak Feller processes*. If $P_t f$ is continuous whenever f is bounded and Borel measurable, then the Markov process is said to be a *strong Feller process*.

We show that we can replace $\mathcal{F}_t^{00}$ in (20.4) by $\mathcal{F}_t^0$.

Proposition 20.2 *Let $(X_t, \mathbb{P}^x)$ be a Markov process and suppose that (20.4) holds. If A is a Borel subset of S, $x \in S$, and $s, t \geq 0$, then*

$$\mathbb{P}^x(X_{s+t} \in A \mid \mathcal{F}_s^0) = \mathbb{P}^{X_s}(X_t \in A), \qquad \mathbb{P}^x\text{-a.s.} \tag{20.5}$$

Proof Since the right-hand side is a function of X_s and hence $\mathcal{F}_s^0$ measurable, we need to show that if $B \in \mathcal{F}_s^0$, then

$$\mathbb{P}^x(X_{s+t} \in A, B) = \mathbb{E}^x\Big[\mathbb{P}^{X_s}(X_t \in A); B\Big]. \tag{20.6}$$

This holds for $B \in \mathcal{F}_s^{00}$ by (20.4). It holds for sets $B \in \mathcal{N}$, the class of null sets, since both sides are 0. Therefore it holds for sets B such that there exists $B_1 \in \mathcal{F}_s^{00}$ with $B \triangle B_1$ being a null set. By linearity it holds for finite disjoint unions of sets of the form just described. The class of such finite disjoint unions is a monotone class that generates $\mathcal{F}_s^0$, and our result follows by the monotone class theorem, Theorem B.2. □

The next step is to go from $\mathcal{F}_s^0$ to $\mathcal{F}_s$.

Proposition 20.3 *Let $(X_t, \mathbb{P}^x)$ be a Markov process and suppose that (20.4) holds. If Assumption 20.1 holds and f is a bounded Borel measurable function, then*

$$\mathbb{E}^x[f(X_{s+t}) \mid \mathcal{F}_s] = \mathbb{E}^{X_s} f(X_t), \qquad \mathbb{P}^x\text{-a.s.} \tag{20.7}$$

It will turn out (see Proposition 20.7 below) that $\mathcal{F}_s^0$ is equal to $\mathcal{F}_s$, but we do not know this yet.

Proof We start with (20.5). By linearity, we have

$$\mathbb{E}^x[f(X_{s+t}) \mid \mathcal{F}_s^0] = \mathbb{E}^{X_s} f(X_t), \qquad \mathbb{P}^x\text{-a.s.,} \tag{20.8}$$

when f is a simple random variable, then by monotone convergence when f is non-negative, and then by linearity again, when f is bounded and Borel measurable. In particular, we have this when f is bounded and continuous.

If $B \in \mathcal{F}_s = \mathcal{F}_{s+}^0$, then $B \in \mathcal{F}_{s+\varepsilon}^0$ for every $\varepsilon > 0$. Hence by (20.8) with s replaced by $s + \varepsilon$, if f is bounded and continuous,

$$\mathbb{E}^x[f(X_{s+t+\varepsilon}); B] = \mathbb{E}^x\left[\mathbb{E}^{X_{s+\varepsilon}} f(X_t); B\right]. \tag{20.9}$$

The right-hand side is equal to

$$\mathbb{E}^x[P_t f(X_{s+\varepsilon}); B];$$

since $P_t f$ is continuous and X_t has paths that are right continuous with left limits, this converges to

$$\mathbb{E}^x[P_t f(X_s); B] = \mathbb{E}^x\left[\mathbb{E}^{X_s} f(X_t); B\right]$$

by dominated convergence. The left-hand side of (20.9) converges, using dominated convergence, the continuity of f, and the fact that X has paths that are right continuous with left limits, to

$$\mathbb{E}^x[f(X_{s+t}); B].$$

We therefore have

$$\mathbb{E}^x[f(X_{s+t}); B] = \mathbb{E}^x\left[\mathbb{E}^{X_s} f(X_t); B\right]. \tag{20.10}$$

A limit argument shows this holds whenever f is bounded and measurable. Since B is an arbitrary event in $\mathcal{F}_s$, that completes the proof. $\qquad\square$

Remark 20.4 In Chapter 16, we discussed the fact that the first time a right continuous process whose jump times are totally inaccessible hits a Borel set is a stopping time, provided the filtration satisfies the usual conditions. Even though the notion of completion of a filtration is a bit different in the context of Markov processes, the result is still true. See Blumenthal and Getoor (1968).

20.2 The Markov property

We start with the Markov property given by Proposition 20.3:

$$\mathbb{E}^x[f(X_{s+t}) \mid \mathcal{F}_s] = \mathbb{E}^{X_s}[f(X_t)], \qquad \mathbb{P}^x\text{-a.s.} \tag{20.11}$$

Since $f(X_{s+t}) = f(X_t) \circ \theta_s$, if we write Y for the random variable $f(X_t)$, we have

$$\mathbb{E}^x[Y \circ \theta_s \mid \mathcal{F}_s] = \mathbb{E}^{X_s} Y, \qquad \mathbb{P}^x\text{-a.s.} \tag{20.12}$$

We wish to generalize this to other random variables Y.

Proposition 20.5 *Let* $(X_t, \mathbb{P}^x)$ *be a Markov process and suppose* (20.11) *holds. Suppose* $Y = \prod_{i=1}^n f_i(X_{t_i-s})$, *where the* f_i *are bounded, Borel measurable, and* $s \leq t_1 \leq \cdots \leq t_n$. *Then* (20.12) *holds.*

Proof We will prove this by induction on n. The case $n = 1$ is (20.11), so we suppose the equality holds for n and prove it for $n + 1$.

Let $V = \prod_{j=2}^{n+1} f_j(X_{t_j - t_1})$ and $h(y) = \mathbb{E}^y V$. By the induction hypothesis,

$$\mathbb{E}^x\left[\prod_{j=1}^{n+1} f_j(X_{t_j}) | \mathcal{F}_s\right] = \mathbb{E}^x\left[\mathbb{E}^x[V \circ \theta_{t_1} | \mathcal{F}_{t_1}] f_1(X_{t_1}) | \mathcal{F}_s\right]$$

$$= \mathbb{E}^x\left[(\mathbb{E}^{X_{t_1}} V) f_1(X_{t_1}) | \mathcal{F}_s\right]$$

$$= \mathbb{E}^x[(h f_1)(X_{t_1}) | \mathcal{F}_s].$$

By (20.11) this is $\mathbb{E}^{X_s}[(h f_1)(X_{t_1 - s})]$. For any y,

$$\mathbb{E}^y[(h f_1)(X_{t_1 - s})] = \mathbb{E}^y[(\mathbb{E}^{X_{t_1 - s}} V) f_1(X_{t_1 - s})]$$

$$= \mathbb{E}^y\left[\mathbb{E}^y[V \circ \theta_{t_1 - s} | \mathcal{F}_{t_1 - s}] f_1(X_{t_1 - s})\right]$$

$$= \mathbb{E}^y[(V \circ \theta_{t_1 - s}) f_1(X_{t_1 - s})].$$

If we replace V by its definition, replace y by X_s, and use the definition of $\theta_{t_1 - s}$, we get the desired equality for $n + 1$ and hence the induction step. □

We now come to the general version of the Markov property. As usual, $\mathcal{F}_\infty = \vee_{t \geq 0} \mathcal{F}_t$. The expression $Y \circ \theta_t$ for general Y may seem puzzling at first. We will give some examples when we get to applications of the strong Markov property in Chapter 21.

Theorem 20.6 *Let $(X_t, \mathbb{P}^x)$ be a Markov process and suppose (20.11) holds. Suppose Y is bounded and measurable with respect to $\mathcal{F}_\infty$. Then*

$$\mathbb{E}^x[Y \circ \theta_s | \mathcal{F}_s] = \mathbb{E}^{X_s} Y, \qquad \mathbb{P}^x\text{-a.s.} \tag{20.13}$$

Proof If in Proposition 20.5 we take $f_j(x) = 1_{A_j}(x)$ for Borel measurable A_j, we have

$$\mathbb{E}^x[1_B \circ \theta_s | \mathcal{F}_s] = \mathbb{E}^{X_s} 1_B \tag{20.14}$$

when $B = \{\omega : \omega(t_1) \in A_1, \ldots, \omega(t_n) \in A_n\}$. It is easy to see that the set of B's for which (20.14) holds is a monotone class. By an argument using the monotone class theorem, (20.14) holds for all B that are measurable with respect to $\mathcal{F}_\infty$. Taking linear combinations, (20.13) holds for Y's that are simple random variables. Using monotone convergence, (20.13) holds for non-negative Y's, and then by linearity for bounded Y's. □

Proposition 20.7 *Let $(X_t, \mathbb{P}^x)$ be a Markov process with respect to $\{\mathcal{F}_t\}$. Let $\mathcal{F}_t^0$ and $\mathcal{F}_t$ be defined by (20.2) and (20.3). Then $\mathcal{F}_t = \mathcal{F}_t^0$ for each $t \geq 0$.*

Proof Let $Y_1 = \prod_{i=1}^n f_i(X_{t_i})$ and $Y_2 = \prod_{j=1}^m g_j(X_{u_j})$, where $t_1 < \cdots < t_n \leq s$ and $0 \leq u_1 < \cdots < u_m$ and the f_j and g_j are bounded Borel measurable functions. Then by Proposition 20.5,

$$\mathbb{E}^x[(Y_1)(Y_2 \circ \theta_s) | \mathcal{F}_s] = Y_1 \mathbb{E}^{X_s} Y_2.$$

Since $\mathbb{E}^{X_s} Y_2$ is a function of X_s, then $(Y_1)(\mathbb{E}^{X_s} Y_2)$ is $\mathcal{F}_s^0$ measurable. Using a monotone class argument, we conclude that if Y is bounded and $\mathcal{F}_\infty$ measurable, then $\mathbb{E}^x[Y | \mathcal{F}_s]$ is $\mathcal{F}_s^0$

measurable. Now apply this to $Y = 1_A$ for $A \in \mathcal{F}_s$ to obtain that $1_A = \mathbb{E}^x[1_A \mid \mathcal{F}_s]$ is $\mathcal{F}_s^0$ measurable. $\qquad\qquad\qquad\qquad\qquad\qquad\qquad\qquad\qquad\qquad\qquad\qquad\qquad\qquad\qquad\qquad\qquad\quad \Box$

The following is known as the *Blumenthal 0–1 law*.

Proposition 20.8 *Let $(X_t, \mathbb{P}^x)$ be a Markov process with respect to $\{\mathcal{F}_t\}$. If $A \in \mathcal{F}_0$, then for each x, $\mathbb{P}^x(A)$ is equal to 0 or 1.*

Proof Suppose $A \in \mathcal{F}_0$. Under $\mathbb{P}^x$, $X_0 = x$, a.s., and then

$$\mathbb{P}^x(A) = \mathbb{E}^{X_0} 1_A = \mathbb{E}^x[1_A \circ \theta_0 \mid \mathcal{F}_0] = 1_A \circ \theta_0 = 1_A \in \{0, 1\}, \qquad \mathbb{P}^x\text{-a.s.}$$

since $1_A \circ \theta_0$ is $\mathcal{F}_0$ measurable. Our result follows because $\mathbb{P}^x(A)$ is a real number and not random. $\qquad\qquad\qquad\qquad\qquad\qquad\qquad\qquad\qquad\qquad\qquad\qquad\qquad\qquad\qquad\qquad\qquad\quad \Box$

20.3 Strong Markov property

Given a stopping time T, recall that the σ-field of events known up to time T is defined to be

$$\mathcal{F}_T = \big\{ A \in \mathcal{F}_\infty : A \cap (T \leq t) \in \mathcal{F}_t \text{ for all } t > 0 \big\}.$$

We define θ_T by $\theta_T(\omega)(t) = \omega(T(\omega) + t)$. Thus, for example, $X_t \circ \theta_T(\omega) = X_{T(\omega)+t}(\omega)$ and $X_T(\omega) = X_{T(\omega)}(\omega)$.

Now we can state the strong Markov property. The notation and definition are admittedly a bit opaque at this stage – be patient until we reach the examples in the next chapter.

Theorem 20.9 *Suppose $(X_t, \mathbb{P}^x)$ is a Markov process with respect to $\{\mathcal{F}_t\}$, that Assumption 20.1 holds, and that T is finite stopping time. If Y is bounded and measurable with respect to $\mathcal{F}_\infty$, then*

$$\mathbb{E}^x[Y \circ \theta_T | \mathcal{F}_T] = \mathbb{E}^{X_T} Y, \qquad \mathbb{P}^x\text{-a.s.}$$

Proof Following the proofs of Section 20.2, it is enough to prove

$$\mathbb{E}^x[f(X_{T+t}) | \mathcal{F}_T] = \mathbb{E}^{X_T} f(X_t) \tag{20.15}$$

for f bounded. We can obtain this by a limit argument if we have (20.15) for f bounded and continuous. Define T_n to be equal to $(k+1)/2^n$ on the event $(k/2^n \leq T < (k+1)/2^n)$.

If $A \in \mathcal{F}_T$, then $A \in \mathcal{F}_{T_n}$. Therefore $A \cap (T_n = k/2^n) \in \mathcal{F}_{k/2^n}$ and we have by the Markov property, Theorem 20.6,

$$\begin{aligned}
\mathbb{E}^x[f(X_{T_n+t}); A, T_n = k/2^n] &= \mathbb{E}^x[f(X_{t+k/2^n}); A, T = k/2^n] \\
&= \mathbb{E}^x[\mathbb{E}^{X_{k/2^n}} f(X_t); A, T_n = k/2^n] \\
&= \mathbb{E}^x[\mathbb{E}^{X_{T_n}} f(X_t); A, T_n = k/2^n].
\end{aligned}$$

Then

$$\mathbb{E}^x[f(X_{T_n+t}); A] = \sum_{k=1}^{\infty} \mathbb{E}^x[f(X_{T_n+t}); A, T_n = k/2^n]$$

$$= \sum_{k=1}^{\infty} \mathbb{E}^x\big[\mathbb{E}^{X_{T_n}} f(X_t); A, T_n = k/2^n\big]$$

$$= \mathbb{E}^x[\mathbb{E}^{X_{T_n}} f(X_t); A].$$

Now let $n \to \infty$. $\mathbb{E}^x[f(X_{T_n+t}); A] \to \mathbb{E}^x[f(X_{T+t}); A)]$ by dominated convergence and the continuity of f and the right continuity of X_t. On the other hand, using the continuity of $P_t f$, $\mathbb{E}^{X_{T_n}} f(X_t) = P_t f(X_{T_n}) \to P_t f(X_T) = \mathbb{E}^{X_T} f(X_t)$. Therefore

$$\mathbb{E}^x[f(X_{T+t}); A] = \mathbb{E}^x[\mathbb{E}^{X_T} f(X_t); A]$$

for all $A \in \mathcal{F}_T$, and hence (20.15) holds. □

Recall that we are restricting our attention to Markov processes whose paths are right continuous with left limits. If we have a Markov process $(X_t, \mathbb{P}^x)$ whose paths are right continuous with left limits, which has shift operators $\{\theta_t\}$, and which satisfies the conclusion of Theorem 20.9, whether or not Assumption 20.1 holds, then we say that $(X_t, \mathbb{P}^x)$ is a *strong Markov process*. A strong Markov process is said to be *quasi-left continuous* if $X_{T_n} \to X_T$, a.s., on $\{T < \infty\}$ whenever T_n are stopping times increasing up to T. Unlike in the definition of predictable stopping times given in Chapter 16, we are not requiring the T_n to be strictly less than T. A *Hunt process* is a strong Markov process that is quasi-left continuous. Quasi-left continuity does not imply left continuity; consider the Poisson process.

Proposition 20.10 *If $(X_t, \mathbb{P}^x)$ is a strong Markov process and Assumption 20.1 holds, then X_t is quasi-left continuous.*

Proof First suppose T is bounded, T_n increases to T, $Y = \lim_{n\to\infty} X_{T_n}$, and f and g are bounded and continuous. If $T_n = T$ for some n, then $\lim_{n\to\infty} g(X_{T_n+t}) = g(X_{T+t})$, and if $T_n < T$ for all n, then $\lim_{n\to\infty} g(X_{T_n+t}) = g(X_{(T+t)-})$, where X_{s-} is the left-hand limit at time s. In either case,

$$\lim_{t\to 0} \lim_{n\to\infty} g(X_{T_n+t}) = g(X_T).$$

Then

$$\mathbb{E}^x[f(Y)g(X_T)] = \lim_{t\to 0} \lim_{n\to\infty} \mathbb{E}^x[f(X_{T_n})g(X_{T_n+t})]$$

$$= \lim_{t\to 0} \lim_{n\to\infty} \mathbb{E}^x[f(X_{T_n})P_t g(X_{T_n})]$$

$$= \lim_{t\to 0} \mathbb{E}^x[f(Y)P_t g(Y)] = \mathbb{E}^x[f(Y)g(Y)].$$

By a limit argument we have

$$\mathbb{E}^x[h(Y, X_T)] = \mathbb{E}^x[h(Y, Y)] \tag{20.16}$$

for all bounded measurable functions h on $\mathcal{S} \times \mathcal{S}$. Now take $h(x, y)$ to be zero if $x = y$ and one otherwise. The right-hand side of (20.16) is 0, so the left-hand side is also.

If T is not bounded, apply the argument in the preceding paragraph to the stopping time $T \wedge M$, where M is a positive real, and then let $M \to \infty$. $\square$

Exercises

20.1 Suppose that $\mathcal{S}$ is a locally compact separable metric space and C_0 is the set of continuous functions on $\mathcal{S}$ that vanish at infinity. To say a continuous function f vanishes at infinity means that given $\varepsilon > 0$ there exists a compact set K such that $|f(x)| < \varepsilon$ if $x \notin K$. Show that if Assumption 20.1 is replaced by the assumptions that $P_t f \in C_0$ whenever $f \in C_0$ and $P_t f \to f$ uniformly as $t \to 0$ whenever $f \in C_0$, then the conclusion of Theorem 20.9 still holds.

20.2 Suppose $(X_t, \mathbb{P}^x)$ is a Markov process with respect to a filtration $\{\mathcal{F}_t\}$. Suppose that $\mathcal{E}_t \subset \mathcal{F}_t$ for each t and that X_t is $\mathcal{E}_t$ measurable for each t. Show that $(X_t, \mathbb{P}^x)$ is a Markov process with respect to the filtration $\{\mathcal{E}_t\}$.

20.3 Give an example of a Markov process that is not a strong Markov process.
      Hint: Let the state space be $[0, \infty)$ and starting from $x \in (0, \infty)$, let X move deterministically at constant speed to the right. Starting at 0, let X wait an exponential length of time, and then begin moving at constant speed to the right.

20.4 Let $(X_t, \mathbb{P}^x)$ be Brownian motion and let $\{\mathcal{F}_t\}$ be the minimal augmented filtration. Suppose $B \in \vee_{t \geq 0} \mathcal{F}_t$ and for some $s > 0$ is of the form $1_B = 1_A \circ \theta_s$. Show that if B is a $\mathbb{P}^x$-null set for some x, then it is a $\mathbb{P}^x$-null set for every x.

20.5 Let P_t be transition probabilities for a Poisson process with parameter λ. These are defined in Exercise 19.3. Show that Assumption 20.1 holds.

20.6 Suppose $(X_t, \mathbb{P}^x)$ is a Markov process with transition probabilities P_t, f is a bounded Borel measurable function, $t_0 > 0$, and we define $M_t = P_{t_0 - t} f(X_t)$ for $t \leq t_0$. Show that $(M_t, t \leq t_0)$ is a $\mathbb{P}^x$-martingale for each x.

20.7 Use the Blumenthal 0–1 law to show that if W is a one-dimensional Brownian motion and $T = \inf\{t > 0 : W_t > 0\}$ is the first time Brownian motion hits $(0, \infty)$, then $\mathbb{P}(T = 0) = 1$.

20.8 Let A be a Borel subset of a metric space $\mathcal{S}$. Let $T_A = \inf\{t : X_t \in A\}$, where $(X_t, \mathbb{P}^x)$ is a strong Markov process. Show that $\mathbb{P}^x(T_A = 0)$ is either 0 or 1 for each x.

20.9 Let $(X_t, \mathbb{P}^x)$ be a strong Markov process and let A be a Borel subset of $\mathcal{S}$. We define A^r by setting $A^r = \{x : \mathbb{P}^x(T_A = 0) = 1\}$, where T_A is the first hitting time of A. Thus A^r is the set of points that are regular for A. Prove that for each x,

$$\mathbb{P}^x(X_{T_A} \in A \cup A^r) = 1.$$

21

Applications of the Markov properties

We give some applications of the Markov property and the strong Markov property. In the first application, we show that d-dimensional Brownian motion is transient if $d \geq 3$. Next we consider estimates on additive functionals. (An example of an additive functional is $A_t = \int_0^t f(X_s)\, ds$, where f is a non-negative function on the state space of the Markov process X.) Third is a sufficient criterion for a Markov process to have continuous paths. Finally, we discuss harmonic functions and show how to solve the classical Dirichlet problem of analysis and partial differential equations.

21.1 Recurrence and transience

Let $W_t = (W_1(t), \ldots, W_d(t))$ be a d-dimensional Brownian motion started at 0 with $d \geq 3$ and let $W_t^x = x + W_t$ be Brownian motion started at x. Let $h(y) = |y|^{2-d}$. A direct calculation of derivatives shows that

$$\Delta h(x) = \sum_{i=1}^{d} \frac{\partial^2 h}{\partial x_i^2}(x) = 0, \qquad x \neq 0.$$

(Noting that

$$\frac{\partial}{\partial y_i}|y| = \frac{\partial}{\partial y_i}(y_1^2 + \cdots + y_d^2)^{1/2} = \frac{y_i}{|y|}$$

helps with the calculation.) By Exercise 9.4, $\langle W_i, W_j \rangle_t$ equals 0 if $i \neq j$ and we saw in Section 9.3 that it equals t if $i = j$. Suppose $r < |x| < R$, and let

$$S = \inf\{t : |W_t^x| \leq r \text{ or } |W_t^x| \geq R\}.$$

S is finite, a.s., because $|W_t^x| \geq |W_1(t)| - |x|$ and $W_1(t)$ exits $[-2R, 2R]$ in finite time by Theorem 7.2. By Itô's formula,

$$h(W_{t \wedge S}^x) = h(W_0^x) + \text{martingale} + \frac{1}{2} \int_0^{t \wedge S} \sum_{i=1}^{d} \frac{\partial^2 h}{\partial x_i^2}(W_s^x)\, ds$$

$$= h(x) + \text{martingale}.$$

Therefore $h(W_{t \wedge S}) - h(x)$ is a martingale started at 0. The function h is equal to r^{2-d} on $\partial B(0, r)$, the boundary of $B(0, r)$, and equal to R^{2-d} on $\partial B(0, R)$, the boundary of $B(0, R)$.

By Corollary 3.17, we deduce

$$\mathbb{P}(W_t^x \text{ hits } B(0, r) \text{ before } B(0, R))$$
$$= \mathbb{P}(h(W_t^x) - h(x) \text{ hits } r^{2-d} - |x|^{2-d} \text{ before } R^{2-d} - |x|^{2-d})$$
$$= \frac{|x|^{2-d} - R^{2-d}}{r^{2-d} - R^{2-d}}.$$

If we let $R \to \infty$ and recall that $2 - d < 0$, we see that

$$\mathbb{P}(W_t^x \text{ ever hits } \partial B(0, r)) = \left(\frac{r}{|x|}\right)^{d-2}. \tag{21.1}$$

We want to use the strong Markov property to go from (21.1) to

$$\lim_{t \to \infty} |W_t^x| = \infty.$$

(There are other ways besides the strong Markov property of showing this.) The first step in doing this is to convert to the Markov process notation. Let $(X_t, \mathbb{P}^x)$ be a Brownian motion. What we have shown is that

$$\mathbb{P}^x(X_t \text{ ever hits } \partial B(0, r)) = \left(\frac{r}{|x|}\right)^{d-2}. \tag{21.2}$$

Let $M > 0$ and let

$$S_1 = \inf\{t : |X_t| \geq 2M\},$$
$$T_1 = \inf\{t > S_1 : |X_t| \leq M\},$$
$$S_2 = \inf\{t > T_1 : |X_t| \geq 2M\},$$
$$T_2 = \inf\{t > S_2 : |X_t| \leq M\},$$

and so on. Another way of writing this is to define

$$S = \inf\{t > 0 : |X_t| \geq 2M\}, \qquad T = \inf\{t > 0 : |X_t| \leq M\},$$

and then to let $S_1 = S$, and for each $i \geq 1$,

$$T_i = S_i + T \circ \theta_{S_i}, \qquad S_{i+1} = T_i + S \circ \theta_{T_i}.$$

Let us explain what is going on. Given a path ω, which is a continuous function from $[0, \infty)$ to $\mathbb{R}^d$, $T \circ \theta_{S_i}$ means to proceed along the path until time S_i, disregard this piece, and then see how long it takes after time S_i to first enter $B(0, M)$. If we add the quantity S_i to $T \circ \theta_{S_i}$, we then get the amount of time for X_t to first enter $B(0, M)$ after time S_i. Thus T_i with the shift notation is the same as $\inf\{t > S_i : X_t \in B(0, M)\}$. The shift notation interpretation of S_{i+1} is similar.

Now we can apply the strong Markov property. Since $T_{i+1} = S_{i+1} + T \circ \theta_{S_{i+1}}$, we can write

$$\mathbb{P}^x(T_{i+1} < \infty) = \mathbb{P}^x(S_{i+1} < \infty, T \circ \theta_{S_{i+1}} < \infty)$$
$$= \mathbb{E}^x\left[\mathbb{P}^x(T \circ \theta_{S_{i+1}} < \infty \mid \mathcal{F}_{S_{i+1}}); S_{i+1} < \infty\right]$$
$$= \mathbb{E}^x\left[\mathbb{P}^{X_{S_{i+1}}}(T < \infty); S_{i+1} < \infty\right].$$

At time S_{i+1}, we have $|X_{S_{i+1}}| = 2M$, and by (21.1)

$$\mathbb{P}^{X_{S_{i+1}}}(T < \infty) = (\tfrac{1}{2})^{d-2}.$$

Therefore

$$\mathbb{P}^x(T_{i+1} < \infty) \leq 2^{2-d}\mathbb{P}^x(S_{i+1} < \infty) \leq 2^{2-d}\mathbb{P}^x(T_i < \infty).$$

The last inequality is simply the fact that $S_{i+1} \geq T_i$. Since $\mathbb{P}^x(T_1 < \infty) \leq 1$, induction tells us that

$$\mathbb{P}^x(T_i < \infty) \leq 2^{(2-d)(i-1)} \to 0$$

as $i \to \infty$. Hence $\mathbb{P}^x(T_i < \infty$ for all $i) = 0$. Since T_i increases as i increases, for almost all ω, T_i will be infinite for i sufficiently large (how large will depend on ω). Hence X_t returns to $B(0, M)$ for a last time, a.s. Since M is arbitrary, this proves that X_t tends to ∞ as $t \to \infty$.

We have thus proved

Proposition 21.1 *If $(X_t, \mathbb{P}^x)$ is a d-dimensional Brownian motion and $d \geq 3$, then $|X_t| \to \infty$ as $t \to \infty$ with $\mathbb{P}^x$-probability one for each x.*

21.2 Additive functionals

Let D be a closed subset of $\mathcal{S}$, let $f : D \to [0, \infty)$, let $S = \tau_D$, and let

$$A = \sup_{x \in D} \mathbb{E}^x \int_0^S f(X_s)\, ds,$$

where $\tau_D = \inf\{t > 0 : X_t \notin D\}$ is the first time X exits D.

Proposition 21.2 *If $A < \infty$, then*

$$\sup_{x \in D} \mathbb{P}^x\left(\int_0^S f(X_s)ds \geq 2kA\right) \leq 2^{-k}. \tag{21.3}$$

This is rather remarkable: as soon as one gets a bound on the expectation, although it must be uniform in x, one gets exponential tails for the distribution. A use of Chebyshev's inequality would only give the bound $(2k)^{-1}$.

Proof Let $B_t = \int_0^{t \wedge S} f(X_s)\, ds$. This is a special case of what is known as an additive functional; see Section 22.3. Let $U_1 = \inf\{t : B_t \geq 2A\}$, and let $U_{i+1} = U_i + U_1 \circ \theta_{U_i}$. To explain this formula, composing ω with θ_{U_i} means we disregard the path before time U_i. Thus $U_1 \circ \theta_{U_i}$ is the length of time after time U_i until B_t has increased an amount $2A$ over its value at U_i. Therefore $U_i + U_1 \circ \theta_{U_i}$ is the $(i + 1)$st time B has increased by $2A$. The event $\mathbb{P}^x(B_S \geq 2kA)$ is bounded by

$$\begin{aligned}
\mathbb{P}^x(U_k \leq S) &= \mathbb{P}^x(U_{k-1} \leq S, U_1 \circ \theta_{U_{k-1}} \leq S \circ \theta_{U_{k-1}}) \\
&= \mathbb{E}^x\left[\mathbb{P}^x(U_1 \circ \theta_{U_{k-1}} \leq S \circ \theta_{U_{k-1}} | \mathcal{F}_{U_{k-1}}); U_{k-1} \leq S\right] \\
&= \mathbb{E}^x\left[\mathbb{P}^{X_{U_{k-1}}}(U_1 \leq S); U_{k-1} \leq S\right].
\end{aligned}$$

If $U_{k-1} \le S$, then $X_{U_{k-1}} \in D$. If $y \in D$,

$$\mathbb{P}^y(U_1 \le S) \le \mathbb{P}^y\left(\int_0^S f(X_s)ds \ge 2A\right) \le \frac{\mathbb{E}^y\int_0^S f(X_s)ds}{2A} \le \tfrac{1}{2}$$

by Chebyshev's inequality. Then

$$\mathbb{P}^x(U_k \le S) \le \tfrac{1}{2}\mathbb{P}^x(U_{k-1} \le S)$$

and (21.3) follows by induction. $\qquad\qquad\square$

We give another proof of Proposition 4.5.

Proposition 21.3 *Let W be a one-dimensional Brownian motion. If T is a finite stopping time and $a < b$, then*

$$\mathbb{P}(W_{T+t} \in [a,b] \mid \mathcal{F}_T) \le \frac{b-a}{\sqrt{2\pi t}}, \qquad a.s.$$

Proof Let $(X_t, \mathbb{P}^x)$ be a one-dimensional Brownian motion. If $y \in \mathbb{R}$, then

$$\mathbb{P}^y(X_t \in [a,b]) = \mathbb{P}^0(X_t \in [a-y, b-y]) \qquad (21.4)$$

$$= \frac{1}{\sqrt{2\pi t}} \int_{a-y}^{b-y} e^{-z^2/2t} \, dz \le \frac{b-a}{\sqrt{2\pi t}}.$$

By the strong Markov property,

$$\mathbb{P}(W_{T+t} \in [a,b] \mid \mathcal{F}_T) = \mathbb{P}^0(X_{T+t} \in [a,b] \mid \mathcal{F}_T) = \mathbb{E}^0[1_{[a,b]}(X_t) \circ \theta_T \mid \mathcal{F}_T]$$

$$= \mathbb{E}^{X_T}[1_{[a,b]}(X_t)] = \mathbb{P}^{X_T}(X_t \in [a,b]).$$

Now use (21.4) with y replaced by X_T. $\qquad\qquad\square$

21.3 Continuity

Let us now come up with a criterion for a Markov process to have continuous paths. We assume we have a strong Markov process $(X_t, \mathbb{P}^x)$ whose paths are right continuous with left limits. Let $d(\cdot, \cdot)$ be the metric for the state space $\mathcal{S}$.

Lemma 21.4 *Let $(X_t, \mathbb{P}^x)$ be a strong Markov process with state space $\mathcal{S}$. For all $x \in \mathcal{S}$ and all $\lambda \ge 0$,*

$$\mathbb{P}^x(\sup_{s \le t} d(X_s, x) \ge \lambda) \le 2 \sup_{s \le t} \sup_{y \in \mathcal{S}} \mathbb{P}^y(d(X_s, X_0) \ge \lambda/2).$$

Note that the left-hand side has the supremum inside while the right-hand side has the suprema outside the probability.

Proof Let us use the notation

$$F(t, \lambda) = \sup_{s \le t} \sup_{y \in \mathcal{S}} \mathbb{P}^y(d(X_s, X_0) \ge \lambda). \qquad (21.5)$$

Write $S = \inf\{t : d(X_t, X_0) \geq \lambda\}$. Then by the strong Markov property,

$$\mathbb{P}^x(\sup_{s \leq t} d(X_s, x) \geq \lambda) \leq \mathbb{P}^x(d(X_t, x) \geq \lambda/2) + \mathbb{P}^x(S < t, d(X_t, X_0) \leq \lambda/2)$$

$$\leq F(t, \lambda/2) + \mathbb{E}^x\Big[\mathbb{P}^{X_S}(d(X_{t-S}, X_0) \geq \lambda/2)\Big]$$

$$\leq 2F(t, \lambda/2); \tag{21.6}$$

see Exercise 21.2. □

Proposition 21.5 *Let $(X_t, \mathbb{P}^x)$ be a strong Markov process. With $F(t, \lambda)$ defined as in (21.5), suppose*

$$\frac{F(t, \lambda)}{t} \to 0 \tag{21.7}$$

as $t \to 0$ for each $\lambda > 0$. Then X_t has continuous paths with $\mathbb{P}^x$-probability one for each x.

For X a Brownian motion, $F(t, \lambda) \leq 2e^{-\lambda^2/8t}$ by Proposition 3.15, and hence $F(t, \lambda)/t \to 0$ as $t \to 0$. Thus Brownian motion satisfies (21.7). On the other hand, (21.7) is not satisfied for the Poisson process; see Exercise 21.3.

Proof Suppose $\lambda, t_0 > 0$ and X has a jump of size larger than 4λ at some time before t_0 with positive probability, that is,

$$\mathbb{P}^x(\sup_{t \leq t_0} d(X_{t-}, X_t) \geq 4\lambda) > 0,$$

where $X_{t-} = \lim_{s \uparrow t, s < t} X_s$. Then for each n there exists $k \leq [t_0 2^n] + 1$ such that

$$\sup_{s,t \in [k/2^n, (k+1)/2^n]} d(X_s, X_t) \geq 4\lambda;$$

$[x]$ is the largest integer less than or equal to x. Therefore there exists $k \leq [t_0 2^n] + 1$ such that

$$\sup_{s \in [k/2^n, (k+1)/2^n]} d(X_s, X_{k/2^n}) \geq 2\lambda.$$

But by Lemma 21.4

$$\mathbb{P}^x(\exists k \leq [t_0 2^n] + 1 : \sup_{k/2^n \leq s \leq (k+1)/2^n} d(X_s, X_{k/2^n}) \geq 2\lambda)$$

$$\leq ([t_0 2^n] + 1) \sup_y \mathbb{P}^y(\sup_{s \leq 2^{-n}} d(X_s, X_0) \geq 2\lambda)$$

$$\leq 2([t_0 2^n] + 1)F(2^{-n}, \lambda)$$

for every n. In the first inequality we used the Markov property at time $k/2^n$ and the fact that there are at most $[t_0 2^n] + 1$ intervals. Letting $n \to \infty$, we see the probability of a jump of size larger than 4λ before time t_0 must be zero. Since λ and t_0 are arbitrary, the paths of X are continuous. □

21.4 Harmonic functions

Suppose $(X_t, \mathbb{P}^x)$ is a continuous Markov process satisfying the strong Markov property, and for each x, the sets of paths are right continuous with left limits with $\mathbb{P}^x$-probability one. Let

D be an open subset of S, and suppose that $\tau_D < \infty$, a.s., with respect to each $\mathbb{P}^x$, where $\tau_D = \inf\{t : X_t \notin D\}$ is the time of the first exit from D. Let f be a bounded measurable function on ∂D, the boundary of D.

Proposition 21.6 *Define*

$$h(x) = \mathbb{E}^x f(X_{\tau_D})$$

and $\mathcal{F}'_s = \mathcal{F}_{s \wedge \tau_D}$. Then for each x, $h(X_{t \wedge \tau_D})$ is a martingale under $\mathbb{P}^x$ with respect to the filtration $\{\mathcal{F}'_t\}$.

Proof   Let $s < t$. Consider a path ω starting at x and continuing until it exits D at time $\tau_D(\omega)$. If we have $u \leq \tau_D$ and we cut off the first u time units of the path, we have a path going from $X_u(\omega)$ and proceeding until it exits D. But note that the point at which it exits will not be changed by cutting off a piece from the beginning of the path. Therefore $X_{\tau_D} \circ \theta_u = X_{\tau_D}$ if $u \leq \tau_D$. Using this,

$$\begin{aligned}
\mathbb{E}^x[h(X_{t \wedge \tau_D}) \mid \mathcal{F}_{s \wedge \tau_D}] &= \mathbb{E}^x\Big[\mathbb{E}^{X_{t \wedge \tau_D}} f(X_{\tau_D}) \mid \mathcal{F}_{s \wedge \tau_D}\Big] \\
&= \mathbb{E}^x[\mathbb{E}^x[f(X_{\tau_D}) \circ \theta_{t \wedge \tau_D} \mid \mathcal{F}_{t \wedge \tau_D}] \mid \mathcal{F}_{s \wedge \tau_D}] \\
&= \mathbb{E}^x[f(X_{\tau_D}) \mid \mathcal{F}_{s \wedge \tau_D}] \\
&= \mathbb{E}^x[f(X_{\tau_D}) \circ \theta_{s \wedge \tau_D} \mid \mathcal{F}_{s \wedge \tau_D}] \\
&= \mathbb{E}^{X_{s \wedge \tau_D}} f(X_{\tau_D}) = h(X_{s \wedge \tau_D}),
\end{aligned}$$

as required. $\square$

This becomes particularly interesting in the case when X_t is a d-dimensional Brownian motion. Suppose D is a bounded domain (i.e., a bounded open subset) in $\mathbb{R}^d$. There exists M such that $D \subset B(0, M)$. We know X_t^1, the first component of X_t is a one-dimensional Brownian motion, and by Theorem 7.2, X_t^1 will exit $[-M, M]$ in finite time, no matter what X_0^1 is. Therefore the time for X_t to exit D will be finite almost surely with respect to each $\mathbb{P}^x$. Take $x \in D$ and take δ smaller than the distance from x to the boundary of D. If $S = \inf\{t : |X_t - x| \geq \delta\}$, the first time X leaves the ball of radius δ about x, then by Proposition 21.6 and optional stopping, we have

$$h(x) = \mathbb{E}^x h(X_S). \tag{21.8}$$

By Exercise 2.3 we know that d-dimensional Brownian motion is rotationally invariant. We conclude from this that the location where a Brownian motion hits the boundary of a ball of radius δ about the starting point must have a uniform distribution. Hence X_S will be uniformly distributed on $\partial B(x, \delta)$. Thus (21.8) can be rewritten as

$$h(x) = \int_{\partial B(x, \delta)} h(y)\, \sigma_{x, \delta}(dy),$$

where $\sigma_{x, \delta}$ is a surface measure on $\partial B(x, \delta)$ normalized to have total mass one. This holds for every δ small enough, and since h is bounded (because f is), it can be shown that h is C^2

in D and is harmonic there:

$$\Delta h(x) = \sum_{i=1}^{d} \frac{\partial^2 h}{\partial x_i^2}(x) = 0;$$

the proof is not obvious – see Bass (1995), Section II.1.

We can use Proposition 21.6 to give a solution to the Dirichlet problem. In the Dirichlet problem one is given a domain in $\mathbb{R}^d$ and a continuous function f on the boundary of D. One wants to find a continuous function h that is harmonic inside D, that is, $\Delta h(x) = 0$ for $x \in D$, and that agrees with f on ∂D. There are domains for which one cannot solve the Dirichlet problem, but a solution can be found provided the domain is moderately nice. We explain how to solve the Dirichlet problem probabilistically; the class of domains where one can do this is the same as the class where one can solve the Dirichlet problem analytically.

Let us say that a point x is *regular* for a Borel subset A if $\mathbb{P}^x(T_A = 0) = 1$, where $T_A = \inf\{t > 0 : X_t \in A\}$. Thus a point x is regular for a set A if starting at x the Brownian motion enters A immediately. For example, a consequence of Theorem 7.2 is that the point 0 is regular for the set $A = (0, \infty)$ when we have a one-dimensional Brownian motion.

Theorem 21.7 *Suppose D is a bounded open domain in $\mathbb{R}^d$ and f is a function on ∂D that is continuous on ∂D. Let $(X_t, \mathbb{P}^x)$ be a d-dimensional Brownian motion and $\tau_D = \inf\{t : X_t \in D^c\}$. If each point of ∂D is regular for D^c, then $h(x) = \mathbb{E}^x f(X_{\tau_D})$ is a solution to the Dirichlet problem.*

The regularity condition says that starting at any point of ∂D, Brownian motion enters D^c immediately. Uniqueness of the solution to the Dirichlet problem is easy, and we do not address this here.

Proof We have already seen in Proposition 21.6 and the remarks immediately following the proof of that proposition that h is harmonic in D. This implies that h is continuous in D. Thus we only need to show that h agrees with f on ∂D.

Our first step is to fix t and ε and to show that the set

$$\{x : \mathbb{P}^x(\tau_D \le t) > 1 - \varepsilon\}$$

is an open set. Let $s < t$, define $\varphi_s(x) = \mathbb{P}^x(\tau_D \le t - s)$, and let

$$w_s(x) = \mathbb{P}^x(X_u \in D^c \text{ for some } u \in [s, t]).$$

By the Markov property at time s,

$$w_s(x) = \mathbb{E}^x \mathbb{P}^{X_s}(X_u \in D^c \text{ for some } u \in [0, t-s]) = \mathbb{E}^x[\mathbb{P}^{X_s}(\tau_D \le t - s)]$$
$$= \mathbb{E}^x \varphi_s(X_s) = (2\pi s)^{-d/2} \int \varphi_s(y) e^{-|x-y|^2/2s} \, dy.$$

By dominated convergence, the last integral is a continuous function of x. If

$$w_0(x) = \mathbb{P}^x(X_u \in D^c \text{ for some } u \in [0, t]),$$

then $w_s(x) \uparrow w_0(x)$, so $\{x : w_0(x) > 1 - \varepsilon\} = \cup_{s \in (0,t)}\{x : w_s(x) > 1 - \varepsilon\}$ is open.

Let $z \in \partial D$. Let $\varepsilon > 0$ and choose η such that $|f(w) - f(z)| < \varepsilon$ if $|w - z| < \eta$ and $w \in \partial D$. Pick t small so that $\mathbb{P}^0(\sup_{s \le t} |X_s| > \eta/2) < \varepsilon$; this is possible because Brownian motion has continuous paths. Because $z \in \partial D$ and every point of ∂D is regular for D^c, $\mathbb{P}^z(\tau_D \leq t) = 1$. Finally choose $\delta < (\eta/2) \wedge \varepsilon$ so that if $|w - z| < \delta$ and $w \in D$, then $\mathbb{P}^w(\tau_D \leq t) > 1 - \varepsilon$.

Now if $|w - z| < \delta$ and $w \in D$, then

$$\mathbb{P}^w(|X_{\tau_D} - z| < \eta) \geq \mathbb{P}^w(\tau_D \leq t, \sup_{s \leq t} |X_s - w| \leq \eta/2)$$

$$\geq \mathbb{P}^w(\tau_D \leq t) - \mathbb{P}^0(\sup_{s \leq t} |X_s| > \eta/2)$$

$$\geq (1 - \varepsilon) - \varepsilon.$$

The set ∂D is a bounded and closed subset of $\mathbb{R}^d$, hence compact, and since f is continuous on ∂D, there exists M such that $|f|$ is bounded by M. If $|w - z| < \delta$ and $w \in D$,

$$|h(w) - f(z)| = |\mathbb{E}^w f(X_{\tau_D}) - f(z)|$$

$$\leq |\mathbb{E}^w[f(X_{\tau_D}); |X_{\tau_D} - z| < \eta] - f(z)\mathbb{P}^w(|X_{\tau_D} - z| < \eta)|$$

$$+ 2M\mathbb{P}^w(|X_{\tau_D} - z| \geq \eta)$$

$$\leq \varepsilon \mathbb{P}^w(|X_{\tau_D} - z| < \eta) + 4M\varepsilon \leq (1 + 4M)\varepsilon.$$

We used the fact that $|f(X_{\tau_D}) - f(z)| < \varepsilon$ if $|X_{\tau_D} - z| < \eta$. Since ε is arbitrary, this proves that $h(w) \to f(z)$ as $w \to z$ inside D. $\qquad\square$

Let us give a sufficient condition for a point to be regular for a domain D. Let $\widetilde{V}_a = \{(x_1, \ldots, x_d) : x_1 > 0, (x_2^2 + \cdots + x_d^2) < a^2 x_1^2\}$. The vertex of $\widetilde{V}_a$ is the origin. A cone V in $\mathbb{R}^d$ is a translation and rotation of $\widetilde{V}_a$ for some a.

The following is known as the *Poincaré cone condition*.

Proposition 21.8 *Suppose there exists a cone V with vertex $y \in \partial D$ such that $V \cap B(y, r) \subset D^c$ for some $r > 0$. Then y is regular for D^c.*

Proof By translation and rotation of the coordinates, we may suppose $y = 0$ and $V = \widetilde{V}_a$ for some a. Then for each t,

$$\mathbb{P}^0(\tau_D \leq t) \geq \mathbb{P}^0(X_t \in D^c) \geq \mathbb{P}^0(X_t \in V \cap B(0, r))$$

$$\geq \mathbb{P}^0(X_t \in V) - \mathbb{P}^0(X_t \notin B(0, r)).$$

By scaling, the last term is $\mathbb{P}^0(X_1 \in V) - \mathbb{P}^0(X_1 \notin B(0, r/\sqrt{t}))$, which converges to

$$\mathbb{P}^0(X_1 \in V) = (2\pi)^{-d/2} \int_V e^{-|z|^2/2} \, dz > 0$$

as $t \to 0$. Observe $\mathbb{P}^0(\tau_D \leq t)$ converges to $\mathbb{P}^0(\tau_D = 0)$. By the Blumenthal 0–1 law (Proposition 20.8), $\mathbb{P}^0(\tau_D = 0) = 1$. $\qquad\square$

Continue to suppose $(X_t, \mathbb{P}^x)$ is a d-dimensional Brownian motion and D is a bounded domain, but now we suppose $d \geq 3$. Define

$$U(x, A) = \mathbb{E}^x \int_0^\infty 1_A(X_s) \, ds, \qquad x \in D.$$

This is the same as the λ-resolvent of 1_A with $\lambda = 0$. We write

$$U(x, A) = \mathbb{E}^x \int_0^\infty 1_A(X_s), ds$$

$$= \int_0^\infty \mathbb{P}^x(X_s \in A) \, ds$$

$$= \int_0^\infty \int_A \frac{1}{(2\pi s)^{d/2}} e^{-|y-x|^2/2s} \, dy \, ds$$

$$= \int_A \int_0^\infty \frac{1}{(2\pi s)^{d/2}} e^{-|y-x|^2/2s} \, ds \, dy.$$

Some calculus shows that the inside integral is equal to $c|x - y|^{2-d}$. If we denote $c|x - y|^{2-d}$ by $u(x, y)$, we then have that

$$U(x, A) = \int_A u(x, y) \, dy. \tag{21.9}$$

The expression $u(x, y)$ is called the *Newtonian potential density*. Note that $u(x, y)$ is a function only of $|x - y|$, it blows up as $|x - y| \to 0$, and tends to 0 as $|x - y| \to \infty$.

If x is in the interior of D, then $u(x, \cdot)$ will be bounded on ∂D. Define $h_x(z) = \mathbb{E}^z u(x, X_{\tau_D})$; we saw above that h_x is harmonic. Now define $g_D(x, y) = u(x, y) - h_x(y)$; this function of two variables is called the *Green's function* or *Green function* for D with pole at x. This is a well-known object in analysis – let us give a probabilistic interpretation. Since $u(x, y)$ is symmetric in x and y, if $A \subset D$ we have

$$\int_A g_D(x, y) \, dx = \int_A u(x, y) \, dx - \int_A \mathbb{E}^y u(x, X_{\tau_D}) \, dx \tag{21.10}$$

$$= \mathbb{E}^y \int_0^\infty 1_A(X_s) \, ds - \mathbb{E}^y \int_A u(x, X_{\tau_D}) \, dx$$

$$= \mathbb{E}^y \int_0^\infty 1_A(X_s) \, ds - \mathbb{E}^y \left[\mathbb{E}^{X_{\tau_D}} \int_0^\infty 1_A(X_s) \, ds \right].$$

Using the strong Markov property and then a change of variables,

$$\mathbb{E}^y \left[\mathbb{E}^{X_{\tau_D}} \int_0^\infty 1_A(X_s) \, ds \right] = \mathbb{E}^y \left[\mathbb{E}^y \left[\int_0^\infty 1_A(X_s) \circ \theta_{\tau_D} \, ds \mid \mathcal{F}_{\tau_D} \right] \right]$$

$$= \mathbb{E}^y \int_0^\infty 1_A(X_s) \circ \theta_{\tau_D} \, ds$$

$$= \mathbb{E}^y \int_0^\infty 1_A(X_{\tau_D+s}) \, ds$$

$$= \mathbb{E}^y \int_{\tau_D}^\infty 1_A(X_s) \, ds.$$

Substituting this in (21.10) we have

$$\int_A g_D(x, y) \, dx = \mathbb{E}^y \int_0^{\tau_D} 1_A(X_s) \, ds.$$

For this reason g_D is sometimes called the *occupation time density* for D.

Exercises

21.1   Suppose $d = 2$, $(X_t, \mathbb{P}^x)$ is a two-dimensional Brownian motion, and $r > 0$. Imitate the argument of Proposition 21.1 but with $h(x) = \log(|x|)$ to show that $\mathbb{P}^x(X_t \text{ hits } B(0, r)) = 1$ when $|x| > r$. Then use the strong Markov property to show that there are times $T_i \to \infty$ with $X_{T_i} \in B(0, r)$. That is, two-dimensional Brownian motion is neighborhood recurrent.

21.2   In the proof of Lemma 21.4, justify each inequality in (21.6).

21.3   Let $(X_t, \mathbb{P}^x)$ be a Poisson process with parameter a and let F be defined by (21.5). Show $F(t, 1/2)/t$ does not converge to 0 as $t \to 0$.

21.4   Suppose $d \geq 3$, $(X_t, \mathbb{P}^x)$ is a d-dimensional Brownian motion, and

$$U f(x) = \mathbb{E}^x \int_0^\infty f(X_s) \, ds.$$

Show that if f is bounded and measurable with compact support, then Uf is continuous and $|Uf(x)| \to 0$ as $|x| \to \infty$. Show that if $f \in C^2$ with compact support, then Uf is C^2. Show that $\frac{1}{2} \Delta U f = -f$.

21.5   Let W_t be a Brownian motion and f a continuous function. Prove that if $f(W_t)$ is a submartingale, then f must be convex.

21.6   Prove the *maximum principle* for harmonic functions. This says that if h is harmonic in a bounded domain D, then

$$\sup_{x \in \overline{D}} |h(x)| \leq \sup_{x \in \partial D} |h(x)|.$$

21.7   If W is a d-dimensional Brownian motion started at 0, find $\mathbb{E}\,T$, where T is the first time W exits the ball of radius r centered at the origin.

   Hint: Use the fact that $|W_t|^2 - dt$ is a martingale.

21.8   Let $f : \mathbb{R} \to \mathbb{R}$ be a bounded function with $|f(x) - f(y)| \leq |x - y|$ for all $x, y \in \mathbb{R}$. Let $D_\varepsilon = \{(x, y) \in \mathbb{R}^2 : f(x) < y < f(x) + \varepsilon\}$ for $\varepsilon \in (0, 1)$. Let $(X_t, \mathbb{P}^x)$ be a two-dimensional Brownian motion and let $\tau_\varepsilon = \inf\{t : X_t \notin D_\varepsilon\}$. Prove that there exists a constant c not depending on ε such that $\mathbb{E}^0 \tau_\varepsilon \leq c\varepsilon^2$.

   Hint: By Exercise 21.7 the expected time for two-dimensional Brownian motion to leave a ball of radius 2ε is less than $c\varepsilon^2$. Then use the strong Markov property repeatedly at the times S_i, where S_i is the first time after time S_{i-1} that Brownian motion has moved at least 2ε from $X_{S_{i-1}}$.

22

Transformations of Markov processes

There are a number of interesting transformations that make new Markov processes out of old. We will look at four: killing, conditioning, changing time, and stopping at a last exit time. These are only a few of the possible transformations.

22.1 Killed processes

One sometimes wants to consider a Markov process up until a stopping time ζ, called the *lifetime* of the process. We affix to our state space $\mathcal{S}$ an isolated point Δ, called the *cemetery* state, and the topology on $\mathcal{S}_\Delta = \mathcal{S} \cup \{\Delta\}$ is the one generated by the collection of open sets of $\mathcal{S}$ together with the set $\{\Delta\}$. We define the *killed process* $\widehat{X}$ by

$$\widehat{X}_t = \begin{cases} X_t, & t < \zeta; \\ \Delta, & t \geq \zeta, \end{cases} \tag{22.1}$$

and we say we *kill* the process X at time ζ. Every function f on $\mathcal{S}$ is defined to be 0 at Δ.

One example of this situation would be to let $\zeta = \tau_D$, where D is a subset of $\mathcal{S}$ and $\tau_D = \inf\{t > 0 : X_t \notin D\}$, the first exit from the set D. Another common occurrence is to let $\zeta = S$, where S is a random variable with an exponential distribution with parameter λ, i.e., $\mathbb{P}(S > t) = e^{-\lambda t}$, such that S is independent of X. A third possibility would be to let $\zeta = \inf\{t : \int_0^t f(X_s)\, ds \geq 1\}$, where f is a non-negative function. The crucial property of ζ is that it be a *terminal time*:

$$\zeta = s + \zeta \circ \theta_s \qquad \text{if } s < \zeta. \tag{22.2}$$

Proposition 22.1 *If $(X_t, \mathbb{P}^x)$ is a strong Markov process and (22.2) holds, then $(\widehat{X}_t, \mathbb{P}^x)$ satisfies the Markov and strong Markov properties.*

Proof As in Section 20.2, we need to show

$$\mathbb{E}^x[f(\widehat{X}_t) \circ \theta_T | \mathcal{F}_T] = \mathbb{E}^{\widehat{X}_T} f(\widehat{X}_t), \qquad \mathbb{P}^x\text{-a.s.}$$

If $A \in \mathcal{F}_T$,

$$\mathbb{E}^x[f(\widehat{X}_t) \circ \theta_T; A] = \mathbb{E}^x[f(X_{t+T}); A, T + t < \zeta].$$

On the other hand,

$$\mathbb{E}^{\widehat{X_T}} f(\widehat{X_t}) = \mathbb{E}^{X_T}[f(X_t); t < \zeta]1_{(T < \zeta)}$$
$$= \mathbb{E}^x[f(X_t) \circ \theta_T; t \circ \theta_T < \zeta \circ \theta_T | \mathcal{F}_T]1_{(T < \zeta)}$$
$$= \mathbb{E}^x[f(X_{t+T}); T + t \circ \theta_T < T + \zeta \circ \theta_T, T < \zeta | \mathcal{F}_T]$$
$$= \mathbb{E}^x[f(X_{t+T}); T + t < \zeta | \mathcal{F}_T],$$

since $T + t \circ \theta_T = T + t$ and $T + \zeta \circ \theta_T = \zeta$ on $(T < \zeta)$. Hence

$$\mathbb{E}^x[\mathbb{E}^{\widehat{X_T}} f(\widehat{X_t}); A] = \mathbb{E}^x[f(X_{t+T}); T + t < \zeta, A],$$

as required.   □

22.2 Conditioned processes

Another type of transformation of a Markov process is by conditioning, also known as *Doob's h-path transform*. To motivate this, let D be a domain in $\mathbb{R}^d$ and let X_t be a Brownian motion killed on exiting the domain. One would like to give a precise meaning to the intuitive notion of Brownian motion conditioned to exit the domain at a certain point. Let h be a positive harmonic function in D (i.e., h is C^2 in D, and $\Delta h = 0$ there) and suppose that h is 0 everywhere on the boundary of D except at one point z. The Poisson kernel for the ball or for the half-space gives examples of such harmonic functions. Then, heuristically, we have by the Markov property at time t,

$$\mathbb{P}^x(X_t \in dy | X_{\tau_D} = z) = \frac{\mathbb{P}^x(X_t \in dy, X_{\tau_D} = z)}{\mathbb{P}^x(X_{\tau_D} = z)}$$
$$= \frac{\mathbb{P}^x(X_t \in dy)\mathbb{P}^y(X_{\tau_D} = z)}{\mathbb{P}^x(X_{\tau_D} = z)}.$$

If $p^0(t, x, dy)$ represents the probability that Brownian motion started at x and killed on leaving D is in dy at time t, we then expect that the analogous probability for Brownian motion conditioned to exit D at z ought to be $h(y)p^0(t, x, dy)/h(x)$. We now make this precise.

Let us look at a strong Markov process X. We say a function h is *invariant* with respect to X if $P_t h(x) = h(x)$ for all t and x, where P_t is the semigroup associated with X. If h is invariant, by the Markov property,

$$\mathbb{E}^x[h(X_t) \mid \mathcal{F}_s] = \mathbb{E}^x[h(X_{t-s}) \circ \theta_s \mid \mathcal{F}_s] = \mathbb{E}^{X_s}h(X_{t-s})$$
$$= P_{t-s}h(X_s) = h(X_s),$$

and so for each x, $h(X_t)$ is a martingale with respect to $\mathbb{P}^x$. Conversely, if $h(X_t)$ is a martingale with respect to $\mathbb{P}^x$ for all x,

$$P_t h(x) = \mathbb{E}^x h(X_t) = h(x)$$

by the definition of martingale, and so h is invariant. In the case of Brownian motion killed on leaving a domain, the invariant functions are thus the harmonic ones.

Now let h be a non-negative invariant function for a strong Markov process X. Letting $M_t = h(X_t)/h(X_0)$, M_t is a non-negative continuous martingale with $M_0 = 1$, $\mathbb{P}^x$-a.s., as long as $h(x) > 0$.

We define the h-path transform of the Markov process X by setting

$$\mathbb{P}_h^x(A) = \mathbb{E}^x[M_t; A], \qquad A \in \mathcal{F}_t. \tag{22.3}$$

Since $M_0 = 1$, $\mathbb{P}_h^x(\Omega) = 1$. Observe that P_h^x gives more mass to paths where $h(X_t)$ is big and less to where it is small. Note the similarity to the Girsanov theorem.

We have the following.

Proposition 22.2 *Suppose* $(X_t, \mathbb{P}^x)$ *is a strong Markov process and that h is non-negative and invariant. Then* $(X_t, \mathbb{P}_h^x)$ *forms a strong Markov process.*

Proof Suppose $A \in \mathcal{F}_s$ and $h(x) \neq 0$. (We leave consideration of the case where $h(x) = 0$ to the reader.) Then

$$\mathbb{E}_h^x[f(X_{t+s}); A] = \frac{\mathbb{E}^x[f(X_{t+s})h(X_{t+s}); A]}{h(x)}$$

$$= \frac{\mathbb{E}^x[\mathbb{E}^{X_s}[f(X_t)h(X_t)]; A]}{h(x)}$$

$$= \mathbb{E}^x\left[\frac{1}{h(X_s)}\mathbb{E}^{X_s}[f(X_t)h(X_t)]h(X_s); A\right]$$

by the Markov property for X. This is equal to

$$\mathbb{E}^x\left[\mathbb{E}_h^{X_s}[f(X_t)]h(X_s); A\right]/h(x) = E_h^x[\mathbb{E}_h^{X_s}f(X_t); A].$$

The Markov property follows from this. The strong Markov property is proved in almost identical fashion. $\square$

Let us consider an example. Let $(X_t, \mathbb{P}^x)$ be a Brownian motion on the non-negative axis killed on first hitting 0. This is the same as a Brownian motion killed on exiting $(0, \infty)$. This will be a strong Markov process. Since the second derivative of the function $h(x) = x$ is 0, then h is harmonic on $(0, \infty)$, and so is invariant for the killed Brownian motion. Let us now condition using the function h to get Brownian motion conditioned to hit infinity before hitting zero.

To identify the resulting process, we argue as follows. Fix x and let $T_\varepsilon = \inf\{t > 0 : X_t < \varepsilon\}$. The Radon–Nikodym derivative of the law of $\mathbb{P}_h^x$ with respect to $\mathbb{P}^x$ on $\mathcal{F}_{t \wedge T_\varepsilon}$ is $M_{t \wedge T_\varepsilon} = h(X_{t \wedge T_\varepsilon})/h(x)$. We can rewrite $M_{t \wedge T_\varepsilon}$ as

$$M_{t \wedge T_\varepsilon} = \exp(\log X_{t \wedge T_\varepsilon} - \log x) = \exp\left(\int_0^{t \wedge T_\varepsilon} \frac{1}{X_s}\, dX_s - \frac{1}{2}\int_0^{t \wedge T_\varepsilon} \left(\frac{1}{X_s}\right)^2 ds\right),$$

using Itô's formula. By the Girsanov theorem, under $\mathbb{P}_h^x$,

$$W_{t \wedge T_\varepsilon} = X_{t \wedge T_\varepsilon} - \int_0^{t \wedge T_\varepsilon} \frac{1}{X_s}\, ds$$

is a martingale. By Exercise 13.2, its quadratic variation is $t \wedge T_\varepsilon$, and so by Exercise 12.3, $W_{t \wedge T_\varepsilon}$ is a Brownian motion stopped at time T_ε. We have

$$X_{t \wedge T_\varepsilon} = x + W_{t \wedge T_\varepsilon} + \int_0^{t \wedge T_\varepsilon} \frac{1}{X_s}\, ds,$$

or X satisfies the stochastic differential equation

$$dX_t = dW_t + \frac{1}{X_t}\, dt$$

for $t \leq T_\varepsilon$. We will see later (Section 24.3) that this is the stochastic differential equation defining the Bessel process of order 3. The same argument shows that Brownian motion killed on exiting $(0, a)$ and then conditioned to hit a before 0 is also a Bessel process of order 3 up until the time of first hitting a.

22.3 Time change

An *additive functional* is an increasing adapted process with $A_0 = 0$, a.s., such that

$$A_t = A_s + A_{t-s} \circ \theta_s \tag{22.4}$$

if $s < t$. The simplest examples are what are known as *classical additive functionals*: $A_t = \int_0^t f(X_r)\, dr$, where f is a non-negative measurable function. We have

$$A_t - A_s = \int_s^t f(X_r)\, dr = \int_0^{t-s} f(X_r)\, dr \circ \theta_s = A_{t-s} \circ \theta.$$

If we have the uniform limit of additive functionals, we again get an additive functional, and thus, for example, the local times L_t^x of a one-dimensional Brownian motion are also additive functionals.

Given a Markov process X and an additive functional A, let

$$B_t = \inf\{u : A_u > t\}$$

and

$$X_t' = X_{B_t}.$$

Let $\mathcal{F}_t' = \mathcal{F}_{B_t}$. Thus X' is a time change of X.

Proposition 22.3 *Let $(X_t, \mathbb{P}^x)$ be a strong Markov process and A_t an additive functional. With B defined as above, $(X_t', \mathbb{P}^x)$ is also a strong Markov process.*

Proof We verify the strong Markov property. Let $\mathcal{F}_t' = \mathcal{F}_{B_t}$. Then if T is a stopping time for $\mathcal{F}_t'$, we have

$$\mathbb{E}^x[f(X_{T+t}') \mid \mathcal{F}_T'] = \mathbb{E}^x[f(X(B_{T+t})) \mid \mathcal{F}_{B_T}].$$

B_T can be seen to be a stopping time with respect to $\{\mathcal{F}_t\}$ and $B_{T+t} = B_t \circ \theta_{B_T}$ where the θ_t are the shift operators, so this is

$$\mathbb{E}^x \mathbb{E}^{X(B_T)} f(X_{B_t}) = \mathbb{E}^x \mathbb{E}^{X_T'} f(X_t').$$

This suffices to show that X_t' is a strong Markov process. $\qquad \square$

22.4 Last exit decompositions

Let A be a Borel set, and let L be the last visit to A:

$$L = \sup\{s : X_s \in A\}.$$

We define L to be 0 if X never hits A. The random time L is not a stopping time, but we can nevertheless kill the process X at time L. It turns out the resulting process Y is the process X conditioned by the function $h(x) = \mathbb{P}^x(T_A < \infty)$. The intuitive meaning of this is that Y is X conditioned to hit the set A.

Let $T = \inf\{t : X_t \in A\}$, and set

$$Y_t = \begin{cases} X_t, & t < L, \\ \Delta, & t \geq L. \end{cases}$$

Let $\mathcal{H}_t = \sigma(Y_s; s \leq t)$.

Proposition 22.4 *If $(X_t, \mathbb{P}^x)$ is a strong Markov process, then $(Y_t, \mathbb{P}^x)$ is a Markov process with respect to $\{\mathcal{H}_t\}$.*

Proof If $B \subset \mathcal{S}$ (so that $\Delta \notin B$), then

$$(Y_t \in B) = (X_t \in B, L > t) = (X_t \in B, T \circ \theta_t < \infty),$$

since L, the last time that X is in A, will be larger than t if and only if X hits A at some time after time t. We conclude that the function $x \to \mathbb{P}^x(Y_t \in B)$ is Borel measurable. Since

$$\mathbb{P}^x(Y_t = \Delta) = \mathbb{P}^x(L \leq t) = 1 - \mathbb{P}^x(L > t) = 1 - \mathbb{P}^x(T \circ \theta_t < \infty),$$

then the function $x \to \mathbb{P}^x(Y_t = \Delta)$ is also Borel measurable.

We need to show that if $C \in \mathcal{H}_s$,

$$\mathbb{E}^x[f(Y_t); C] = \mathbb{E}^x[Q_{t-s}f(Y_s); C], \tag{22.5}$$

where f is bounded and measurable, $h(x) = \mathbb{P}^x(L > 0)$, and

$$Q_t g(x) = \frac{1}{h(x)} P_t(gh)(x)$$

when $h(x) \neq 0$. (Set $Q_t g(x) = 0$ if $h(x) = 0$.)

It suffices to show (22.5) when $C = (Y_{r_1} \in B_1, \ldots, Y_{r_n} \in B_n)$ for $r_1 \leq \cdots \leq r_n \leq s$ and the $B_1, \ldots, B_n$ are Borel sets. If we set

$$C_s = (X_{r_1} \in B_1, \ldots, X_{r_n} \in B_n),$$

then $C_s \in \mathcal{F}_s$, $C \cap (L > s) = C_s \cap (L > s)$, and $C \cap (L > t) = C_s \cap (L > t)$.

We start with

$$\mathbb{E}^x[f(Y_t); C] = \mathbb{E}^x[f(X_t); C, L > t] = \mathbb{E}^x[f(X_t); C_s, L > t]$$
$$= \mathbb{E}^x[f(X_t); C_s, L \circ \theta_t > 0].$$

Conditioning on $\mathcal{F}_t$, this is equal to

$$\mathbb{E}^x[f(X_t)\mathbb{P}^{X_t}(L > 0); C_s] = \mathbb{E}^x[f(X_t)h(X_t); C_s].$$

Conditioning on $\mathcal{F}_s$, this in turn is equal to

$$\mathbb{E}^x[P_{t-s}(fh)(X_{t-s}); C_s] = \mathbb{E}^x[h(X_s)Q_{t-s}f(X_s); C_s] \qquad (22.6)$$
$$= \mathbb{E}^x[\mathbb{P}^{X_s}(L > 0)Q_{t-s}f(X_s); C_s]$$
$$= \mathbb{E}^x[Q_{t-s}f(X_s); C_s, L \circ \theta_s > 0],$$

where we used the Markov property for the last equality. Continuing, we have that the last line of (22.6) is equal to

$$\mathbb{E}^x[Q_{t-s}f(X_s); C_s, L > s] = \mathbb{E}^x[Q_{t-s}f(X_s); C, L > s]$$
$$= \mathbb{E}^x[Q_{t-s}f(Y_s); C],$$

as desired.   □

We can also look at X_{L+t}, where L is as above. This new process is again a strong Markov process, and this time is the process X conditioned by the function $h(x) = \mathbb{P}^x(T_A = \infty)$. The intuitive meaning of this is that X_{L+t} is X conditioned never to hit A. Since we are looking at the process after the last visit to A, this is entirely plausible. For a proof of the Markov property of X_{L+t}, see Meyer *et al.* (1972).

Exercises

22.1 Let $(X_t, \mathbb{P}^x)$ be a one-dimensional Brownian motion, L_t^x the local time of Brownian motion at x, and m a positive finite measure on $\mathbb{R}$. Show that $A_t = \int L_t^x\, m(dx)$ is an additive functional.

22.2 We consider the *space-time process*. Let $V_t = V_0 + t$. The process V_t is simply the process that increases deterministically at unit speed. Thus V_t can represent time. If $(X_t, \mathbb{P}^x)$ is a Markov process, show that $((X_t, V_t), \mathbb{P}^{(x,v)})$ is also a Markov process. Is $((X_t, V_t), \mathbb{P}^{(x,v)})$ necessarily a strong Markov process if $(X_t, \mathbb{P}^x)$ is a strong Markov process?

For some applications, one lets $V_t = V_0 - t$, and one thinks of time running backwards. Space-time processes are useful when considering parabolic partial differential equations.

22.3 Suppose $(X_t, \mathbb{P}^x)$ is a strong Markov process and f is a non-negative invariant function for $(X_t, \mathbb{P}^x)$. Write $\mathbb{Q}^x$ for $\mathbb{P}_f^x$. Suppose g is a non-negative invariant function for $(X_t, \mathbb{Q}^x)$. Show that fg is a non-negative invariant function for $(X_t, \mathbb{P}^x)$ and that $\mathbb{Q}_g^x = \mathbb{P}_{fg}^x$.

22.4 Suppose A and B are additive functionals for a Markov process and A and B have continuous paths. Prove that if $\mathbb{E}^x A_t = \mathbb{E}^x B_t$ for all x and t, then

$$\mathbb{P}^x(A_t \neq B_t \text{ for some } t \geq 0) = 0$$

for all x.
  Hint: Show $A_t - B_t$ is a martingale.

22.5 Suppose A and B are additive functionals with continuous paths and suppose $\mathbb{E}^x A_\infty = \mathbb{E}^x B_\infty < \infty$ for each x. Show

$$\mathbb{P}^x(A_t \neq B_t \text{ for some } t \geq 0) = 0$$

for each x.
  Hint: If $f(x) = \mathbb{E}^x A_\infty$, then

$$\mathbb{E}^x[A_\infty \mid \mathcal{F}_t] - A_t = \mathbb{E}^{X_t} A_\infty = f(X_t),$$

and similarly with B in place of A. Then $A - B$ is a $\mathbb{P}^x$ martingale for each x.

22.6 Let A be an additive functional with continuous paths. Suppose there exists $K > 0$ such that $\mathbb{E}^x A_\infty \leq K$ for each x. Prove that there exists a constant c depending only on K such that

$$\mathbb{E}\, e^{cA_\infty} < \infty, \qquad x \in \mathcal{S}.$$

22.7 Here is an argument that the law of a Brownian motion conditioned to have a maximum at a certain level is a Bessel process of order 3.

Let W be a one-dimensional Brownian motion killed on hitting 0. Let $S_t = \sup_{s \leq t} W_s$ be the maximum. By Exercise 19.1, $X = (W, S)$ is a Markov process. Determine the law of X for $t \leq L$, where L is the last time X hits the diagonal. To define L more precisely, let $D = \{(w, s) : w = s, w > 0\}$ and $L = \sup\{t \geq 0 : X_t \in D\}$. L is finite, a.s., because W will hit 0 in finite time with probability one.

Notes

Markov processes are in some sense supposed to have the property that the past and the future are independent given the present. From this point of view, one might hope that a Markov process run backwards is again a Markov process. This is, more or less, the case; see Chung and Walsh (1969) or Rogers and Williams (2000a).

23

Optimal stopping

A nice application of Markov process theory is optimal stopping. Suppose we have a *reward function* $g \geq 0$ and we want to find the stopping time T that maximizes the value of $\mathbb{E}^x g(X_T)$ and we also want to find the value of this expectation. This is the *optimal stopping problem.*

An important example of an optimal stopping problem is pricing the American put. (See Chapter 28 for more on options.) A European put is an option to sell a share of stock at a fixed price K at a certain time t_0. If at time t_0 the price S_{t_0} of the stock is lower than K, one can make a profit by buying a share of stock on the stock exchange for S_{t_0} dollars, exercising the put (which means selling a share of stock for K dollars), and taking home a profit of $K - S_{t_0}$. If the price of the stock is above K at time t_0, it would be silly to exercise the put, and thus the put is worthless. An American put is almost the same, but one has the option to sell a share of stock at price K at any time before time t_0. An American put is more valuable than a European put because if one exercises the option early, that is, sells the share of stock before time t_0, then one can put the money in a risk-free asset such as a bond or in the bank and earn interest on the money. When should one exercise an American put to maximize the expected return? One cannot look into the future, so the time should be a stopping time. The stopping time should depend on the stock price, the exercise price, and also the time until time t_0. Thus one is in the optimal stopping context with $X_t = (t, S_t)$, where S_t is the stock price, and one wants to find a stopping time T that maximizes a certain reward function.

23.1 Excessive functions

A solution to the optimal stopping problem can be given in the Markov case through the use of excessive functions. A non-negative function f is *excessive* for a Markov process X if $P_t f(x) \leq f(x)$ for all t and x and $P_t f(x)$ increases up to $f(x)$ pointwise as $t \to 0$. Here P_t is the semigroup associated with the Markov process X. If $g \geq 0$, define

$$Ug(x) = \int_0^\infty P_s g(x)\,ds = \mathbb{E}^x \int_0^\infty g(X_s)\,ds. \tag{23.1}$$

When $g \geq 0$, Ug is excessive. To see this, using the semigroup property and a change of variables,

$$P_t f(x) = P_t \left(\int_0^\infty P_s g(x)\,ds \right) = \int_0^\infty P_{s+t} g(x)\,ds$$
$$= \int_t^\infty P_s g(x)\,ds.$$

This is certainly less than the integral from 0 to ∞, hence is less than $f(x)$, and $P_t f(x)$ increases up to $f(x)$ by monotone convergence. (It is possible that f is infinite for some or all x.)

The theory of excessive functions is an important part of Markov process theory and we refer the reader to Blumenthal and Getoor (1968), a book which has inspired a generation of Markov process theorists.

We have the following.

Lemma 23.1 *If f is excessive, there exist functions $g_n \geq 0$ such that Ug_n increases up to f, where Ug_n is defined by (23.1).*

Proof Let $g_n = n(f - P_{1/n}f)$. Since f is excessive, then $g_n \geq 0$. We have

$$
\begin{aligned}
Ug_n &= n \int_0^\infty P_s f \, ds - n \int_0^\infty P_{s+(1/n)} f \, ds \\
&= n \int_0^{1/n} P_s f \, ds,
\end{aligned}
$$

which is less than f and increases to f. $\qquad\qquad\square$

Next we have

Proposition 23.2 *(1) If f is excessive, T is a finite stopping time, and $h(x) = \mathbb{E}^x f(X_T)$, then h is excessive.*

(2) If f is excessive and T is a finite stopping time, then $f(x) \geq \mathbb{E}^x f(X_T)$.

(3) If f is excessive, then $f(X_t)$ is a supermartingale

Proof (1) First suppose $f = Ug$ for some non-negative function g. Then

$$
\begin{aligned}
h(x) &= \mathbb{E}^x Ug(X_T) = \mathbb{E}^x \mathbb{E}^{X_T} \int_0^\infty g(X_s) \, ds \qquad\qquad (23.2) \\
&= \mathbb{E}^x \int_0^\infty g(X_{s+T}) \, ds = \mathbb{E}^x \int_T^\infty g(X_s) \, ds
\end{aligned}
$$

by the strong Markov property and a change of variables. The same argument shows that

$$
P_t h(x) = \mathbb{E}^x h(X_t) = \mathbb{E}^x \mathbb{E}^{X_t} \int_T^\infty g(X_s) \, ds = \mathbb{E}^x \int_{T+t}^\infty g(X_s) \, ds.
$$

This is less than $\mathbb{E}^x \int_T^\infty g(X_s) \, ds = h(x)$ and increases up to $h(x)$ as $t \downarrow 0$.

Now let f be excessive but not necessarily of the form Ug. In the paragraph above, replace g by the g_n that were defined in Lemma 23.1 to conclude

$$
P_t h(x) = \lim_{n \to \infty} P_t Ug_n(x) \leq \lim_{n \to \infty} Ug_n(x) = h(x).
$$

That $P_t h$ increases up to h is proved similarly; there is no difficulty interchanging the limit as n tends to infinity and the limit as t tends to 0 since $P_t Ug_n$ increases both as n increases and as t decreases.

(2) As in the proof of (1), it suffices to consider the case where $f = Ug$ and then take limits. By (23.2),

$$\mathbb{E}^x Ug(X_T) = \mathbb{E}^x \int_T^\infty g(X_s)\, ds \leq \mathbb{E}^x \int_0^\infty g(X_s)\, ds = Ug(x).$$

(3) By the Markov property,

$$\mathbb{E}^x[f(X_t) \mid \mathcal{F}_s] = \mathbb{E}^{X_s} f(X_{t-s}) = P_{t-s} f(X_s) \leq f(X_s).$$

The proof is complete. □

By Proposition 23.2, $f(X_t)$ is a supermartingale and therefore has left and right limits along the dyadic rationals. We could take a version of $f(X_t)$ that is right continuous, but there is the danger that doing so would result in a version of X that is not right continuous with left limits. We want to have X fixed and then conclude that $f(X_t)$ is right continuous with left limits without needing to take a version.

Proposition 23.3 *Let $(X_t, \mathbb{P}^x)$ be a strong Markov process. If f is excessive, then for each x, $f(X_t)$ is right continuous with left limits $\mathbb{P}^x$ almost surely.*

For a proof, we refer the reader to Blumenthal and Getoor (1968), Theorem II.2.12 or to Exercise 23.8.

Given a function g, the function G is an *excessive majorant* for g if G is excessive and $G \geq g$ pointwise. G is the *least excessive majorant* for g if (1) G is an excessive majorant, and (2) if $\widetilde{G}$ is any other excessive majorant, then $G \leq \widetilde{G}$ pointwise.

It turns out, which we will prove below, that an optimal stopping time is to stop the first time X_t leaves the set where $g(x) < G(x)$. Therefore it is important to be able to calculate the least excessive majorant of a function.

Here is one method of constructing the least excessive majorant. We say a function $f : \mathcal{S} \to \mathbb{R}$ is *lower semicontinuous* if $\{x : f(x) > a\}$ is an open set for every real number a. See Exercise 23.9 for information about lower semicontinuous functions.

Proposition 23.4 *Suppose that g is non-negative, bounded, and continuous and that Assumption 20.1 holds. Let $g_0 = g$, let $T_n = \{k/2^n : 0 \leq k \leq n2^n\}$, and define*

$$g_n(x) = \max_{t \in T_n} P_t g_{n-1}(x)$$

for $n = 1, 2, \ldots$ Then $g_n(x)$ increases pointwise to $G(x)$, the least excessive majorant of g.

Proof Since $g_n(x) \geq P_0 g_{n-1}(x) = \mathbb{E}^x g_{n-1}(X_0) = g_{n-1}(x)$, the sequence $g_n(x)$ is increasing. Call the limit $H(x)$.

We first show H is lower semicontinuous. If g_{n-1} is bounded and continuous, then $P_t g_{n-1}$ is bounded and continuous for each t by Assumption 20.1. Since the maximum of a finite number of continuous functions is continuous, then g_n is bounded and continuous. By an induction argument, each g_n is continuous. By Exercise 23.9, H is lower semicontinuous.

We next show that H is excessive. If $t \in T_m$ and $n \geq m$, then

$$H(x) \geq g_n(x) \geq P_t g_{n-1}(x) = \mathbb{E}^x g_{n-1}(X_t).$$

Letting n tend to infinity, $H(x) \geq \mathbb{E}^x H(X_t)$ if $t \in T_m$ for some m. Now take $t_k \in \cup_m T_m$ with $t_k \to t$. Since H is lower semicontinuous, then using Exercise 23.9 and Fatou's lemma,

$$H(x) \geq \liminf_{k \to \infty} \mathbb{E}^x H(X_{t_k}) \geq \mathbb{E}^x [\liminf_{k \to \infty} H(X_{t_k})] \geq \mathbb{E}^x H(X_t).$$

If $a \in \mathbb{R}$, let $E_a = \{y : H(y) > a\}$, which is open. If $a < H(x)$, then

$$P_t H(x) = \mathbb{E}^x H(X_t) \geq a \mathbb{P}^x (X_t \in E_a) \to a$$

as $t \to 0$. Therefore $\liminf_{t \to 0} P_t H(x) \geq a$ for all $a < H(x)$, hence

$$\liminf_{t \to 0} P_t H(x) \geq H(x),$$

and we conclude $P_t H(x) \to H(x)$ as $t \to 0$. Thus H is excessive.

Suppose now that F is excessive and $F \geq g$ pointwise. If $F \geq g_{n-1}$, then $F(x) \geq P_t F(x) \geq P_t g_{n-1}(x)$ for every $t \in T_n$, hence $F(x) \geq g_n(x)$. By an induction argument, $F(x) \geq g_n(x)$ for all n, hence $F(x) \geq H(x)$. Therefore H is the least excessive majorant of g. $\square$

In one case, at least, finding the least excessive majorant is easy. Suppose we have a one-dimensional Brownian motion killed on leaving an interval $[a, b]$ and a non-negative function g defined on $[a, b]$. Then the least excessive majorant is the smallest concave function G that is larger than or equal to g everywhere. To see this, if G is the smallest concave function, by Jensen's inequality

$$P_t G(x) = \mathbb{E}^x G(X_t) \leq G(\mathbb{E}^x X_t) \leq G(x).$$

Because G is concave, it is continuous, and so $P_t G(x) = \mathbb{E}^x G(X_t) \to G(x)$ as $t \to 0$. Therefore G is excessive. If $\widetilde{G}$ is another excessive function larger than g and $a \leq c < x < d \leq b$, we have $\widetilde{G}(x) \geq \mathbb{E}^x \widetilde{G}(X_S)$, where S is the first time the process leaves $[c, d]$ by Proposition 23.2(1). Since X is equal to a Brownian motion up to time S, we know the exact distribution of X_S; see Proposition 3.16. Therefore

$$\widetilde{G}(x) \geq \mathbb{E}^x \widetilde{G}(X_S) = \frac{d-x}{d-c} \widetilde{G}(c) + \frac{x-c}{d-c} \widetilde{G}(d).$$

Rearranging this inequality shows that $\widetilde{G}$ is concave. Recall that the minimum of two concave functions is concave, so $G \wedge \widetilde{G}$ is a concave function larger than g that is less than or equal to G. But G is the smallest concave function larger than or equal to g, hence $G = G \wedge \widetilde{G}$, or $G \leq \widetilde{G}$. Thus G is the least excessive majorant of g.

23.2 Solving the optimal stopping problem

Now let us turn to proving that an optimal stopping time can be given in terms of the least excessive majorant. For simplicity we will suppose that g is non-negative, continuous, and bounded. We will assume that our Markov process and g are such that a least excessive majorant G exists. Let g^* be the *optimal reward*:

$$g^*(x) = \sup\{\mathbb{E}^x g(X_T) : T \text{ a stopping time}\}.$$

Let $D = \{x : g(x) < G(x)\}$, the *continuation region* and let $\tau_D = \inf\{t : X_t \notin D\}$.

Theorem 23.5 *Let $(X_t, \mathbb{P}^x)$ be a strong Markov process and g, g^*, G, and D as above. If $\tau_D < \infty$, $\mathbb{P}^x$-a.s., then $g^*(x) = G(x) = \mathbb{E}^x g(X_{\tau_D})$.*

In other words, an optimal stopping time is to stop the first time the process hits $\{x : G(x) = g(x)\}$.

Proof Let $D_\varepsilon = \{x : g(x) < G(x) - \varepsilon\}$, and write τ_ε for τ_{D_ε}. Let $H_\varepsilon(x) = \mathbb{E}^x[G(X_{\tau_\varepsilon})]$, which is excessive by Proposition 23.2(2).

The first step of the proof is to prove (23.3) below. Second, we prove $G(x) \leq g^*(x)$. The third step is to prove that $G(x) = g^*(x)$ and the fourth that $g^*(x) = \mathbb{E}^x g(X_{\tau_D})$.

Step 1. Let $\varepsilon > 0$. We claim

$$g(x) \leq H_\varepsilon(x) + \varepsilon, \qquad x \in D. \tag{23.3}$$

To prove this, we suppose not, that is, we let

$$b = \sup_{x \in D}(g(x) - H_\varepsilon(x))$$

and suppose $b > \varepsilon$. Choose $\eta < \varepsilon$, and then choose x_0 such that

$$g(x_0) - H_\varepsilon(x_0) \geq b - \eta. \tag{23.4}$$

Since $H_\varepsilon + b$ is an excessive majorant of g by the definition of b, and G is the least excessive majorant, then

$$G(x_0) \leq H_\varepsilon(x_0) + b. \tag{23.5}$$

From (23.4) and (23.5) we conclude

$$G(x_0) \leq g(x_0) + \eta. \tag{23.6}$$

By the Blumenthal 0–1 law (Proposition 20.8), either τ_ε is strictly positive with $\mathbb{P}^{x_0}$ probability one or else zero with $\mathbb{P}^{x_0}$ probability one. In the first case, for each $t > 0$,

$$\begin{aligned}
g(x_0) + \eta &\geq G(x_0) \\
&\geq \mathbb{E}^x[G(X_{t \wedge \tau_\varepsilon})] \\
&\geq \mathbb{E}^{x_0}[g(X_t) + \varepsilon; \tau_\varepsilon > t].
\end{aligned}$$

The first inequality is (23.6), the second is due to G being excessive, and the third because $G > g + \varepsilon$ up until the time τ_ε. If we let $t \to 0$ and use the fact that g is continuous, we get $g(x_0) + \eta \geq g(x_0) + \varepsilon$, a contradiction to the way we chose η.

In the second case, where $\tau_\varepsilon = 0$ with $\mathbb{P}^{x_0}$-probability one, we have

$$H_\varepsilon(x_0) = \mathbb{E}^{x_0} G(X_{\tau_\varepsilon}) = \mathbb{E}^{x_0} G(X_0) = G(x_0) \geq g(x_0) \geq H_\varepsilon(x_0) + b - \eta,$$

a contradiction since we chose $\eta < b$.

In either case we reach a contradiction, so (23.3) must hold.

Step 2. A conclusion we reach from (23.3) is that $H_\varepsilon + \varepsilon$ is an excessive majorant of g. Therefore

$$\begin{aligned}
G(x) &\leq H_\varepsilon(x) + \varepsilon \\
&= \mathbb{E}^x[G(X_{\tau_\varepsilon})] + \varepsilon \\
&\leq \mathbb{E}^x[g(X_{\tau_\varepsilon}) + \varepsilon] + \varepsilon \\
&\leq g^*(x) + 2\varepsilon.
\end{aligned} \tag{23.7}$$

The first inequality holds because G is the least excessive majorant, the second inequality because $g(X_{\tau_\varepsilon}) + \varepsilon = G(X_{\tau_\varepsilon})$ by the definition of τ_ε, and the third by the definition of g^*. Since ε is arbitrary, we see that $G(x) \le g^*(x)$.

Step 3. For any stopping time T, because G is excessive and majorizes g,

$$G(x) \ge \mathbb{E}^x G(X_T) \ge \mathbb{E}^x g(X_T).$$

Taking the supremum over all stopping times T, $G(x) \ge g^*(x)$, and therefore $G(x) = g^*(x)$.

Step 4. Because τ_D is finite almost surely, the continuity of g tells us that $\mathbb{E}^x g(X_{\tau_\varepsilon}) \to \mathbb{E}^x g(X_{\tau_D})$ as $\varepsilon \to 0$. By the definition of g^*, we know that $\mathbb{E}^x g(X_{\tau_\varepsilon}) \le g^*(x)$.

On the other hand, by the definitions of τ_ε and H_ε,

$$\mathbb{E}^x g(X_{\tau_\varepsilon}) = \mathbb{E}^x G(X_{\tau_\varepsilon}) - \varepsilon = H_\varepsilon(x) - \varepsilon.$$

By the first inequality in (23.7), the right-hand side is greater than or equal to $G(x) - 2\varepsilon = g^*(x) - 2\varepsilon$. Letting $\varepsilon \to 0$ we obtain

$$\mathbb{E}^x g(X_{\tau_D}) \ge g^*(x)$$

as desired. $\qquad\square$

The following two corollaries are useful in applications.

Corollary 23.6 *Suppose there exists a Borel set A such that h is an excessive majorant of g, where $h(x) = \mathbb{E}^x g(X_{\tau_A})$ and $\tau_A = \inf\{t : X_t \notin A\}$. Then $g^*(x) = h(x)$.*

Proof Let G be the least excessive majorant of g. Then $h(x) \ge G(x)$. However,

$$h(x) = \mathbb{E}^x g(X_{\tau_A}) \le \sup_T \mathbb{E}^x g(X_T) = g^*(x) = G(x)$$

by Theorem 23.5. $\qquad\square$

Corollary 23.7 *Suppose g is continuous and G, the least excessive majorant of g, is lower semicontinuous. Let D be the continuation region, suppose $\tau_D < \infty$, a.s., and let $h(x) = \mathbb{E}^x g(X_{\tau_D})$. If $h \ge g$, then $h = g^*$.*

Proof Note $D = \{x : g(x) < G(x)\} = \cup_{a<b}[(g(x) < a) \cap (G(x) > b)]$, where the union is over all pairs of real numbers $a < b$. Since G is lower semicontinuous and g is continuous, then D is open. This implies $X_{\tau_D} \notin D$, and so $g(X_{\tau_D}) \ge G(X_{\tau_D})$, a.s. Since $g \le G$, we see that

$$h(x) = \mathbb{E}^x g(X_{\tau_D}) = \mathbb{E}^x G(X_{\tau_D}).$$

Since G is excessive, then h is also excessive by Proposition 23.2. Therefore h is an excessive majorant of g and we can apply Corollary 23.6. $\qquad\square$

Exercises

23.1 Show that if f is excessive, then $1 - e^{-f}$ is excessive. Thus, for some purposes it is enough to look at bounded excessive functions.

23.2 Show that if f and g are excessive, then $f \wedge g$ is excessive.

23.3   Let A_t be an additive functional (defined in (22.4)) and let $f(x) = \mathbb{E}^x A_\infty$. Show that f is excessive.

23.4   Let f be an excessive function for a strong Markov process $(X_t, \mathbb{P}^x)$. Let $\varepsilon > 0$ and $S_1 = \inf\{t : |f(X_t) - f(X_0)| \geq \varepsilon\}$. Let $S_{i+1} = S_i + S_1 \circ \theta_{S_i}$. Prove that $f(X_{S_i})$ is a supermartingale with respect to the σ-fields $\mathcal{F}_{S_i}$ and with respect to $\mathbb{P}^x$ for each x.

23.5   For each n, let A_t^n be an additive functional with continuous paths and suppose that $f_n(x) = \mathbb{E}^x A_\infty^n$ is finite for every x. Suppose A_t is a continuous additive functional with $f(x) = \mathbb{E}^x A_\infty$ also finite for each x. Suppose f_n converges to f uniformly. Prove that for each x, with $\mathbb{P}^x$-probability one, A_t^n converges to A_t, uniformly over $t \geq 0$.

   Hint: Use Proposition 9.11.

23.6   Suppose f is bounded and excessive, $\lambda \geq 0$, and $A = \{y : f(y) \leq \lambda\}$. Prove that if $x \in A^r$ (i.e., x is regular for A), then $f(x) \leq \lambda$.

   Hint: Use the optional section theorem (Theorem 16.12) to find stopping times T_m whose graphs are contained in $\{(t, \omega) : t \leq 1/m, f(X_t) \leq \lambda\}$ with $\mathbb{P}^x$-probability at least $1 - (1/m)$.

   If the g_n are as in Lemma 23.1, write

   $$Ug_n(x) = \mathbb{E}^x \int_0^{T_m} g_n(X_s)\, ds + \mathbb{E}^x U g_n(X_{T_m})$$

   $$\leq \mathbb{E}^x \int_0^{T_m} g_n(X_s)\, ds + \mathbb{E}^x f(X_{T_m})$$

   $$\leq \mathbb{E}^x \int_0^{T_m} g_n(X_s)\, ds + \lambda + \|f\|_\infty/m.$$

   Let $m \to \infty$, then $n \to \infty$.

23.7   Suppose f is bounded and excessive, $\lambda \geq 0$, and $B = \{y : f(y) \geq \lambda\}$. Prove that if $x \in B^r$, then $f(x) \leq \lambda$.

   Hint: Use the optional stopping theorem as in Exercise 23.6 to find stopping times R_m analogous to the T_m. Write

   $$f(x) \geq \mathbb{E}^x f(X_{R_m}) \geq \lambda - \|f\|_\infty/m,$$

   and then let $m \to \infty$.

23.8   (1) Suppose f is bounded and excessive, $x \in S$, $\varepsilon > 0$, and $C = \{y : |f(y) - f(x)| \geq \varepsilon\}$. Use Exercises 23.6 and 23.7 to show that if $z \in C^r$, then $|f(z) - f(x)| \geq \varepsilon$.

   (2) Let f, ε, and x be as in (1) and set $S = \inf\{t > 0 : |f(X_t) - f(x)| \geq \varepsilon\}$. Use Exercise 20.9 to show that $|f(X_S) - f(x)| \geq \varepsilon$ with $\mathbb{P}^x$-probability one.

   (3) Let f, ε, x, and S be as in (2). Define $S = 0$ and $S_{i+1} = S_i + S \circ \theta_{S_i}$. By Exercise 23.4, $f(X_{S_i})$ is a positive supermartingale. Use Corollary A.36 to show $S_i \to \infty$, $\mathbb{P}^x$-a.s. Deduce that with $\mathbb{P}^x$-probability one, $f(X_t)$ has paths that are right continuous with left limits.

   (4) Use Exercise 23.1 to show that if f is excessive but not necessarily bounded, then $f(X_t)$ has paths that are right continuous with left limits.

23.9   (1) Show that every continuous function is lower semicontinuous.

   (2) Show that if f is lower semicontinuous and $x \in S$, then

   $$\liminf_{y \to x} f(y) \geq f(x).$$

   (3) Show that if f_n is a sequence of continuous functions increasing to f, then f is lower semicontinuous.

23.10 Suppose g is non-negative, bounded, and continuous, and Assumption 20.1 holds. Let $g_0 = g$ and define $g_n(x) = \sup_{t \geq 0} P_t g_{n-1}(x)$ for $n \geq 1$. Prove that g_n increases to the least excessive majorant of g.

Notes

See Øksendal (2003) for further information on optimal stopping.

Exercise 23.3 shows that $\mathbb{E}^x A_\infty$ is an excessive function if A is an additive functional. To a large extent the converse is true: given an excessive function f and some mild conditions, there exists an additive functional A such that $f(x) = \mathbb{E}^x A_\infty$ for all x. The proof is a modification of the Doob–Meyer decomposition of $f(X_t)$ that takes into account the fact there is a family of probabilities instead of just one; see Blumenthal and Getoor (1968).

The optimal stopping problem involving American puts has a theoretical solution: look at the least excessive majorant for a certain reward function. The reward function is not just $(K - s)^+$ because the interest earned on the money obtained after the sale of a share of stock needs to be taken into account. Moreover, the excessive functions here are relative to the space-time process (S_t, t), not those relative to S_t. Finding a satisfactory solution to this optimal stopping problem is still open and is important.

Stochastic differential equations

Stochastic differential equations are used in modeling a wide variety of physical and economic situations, and are one of the main reasons for the interest in stochastic integrals.

We consider stochastic differential equations (SDEs) of the form

$$dX_t = \sigma(X_t)\, dW_t + b(X_t)\, dt,$$

where σ and b are real-valued functions and W is a one-dimensional Brownian motion. We also consider multidimensional analogs of this equation. If X represents the position of a particle, the $\sigma(X_t)\, dW_t$ term says that the particle X diffuses like a multiple of Brownian motion, but how strong the diffusivity is depends on the location of the particle. The $b(X_t)\, dt$ term represents a push in one direction or another, the size of the push depending on the location of the particle.

24.1 Pathwise solutions of SDEs

Let W_t be a one-dimensional Brownian motion with respect to a filtration $\{\mathcal{F}_t\}$ satisfying the usual conditions; see Chapter 1. We want to consider SDEs of the form

$$dX_t = \sigma(X_t)\, dW_t + b(X_t)\, dt, \qquad X_0 = x_0. \tag{24.1}$$

This means that X_t satisfies the equation

$$X_t = x_0 + \int_0^t \sigma(X_s)\, dW_s + \int_0^t b(X_s)\, ds, \qquad t \geq 0. \tag{24.2}$$

Here σ and b are Borel measurable functions, the first integral in (24.2) is a stochastic integral with respect to the Brownian motion W_t, and (24.2) holds almost surely, that is, we can find versions of $\int_0^t \sigma(X_s)\, dW_s$ such that for almost all ω, (24.2) holds for all t. In order to be able to define the stochastic integral, we require that any solution X_t to (24.2) be adapted to the filtration $\{\mathcal{F}_t\}$. If X satisfies (24.2), then X will automatically have continuous paths. We want to consider existence and uniqueness of solutions to the equation (24.2).

Definition 24.1 A stochastic process X will be a *pathwise solution* to (24.1) if X is adapted to the filtration $\{\mathcal{F}_t\}$ and (24.2) holds almost surely, where the null set does not depend on t. We say the solution to (24.1) is *pathwise unique* if whenever X_t' is another solution, then

$$\mathbb{P}(X_t \neq X_t' \text{ for some } t \geq 0) = 0. \tag{24.3}$$

Sometimes pathwise uniqueness is used for a slightly stronger concept: one can let W be a Brownian motion with respect to each of two filtrations $\{\mathcal{F}_t\}$ and $\{\mathcal{F}_t'\}$, which are possibly different, and one can let X'_t be adapted to $\{\mathcal{F}'_t\}$. One then requires (24.3) to hold. We won't need to use this modification of the definition, and in any case our proof of uniqueness will be equally valid in this situation.

The function σ in (24.1) is called the *diffusion coefficient* and the function b is called the *drift coefficient*. σ tells us the intensity of the noise at a point, and b tells us if there is a push in any direction at a given point.

We will suppose that σ and b are *Lipschitz functions*: there exists a constant c such that

$$|\sigma(x) - \sigma(y)| \le c|x - y|, \qquad |b(x) - b(y)| \le c|x - y|. \tag{24.4}$$

We also suppose for now that σ and b are bounded.

Theorem 24.2 *Suppose σ and b are bounded Lipschitz functions. Then there exists a pathwise solution to* (24.1) *and this solution is pathwise unique.*

Proof Existence. Let $X_0(t) = x_0$ for all t and define $X_i(t)$ recursively by

$$X_{i+1}(t) = x_0 + \int_0^t \sigma(X_i(s))\, dW_s + \int_0^t b(X_i(s))\, ds. \tag{24.5}$$

Note that $X_0(t)$ is trivially adapted to $\{\mathcal{F}_t\}$, and an induction argument shows that X_i is adapted to $\{\mathcal{F}_t\}$ for each i.

Fix t_0. We will show existence (and uniqueness) up to time t_0; since t_0 is arbitrary, this will achieve the theorem.

Since $(x + y)^2 \le 2x^2 + 2y^2$, then

$$\mathbb{E}\sup_{r \le t} |X_{i+1}(r) - X_i(r)|^2 = \mathbb{E}\left[\sup_{r \le t}\left(\int_0^r [\sigma(X_i(s)) - \sigma(X_{i-1}(s))]\, dW_s\right.\right.$$
$$\left.\left. + \int_0^r [b(X_i(s)) - b(X_{i-1}(s))]\, ds\right)^2\right]$$
$$\le 2\mathbb{E}\left[\sup_{r \le t}\left(\int_0^r [\sigma(X_i(s)) - \sigma(X_{i-1}(s))]\, dW_s\right)^2\right]$$
$$+ 2\mathbb{E}\left[\sup_{r \le t}\left(\int_0^r [b(X_i(s)) - b(X_{i-1}(s))]\, ds\right)^2\right].$$

By Doob's inequalities (Theorem 3.6) and the fact that σ is a Lipschitz function, the first term after the inequality is bounded by

$$8\mathbb{E}\left[\left(\int_0^t [\sigma(X_i(s)) - \sigma(X_{i-1}(s))]\, dW_s\right)^2\right] = 8\mathbb{E}\int_0^t [\sigma(X_i(s)) - \sigma(X_{i-1}(s))]^2\, ds$$
$$\le c\mathbb{E}\int_0^t |X_i(s) - X_{i-1}(s)|^2\, ds.$$

By the Cauchy–Schwarz inequality, the fact that $t \le t_0$, and the fact that b is a Lipschitz function, the second term is bounded by

$$2\mathbb{E}\left(\int_0^t |b(X_i(s)) - b(X_{i-1}(s))|\, ds\right)^2 \le 2t_0\mathbb{E}\int_0^t |b(X_i(s)) - b(X_{i-1}(s))|^2\, ds$$
$$\le c\mathbb{E}\int_0^t |X_i(s) - X_{i-1}(s)|^2\, ds.$$

Therefore

$$\mathbb{E} \sup_{r \leq t} |X_{i+1}(r) - X_i(r)|^2 \leq c\mathbb{E} \int_0^t |X_i(s) - X_{i-1}(s)|^2 \, ds. \tag{24.6}$$

Let $g_i(t) = \mathbb{E} \sup_{r \leq t} |X_i(r) - X_{i-1}(r)|^2$. Thus provided we choose A big enough, $g_1(t) \leq A$ for $t \leq t_0$ and

$$g_{i+1}(t) \leq A \int_0^t g_i(s) \, ds, \qquad t \leq t_0.$$

(Clearly $|X_{i+1}(t) - X_i(t)|^2 \leq \sup_{r \leq t} |X_{i+1}(r) - X_i(r)|^2$.) Thus

$$g_2(t) \leq A \int_0^t g_1(s) \, ds \leq A \int_0^t A \, ds = A^2 t,$$

$$g_3(t) \leq A \int_0^t g_2(s) \, ds \leq A \int_0^t A^2 s \, ds = A^3 t^2 / 2,$$

and continuing by induction,

$$g_i(t) \leq A^i t^{i-1} / (i-1)!$$

Exercise 24.1 asks you to show that if we define

$$\|Y\|_t = (\mathbb{E} \sup_{r \leq t} |Y_r|^2)^{1/2} \tag{24.7}$$

when Y is a stochastic process, then $\|Y\|_t$ is a norm and the corresponding metric is complete. Hence

$$(\mathbb{E} \sup_{r \leq t_0} |X_n(s) - X_m(s)|^2)^{1/2} = \|X_n - X_m\|_{t_0}$$

$$\leq \sum_{i=m}^{n-1} \|X_{i+1} - X_i\|_{t_0}$$

$$\leq \sum_{i=m}^{n-1} (g_i(t_0))^{1/2}$$

can be made small by taking m, n large. (We use the ratio test to show that the sum $\sum (A^i t_0^{i-1}/(i-1)!)^{1/2}$ converges.) We have $\mathbb{E} X_0(t)^2 < \infty$. By the completeness of $\| \cdot \|_{t_0}$ there exists X_t such that $\mathbb{E} \sup_{s \leq t_0} |X_n(s) - X_s|^2 \to 0$ as $n \to \infty$. This implies there exists a subsequence $\{n_j\}$ such that $\sup_{s \leq t_0} |X_{n_j}(s) - X_s|^2 \to 0$ almost surely; since each X_{n_j} is continuous in t, then X_t is also. Taking a limit in (24.5) as $n \to \infty$ shows X_t satisfies (24.2).

Uniqueness. Suppose X_t and X_t' are two solutions to (24.2). Let

$$g(t) = \mathbb{E} \sup_{r \leq t} |X_r - X_r'|^2.$$

Very similarly to the existence proof, $\mathbb{E} \sup_{r \le t} |X_r|^2 < \infty$, the same with X replaced by X', and

$$\mathbb{E} \sup_{r \le t} |X_r - X'_r|^2 \le 2\mathbb{E} \left[\sup_{r \le t} \left(\int_0^r [\sigma(X_s) - \sigma(X'_s)] \, dW_s \right)^2 \right]$$

$$+ 2\mathbb{E} \left[\sup_{r \le t} \left(\int_0^r [b(X_s) - b(X'_s)] \, ds \right)^2 \right]$$

$$\le c\mathbb{E} \int_0^t |X_s - X'_s|^2 \, ds.$$

Therefore there exists $A > 0$ such that $g(t)$ is bounded by A and $g(t) \le A \int_0^t g(s) \, ds$.

Then $g(t) \le A \int_0^t A \, ds = A^2 t$, $g(t) \le A \int_0^t A^2 s \, ds = A^3 t^2/2$, etc. Thus we have $g(t) \le A^i t^{i-1}/(i-1)!$ for all i, which is only possible if $g(t) = 0$. This implies that $X_t = X'_t$ for all $t \le t_0$, except for a null set. $\square$

We also want to consider the SDE (24.1) when σ and b are Lipschitz functions, but not necessarily bounded. Note $|\sigma(x)| \le |\sigma(0)| + c|x|$, so that $|\sigma(x)|$ is less than or equal to $c(1 + |x|)$, and the same for b.

Theorem 24.3 *Suppose σ and b are Lipschitz functions, but not necessarily bounded. Then there exists a pathwise solution to (24.1) and this solution is pathwise unique.*

Proof Let σ_n and b_n be bounded Lipschitz functions that agree with σ and b, respectively, on $[-n, n]$. Let X_n be the unique pathwise solution to (24.1) with σ and b replaced by σ_n and b_n, respectively. Let $T_n = \inf\{t : |X_n(t)| \ge n\}$. We claim $X_n(t) = X_m(t)$ if $t \le T_n \wedge T_m$; to prove this, let $g(t) = \mathbb{E} \sup_{s \le t \wedge T_n \wedge T_m} |X_n(s) - X_m(s)|^2$, and proceed as in the uniqueness part of the proof of Theorem 24.2. We then have existence and uniqueness of the SDE for $t \le T_n$ for each n.

To complete the proof, it suffices to show $T_n \to \infty$. Let

$$h_n(t) = \mathbb{E} \sup_{s \le t \wedge T_n} |X_n(s)|^2.$$

Then

$$h_n(t) \le c|x_0|^2 + c\mathbb{E} \left(\int_0^t \sigma_n(X_n(s)) \, dW_s \right)^2 + c\mathbb{E} \int_0^t b_n(X_n(s))^2 \, ds$$

$$\le c|x_0|^2 + c\mathbb{E} \int_0^t \sigma_n(X_n(s))^2 \, ds + ct_0 \mathbb{E} \int_0^t b_n(X_n(s))^2 \, ds$$

$$\le c|x_0|^2 + c + c\mathbb{E} \int_0^t |X_n(s)|^2 \, ds$$

$$\le c + c \int_0^t h_n(s) \, ds,$$

using estimates very similar to those of the proof of Theorem 24.2. By Exercise 24.2, $h_n(t) \le ce^{ct}$ if $t \le t_0$. Note the constant c can be chosen to be independent of n. Then

$$\mathbb{P}(T_n < t_0) = \mathbb{P}(\sup_{s \le t_0} |X_n(s)| \ge n) \le \frac{\mathbb{E} \sup_{s \le t_0} |X_n(s)|^2}{n^2} \le \frac{h_n(t_0)}{n^2} \to 0$$

as $n \to \infty$. Since t_0 is arbitrary, $T_n \to \infty$, a.s. $\square$

Although we considered one-dimensional SDEs for simplicity, the same arguments apply when we have higher-dimensional SDEs. Let

$$W = (W^1, \ldots, W^d)$$

be a d-dimensional Brownian motion, let $\sigma_{ij}(x)$ be bounded Lipschitz functions for $i = 1, \ldots, n$ and $j = 1, \ldots, d$, and let $b_i(x)$ be bounded Lipschitz functions for $i = 1, \ldots, n$. Consider the system of equations

$$dX_t^i = \sum_{j=1}^d \sigma_{ij}(X_t)\, dW_t^j + b_i(X_t)\, dt, \qquad i = 1, \ldots, n. \tag{24.8}$$

This is frequently written in matrix form

$$dX_t = \sigma(X_t)\, dW_t + b(X_t)\, dt \tag{24.9}$$

where we view $X = (X^1, \ldots, X^n)$ as a $n \times 1$ matrix, $b = (b_1, \ldots, b_n)$ as a $n \times 1$ matrix-valued function, W as a $d \times 1$ matrix, and σ as a $n \times d$ matrix-valued function. We have existence and uniqueness to the system (24.8). Exercise 24.5 asks you to prove this in the case when $n = d$, although there is nothing at all special about requiring $n = d$.

24.2 One-dimensional SDEs

Although our proof of pathwise existence and uniqueness was for SDEs in one dimension, as is pointed out in Exercise 24.5, almost the same proof works in higher dimensions. In this section we look at a pathwise uniqueness result that is valid only for SDEs on $\mathbb{R}$. The equation we look at is the same as the one in the last section, namely,

$$X_t = x_0 + \int_0^t \sigma(X_s)\, dW_s + \int_0^t b(X_s)\, ds. \tag{24.10}$$

Theorem 24.4 *Suppose b is bounded and Lipschitz. Suppose there exists a continuous function $\rho : [0, \infty) \to [0, \infty)$ such that $\rho(0) = 0$,*

$$\int_0^\varepsilon \rho^{-2}(u)\, du = \infty \tag{24.11}$$

for all $\varepsilon > 0$, and σ is bounded and satisfies

$$|\sigma(x) - \sigma(y)| \leq \rho(|x - y|)$$

for all x and y. Then the solution to (24.10) is pathwise unique.

For an example, let $b(x) = 0$ for all x, and let σ be Hölder continuous of order α, that is, there exists c such that $|\sigma(x) - \sigma(y)| \leq c|x - y|^\alpha$. Then we take $\rho(x) = x^\alpha$, and the integral condition in the theorem is satisfied if and only if $\alpha \geq 1/2$. If (24.11) holds for all $\varepsilon > 0$, we say the *Yamada–Watanabe condition* holds.

Instead of proving this theorem right away and then essentially repeating the proof to give a comparison theorem, we will state and prove a comparison theorem (Theorem 24.5) and then obtain Theorem 24.4 as a corollary of Theorem 24.5.

We only prove the uniqueness of the solution to (24.10) here. The existence is a consequence of some measure-theoretic magic; see Revuz and Yor (1999), Theorem IX.1.7.

Theorem 24.5 *Suppose σ satisfies the conditions in Theorem 24.4. Suppose X_t satisfies (24.10) with b a Lipschitz function. Suppose Y_t is a continuous semimartingale satisfying*

$$Y_t \geq Y_0 + \int_0^t \sigma(Y_s)\, dW_s + \int_0^t B(Y_s)\, ds,$$

where B is a Borel measurable function and $B(z) \geq b(z)$ for all z. If $Y_0 \geq x$, a.s., then $Y_t \geq X_t$ almost surely for all t.

Proof Let $a_n \downarrow 0$ be selected so that

$$\int_{a_n}^{a_{n-1}} (\rho(u))^{-2}\, du = n.$$

This can be done inductively. Choose a_0 arbitrarily. Since $\int_r^{a_0} \rho(x)^{-2}\, dx$ increases to infinity as $r \to 0$, we can choose a_1 such that $\int_{a_1}^{a_0} \rho(x)^{-2}\, dx = 1$; in a similar manner we choose $a_2, a_3, \ldots$. Let h_n be continuous, supported in (a_n, a_{n-1}), $0 \leq h_n(u) \leq 2/n\rho^2(u)$, and $\int_{a_n}^{a_{n-1}} h_n(u)\, du = 1$ for each n. The idea here is to start with the function $(1+\varepsilon)1_{(a_n, a_{n-1})}(u)/(n\rho(u)^2)$ for some small ε, and then modify this near a_n and a_{n-1} to get a function that is continuous, is supported in (a_n, a_{n-1}), and integrates to 1. Let f_n be such that $f_n(0) = f_n'(0) = 0$ and $f_n'' = h_n$. Note

$$f_n'(u) = \int_0^u f_n''(s)\, ds = \int_0^u h_n(s)\, ds \leq 1$$

and $f_n'(u) \geq 0$, so $0 \leq f_n'(u) \leq 1$ and $f_n'(u) = 1$ if $u \geq a_{n-1}$. Hence $f_n(u) \uparrow u$ as $n \to \infty$ for each $u \geq 0$.

Since $x \leq y$, then $f_n(x - y) = 0$, and we have by Itô's formula

$$f_n(X_t - Y_t) = \text{martingale} + \int_0^t f_n'(X_s - Y_s)[b(X_s) - B(Y_s)]\, ds \tag{24.12}$$

$$+ \frac{1}{2}\int_0^t f_n''(X_s - Y_s)[\sigma(X_s) - \sigma(Y_s)]^2\, ds.$$

We take expectations of both sides. The martingale term has 0 expectation. The final term on the right-hand side is bounded in expectation by

$$\frac{1}{2}\mathbb{E}\int_0^t \frac{2}{n(\rho|X_s - Y_s|)^2}(\rho|X_s - Y_s|)^2\, ds \leq \frac{t}{n}$$

by the assumptions on σ and the bound on $f_n'' = h_n$, and so goes to 0 as $n \to \infty$. The expectation of the second term on the right of (24.12) is bounded above by

$$\mathbb{E}\int_0^t f_n'(X_s - Y_s)[b(X_s) - b(Y_s)]\, ds + \mathbb{E}\int_0^t f_n'(X_s - Y_s)[b(Y_s) - B(Y_s)]\, ds$$

$$\leq c\mathbb{E}\int_0^t (1_{[0,\infty)}(X_s - Y_s))\, |X_s - Y_s|\, ds$$

$$= c\mathbb{E}\int_0^t (X_s - Y_s)^+\, ds.$$

Letting $n \to \infty$,

$$\mathbb{E}\,(X_t - Y_t)^+ \le c \int_0^t \mathbb{E}\,(X_s - Y_s)^+ \, ds.$$

If we set $g(t) = \mathbb{E}\,(X_t - Y_t)^+$, we have

$$g(t) \le c \int_0^t g(s) \, ds,$$

and by Exercise 24.2 we conclude $g(t) = 0$ for each t. Using the continuity of the paths of X_t and Y_t completes the proof. □

We now prove Theorem 24.4.

Proof of Theorem 24.4 Suppose X and X' are two solutions to (24.10). Then by Theorem 24.5 with $Y = X'$ and $B = b$, we have $X_t \le X_t'$ for all t. Applying this argument with X and X' reversed yields $X_t' \le X_t$ for all t, which completes the proof. □

24.3 Examples of SDEs

Ornstein–Uhlenbeck process
The Ornstein–Uhlenbeck process is the solution to the SDE

$$dX_t = dW_t - \frac{X_t}{2}\, dt, \qquad X_0 = x. \tag{24.13}$$

The existence and uniqueness follow by Theorem 24.3. Note that the drift coefficient is not bounded, so Theorem 24.2 is not sufficient. The process behaves like a Brownian motion, with a drift that pushes the process towards the origin; the farther the process gets from the origin, the stronger the push.

The equation (24.13) can be solved explicitly. Rearranging, multiplying by $e^{t/2}$, and using the product rule,

$$d[e^{t/2}X_t] = e^{t/2}\, dX_t + e^{t/2}\frac{X_t}{2} dt = e^{t/2}\, dW_t,$$

so

$$e^{t/2}X_t = X_0 + \int_0^t e^{s/2}\, dW_s,$$

or

$$X_t = e^{-t/2}x + e^{-t/2}\int_0^t e^{s/2}\, dW_s. \tag{24.14}$$

We used here the fact that the martingale part of the semimartingale $Z_t = e^{t/2}$ is zero, and therefore $\langle Z, W \rangle_t = 0$. By Exercise 24.6, X_t is a Gaussian process and the distribution of X_t is that of a normal random variable with mean $e^{-t/2}x$ and variance equal to $e^{-t} \int_0^t (e^{s/2})^2\, ds = 1 - e^{-t}$.

If we let $Y_t = \int_0^t e^{s/2}\, dW_s$ and $V_t = Y_{\log(t+1)}$, then Y_t is a mean-zero continuous Gaussian process with independent increments, and hence so is V_t. Since

$$\mathrm{Var}\,(V_u - V_t) = \int_{\log(t+1)}^{\log(u+1)} e^s\, ds = u - t,$$

then V_t is a Brownian motion. Hence

$$X_t = e^{-t/2}x + e^{-t/2}V(e^t - 1).$$

This representation of an Ornstein–Uhlenbeck process in terms of a Brownian motion is useful for, among other things, calculating the exit probabilities of a square root boundary.

Linear equations
We consider the linear equation

$$dX_t = AX_t \, dW_t + BX_t \, dt, \qquad X_0 = x_0, \tag{24.15}$$

where A and B are constants. One place this comes up is in models of stock prices in financial mathematics; see Chapter 28. We have pathwise existence and uniqueness by Theorem 24.3; here both the diffusion and drift coefficients are unbounded.

We will give a candidate for the solution, and verify that it solves (24.15). By the pathwise uniqueness, this will then be the only solution. Our candidate is

$$X_t = x_0 e^{AW_t + (B - A^2/2)t}.$$

To verify that this is a solution, we use Itô's formula with the process $AW_t + (B - A^2/2)t$ and the function e^x:

$$X_t = x_0 + \int_0^t e^{AW_s + (B - A^2/2)s} A \, dW_s + \int_0^t e^{AW_s + (B - A^2/2)s}(B - A^2/2) \, ds$$

$$+ \frac{1}{2} \int_0^t e^{AW_s + (B - A^2/2)s} A^2 \, ds$$

$$= x_0 + \int_0^t e^{AW_s + (B - A^2/2)s} A \, dW_s + \int_0^t e^{AW_s + (B - A^2/2)s} B \, ds$$

$$= x_0 + \int_0^t AX_s \, dW_s + \int_0^t BX_s \, ds.$$

Let us summarize our discussion.

Proposition 24.6 *The unique pathwise solution to*

$$dX_t = AX_t \, dW_t + BX_t \, dt$$

is

$$X_t = X_0 e^{AW_t + (B - A^2/2)t}.$$

If we write $Z_t = AW_t + Bt$, then (24.15) becomes

$$dX_t = X_t \, dZ_t, \qquad X_0 = x_0. \tag{24.16}$$

The equation (24.16) makes sense for arbitrary continuous semimartingales Z, and by using Itô's formula as above, one can see that a solution is $X_t = x_0 e^{Z_t - \langle Z \rangle_t/2}$.

Bessel processes

We consider Bessel processes and the squares of Bessel processes. The reason for the name is that these processes turn out to be Markov processes and the infinitesimal generator of the semigroup (see Chapter 37) is related to Bessel's equation, a type of differential equation.

A Bessel process of order $\nu \geq 2$ is defined to be a solution of the SDE

$$dX_t = dW_t + \frac{\nu - 1}{2X_t} dt, \qquad X_0 = x. \tag{24.17}$$

Bessel processes of order $0 \leq \nu < 2$ can also be defined using (24.17), but only up until the first time the process X reaches 0; some extra information needs to be given as to what the process does at 0. The square of a Bessel process of order $\nu \geq 0$ is defined to be the solution to the SDE

$$dY_t = 2\sqrt{|Y_t|} \, dW_t + \nu \, dt, \qquad Y_0 = y. \tag{24.18}$$

There is no difficulty defining the square of a Bessel process for $0 \leq \nu < 2$.

By Theorem 24.4 we have pathwise uniqueness for the solution to (24.18), because $|\,|y|^{1/2} - |x|^{1/2}| \leq |y - x|^{1/2}$, and we can thus take $\rho(u) = 2u^{1/2}$ in Theorem 24.4. The solution to (24.18) when $\nu = 0$ and $y = 0$ is clearly $Y_t = 0$ for all t. By Theorem 24.5 with $b(x) = \nu$ and $B(x) = 0$, we see that the solution to (24.18) is greater than or equal to 0 for all t. We may thus omit the absolute value in (24.18) and rewrite it as

$$dY_t = 2\sqrt{Y_t} \, dW_t + \nu \, dt, \qquad Y_0 = y. \tag{24.19}$$

If we apply Itô's formula to the solution Y_t of (24.19) with the function $\sqrt{x}$, we see that $X_t = \sqrt{Y_t}$ solves (24.17) for t up until the first time Y reaches 0; the function $\sqrt{x}$ is twice continuously differentiable as long as we stay away from 0. We will see shortly that the square of a Bessel process started away from 0 never hits 0 if and only if $\nu \geq 2$.

Using Itô's formula with a d-dimensional process W_t and the function $|x|^2$ shows that the square of the modulus of a d-dimensional Brownian motion is the square of a Bessel process of order d; this is Exercise 24.7.

Bessel processes have the same scaling properties as Brownian motion. That is, if X_t is a Bessel process of order ν started at x, then $aX_{a^{-2}t}$ is a Bessel process of order ν started at ax. In fact, from (24.17),

$$d(aX_{a^{-2}t}) = a \, dW_{a^{-2}t} + a^2 \frac{\nu - 1}{2aX_{a^{-2}t}} d(a^{-2}t),$$

and the assertion follows from the uniqueness of the solution to (24.17) and the fact that $aW(a^{-2}t)$ is again a Brownian motion.

Bessel processes are useful for comparison purposes, and so the following is worthwhile.

Proposition 24.7 *Suppose Y_t is the square of a Bessel process of order ν. Suppose $Y_0 = y$. The following hold with probability one.*
(1) If $\nu > 2$ and $y > 0$, Y_t never hits 0.
(2) If $\nu = 2$ and $y > 0$, Y_t hits every neighborhood of 0, but never hits the point 0.
(3) If $0 < \nu < 2$, Y_t hits 0.
(4) If $\nu = 0$, then Y_t hits 0. If started at 0, then Y_t remains at 0 forever.

When we say that Y_t hits 0, we consider only times $t > 0$. We define $T_0 = \inf\{t > 0 : Y_t = 0\}$ and say that Y_t hits 0 if $T_0 < \infty$.

Proof We prove (2). An application of Itô's formula with the process being the square of a Bessel process of order 2 and the function being $\log x$ shows that $\log Y_t$ is a martingale up until the first hitting time of 0; cf. Exercise 21.1. The quadratic variation of $\log Y_t$ is $\int_0^t Y_s^{-2} \, ds$ for t less than the hitting time of 0. Suppose $0 < a < y < b$.

We claim that Y_t leaves the interval $[a, b]$, a.s. If not, $\langle \log Y \rangle_t \geq b^{-2} t \to \infty$ as $t \to \infty$. Since $\log Y_t$ is a martingale, it is a time change of Brownian motion, and Brownian motion leaves $[\log a, \log b]$ with probability one, a contradiction.

Then by Corollary 3.17,

$$\mathbb{P}(Y_t \text{ hits } a \text{ before } b) = \frac{\log b - \log y}{\log b - \log a}. \tag{24.20}$$

Letting $b \to \infty$, we see that $\mathbb{P}(Y_t \text{ hits } a) = 1$, and since a is arbitrary, Y_t hits every neighborhood of 0. If in (24.20) we hold b fixed instead and let $a \to 0$, we see $\mathbb{P}(Y_t \text{ hits } 0 \text{ before } b) = 0$; since b is arbitrary, this proves that Y_t never hits the point 0.

Parts (1), (3), and (4) are similar, but instead of $\log |x|$ we use $|x|^{(2-\nu)/2}$. The details are left as Exercise 24.8. □

Exercises

24.1 Show that $\| \cdot \|_t$ defined by (24.7) gives rise to a complete normed linear space.

24.2 Suppose $g(t)$ is non-negative and bounded on each finite subinterval of $[0, \infty)$. Suppose there exist constants A and B such that

$$g(t) \leq A + B \int_0^t g(s) \, ds \tag{24.21}$$

for each $t \geq 0$. Prove that $g(t) \leq A e^{Bt}$ for all $t \geq 0$. This result is known as *Gronwall's lemma*.
   Hint: Write

$$g(t) \leq A + B \int_0^t \left[A + B \int_0^s g(r) \, dr \right] ds,$$

use (24.21) to substitute for $g(r)$, and iterate.

24.3 The starting point in (24.1) can be random. Suppose Y is a random variable that is measurable with respect to $\mathcal{F}_0$, Y is square integrable, and σ and b are bounded and Lipschitz. Prove pathwise existence and uniqueness for the equation

$$X_t = Y + \int_0^t \sigma(X_s) \, dW_s + \int_0^t b(X_s) \, ds.$$

24.4 The functions σ and b in (24.1) can depend on time as well as space. Suppose $\sigma : [0, \infty) \times \mathbb{R} \to \mathbb{R}$, $b : [0, \infty) \times \mathbb{R} \to \mathbb{R}$ are bounded and uniformly Lipschitz in the second variable: there exists c independent of s such that $|\sigma(s, x) - \sigma(s, y)| \leq c|x - y|$ and similarly for b. Prove pathwise existence and uniqueness for the equation

$$X_t = x_0 + \int_0^t \sigma(s, X_s) \, dW_s + \int_0^t b(s, X_s) \, ds.$$

24.5 Here is a multidimensional analog of (24.1). Suppose the functions $\sigma_{ij} : \mathbb{R}^d \to \mathbb{R}$, $1 \le i, j \le d$, are bounded and Lipschitz, and $b_i : \mathbb{R}^d \to \mathbb{R}$, $i = 1, \ldots, d$, are bounded and Lipschitz, W^j are independent one-dimensional Brownian motions, $x_0 = (x_0^{(1)}, \ldots, x_0^{(d)})$, and $X_t = (X_t^{(1)}, \ldots, X_t^{)d})$ satisfies

$$X_t^{(i)} = x_0^{(i)} + \int_0^t \sum_{j=1}^d \sigma_{ij}(X_s)\, dW_s^j + \int_0^t b_i(X_s)\, ds \tag{24.22}$$

for $i = 1, \ldots, d$. Prove pathwise existence and uniqueness for this system of equations.

24.6 Suppose f and g map $[0, \infty) \to \mathbb{R}$ with $\int_0^\infty f(t)^2\, dt < \infty$ and $\int_0^\infty g(t)^2\, dt < \infty$. Show that $\int_0^\infty f(t)\, dW_t$ is a mean zero Gaussian random variable, the same with f replaced by g, and

$$\mathrm{Cov}\left(\int_0^\infty f(t)\, dW_t, \int_0^\infty g(t)\, dW_t \right) = \int_0^\infty f(t)g(t)\, dt.$$

Hint: Approximate f and g by piecewise constant deterministic functions.

24.7 Show that if W_t is a d-dimensional Brownian motion, then $|W_t|^2$ is the square of a Bessel process of order d.

24.8 Prove (1), (3), and (4) of Proposition 24.7.

24.9 Let X be the solution to $dX_t = \sigma(X_t)\, dW_t + b(X_t)\, dt$, where W is a one-dimensional Brownian motion, σ and b are Lipschitz continuous real-valued functions, and $|\sigma(x)| \le c(1 + |x|)$ and $|b(x)| \le c(1 + |x|)$. Let $t_0 > 0$. Prove that if $p \ge 2$, then

$$\mathbb{E}\left[\sup_{s \le t_0} |X_s|^p \right] \le c(1 + |x_0|^p).$$

24.10 Let W be a one-dimensional Brownian motion and let X_t^x be the solution to

$$dX_t = \sigma(X_t)\, dW_t + b(X_t)\, dt, \qquad X_0 = x.$$

Suppose σ and b are C^∞ functions and that σ and b and all their derivatives are bounded. Show that for each t the map $x \to X_t^x$ is continuous in x with probability one. Show that the map is differentiable in x.

24.11 Suppose $A(t)$ and $B(t)$ are deterministic functions of t. Find an explicit solution to the one-dimensional SDE

$$dX_t = A(t)\, dW_t + B(t)\, dt, \qquad X_0 = x.$$

Notes

If one wants to have a stochastic differential equation with jumps, besides a Brownian motion, one integrates with respect to a Poisson point process, which is defined in Chapter 18. Using the notation of that chapter, one considers the stochastic differential equation

$$dX_t = \sigma(X_{t-})\, dW_t + b(X_{t-})\, dt$$

$$+ \int_S F(X_{t-}, z)\, (\mu(dt, dz) - \nu(dt, dz)), \qquad X_0 = x_0,$$

which means that we want a solution to

$$X_t = x_0 + \int_0^t \sigma(X_{s-})\,dW_s + \int_0^t b(X_{s-})\,ds$$
$$+ \int_0^t \int_S F(X_{s-}, z)\,(\mu(ds, dz) - \nu(ds, dz)).$$

There is pathwise existence and uniqueness to this SDE provided F satisfies a suitable Lipschitz-like condition; see Skorokhod (1965).

25

Weak solutions of SDEs

In Chapter 24 we considered SDEs of the form

$$dX_t = \sigma(X_t)\,dW_t + b(X_t)\,dt, \tag{25.1}$$

where W is a Brownian motion and σ and b are Lipschitz functions, or in one dimension, where σ has a modulus of continuity satisfying an integral condition. When the coefficients σ and b fail to be sufficiently smooth, it is sometimes the case that (25.1) may not have a pathwise solution at all, or it may not be unique. We define another notion of existence and uniqueness that is useful.

Definition 25.1 A *weak solution* $(X, W, \mathbb{P})$ to (25.1) exists if there exists a probability measure $\mathbb{P}$ and a pair of processes (X_t, W_t) such that W_t is a Brownian motion under $\mathbb{P}$ and (25.1) holds. There is *weak uniqueness* holding for (25.1) if whenever $(X, W, \mathbb{P})$ and $(X', W', \mathbb{P}')$ are two weak solutions, then the joint law of (X, W) under $\mathbb{P}$ and the joint law of (X', W') under $\mathbb{P}'$ are equal. When this happens, we also say that the solution to (25.1) is *unique in law*.

Let us discuss the relationship between weak solutions and pathwise solutions. If the solution to (25.1) is pathwise unique, then weak uniqueness holds. For a proof of this result under very general hypotheses, see Revuz and Yor (1999), theorem IX.1.7. In the case that σ and b are Lipschitz functions, the proof is much simpler.

Proposition 25.2 *Suppose σ and b are bounded Lipschitz functions and $x_0 \in \mathbb{R}^d$. Then weak uniqueness holds for* (25.1).

Proof For notational simplicity we consider the case of dimension one. Suppose $(X, W, \mathbb{P})$ and $(X', W', \mathbb{P}')$ are two weak solutions to (25.1). Let $X_0(t) = x_0$ and define $X_{i+1}(t)$ by

$$X_{i+1}(t) = x_0 + \int_0^t \sigma(X_i(s))\,dW_s + \int_0^t b(X_i(s))\,ds. \tag{25.2}$$

We saw by the proof of Theorem 24.2 that the limit of the X_i exists, uniformly over finite time intervals, and solves (25.1), and the solution is pathwise unique. Since X also solves (25.1), we conclude that X_i converges (uniformly over finite time intervals) to X, a.s., with respect to $\mathbb{P}$. Similarly, if we let $X_0'(t) = x_0$ and define $X_{i+1}'(t)$ by

$$X_{i+1}'(t) = x_0 + \int_0^t \sigma(X_i'(s))\,dW_s' + \int_0^t b(X_i'(s))\,ds, \tag{25.3}$$

then X_i' converges, uniformly over finite time intervals, to X'.

Now since W is a Brownian motion under $\mathbb{P}$ and W' is a Brownian motion under $\mathbb{P}'$, then the law of (X_0, W) under $\mathbb{P}$ equals the law of (X_0', W') under $\mathbb{P}'$. By (25.2) and (25.3), the law of (X_1, W) under $\mathbb{P}$ equals the law of (X_1', W') under $\mathbb{P}'$, and iterating, the law of (X_i, W) under $\mathbb{P}$ equals the law of (X_i', W') under $\mathbb{P}'$. Passing to the limit, the law of (X, W) under $\mathbb{P}$ equals the law of (X', W') under $\mathbb{P}'$. $\qquad\square$

We now give an example to show that weak uniqueness might hold even if pathwise uniqueness does not. Let $\sigma(x)$ be equal to 1 if $x \geq 0$ and -1 otherwise. We take b to be identically 0. We consider solutions to

$$X_t = \int_0^t \sigma(X_s) \, dW_s. \tag{25.4}$$

Weak uniqueness holds since if W is a Brownian motion under $\mathbb{P}$, then X_t must be a martingale, and the quadratic variation of X is $d\langle X\rangle_t = \sigma(X_t)^2 \, dt = dt$; by Lévy's theorem (Theorem 12.1), X_t is a Brownian motion. Given a Brownian motion X_t and letting $W_t = \int_0^t \frac{1}{\sigma(X_s)} \, dX_s$, then again by Lévy's theorem, W_t is a Brownian motion and $X_t = \int_0^t \sigma(X_s) \, dW_s$; thus weak solutions exist.

On the other hand, pathwise uniqueness does not hold. To see this, let $Y_t = -X_t$. We have

$$Y_t = \int_0^t \sigma(Y_s) \, dW_s - 2 \int_0^t 1_{\{0\}}(X_s) \, dW_s. \tag{25.5}$$

The second term on the right has quadratic variation $4 \int_0^t 1_{\{0\}}(X_s) \, ds$; this is 0 almost surely because we showed in Exercise 11.1 that the amount of time Brownian motion spends at 0 has Lebesgue measure 0. Therefore Y is another pathwise solution to (25.4).

This example is not satisfying because one would like σ to be positive and even continuous if possible. Such examples exist, however. For each $\beta < 1/2$, Barlow (1982) has constructed functions σ that are Hölder continuous of order β and bounded above and below by positive constants and for which

$$dX_t = \sigma(X_t) \, dW_t, \qquad X_0 = x_0, \tag{25.6}$$

has a unique weak solution but no pathwise solution exists.

Let us show how the technique of time change can be used to study weak uniqueness. We consider the SDE (25.6).

Proposition 25.3 *If σ is Borel measurable and there exist $c_2 > c_1 > 0$ such that $c_1 \leq \sigma(x) \leq c_2$ for all x, then weak existence and weak uniqueness hold for (25.6).*

Proof We consider only uniqueness, leaving existence as Exercise 25.1. Suppose $(X, W, \mathbb{P})$ and $(X', W', \mathbb{P}')$ are two weak solutions. Then X_t is a martingale, and as in Section 12.2, if we set

$$A_t = \int_0^t \sigma(X_s)^2 \, ds, \qquad \tau_t = \inf\{s : A_s \geq t\},$$

then $M_t = X_{\tau_t}$ is a Brownian motion under $\mathbb{P}$. Define A', τ', and M' analogously. The law of M under $\mathbb{P}$ is that of a Brownian motion, as is that of M' under $\mathbb{P}'$.

Now let

$$B_t = \int_0^t \frac{1}{\sigma(M_s)^2} \, ds, \qquad \rho_t = \inf\{s : B_s \geq t\}. \tag{25.7}$$

Since M_t is a Brownian motion and σ is bounded above and below by positive constants, then B_t is continuous, strictly increasing, and increases to infinity as $t \to \infty$, and the same is therefore true of ρ_t. By a change of variables,

$$B_t = \int_0^t \frac{1}{\sigma(X_{\tau_s})^2} \, ds = \int_0^{\tau_t} \frac{1}{\sigma(X_u)^2} \, dA_u$$

$$= \int_0^{\tau_t} \frac{1}{\sigma(X_u)^2} \sigma(X_u)^2 \, du = \tau_t.$$

Therefore $M_{\rho_t} = X_{\tau(\rho_t)} = X_t$. We have the analogous formulas with primes.

The law of M under $\mathbb{P}$ equals the law of M' under $\mathbb{P}'$ since both are Brownian motions, so by (25.7) the law of (M, B) under $\mathbb{P}$ equals the law of (M', B') under $\mathbb{P}'$, and consequently the law of (M, ρ) under $\mathbb{P}$ equals the law of (M', ρ') under $\mathbb{P}'$. Since $X_t = M_{\rho_t}$ and similarly for X', we conclude the law of X under $\mathbb{P}$ equals the law of X' under $\mathbb{P}'$. Finally, since $W_t = \int_0^t \frac{1}{\sigma(X_s)} \, dX_s$ and similarly for W', the joint law of (X, W) under $\mathbb{P}$ equals the joint law of (X', W') under $\mathbb{P}'$. $\qquad \square$

We point out that in the above proof it is essential that one can reconstruct X from M in a measurable way.

We now use the Girsanov theorem to prove weak uniqueness for (25.1).

Proposition 25.4 *Suppose σ and b are measurable and bounded above and σ is bounded below by a positive constant. Then weak existence and uniqueness holds for (25.1).*

Proof We prove the weak uniqueness, leaving it as Exercise 25.2 to prove existence. Define $\{\mathcal{F}_t\}$ to be the minimal augmented filtration generated by X,

$$M_t = \exp\left(-\int_0^t \frac{b}{\sigma}(X_s) \, dW_s - \frac{1}{2} \int_0^t \left(\frac{b}{\sigma}(X_s)\right)^2 ds \right),$$

and $\mathbb{Q}$ the probability measure defined by $\mathbb{Q}(A) = \mathbb{E}_{\mathbb{P}}[M_t; A]$ if $A \in \mathcal{F}_t$. By Theorem 13.3, under $\mathbb{Q}$, the process $\widetilde{W}_t = W_t + \int_0^t (b/\sigma)(X_s) \, ds$ is a Brownian motion, and

$$dX_t = \sigma(X_t)\left(dW_t + \frac{b}{\sigma}(X_t) \, dt\right) = \sigma(X_t) \, d\widetilde{W}_t.$$

Define M', $\mathbb{Q}'$, and $\widetilde{W}'$ analogously. By Proposition 25.3 the law of $(X, \widetilde{W})$ under $\mathbb{Q}$ is equal to the law of $(X', \widetilde{W}')$ under $\mathbb{Q}'$. Let $n \geq 1$, $t_1 < \cdots < t_n$, and let $A_1, \ldots, A_n$ be Borel subsets of $\mathbb{R}$. Set $B = \{X_{t_1} \in A_1, \ldots, X_{t_n} \in A_n\}$ and define B' analogously. We have

$$\mathbb{P}(B) = \int_B \frac{d\mathbb{P}}{d\mathbb{Q}} \, d\mathbb{Q} = \int_B \exp\left(\int_0^t \frac{b}{\sigma}(X_s) \, dW_s + \frac{1}{2} \int_0^t \left(\frac{b}{\sigma}(X_s)\right)^2 ds \right) d\mathbb{Q}$$

$$= \int_B \exp\left(\int_0^t \frac{b}{\sigma}(X_s) \, d\widetilde{W}_s - \frac{1}{2} \int_0^t \left(\frac{b}{\sigma}(X_s)\right)^2 ds \right) d\mathbb{Q}.$$

Using the analogous formula for $\mathbb{P}'(B')$ and the fact that the law of $(X, \widetilde{W})$ under $\mathbb{Q}$ is the same as that of $(X', \widetilde{W}')$ under $\mathbb{Q}'$, we see that $\mathbb{P}(B) = \mathbb{P}'(B')$; thus the finite-dimensional distributions of X under $\mathbb{P}$ and of X' under $\mathbb{P}'$ are the same. Since both X and X' are continuous processes, we conclude from Theorem 2.6 that the law of X under $\mathbb{P}$ equals the law of X' under $\mathbb{P}'$. Defining $Y_t = X_t - \int_0^t b(X_s)\, ds$ and similarly for Y', the joint law of (X, Y) under $\mathbb{P}$ equals the joint law of (X', Y') under $\mathbb{P}'$. Finally, $W_t = \int_0^t \frac{1}{\sigma(X_s)}\, dY_s$ and similarly for W', so we obtain our conclusion. $\qquad\square$

The procedure of using the Girsanov theorem to get rid of the drift also works in higher dimensions. However the time change procedure of Proposition 25.3 is not nearly as useful in higher dimensions as in one dimension. The question of weak uniqueness for the system of equations in Exercise 24.5 is quite an interesting one; see Bass (1997) and Stroock and Varadhan (1977).

Exercises

25.1 Show weak existence holds under the hypotheses of Proposition 25.3.

25.2 Show weak existence holds under the hypotheses of Proposition 25.4.

25.3 Here is an example of an SDE where weak uniqueness does not hold. Suppose W is a one-dimensional Brownian motion and $\alpha \in (0, \frac{1}{2})$. Let $\sigma(x) = 1 \wedge |x|^\alpha$. Find two solutions to

$$dX_t = \sigma(X_t)\, dW_t$$

that are not equal in law.

Hint: One is the solution that is identically zero. The other can be constructed by time changing a Brownian motion by the inverse of the increasing process

$$2 \int_0^t (1 \wedge |X_s|^{2\alpha})^{-1}\, ds.$$

25.4 (1) Suppose a_s and b_s are bounded predictable processes with a_s bounded below by a positive constant. Let W be a one-dimensional Brownian motion. Suppose Y is a one-dimensional semimartingale such that

$$dY_t = a_t Y_t\, dW_t + b_t\, dt, \qquad Y_0 = 0.$$

Prove that if $t_0 > 0$ and $\varepsilon > 0$, there exists a constant $c > 0$ depending only on t_0, ε, and the bounds on a_s and b_s such that

$$\mathbb{P}(\sup_{s \le t_0} |Y_s| < \varepsilon) > c.$$

(2) Now let W be d-dimensional Brownian motion, let $x \in \mathbb{R}^d$, and let σ be a $d \times d$ matrix-valued function that is bounded and such that $\sigma\sigma^T(x)$ is positive definite, uniformly in x. That is, there exists $\Lambda > 0$ such that for all x,

$$\sum_{i,j=1}^d y_i y_j (\sigma(x)\sigma^T(x))_{ij} \ge \Lambda \sum_{i=1}^d y_i^2, \qquad (y_1, \ldots, y_d) \in \mathbb{R}^d.$$

Let b be a $d \times 1$ matrix-valued function that is bounded. Let X be the solution to

$$dX_t = \sigma(X_t)\, dW_t + b(X_t)\, dt, \qquad X_0 = x.$$

Use Itô's formula to find an equivalent expression for $|X_t - x|^2$. Then use (1) to prove that if $t_0 > 0$ and $\varepsilon > 0$, there exists a constant $c > 0$ not depending on x such that

$$\mathbb{P}^x(\sup_{s \leq t_0} |X_s - x| < \varepsilon) > c.$$

25.5 This is the *support theorem* for solutions to SDEs. Let X, x, ε, and t_0 be as in (2) of Exercise 25.4. Suppose $\psi : [0, t_0] \to \mathbb{R}^d$ is a continuous function with $\psi(0) = x$. Use the Girsanov theorem to prove that there exists $c > 0$ such that

$$\mathbb{P}^x(\sup_{s \leq t_0} |X_s - \psi(s)| < \varepsilon) > c.$$

25.6 Suppose weak uniqueness holds for the one-dimensional stochastic differential equation

$$dX_t = \sigma(X_t) \, dW_t, \qquad X_0 = x, \tag{25.8}$$

where W is a one-dimensional Brownian motion. Suppose also that there exists a process X' that is adapted to the minimal augmented filtration of W with $X_0' = x$ and $dX_t' = \sigma(X_t') \, dW_t$. Prove that pathwise uniqueness holds for (25.8).

Hint: Show there exists a measurable map F from $C[0, \infty) \to C[0, \infty)$ such that $X' = F(W)$. If X'' is another solution to (25.8), then weak uniqueness shows that the laws of (X'', W) and (X', W) are equal, hence $X'' = F(W) = X'$.

26

The Ray–Knight theorems

The local time of Brownian motion, L_t^x, is parameterized by space and time: x and t. Ray and Knight independently discovered that at certain stopping times T, the process $x \to L_T^x$ is a Markov process.

Times that work are (1) the first time local time at 0 reaches a level r; (2) an exponential random variable T that is independent of the Brownian motion; and (3) the first time T that Brownian motion reaches the level one. We will prove the version of the Ray–Knight theorems in the last case. We will show that if W is a Brownian motion with local times L_t^x and

$$T = \inf\{t > 0 : W_t = 1\},$$

then the process L_T^{1-x} indexed by x has the same law as the square of a Bessel process of order 2. We will see in Chapter 39 that the square of a Bessel process is a Markov process.

We will use the following lemma.

Lemma 26.1 *Suppose $X_t^{(j)}$, $j = 1, 2$, are two continuous processes such that*

$$\mathbb{E} \exp\left(-\int_0^1 f(s)X_s^{(1)}\, ds\right) = \mathbb{E} \exp\left(-\int_0^1 f(s)X_s^{(2)}\, ds\right)$$

whenever f is a non-negative continuous function with support in $(0, 1)$. Then the laws of $\{X_t^{(j)}; 0 \le t \le 1\}$, $j = 1, 2$, are equal.

Proof Let φ be a non-negative continuous function with support in $[0, 1]$ such that $\int_0^1 \varphi(x)\, dx = 1$, and let $\varphi_\varepsilon(x) = \varepsilon^{-1}\varphi(x/\varepsilon)$, so that the sequence $\{\varphi_\varepsilon\}$ is an approximation to the identity. If g is a continuous function and $t \ne 0$, then $\int g(s)\varphi_\varepsilon(s - t)\, ds \to g(t)$. Now let $t_1, \ldots, t_n \in (0, 1)$, $a_1, \ldots, a_n > 0$, and set $f_\varepsilon(x) = \sum_{i=1}^n a_i \varphi_\varepsilon(x - t_i)$. Using the hypothesis and letting $\varepsilon \to 0$, we obtain

$$\mathbb{E} \exp\left(-\sum_{i=1}^n a_i X_{t_i}^{(1)}\right) = \mathbb{E} \exp\left(-\sum_{i=1}^n a_i X_{t_i}^{(2)}\right).$$

The left-hand side is the joint Laplace transform of $(X_{t_1}^{(1)}, \ldots, X_{t_n}^{(1)})$ and the right-hand side is the same for $X^{(2)}$. By the uniqueness of the Laplace transform, the finite-dimensional distributions of $X^{(1)}$ and $X^{(2)}$ are equal. Both processes have continuous paths, and the conclusion now follows from Theorem 2.6. $\qquad \square$

Let B_t be a Brownian motion, not necessarily the same as W_t, and let Z_t be the non-negative solution to

$$dZ_t = 2\sqrt{Z_t}\,dB_t + 2\,dt, \qquad Z_0 = 0,\ 0 \le t \le 1. \tag{26.1}$$

The solution to this equation is unique by Theorem 24.4, and Z_t is the square of a Bessel process of order 2.

Theorem 26.2 *The processes* $\{L_T^{1-x}; 0 \le x \le 1\}$ *and* $\{Z_x; 0 \le x \le 1\}$ *have the same law.*

Proof Let $f \ge 0$ be a continuous function whose support $[a, b]$ is a subset of $(0, 1)$. Let F be the solution to

$$F''(x) = 2F(x)f(x), \qquad F(1) = 1, \quad F'(1) = 0;$$

see Exercise 26.1. Define $g(x) = f(1 - x)$ and $G(x) = F(1 - x)$, so that $G'' = 2Gg$, $G'(0) = 0$, and $G(0) = 1$. We will show

$$\mathbb{E}\exp\left(-\int_0^1 f(x)L_T^{1-x}\,dx\right) = \mathbb{E}\exp\left(-\int_0^1 f(t)Z_t\,dt\right), \tag{26.2}$$

and then apply Lemma 26.1.

The left-hand side of (26.2) is equal to

$$\mathbb{E}\exp\left(-\int_0^1 f(1-x)L_T^x\,dx\right) = \mathbb{E}\exp\left(-\int_0^1 g(x)L_T^x\,dx\right)$$

$$= \mathbb{E}\exp\left(-\int_0^T g(X_s)\,ds\right),$$

where the last equality follows from the occupation time formula (Theorem 14.4). Let

$$M_t = G(W_t)e^{-\int_0^t g(W_s)\,ds}.$$

By Itô's formula and the product formula,

$$dM_t = -G(W_t)g(W_t)e^{-\int_0^t g(W_s)\,ds}\,dt + G'(W_t)e^{-\int_0^t g(W_s)\,ds}\,dW_t$$

$$+ \tfrac{1}{2}G''(W_t)e^{-\int_0^t g(W_s)\,ds}\,dt$$

$$= G'(W_t)e^{-\int_0^t g(W_s)\,ds}\,dW_t,$$

since $\tfrac{1}{2}G'' - Gg = 0$. Therefore M_t is a martingale. Since G is bounded on $(-\infty, 1]$, then $M_{t\wedge T}$ is bounded and we then have

$$1 = G(0) = \mathbb{E}\,M_0 = \mathbb{E}\,M_T = \mathbb{E}\,G(1)e^{-\int_0^T g(W_s)\,ds},$$

so

$$\mathbb{E}\exp\left(-\int_0^T g(W_s)\,ds\right) = \frac{1}{G(1)}. \tag{26.3}$$

Now look at the right-hand side of (26.2). Let

$$N_t = \frac{1}{F(t)}\exp\left(Z_t\frac{F'(t)}{2F(t)} - \int_0^t f(s)Z_s\,ds\right).$$

Let

$$Y_t = Z_t \frac{F'(t)}{2F(t)},$$

so using (26.1),

$$dY_t = Z_t \frac{2F(t)F''(t) - 2F'(t)^2}{4F(t)^2} \, dt + 2 \frac{F'(t)}{2F(t)} \sqrt{Z_t} \, dB_t + 2 \frac{F'(t)}{2F(t)} \, dt.$$

If

$$X_t = Y_t - \int_0^t f(s) Z_s \, ds,$$

then the martingale part of X is

$$\int_0^t \frac{F'(s)}{F(s)} \sqrt{Z_s} \, dB_s,$$

and hence

$$d\langle X \rangle_t = \left(\frac{F'(t)}{F(t)} \right)^2 Z_t \, dt.$$

By Itô's formula and the product formula and using $F'' = 2Ff$,

$$\begin{aligned}
dN_t &= -\frac{F'(t)}{F(t)^2} e^{X_t} \, dt + \frac{1}{F(t)} e^{X_t} \left\{ Z_t \frac{F''(t)}{2F(t)} \, dt - Z_t \frac{F'(t)^2}{2F(t)^2} \, dt \right. \\
&\quad \left. + \frac{F'(t)}{F(t)} \sqrt{Z_t} \, dB_t + \frac{F'(t)}{F(t)} \, dt - f(t) Z_t \, dt \right\} \\
&\quad + \frac{1}{2} \frac{1}{F(t)} e^{X_t} \frac{F'(t)^2}{F(t)^2} Z_t \, dt \\
&= \frac{F'(t)}{F(t)} \sqrt{Z_t} \, dB_t.
\end{aligned}$$

Observe that F is continuous and positive on $[0, 1]$, hence bounded below on $[0, 1]$ by a positive constant. Also F' is bounded above on $[0, 1]$. We see that $N_{t \wedge 1}$ is a martingale. Then $\mathbb{E} N_0 = 1/F(0) = 1/G(1)$, while

$$\begin{aligned}
\mathbb{E} N_1 &= \frac{1}{F(1)} \mathbb{E} \exp \left(Z_1 \frac{F'(1)}{2F(1)} - \int_0^1 f(s) Z_s \, ds \right) \\
&= \mathbb{E} \, e^{-\int_0^1 f(s) Z_s \, ds}.
\end{aligned}$$

Therefore

$$\mathbb{E} \exp \left(-\int_0^1 f(s) Z_s \, ds \right) = \mathbb{E} N_1 = \mathbb{E} N_0 = \frac{1}{G(1)}.$$

Combining with (26.3), we conclude the two sides of (26.2) are equal. $\qquad \square$

You may wonder how the function F was arrived at. Exercises 26.2 and 26.3 may shed some light on this.

Exercises

26.1   Suppose f is a non-negative continuous function whose support $[a, b]$ is a subset of $(0, 1)$. Show that there is a unique solution to the ordinary differential equation $F''(x) = 2F(x)f(x)$, $F(1) = 1, F'(1) = 0$, that F is everywhere positive, and F is bounded on $[0, \infty)$.

Hint: Since f is zero in (b, ∞), then F'' is zero there, and hence is of the form $F(x) = Ax + B$ for some A and B for $x \geq b$. Since $F'(1) = 0$, conclude that A is 0.

26.2   Suppose X_t is a solution to the one-dimensional SDE

$$dX_t = \sigma(X_t)\, dW_t + b(X_t)\, dt.$$

Suppose σ and b are bounded and continuous and f is a bounded and continuous function. What ordinary differential equation must $H(x)$ satisfy (in terms of σ, b, and f) in order that

$$M_t = H(X_t)e^{\int_0^t f(X_s)\, ds}$$

be a martingale?

26.3   Suppose X_t is a solution to the one-dimensional SDE

$$dX_t = \sigma(X_t)\, dW_t + b(X_t)\, dt.$$

Suppose σ and b are bounded and continuous and f is a bounded and continuous function. What partial differential equation must $K(x, t))$ satisfy (in terms of σ, b, and f) in order that

$$N_t = K(X_t, t)e^{\int_0^t f(s)X_s\, ds}$$

be a martingale?

26.4   Let W be a Brownian motion and L_t^y the local times at level y. Prove that local times at a fixed time t are not a Markov process. That is, let $t > 0$ be fixed and show that $(L_t^y, y \geq 0)$ is not a Markov process in the variable y.

26.5   Let S be the first time two-dimensional Brownian motion exits the unit ball and let $\psi(\lambda) = \mathbb{P}^0(S > \lambda)$. If W is a one-dimensional Brownian motion with local times L_t^x and $T = \inf\{t > 0 : W_t = 1\}$, find the distribution of $Y = \sup_{0 \leq x \leq 1} L_T^x$ in terms of ψ, i.e., write $\mathbb{P}(Y \leq \lambda)$ in terms of the function ψ.

26.6   Suppose $x \in (0, 1)$. With W and T as in Exercise 26.5, find the distribution of L_T^x.

26.7   Let W be a one-dimensional Brownian motion with local times L_t^x. Let $T_r = \inf\{t > 0 : L_t^0 = r\}$. The law of the process $x \to L_{T_r}^x$ can be described as follows:

(1) The law of $\{L_{T_r}^x, x \geq 0\}$ is the same as the law of $\{X_x, x \geq 0\}$ started at r, where X is the square of a Bessel process of order 0.

(2) The law of $\{L_{T_r}^{-x}, x \geq 0\}$ is also the same as the law of $\{X_x, x \geq 0\}$ started at r, where X is the square of a Bessel process of order 0.

(3) The processes $\{L_{T_r}^x, x \geq 0\}$ and $\{L_{T_r}^{-x}, x \geq 0\}$ are independent of each other.

This is proved in Revuz and Yor (1999), Section XI.2, or for a challenge, try to prove (1) for yourself using the techniques of this chapter. Using this description of $L_{T_r}^x$, find the distribution of $L_{T_r}^* = \sup_x L_{T_r}^x$.

Notes

There are several other proofs of the Ray–Knight theorems. One by Walsh (Rogers and Williams, 2000b; Walsh, 1978) uses excursion theory. In the next chapter we will indicate some ideas used in that proof.

27

Brownian excursions

The paths of a Brownian motion W_t are continuous, so the zero set $Z(\omega) = \{t : W_t(\omega) = 0\}$ is a closed set. The complement of $Z(\omega)$ is an open subset of the reals, hence is the countable union of disjoint open intervals. If (a, b) is one of those intervals (depending on ω, of course), then $\{W_t(\omega) : a \leq t \leq b\}$ is a continuous function of t that is zero at $t = a$ and $t = b$ but is never 0 for any $t \in (a, b)$. We call this piece of the path of $W_t(\omega)$ an *excursion*.

To be more formal, let $\mathcal{E}$ be the collection of continuous functions f with domain $[0, \infty)$ such that the following hold: there exists a positive real σ_f such that $f(0) = 0$, $f(\sigma_f) = 0$, $f(t) \neq 0$ if $t \in (0, \sigma_f)$, and $f(t) = 0$ if $t > \sigma_f$. We make $\mathcal{E}$ into a metric space by furnishing it with the supremum norm. Given a Borel subset A of $\mathcal{E}$, we say that the Brownian motion W has had an excursion in A by time t if there exists a time u and a function $f \in A$ such that $u + \sigma_f \leq t$ and $W_{u+s}(\omega) = f(s)$ for all $s \leq \sigma_f$. Let $K_t(A)$ be the number of excursions of W in A by time t. Let L_t^0 be Brownian local time at 0, and let

$$T_r = \inf\{t > 0 : L_t^0 \geq r\} \tag{27.1}$$

be the inverse of Brownian local time at zero.

Set

$$N_r(A) = K_{T_r}(A).$$

Although $N_t(A)$ might be identically infinite for some sets A, it will be finite for others. For example, let $\delta > 0$ and suppose that every function in A has a supremum greater than δ. The continuity of the paths of W implies that $N_t(A)$ is finite for every t.

The main result of this section is the following.

Theorem 27.1 $N_t(\cdot)$ *is a Poisson point process.*

Proof If $N_t(B)$ is not infinite, then it has right-continuous paths that increase at most 1 at any given time. The main step will be to show that $N_t(B)$ has stationary increments and $N_t(B) - N_s(B)$ is independent of the σ-field generated by the random variables

$$\{N_r(A) : r \leq s, A \text{ a Borel subset of } \mathcal{E}\}.$$

If $r_1 \leq \cdots \leq r_n \leq s < t$, $k \geq 0$, $j_1, \ldots, j_n \geq 0$, and B and $A_1, \ldots, A_n$ are Borel subsets of $\mathcal{E}$, then

$$\mathbb{P}(N_t(B) - N_s(B) = k; N_{r_1}(A_1) = j_1, \ldots N_{r_n}(A_n) = j_n) \tag{27.2}$$
$$= \mathbb{P}(K_{T_t}(B) - K_{T_s}(B) = k; K_{T_{r_1}}(A_1) = j_1, \ldots, K_{T_{r_n}}(A_n) = j_n)$$
$$= \mathbb{E}\left[\mathbb{P}^{W_{T_s}}(K_{T_{t-s}}(B) - K_{T_0}(B) = k); K_{T_{r_1}}(A_1) = j_1, \ldots, K_{T_{r_n}}(A_n) = j_n\right],$$

where we used the strong Markov property at time T_s. Since T_s is the first time that local time of Brownian motion at 0 exceeds s and L_t^0 increases only when W is at 0, then at time T_s the process W is at 0, so $W_{T_s} = 0$. Therefore the last expression in (27.2) equals

$$\mathbb{P}^0(K_{T_{t-s}}(B) - K_0(B) = k)\mathbb{P}(K_{T_{r_1}}(A_1) = j_1, \ldots, K_{T_{r_n}}(A_n) = j_n),$$

which can be rewritten as

$$\mathbb{P}^0(N_{t-s}(B) - N_0(B) = k)\mathbb{P}(N_{r_1}(A_1) = j_1, \ldots, N_{r_n}(A_n) = j_n).$$

This shows that the law of $N_t(B) - N_s(B)$ is the same as the law of $N_{t-s}(B) - N_0(B)$ and is independent of $\sigma(N_r(A) : r \leq s, A \subset \mathcal{E})$, which is what we wanted.

Observe that $N_t(B)$ is constant except for jumps of size one. By Proposition 5.4, $N_t(B)$ is a Poisson process. It is clear that $N_t(B)$ is a measure in B, which completes the proof. □

Let $m(A) = \mathbb{E}^0 N_1(A)$. The measure A is called the *excursion measure*. We can say a few things about m.

Proposition 27.2 *If*

$$A = \{f \in \mathcal{E} : \sup_t |f(t)| > a\},$$

then $m(A) = 1/a$.

Proof Let $U = \inf\{t : |W_t| = a\}$ and $V = \inf\{t > U : W_t = 0\}$. Since $|W_t| - L_t^0$ is a martingale by Theorem 14.1, then $\mathbb{E}^0|W_{t \wedge U}| = \mathbb{E}^0 L_{t \wedge U}^0$. Letting $t \to \infty$ and using dominated convergence on the left and monotone convergence on the right,

$$\mathbb{E} L_U^0 = \mathbb{E}^0|W_U| = a.$$

Set $R = \inf\{r : N_r(A) = 1\}$. Because $N_r(A)$ is a Poisson process, then R is an exponential random variable with parameter $\mathbb{E} N_1(A) = m(A)$. It therefore suffices to show $\mathbb{E}^0 R = a$; see (A.9).

We have $R = \inf\{r : K_{T_r}(A) = 1\}$, and because K can only increase at times when $W_t = 0$, then

$$R = \inf\{L_t^0 : K_t(A) = 1\}.$$

Now $K_t(A)$ will first equal one when $t = V$. But because local time at 0 does not increase when W is not at 0, $L_V^0 = L_U^0$. Therefore

$$\mathbb{E}^0 R = \mathbb{E}^0 L_V^0 = \mathbb{E}^0 L_U^0 = a.$$

We conclude that $m(A) = 1/a$. □

By symmetry, if $B = \{f \in \mathcal{E} : \sup_t f(t) > a\}$, then $m(B) = 1/(2a)$. One can say more about m. Consider those excursions whose maximum is some fixed value b. Starting at any point other than 0, the excursion can be viewed as a Brownian motion killed at 0 and conditioned to have maximum b. Such a path can be decomposed into the part before the maximum, which is a Brownian motion conditioned to hit b before 0, and the part after the maximum, which is Brownian motion conditioned to hit 0 before b. The former can be shown to have the same law as a three-dimensional Bessel process, up until it hits the level b (see the example in Section 22.2), and the latter the same law as $b - X_t$, where X_t is also a three-dimensional Bessel process up until it hits the level b. Moreover, the part of the path before the maximum can be taken to be independent of the part of the path after the maximum. See Rogers and Williams (2000b) for details.

Let us briefly revisit the Ray–Knight theorems and indicate how Brownian excursions can be used to obtain information about local times at different levels. Fix r and let $T_r = \inf\{t > 0 : L_t^0 \geq r\}$. If $x > 0$ and $y_1, \ldots, y_n < 0$, then the local time at x is a function of the excursions from 0 that hit x and the local times at $y_1, \ldots, y_n$ are functions of the excursions that go below zero. Since the set of excursions that take positive values and those that take negative values are independent, then $L_{T_r}^x$ should be independent of $L_{T_r}^{y_1}, \ldots, L_{T_r}^{y_n}$. To find the distribution of $L_{T_r}^x$, there are a Poisson number of excursions that reach the level x. Each excursion that reaches x contributes an amount to the local time at x that is an exponential random variable; see Exercise 27.1. After proving some additional independence, namely, that the amount each excursion contributes to local time at x is independent of the amount any other excursion contributes and that the amount contributed by an excursion is independent of the number of excursions reaching x, we see that $L_{T_r}^x$ should have the same distribution as a Poisson number of independent exponential random variables.

Exercises

27.1 Let W be a Brownian motion, $x > 0$, and $T = \inf\{t > 0 : W_t = x\}$. If L_t^x is the local time at x, show that the distribution of L_T^x is an exponential random variable. Determine the parameter of this exponential random variable.

27.2 Let W be a one-dimensional Brownian motion. This exercise asks you to prove that the normalized number of downcrossings by time t converges to local time at 0. If $a > 0$, let $S_0 = 0$, $T_0 = \inf\{t : W_t = a\}$, and for $i \geq 1$,

$$S_i = \inf\{t > T_{i-1} : W_t = 0\}, \qquad T_i = \inf\{t > S_i : W_t = a\}.$$

Then $D_t(a)$, the number of downcrossings up to time t, is defined to be $\sup\{k : S_k \leq t\}$. Prove that there exists a constant c such that

$$\lim_{a \to 0} a D_t(a) = c L_t^0, \qquad \text{a.s.,}$$

where L_t^0 is local time at 0 of W. Determine c.
Hint: Use Exercise 18.5.

27.3 Let $(X_t, \mathbb{P}^x)$ be a Brownian motion.
(1) Use the reflection principle to find

$$\mathbb{P}^0(X_s > -a \text{ for all } s \leq r).$$

This is the same as $\mathbb{P}^a(T_0 > r)$, where T_0 is the first time the Brownian motion hits 0.

(2) Let

$$A(a, r) = \{f \in \mathcal{E} : \sup f(t) > a, \sigma_f > r\},$$

$$B(r) = \{f \in \mathcal{E} : \sigma_f > r, \sup f(t) > 0\},$$

and

$$C(a) = \{f \in \mathcal{E} : \sup f(t) > a\}.$$

Prove that

$$m(B(r)) = \lim_{a \to 0} m(A(a, r)) = \lim_{a \to 0} [m(C(a)) \times \mathbb{P}^a(T_0 > r)]$$

and use this and (1) to compute $m(B(r))$. By symmetry, $m(\{f \in \mathcal{E} : \sigma_f > r\})$ will be twice the value of $m(B(r))$.

27.4 Let W be a Brownian motion. Let $E_t(r)$ be the number of excursions of length larger than r that have been completed by time t. An excursion of length larger than r means that $\sigma_f > r$. Show that there exists a constant c such that

$$\lim_{r \to 0} \sqrt{r} E_t(r) = c L_t^0, \qquad \text{a.s.}$$

Determine c.

One interesting point here is that this shows that L_t^0 is determined entirely by the zero set $Z(\omega) = \{t : W_t(\omega) = 0\}$.

27.5 Let $\delta > 0$ and $A_\delta = \{f \in \mathcal{E} : \sup_t |f(t)| > \delta\}$. Let $S_1 = \inf\{t : K_t(A_\delta) = 1\}$ and $S_2 = \inf\{t > S_1 : K_t(A_\delta) = 2\}$. Thus S_1 and S_2 are the times the first and second excursions in A_δ have been completed. Let $Y_1(t)$ be the excursion completed at time S_1 and define $Y_2(t)$ similarly. To be more precise, if $R_1 = \sup\{t < S_1 : W_t = 0\}$, then $Y_1(s) = W_{R_1+s}$ if $s \leq S_1 - R_1$ and $Y_1(s)$ is equal to 0 for all $s \geq S_1 - R_1$.

Prove that Y_1 and Y_2 are independent.

Hint: Use the strong Markov property at time S_1.

Notes

Besides its use in the Ray–Knight theorems (Rogers and Williams, 2000b), excursion theory is useful in many other contexts. See Rogers and Williams (2000b) for applications to Skorokhod embedding and to the arc sine law.

28

Financial mathematics

A European call option is the option to buy a share of stock at a given price at some particular time in the future. For example, I might buy a call option to purchase one share of Company X for $40 three months from today. When the three months is up, I check the price of Company X. If, say, it is $35, then my option is worthless, because why would I buy a share for $40 using the option when I could buy it on the open market for $35? But if three months from now, the share price is, say, $45, then I can exercise my option, which means I buy a share for $40, and I can then turn around immediately and sell that share for $45 and make a profit of $5. Thus, today, there is a potential for a profit if I have a call option, and so I should pay something to purchase that option. A significant part of financial mathematics is devoted to the question of what is the fair price I should pay for a call option.

Options originated in the commodities market, where farmers wanted to hedge their risks. Since then many types of options have been developed (options are also known as derivatives), and the amount of money invested in options has for the past several years exceeded the amount of money invested in stocks.

In 1973 Black and Scholes, using the reasonable principle that you can't get something for nothing, came up with a convincing formula for the price of an option. This chapter gives two derivations of the Black–Scholes formula, proves the fundamental theorem of finance, and finishes by considering a stochastic control problem. The Black–Scholes formula is a beautiful example of applied stochastic processes.

28.1 Finance models

Let W_t be a Brownian motion. We assume that S_t is the price of a stock or other risky security. If we have $2,000 and we buy 100 shares in a stock that sells for $20 per share and it goes up $2, or if we buy 10 shares in a stock selling for $200 per share that goes up $20, we are equally happy; it is the percentage increase that matters. With this in mind, we assume that S_t satisfies

$$dS_t = \mu S_t \, dt + \sigma S_t \, dW_t. \tag{28.1}$$

This is plausible, since then $dS_t/S_t = \mu \, dt + \sigma \, dW_t$, that is, we are assuming the relative change in price is a multiple of Brownian motion with drift. The quantity μ is known as the *mean rate of return* and σ is called the *volatility*. The solution to this SDE is

$$S_t = S_0 e^{\sigma W_t + (\mu - (\sigma^2/2))t} \tag{28.2}$$

by Proposition 24.6.

We also assume the existence of a bond with price B_t, which is assumed to be riskless, and the equation for B_t is

$$dB_t = rB_t \, dt,$$

which implies

$$B_t = B_0 e^{rt}.$$

Suppose at time t one buys A shares of stock. The cost is AS_t. If one sells the shares at time $t + h$, one receives AS_{t+h}, and the net gain is $A(S_{t+h} - S_t)$. One can also sell short, i.e., let A be negative. The formula for the gain is the same.

Suppose at time t_i one holds A_i shares, up until time t_{i+1}. The total net gain over the whole period t_0 to t_n is $\sum_{i=0}^{n-1} A_i(S_{t_{i+1}} - S_{t_i})$. This is the same as the stochastic integral $\int_0^t a_t \, dS_t$ if a_t equals A_i when $t_i \leq t < t_{i+1}$.

One should allow A_i to depend on the entire past $\mathcal{F}_{t_i}$. Idealizing, one allows continuous trading, and if a_s is the number of shares held at time s, the net gain through trading the stock is $\int_0^t a_s \, dS_s$. One has a similar net gain of $\int_0^t b_s \, dB_s$ when trading bonds if b_s is the number of bonds held at time s.

Although a_t can depend on the entire past $\mathcal{F}_t$, one does not want to let a_t depend on the future. This helps explain why the class of predictable integrands is the appropriate one to use.

The pair (a, b) is called a *trading strategy*. Set

$$V_t = a_t S_t + b_t B_t, \tag{28.3}$$

the amount of wealth one has at time t. The strategy is *self-financing* if

$$V_t = V_0 + \int_0^t a_s \, dS_s + \int_0^t b_s \, dB_s \tag{28.4}$$

for all t. The first integral represents the net gain from trading in the stock, the second integral the net gain from trading in the bond, and (28.4) says that one's wealth at time t is equal to what one starts with plus what one has realized through trading in the stock and bond. We assume throughout that there are no transaction costs (i.e., no brokerage fees).

A European call gives the buyer the option of buying a share of the stock at a fixed time t_E at price K. The time t_E is called the *exercise time*. After time t_E, the option has expired and is worthless.

What is the option worth? At time t_E, if $S_{t_E} \leq K$, the option is worth nothing, for who would pay K dollars for a share of stock when it sells for S_{t_E} dollars? If $S_{t_E} > K$, one can use the option to buy a share of the stock at price K and immediately sell it at price S_{t_E}, to make a profit of $S_{t_E} - K$. Thus the value of the option at time t_E is $(S_{t_E} - K)^+$. An important question is: how much should the option sell for? What is a fair price for the option at time 0?

There are a myriad of types of options. The American call is almost the same as the European call, except that one is allowed to buy a share of the stock at price K at any time in the interval $[0, t_E]$. The European put gives the buyer the option to sell a share of the stock at price K at time t_E, while the American put gives the buyer the option to sell a share at price K anytime before time t_E.

28.2 Black–Scholes formula

In 1973 Black and Scholes came up with their formula for the price of a European call. We will give two derivations of this formula.

Derivation 1. First of all, the interest rate r on the bond may be considered to be the same as the rate of inflation. Thus the value of the option $(S_{t_E} - K)^+$ in today's dollars is

$$C = e^{-rt_E}(S_{t_E} - K)^+. \tag{28.5}$$

In this first derivation we work in today's dollars. Therefore the present-day value of the stock is $P_t = e^{-rt}S_t$. Note $P_0 = S_0$ and the present-day value of our option at time t_E is then

$$C = e^{-rt_E}(S_{t_E} - K)^+ = (P_{t_E} - e^{-rt_E}K)^+. \tag{28.6}$$

By the product formula,

$$\begin{aligned} dP_t &= e^{-rt}\,dS_t - re^{-rt}S_t\,dt \\ &= e^{-rt}\sigma S_t\,dW_t + e^{-rt}\mu S_t\,dt - re^{-rt}S_t\,dt \\ &= \sigma P_t\,dW_t + (\mu - r)P_t\,dt. \end{aligned}$$

The solution to this stochastic differential equation (see Proposition 24.6) is

$$P_t = P_0 e^{\sigma W_t + (\mu - r - \sigma^2/2)t}.$$

Also, the net gain or loss in present-day dollars when holding a_s shares of stock at time s is $\int_0^t a_s\,dP_s$.

Define $\mathbb{Q}$ on $\mathcal{F}_{t_E}$ by

$$d\mathbb{Q}/d\mathbb{P} = M_{t_E} = \exp\left(-\frac{\mu - r}{\sigma}W_{t_E} - \frac{(\mu - r)^2}{2\sigma^2}t_E\right).$$

Under $\mathbb{Q}$, $\widetilde{W}_t = W_t + \frac{\mu - r}{\sigma}t$ is a Brownian motion by the Girsanov theorem.

Now

$$dP_t = \sigma P_t\,dW_t + (\mu - r)P_t\,dt = \sigma P_t\left(dW_t + \frac{\mu - r}{\sigma}\,dt\right) = \sigma P_t\,d\widetilde{W}_t.$$

Therefore under $\mathbb{Q}$, P_t is a martingale since stochastic integrals with respect to martingales are martingales. The solution to the SDE

$$dP_t = \sigma P_t\,d\widetilde{W}_t$$

is

$$P_t = P_0 e^{\sigma \widetilde{W}_t - (\sigma^2/2)t}, \tag{28.7}$$

so P_t and $\widetilde{W}_t$ have the same filtration.

C is $\mathcal{F}_{t_E}$ measurable. By the martingale representation theorem (Theorem 12.3), there exists an adapted process A_s such that

$$C = \mathbb{E}_{\mathbb{Q}}C + \int_0^{t_E} A_s\,d\widetilde{W}_s = \mathbb{E}_{\mathbb{Q}}C + \int_0^{t_E} D_s\,dP_s,$$

where $D_s = A_s/(\sigma P_s)$.

Therefore, if one follows the trading strategy of buying and selling the stock S_t, where one holds D_s shares of stock at time s, one can obtain $C - \mathbb{E}_\mathbb{Q}C$ dollars at time t_E. Or, starting with $\mathbb{E}_\mathbb{Q}C$ dollars and buying and selling stock, one can get the identical output as C, almost surely. A standard assumption in finance is that of no arbitrage, which means you cannot make a profit without taking some risk. To avoid riskless profits, C must sell for $\mathbb{E}_\mathbb{Q}C$.

To explain this in more detail, suppose you could sell the European call for C' dollars. If $C' > \mathbb{E}_\mathbb{Q}C$, you could sell a call for C' dollars, use the money and invest in the trading strategy of holding D_s shares of stock at time s, and at time t_E have $C' + C - \mathbb{E}_\mathbb{Q}C$ worth of stocks and options. The buyer of the option decides whether to exercise the option, and it costs you C dollars to meet that obligation. With probability one, you have gained $C' - \mathbb{E}_\mathbb{Q}C$ dollars, a riskless profit. If $C' < \mathbb{E}_\mathbb{Q}C$, simply reverse the roles of buying and selling. The only way to avoid making a riskless profit is if $C' = \mathbb{E}_\mathbb{Q}C$.

To find $\mathbb{E}_\mathbb{Q}C$, using (28.6) and (28.7) we write

$$\mathbb{E}_\mathbb{Q}C = \mathbb{E}_\mathbb{Q}[(S_0 e^{\sigma \widetilde{W}_{t_E} - \sigma^2 t_E/2} - e^{-rt_E}K)^+] \tag{28.8}$$
$$= \frac{1}{\sqrt{2\pi t_E}} \int (S_0 e^{\sigma y - \sigma^2 t_E/2} - e^{-rt_E}K)^+ e^{-y^2/2t_E}\, dy,$$

which is the Black–Scholes formula. One can, if one wishes, perform some calculations to find alternate expressions for the right-hand side.

It is noteworthy that μ does not appear in (28.8)! You and I might have different opinions as to what μ, the mean rate of return, is equal to, but we should agree on the price of the call. This was a shock to economists when this was first discovered. The value of σ, the volatility, does enter into the formula.

Until we evaluated $\mathbb{E}_\mathbb{Q}C$ in (28.8), the actual form of C was unimportant. For any type of option expiring at time t_E, Derivation 1 tells us that its price at time zero should be its expectation under $\mathbb{Q}$.

Derivation 2. In this approach, which is the one used by Black and Scholes, we use the actual values of the securities, not the present-day values. Let V_t be the value of the option at time t and assume

$$V_t = f(S_t, t_E - t) \tag{28.9}$$

for all t, where f is some function that is sufficiently smooth. We also want $V_{t_E} = (S_{t_E} - K)^+$.

Recall the multivariate version of Itô's formula (Theorem 11.2). We apply this with $d = 2$ and $X_t = (S_t, t_E - t)$. From (28.1),

$$\langle S \rangle_t = \sigma^2 S_t^2\, dt,$$

$\langle t_E - t \rangle_t = 0$ since $t_E - t$ is of bounded variation and hence has no martingale part, and $\langle S, t_E - t \rangle_t = 0$. Also, $d(t_E - t) = -dt$. Then

$$V_t - V_0 = f(S_t, t_E - t) - f(S_0, t_E) \tag{28.10}$$
$$= \int_0^t f_x(S_u, t_E - u)\, dS_u - \int_0^t f_t(S_u, t_E - u)\, du$$
$$+ \tfrac{1}{2}\int_0^t \sigma^2 S_u^2 f_{xx}(S_u, t_E - u)\, du.$$

Here f_x is the partial derivative with respect to x, the first variable, f_{xx} is the second partial derivative with respect to x, and f_t is the partial derivative with respect to t, the second variable. On the other hand,

$$V_t - V_0 = \int_0^t a_u \, dS_u + \int_0^t b_u \, dB_u. \tag{28.11}$$

By (28.3) and (28.9),

$$b_t = \frac{V_t - a_t S_t}{B_t} = \frac{f(S_t, t_{Et} - t) - a_t S_t}{B_t}. \tag{28.12}$$

Also, recall $B_t = B_0 e^{rt}$. Comparing (28.10) with (28.11), we must therefore have

$$a_t = f_x(S_t, t_E - t) \tag{28.13}$$

and

$$- f_t(S_t, t_E - t) + \tfrac{1}{2}\sigma^2 S_t^2 f_{xx}(S_t, t_E - t) = b_t B_0 r e^{rt}. \tag{28.14}$$

Substituting for b_t using (28.12),

$$r[f(S_t, t_E - t) - S_t f_x(S_t, t_E - t)] \tag{28.15}$$
$$= -f_t(S_t, t_E - t) + \tfrac{1}{2}\sigma^2 S_t^2 f_{xx}(S_t, t_E - t)$$

for almost all t and all S_t. Since S_t is a continuous process, (28.15) leads to the parabolic partial differential equation (PDE)

$$f_t = \tfrac{1}{2}\sigma^2 x^2 f_{xx} + rx f_x - rf, \qquad (x, s) \in (0, \infty) \times [0, t_E),$$

and

$$f(x, 0) = (x - K)^+.$$

Solving this equation for f, $f(x, t_E)$ tells us what V_0 should be, i.e., the cost of setting up the equivalent portfolio. This partial differential equation can be solved and the solution is the Black–Scholes formula. Equation (28.13) shows what the trading strategy should be.

Let us now briefly discuss American calls. Recall that these are ones where the holder can buy the security at price K at any time up to time t_E. Since the holder of an American call can always wait up to time t_E, which is equivalent to having a European call, the value of an American call should always be at least as large as the value of the corresponding European call.

Suppose one exercises an American call early. If $S_{t_E} > K$ and one exercised early, at time t_E one has one share of stock, for which one paid K, and one has a profit of $(S_{t_E} - K)$. However, because one purchased the stock before time t_E, one lost the interest $K e^{r(t_E - t)}$ that would have accrued by waiting to exercise the option. (We are supposing $r \geq 0$.) Thus in this case it would have been better to wait until time t_E to exercise the option.

On the other hand, if $S_{t_E} < K$, exercising the option early would mean that one has lost $|S_{t_E} - K|$, whereas for the European option, one would have not exercised at all, and lost nothing (other than the price of the option).

In either case, exercising early gains nothing, hence the price of an American call should be the same as that of a European call.

One can equally well price the European put, the option to sell a share of stock at price K at time t_E, by either Derivation 1 or Derivation 2 of the Black–Scholes formula. However this analysis breaks down for American puts (sell a share of stock anytime up to time t_E), because in this case one gains by selling early: one can earn interest on the money received.

28.3 The fundamental theorem of finance

In the preceding section, we showed there was a probability measure $\mathbb{Q}$ under which P_t was a martingale. This is true very generally. Let S_t be the price of a security in present-day dollars. We will suppose S_t is a continuous semimartingale, and can be written $S_t = M_t + A_t$.

The *NFLVR condition* ("no free lunch with vanishing risk") is that one cannot find fixed positive real numbers $t_0, \varepsilon, b > 0$, and predictable processes H_n with $\int_0^{t_0} |H_n(s)| \, |dA_s| + \int_0^{t_0} H_n^2 \, d\langle M \rangle_s < \infty$, a.s., for each n such that

$$\int_0^{t_0} H_n(s) \, dS_s > -\frac{1}{n}, \qquad \text{a.s.,}$$

for all n and

$$\mathbb{P}\left(\int_0^{t_0} H_n(s) \, dS_s > b \right) > \varepsilon.$$

Here t_0, b, ε do not depend on n. The condition says that one can with positive probability ε make a profit of b and with a loss no larger than $1/n$. $\mathbb{Q}$ is an *equivalent martingale measure* if $\mathbb{Q}$ is a probability measure, $\mathbb{Q}$ is equivalent to $\mathbb{P}$ (which means they have the same null sets), and S_t is a local martingale under $\mathbb{Q}$.

Theorem 28.1 *If S_t is a continuous semimartingale and the NFLVR condition holds, then there exists an equivalent martingale measure $\mathbb{Q}$.*

Proof Let us prove first of all that dA_t is absolutely continuous with respect to $d\langle M \rangle_t$. We suppose not and obtain a contradiction. Consider the measures μ_A and $\mu_{\langle M \rangle}$ on the predictable σ-field defined by

$$\mu_A(D) = \mathbb{E} \int_0^\infty 1_D \, dA_t, \qquad \mu_{\langle M \rangle}(D) = \mathbb{E} \int_0^\infty 1_D \, d\langle M \rangle_t. \tag{28.16}$$

Since A is of bounded variation and continuous, it is a predictable process, and we can write $A_t = B_t - C_t$, where B and C are continuous increasing processes and where μ_B and μ_C are mutually singular measures on the predictable σ-field; we define μ_B and μ_C analogously to (28.16). To give a few more details on how to do this, we write $A_t = B'_t - C'_t$, where B' and C' are continuous increasing processes, we find non-negative predictable processes b_t and c_t such that $B'_t = \int_0^t b_t \, d(B'_t + C'_t)$ and $C'_t = \int_0^t c_t \, d(B'_t + C'_t)$, and then let $B_t = \int_0^t (b_t - (b_t \wedge c_t)) \, d(B'_t + C'_t)$ and $C_t = \int_0^t (c_t - (b_t \wedge c_t)) \, d(B'_t + C'_t)$. We leave it to the reader to check that B and C are the desired processes. Since μ_B and μ_C are mutually singular, there exists a set E in the predictable σ-field such that $\mu_B(D) = \mu_B(D \cap E)$ and $\mu_C(D) = \mu_C(D \cap E^c)$ for all sets D in the predictable σ-field.

If μ_A is not absolutely continuous with respect to $\mu_{\langle M \rangle}$, then at least one of μ_B and μ_C is not absolutely continuous. We assume that μ_B is not, for otherwise we can look at $-S_t$ instead of S_t. Therefore there exists a predictable set F and a fixed time t_0 such that $\int_0^{t_0} 1_F \, dB_s$ is almost surely non-negative, is strictly positive with positive probability, and $\int_0^{t_0} 1_F \, d\langle M \rangle_s = 0$. We can replace F by $F \cap E$ and so assume that $F \subset E$, and hence $\mu_C(F) = \mu_C(F \cap E^c) = 0$. Then

$$\int_0^{t_0} 1_F \, dS_s = \int_0^{t_0} 1_F \, dM_s + \int_0^{t_0} 1_F \, dB_s + \int_0^{t_0} 1_F \, dC_s.$$

The stochastic integral term is 0 because $\int_0^{t_0} (1_F)^2 \, d\langle M \rangle_s = 0$. The integral with respect to C_s is zero because $\mu_C(F) = 0$. We then have the NFLVR condition violated with $H_n = 1_F$ for all n. Hence absolute continuity is established, and by the Radon–Nikodym theorem, $A_t = \int_0^t h_s \, d\langle M \rangle_s$ for some predictable process h_s.

Our next goal is to show $\int_0^t h_s^2 \, d\langle M \rangle_s < \infty$ for all t. Let

$$U = \inf \left\{ t : \int_0^t h_s^2 \, d\langle M \rangle_s = \infty \right\}.$$

On $(U < \infty)$ there are two possibilities:

(1) $\int_0^t h_s^2 \, d\langle M \rangle_s < \infty$ if $t < U$ but $\int_0^U h_s^2 \, d\langle M \rangle_s = \infty$, and

(2) $\int_0^U h_s^2 \, d\langle M \rangle_s < \infty$ but $\int_U^{U+\varepsilon} h_s^2 \, d\langle M \rangle_s = \infty$ for all ε.

(For a real variable analog, consider the two functions $f_1(t) = \int_{-1}^t \frac{1}{|x|} \, dx$ and $f_2(t) = \int_{-1}^t 1_{(x>0)} \frac{1}{x} \, dx$ at $t = 0$.)

Let us investigate case (1) and show that it cannot happen. Choose a fixed time t_0 such that $\mathbb{P}(U < t_0) > 0$. Let

$$R_1 = R_1(n) = \inf \left\{ t : \int_0^t h_s^2 \, d\langle M \rangle_s \geq n^4 \right\} \wedge U \wedge t_0.$$

We suppose

$$\inf_n \mathbb{P}(R_1(n) < U \wedge t_0) > 0 \tag{28.17}$$

and obtain a contradiction. Let $H_t = h_t 1_{[0,R_1]}/n^4$. Then

$$\int_0^{t_0} H_s \, dA_s = \int_0^{R_1} \frac{h_s^2}{n^4} \, d\langle M \rangle_s \geq 1$$

on $(R_1 < U < t_0)$. On the other hand,

$$\mathbb{E} \left(\sup_{t \leq t_0} \left| \int_0^t H_s \, dM_s \right| \right)^2 \leq 4\mathbb{E} \int_0^{t_0} H_s^2 \, d\langle M \rangle_s \leq \frac{4}{n^8} n^4 = \frac{4}{n^4}$$

by Doob's inequalities. Therefore

$$\mathbb{P} \left(\sup_{t \leq t_0} \left| \int_0^t H_s \, dM_s \right| > \frac{1}{n} \right) \leq \frac{\mathbb{E} \sup_{t \leq t_0} \left| \int_0^t H_s \, dM_s \right|^2}{n^{-2}} \leq \frac{4/n^4}{n^{-2}} = \frac{4}{n^2}.$$

Let

$$R_2 = R_2(n) = \inf \left\{ t : \left| \int_0^t H_s \, dM_s \right| \geq 1/n \right\}$$

and let $\widetilde{H}_t = H_t 1_{[0,R_2]}$. We then have

$$\mathbb{P}(R_2 < R_1) \leq \mathbb{P}(R_2 \leq t_0) \leq 4/n^2,$$

$$\int_0^{t_0} \widetilde{H}_s \, dS_s = \int_0^{R_2} \widetilde{H}_s \, dM_s + \int_0^{R_2} \widetilde{H}_s \, dA_s$$

$$\geq -\frac{1}{n} + \int_0^{R_2} \frac{h_s^2}{n^4} \, d\langle M \rangle_s \geq -1/n$$

almost surely, and

$$\mathbb{P}\left(\int_0^{t_0} \widetilde{H}_s \, dS_s \geq \frac{1}{2}\right) \geq \mathbb{P}(R_1 < U < t_0) - \mathbb{P}(R_2 < R_1)$$

$$\geq \mathbb{P}(R_1 < U < t_0) - \frac{4}{n^2}.$$

We do this for each n, and thus obtain a contradiction to the NFLVR condition, so (28.17) cannot hold.

Case (2) is similar: choose δ_n such that $\int_{U+\delta_n}^{U+1} h_s^2 \, d\langle M \rangle_s \geq n^4$ with positive probability, let $H_t = h_t 1_{[U+\delta,U+1]}/n^4$, and proceed as above. We leave the details as Exercise 28.3.

We thus have $\int_0^t h_s^2 \, d\langle M \rangle_s < \infty$, a.s., for each t. Consequently the quantity $\sup_{s \leq t} |\int_0^s h_r \, dM_r|$ is also finite. Let

$$V_n = \inf\left\{t : \left|\int_0^t h_s \, dM_s\right| \geq n \text{ or } \int_0^t h_s^2 \, d\langle M \rangle_s \geq n\right\}.$$

We conclude $V_n \uparrow \infty$.

Define $\mathbb{Q}$ on $\mathcal{F}_{V_n}$ by

$$d\mathbb{Q}/d\mathbb{P} = \exp\left(-\int_0^{V_n} h_s \, dM_s - \frac{1}{2}\int_0^{V_n} h_s^2 \, d\langle M \rangle_s\right).$$

The exponent is bounded, so $\mathbb{Q}$ is well defined. Under $\mathbb{Q}$, if $t \leq V_n$, then

$$M_t - \left\langle -\int_0^{\cdot} h_s \, dM_s, M \right\rangle_t = M_t + \int_0^t h_s \, d\langle M_s \rangle = M_t + A_t$$

is a martingale by the Girsanov theorem (Exercise 13.5). Therefore $S_t = M_t + A_t$ is a local martingale.

Finally, $e^{-\int_0^t h_s \, dM_s - \frac{1}{2}\int_0^t h_s^2 \, d\langle M \rangle_s}$ is never zero nor infinite, so $\mathbb{Q}$ is equivalent to $\mathbb{P}$. $\square$

Let us give two examples to clarify the proof. Let C be the standard Cantor set and let $g(t)$ be the Cantor function. Suppose $S_t = W_t + g(t)$, where W is a Brownian motion. We then let $H_t = 1_C(t)$. Since the Cantor function increases only on the Cantor set, $\int_0^1 H_s \, dg(s) = 1$. Since the Cantor set has Lebesgue measure 0, then $\int_0^1 H_s^2 \, ds = 0$. But this is the quadratic variation of $\int_0^1 H_s \, dW_s$, so this stochastic integral is also 0. It follows that

$$\int_0^1 H_s \, dS_s = \int_0^1 H_s \, dW_s + \int_0^1 H_s \, dg(s) = 1,$$

which says that with the trading strategy H we make a profit of 1 almost surely, that is, without any risk. Therefore the NFLVR condition is violated. This example indicates why we must have dA_t absolutely continuous with respect to $d\langle M\rangle_t$.

Suppose now that W is a Brownian motion and $S_t = W_t + \int_0^t H_s\, ds$ with H_s bounded. Let

$$M_t = e^{-\int_0^t H_s\, dW_s - \frac{1}{2}\int_0^t H_s^2\, ds},$$

and define $\mathbb{Q}$ on $\mathcal{F}_1$ by $d\mathbb{Q}/d\mathbb{P} = M_1$. By the Girsanov theorem, $S_t = W_t + \int_0^t H_s\, ds$ is a martingale under $\mathbb{Q}$. This example shows that if the Radon–Nikodym derivative of dA_t with respect to $d\langle M\rangle_t$ is not too bad, we can apply the Girsanov theorem.

28.4 Stochastic control

The theory of stochastic control, which includes a study of the Hamilton–Jacobi–Bellman (HJB) equation and requires some knowledge of partial differential equations, is beyond the scope of this book. However, we can consider one simple useful example. Suppose we have available to us a stock which satisfies the SDE

$$dS_t = \sigma S_t\, dW_t + \mu S_t\, dt,$$

where W_t is a Brownian motion, and a risk-free asset which satisfies the equation

$$dB_t = r B_t\, dt.$$

We want to put a proportion u of our wealth Z_t into the stock and the remainder into the risk-free asset. We will restrict $0 \le u \le 1$, so that we do not borrow nor have short selling. Also, we take $\mu > r$, for if the mean rate of return on the stock is less than the risk-free rate, we simply put all our money in the risk-free asset. How do we choose u in order to maximize our return?

First of all, what do we mean by maximizing our return? Typically one chooses ahead of time a deterministic function U, called the utility function, and one wants to maximize $\mathbb{E}\, U(Z_{t_0})$ at some fixed time t_0. Usually utility functions are taken to be increasing and concave. The function is chosen to be increasing because more money is considered better. It is chosen concave because one assumes that twice the amount of money will give increased pleasure, but not twice as much pleasure.

Let us work out the optimal control problem when $U(x) = x^p$ for some $p \in (0, 1)$. If Z_t (depending on u) is our wealth, we have $Z_t = S_t + B_t$ and $S_t = u Z_t$, $B_t = (1 - u)Z_t$. We will allow u to depend on t and ω, but our answer will turn out to be deterministic and independent of t, i.e., u is a constant.

We have seen (Proposition 24.6) that

$$S_t = S_0 e^{\sigma W_t - \sigma^2 t/2 + \mu t}$$

and $\langle S\rangle_t = \sigma^2 S_t^2\, dt$ and that the equation for B_t has the solution

$$B_t = B_0 e^{rt}.$$

Therefore neither S_t nor B_t can ever be 0 or negative, and so $Z_t > 0$ for all t. Applying Itô's formula to Z_t^p and noting that $\langle Z \rangle_t = \langle S \rangle_t$, we have

$$
\begin{aligned}
dZ_t^p &= pZ_t^{p-1} \, dZ_t + \tfrac{1}{2}p(p-1)Z_t^{p-2} \, d\langle Z \rangle_t \\
&= pZ_t^{p-1}\sigma S_t \, dW_t + pZ_t^{p-1}\mu S_t \, dt + pZ_t^{p-1} r B_t \, dt \\
&\quad + \tfrac{1}{2}p(p-1)Z_t^{p-2}\sigma^2 S_t^2 \, dt \\
&= puZ_t^p \sigma \, dW_t + puZ_t^p \mu \, dt + p(1-u)rZ_t^p \, dt \\
&\quad + \tfrac{1}{2}p(p-1)Z_t^p \sigma^2 u^2 \, dt.
\end{aligned}
$$

Therefore

$$
\mathbb{E} \, Z_{t_0}^p = \mathbb{E} \, Z_0^p + p\mathbb{E} \int_0^{t_0} Z_t^p [u\mu + (1-u)r + \tfrac{1}{2}(p-1)\sigma^2 u^2] \, dt.
$$

This will be largest if the expression

$$
F(u) = u\mu + (1-u)r + \tfrac{1}{2}(p-1)\sigma^2 u^2
$$

is largest, which by elementary calculus is largest when

$$
u = \frac{\mu - r}{(1-p)\sigma^2}.
$$

Exercises

28.1 Let

$$
\Phi(x) = \frac{1}{\sqrt{2\pi}} \int_{-\infty}^x e^{-y^2/2} \, dy,
$$

the *cumulative normal distribution function*. Rewrite the Black–Scholes formula for the value of a European call in terms of Φ. This is the way the Black–Scholes formula is written in finance books.

28.2 A European put that gives one the option to sell a share of stock at price K at time t_E has value $(K - S_{t_E})^+$ at time t_E. Find the present-day value of the European put at time 0.

28.3 Carry out the details of the proof of Theorem 28.1 for Case 2.

28.4 If the utility function in Section 28.4 is $U(x) = \log x$ instead of $U(x) = x^p$, what is the optimal choice for u?

28.5 Let $a, b > 0$, let Y_i be i.i.d. random variables that take only the values b and $-a$, and let $S_n = \sum_{i=1}^n Y_i$. Show that if $\mathbb{P}(Y_1 = b) > 0$ and $\mathbb{P}(Y_1 = -a) > 0$, there exists a probability measure $\mathbb{Q}$ equivalent to $\mathbb{P}$ under which S_n is a martingale. Describe the Radon–Nikodym derivative of $\mathbb{Q}$ with respect to $\mathbb{P}$.

28.6 Suppose the interest rate r is equal to 0 and an option V has payoff

$$
\sup_{s \leq t_e} S_s
$$

at time t_e. What is the price of V at time 0?

28.7 Suppose the interest rate r is equal to 0. Let U be the option that pays off $-\inf_{s \leq t_e} S_s$ at time t_e. What is the price of U at time 0?

   If V is as in Exercise 28.6, then $U + V$ is the option that pays off the maximum of the stock price minus the minimum of the stock price, in other words, "buy low, sell high." Naturally such an option would be expensive. It is remarkable that there exists a trading strategy that can duplicate this payoff, even though the times when the maximum and minimum occur are not stopping times.

29

Filtering

Stochastic filtering is a nice example of nontrivial interesting mathematics that is extremely useful. For example, it has been used extensively in NASA's space program.

The method we use is called the *innovations approach* to filtering, and uses Lévy's theorem, the martingale representation theorem, and other results from stochastic calculus.

We will start with a fairly general model, except for simplicity we will assume our observation process is one-dimensional. The extension to the d-dimensional case is mostly routine. Later on we will look at a specific model, the linear model, where one can give fairly explicit solutions to the filtering equation for real-life problems.

29.1 The basic model

We start with a probability space $(\Omega, \mathcal{F}, \mathbb{P})$, together with a filtration $\{\mathcal{F}_t\}$ satisfying the usual conditions. In filtering theory, there are a number of filtrations present, and we will need to be careful about which ones are which.

We have a *signal process* X_t taking values in a complete separable metric space $\mathcal{S}$ and we let $\{\mathcal{F}_t^X\}$ be the minimal augmented filtration generated by X. We have a function f mapping $\mathcal{S}$ to the reals, we suppose $\mathbb{E}\,|f(X_t)|^2 < \infty$ for all t, and we suppose that there exists a process A_s adapted to the filtration $\{\mathcal{F}_t^X\}$ such that

$$M_t = f(X_t) - f(X_0) - \int_0^t A_s\, ds$$

is a martingale with respect to the filtration $\{\mathcal{F}_t^X\}$. Next we discuss the observation process. Let W_t be a one-dimensional Brownian motion with respect to the filtration $\{\mathcal{F}_t^X\}$, let h_t be a real-valued process adapted to $\{\mathcal{F}_t^X\}$, and suppose

$$Z_t = W_t + \int_0^t h_s\, ds. \tag{29.1}$$

The process Z_t is called the *observation process* and is what we observe. Let $\{\mathcal{F}_t^Z\}$ be the filtration generated by the process Z. In practice one does not necessarily want to assume that $\{\mathcal{F}_t^Z\}$ is right continuous, but let us assume that it is for simplicity. Requiring the filtration to be complete is not a serious issue.

For an example, suppose that $dX_t = \sigma(X_t)\, d\overline{W}_t + b(X_t)\, dt$ as in Chapter 24, where $\overline{W}_t$ is a d-dimensional Brownian motion and σ and b are matrix valued, and suppose $f \in C^2(\mathbb{R}^d)$ is bounded or has linear growth. Then Itô's formula shows that such an f will satisfy our

229

assumptions. In this case h_s in (29.1) is of the form $g(X_s)$ for a particular function g; see Section 39.3.

The goal of filtering is to get the best estimate of $f(X_t)$ from the observations $\{Z_t\}$. We want to find the best estimate for $f(X_t)$ in the following sense. We want to minimize the mean square error $\mathbb{E}\,|f(X_t) - Y|^2$ over all random variables Y that are $\mathcal{F}_t^Z$ measurable, i.e., over all random variables that can be determined by the observations up to time t. The rationale is that since $\mathcal{F}_t^Z$ is the information we have observed up to time t, we want our estimate to be $\mathcal{F}_t^Z$ measurable, and among all random variables that are $\mathcal{F}_t^Z$ measurable, we want the one closest to $f(X_t)$ in L^2 norm, which means we minimize the mean square error.

Lemma 29.1 *The best mean square error estimate of $f(X_t)$ over the class of $\mathcal{F}_t^Z$ measurable random variables is*

$$Y = \mathbb{E}\,[f(X_t) \mid \mathcal{F}_t^Z].$$

Proof By our assumptions on f, the random variable $V = f(X_t)$ is in $L^2(\mathbb{P})$. Let Y be the best mean square estimator. The collection $\mathcal{M}$ of L^2 random variables which are $\mathcal{F}_t^Z$ measurable is a linear subspace of L^2, and the element of a Hilbert space that minimizes the distance from V to this subspace $\mathcal{M}$ is the projection onto $\mathcal{M}$. Therefore Y is the projection of V onto $\mathcal{M}$. Hence $V - Y$ is orthogonal (in the L^2 sense) to every element of $\mathcal{M}$. In particular, if $E \in \mathcal{F}_t^Z$,

$$\mathbb{E}\,[(V - Y)1_E] = 0,$$

which implies $\mathbb{E}\,[V; E] = \mathbb{E}\,[Y; E]$. This holds for every $E \in \mathcal{F}_t^Z$ and Y is $\mathcal{F}_t^Z$ measurable, hence $Y = \mathbb{E}\,[V \mid \mathcal{F}_t^Z]$. $\quad\square$

Given any process H_t that is $\{\mathcal{F}_t\}$ adapted, we use the notation $\widehat{H}_t = \mathbb{E}\,[H_t \mid \mathcal{F}_t^Z]$. We will look at expressions like $\int_0^t \widehat{H}_s\,ds$, and you might wonder about the joint measurability of $\widehat{H}$ in ω and t, since $\widehat{H}_t$ is only defined almost surely for each t. The way to deal with this is to let $\widehat{H}_t$ be the optional projection of H with respect to the optional σ-field generated by $\{\mathcal{F}_t^Z\}$; see (16.8) in Chapter 16.

29.2 The innovation process

We next define the *innovation process*

$$N_t = Z_t - \int_0^t \widehat{h}_s\,ds. \tag{29.2}$$

(Following our convention on notation, $\widehat{h}_s = \mathbb{E}\,[h_s \mid \mathcal{F}_s^Z]$.) Note that although N_t is $\mathcal{F}_t^Z$ measurable, we cannot determine it from (29.2) because it contains the unknown $\widehat{h}_s$ on the right-hand side.

Proposition 29.2 *N_t is a Brownian motion with respect to the filtration $\{\mathcal{F}_t^Z\}$.*

Proof We will show that N_t is a continuous martingale with respect to the filtration $\{\mathcal{F}_t^Z\}$ whose quadratic variation is t, and then our result follows from Lévy's theorem (Theorem 12.1). That N_t is continuous is obvious, and $\langle N \rangle_t = \langle Z \rangle_t = \langle W \rangle_t = t$ from the definitions of Z and W. Thus we need to show that N is a martingale with respect to $\{\mathcal{F}_t^Z\}$.

If $r \geq s$, we have

$$\mathbb{E}[\widehat{h}_r \mid \mathcal{F}_s^Z] = \mathbb{E}[\mathbb{E}[h_r \mid \mathcal{F}_r^Z] \mid \mathcal{F}_s^Z] = \mathbb{E}[h_r \mid \mathcal{F}_s^Z]. \tag{29.3}$$

Then using Exercise 29.1,

$$\mathbb{E}[N_t - N_s \mid \mathcal{F}_s^Z] = \mathbb{E}[Z_t - Z_s \mid \mathcal{F}_s^Z] - \int_s^t \mathbb{E}[\widehat{h}_r \mid \mathcal{F}_s^Z] \, dr \tag{29.4}$$

$$= \mathbb{E}[W_t - W_s \mid \mathcal{F}_s^Z] + \int_s^t \mathbb{E}[h_r - \widehat{h}_r \mid \mathcal{F}_s^Z] \, dr$$

$$= \mathbb{E}[\mathbb{E}[W_t - W_s \mid \mathcal{F}_s^X] \mid \mathcal{F}_s^Z] = 0,$$

since $\mathcal{F}_s^Z \subset \mathcal{F}_s^X$. $\qquad\qquad\qquad\qquad\qquad\qquad\qquad\qquad\qquad\qquad\qquad\qquad \square$

29.3 Representation of $\mathcal{F}^Z$-martingales

In this section we prove that if Y_t is a martingale with respect to $\{\mathcal{F}_t^Z\}$, then Y can be represented as a stochastic integral with respect to N. This is not an immediate consequence of Theorem 12.3 because we do not know that N_t generates $\{\mathcal{F}_t^Z\}$; the filtration generated by N could conceivably be strictly smaller than the one generated by Z.

Theorem 29.3 *Suppose Y_t is a square integrable martingale with respect to $\{\mathcal{F}_t^Z\}$. Let $\mathcal{P}^Z$ be the predictable σ-field defined on $[0, \infty) \times \Omega$ in terms of $\{\mathcal{F}_t^Z\}$. Then there exists H_s which is $\mathcal{P}^Z$ measurable and with $\mathbb{E}\int_0^\infty H_s^2 \, ds < \infty$ such that*

$$Y_t = Y_0 + \int_0^t H_s \, dN_s \tag{29.5}$$

for all t.

To clarify, $\mathcal{P}^Z$ is the σ-field generated by all bounded left-continuous processes that are adapted to $\{\mathcal{F}_t^Z\}$.

Proof First let us treat the case where $\int_0^t \widehat{h}_s \, dN_s$, $\int_0^t |\widehat{h}_s|^2 \, ds$, and Y_t are each bounded. Define $\mathbb{Q}$ on $\mathcal{F}_t^Z$ by $d\mathbb{Q}/d\mathbb{P} \mid_{\mathcal{F}_t^Z} = M_t$, where

$$M_t = \exp\left(-\int_0^t \widehat{h}_s \, dN_s - \tfrac{1}{2}\int_0^t |\widehat{h}_s|^2 \, ds\right).$$

Then by the Girsanov theorem (Theorem 13.3)

$$Z_t = N_t + \int_0^t \widehat{h}_s \, ds$$

is a martingale under $\mathbb{Q}$ with respect to $\{\mathcal{F}_t^Z\}$. Since $\langle Z \rangle_t = \langle N \rangle_t = \langle W \rangle_t = t$, then Z is a Brownian motion under $\mathbb{Q}$ with respect to $\{\mathcal{F}_t^Z\}$.

Let $\widetilde{Y}_t = M_t^{-1} Y_t$. If $A \in \mathcal{F}_s^Z$, then $A \in \mathcal{F}_s^X$ and

$$\mathbb{E}_\mathbb{Q}[\widetilde{Y}_t; A] = \mathbb{E}_\mathbb{P}[M_t(M_t^{-1}Y_t); A] = \mathbb{E}_\mathbb{P}[Y_t; A] = \mathbb{E}_\mathbb{P}[Y_s; A]$$

$$= \mathbb{E}_\mathbb{P}[M_s(M_s^{-1}Y_s); A] = \mathbb{E}_\mathbb{Q}[\widetilde{Y}_s; A].$$

Therefore $\widetilde{Y}_t$ is a martingale under $\mathbb{Q}$ with respect to $\{\mathcal{F}_t^Z\}$. By the martingale representation theorem (Theorem 12.3) there exists $K_s \in \mathcal{P}^Z$ such that

$$\widetilde{Y}_t = \widetilde{Y}_0 + \int_0^t K_s \, dZ_s = \widetilde{Y}_0 + \int_0^t K_s \, dN_s + \int_0^t K_s \widehat{h}_s \, ds.$$

On the other hand, $dM_t = -M_t \widehat{h}_t \, dN_t$ and $Y_t = M_t \widetilde{Y}_t$. We have $d\langle M, Y \rangle_t = -M_t \widehat{h}_t K_t \, dt$. By the product formula,

$$Y_t = M_0 \widetilde{Y}_0 + \int_0^t \widetilde{Y}_s \, dM_s + \int_0^t M_s \, d\widetilde{Y}_s + \langle M, \widetilde{Y} \rangle_t$$

$$= Y_0 - \int_0^t \widetilde{Y}_s M_s \widehat{h}_s \, dN_s + \int_0^t K_s M_s \, dN_s + \int_0^t K_s \widehat{h}_s M_s \, ds - \int_0^t M_s \widehat{h}_s K_s \, ds,$$

which is of the desired form if we set $H_s = K_s M_s - \widetilde{Y}_s M_s \widehat{h}_s$.

In the general case, let

$$T_K = \inf \left\{ t : \left| \int_0^t \widehat{h}_s \, dN_s \right| + \int_0^t |\widehat{h}_s|^2 \, ds + |Y_t| \ge K \right\}.$$

We apply the above argument to $Y_{t \wedge T_K}$ and use Exercise 29.3 to get

$$Y_{t \wedge T_K} = Y_0 + \int_0^t H_s^K \, dN_s,$$

where H_s^K is predictable with respect to the σ-fields $\{\mathcal{F}_{t \wedge T_K}^Z\}$ and is 0 from time T_K on. Since Y_t is square integrable, $Y_{T_K} \to Y_\infty$ almost surely and in $L^2(\mathbb{P})$ as $K \to \infty$, and

$$\mathbb{E}\left[\int_0^\infty |H_s^K - H_s^L|^2 \, ds \right] = \mathbb{E}\left[|Y_{T_K} - Y_{T_L}|^2 \right] \to 0$$

as $K, L \to \infty$. Using the completeness of L^2, there exists H_s such that $\mathbb{E} \int_0^\infty H_s^2 \, ds < \infty$ and $\mathbb{E} \int_0^\infty |H_s - H_s^K|^2 \, ds \to 0$ as $K \to \infty$. It is routine to check that H_s is $\mathcal{P}^Z$ measurable and that (29.5) holds. $\qquad\square$

29.4 The filtering equation

We now derive the general filtering equation. First we need a lemma.

Lemma 29.4 *If $Y_t - \int_0^t H_s \, ds$ is a martingale with respect to $\{\mathcal{F}_t^X\}$, then $\widehat{Y}_t - \int_0^t \widehat{H}_s \, ds$ is a martingale with respect to $\{\mathcal{F}_t^Z\}$.*

Proof Since $\mathcal{F}_s^Z \subset \mathcal{F}_s^X$,

$$\mathbb{E}\left[\widehat{Y}_t - \widehat{Y}_s - \int_s^t \widehat{H}_r\,dr \mid \mathcal{F}_s^Z\right]$$

$$= \mathbb{E}\left[\mathbb{E}\,[Y_t \mid \mathcal{F}_t^Z] - \mathbb{E}\,[Y_s \mid \mathcal{F}_s^Z] - \int_s^t \mathbb{E}\,[H_r \mid \mathcal{F}_r^Z]\,dr \mid \mathcal{F}_s^Z\right]$$

$$= \mathbb{E}\left[Y_t - Y_s - \int_s^t H_r\,dr \mid \mathcal{F}_s^Z\right]$$

$$= \mathbb{E}\left[\mathbb{E}\left[Y_t - Y_s - \int_s^t H_r\,dr \mid \mathcal{F}_s^X\right] \mid \mathcal{F}_s^Z\right] = 0.$$

The first equality is proved in a fashion similar to the one you were asked to prove in Exercise 29.1. $\quad\square$

Here is the filtering equation.

Theorem 29.5 *Let $M_t = f(X_t) - f(X_0) - \int_0^t A_s\,ds$ be a martingale with respect to $\{\mathcal{F}_t^X\}$ and write F_s for $f(X_s)$. Suppose $\langle M, W\rangle_t = \int_0^t D_s\,ds$. Then*

$$\widehat{F}_t = \widehat{F}_0 + \int_0^t \widehat{A}_s\,ds + \int_0^t (\widehat{F_s h_s} - \widehat{F}_s\widehat{h}_s + \widehat{D}_s)\,dN_s. \tag{29.6}$$

Proof By Lemma 29.4,

$$L_t = \widehat{F}_t - \widehat{F}_0 - \int_0^t \widehat{A}_s\,ds \tag{29.7}$$

is a martingale with respect to $\{\mathcal{F}_t^Z\}$ and by Theorem 29.3, there exists H_s such that

$$L_t = \int_0^t H_s\,dN_s. \tag{29.8}$$

By the product formula

$$F_t Z_t = \int_0^t F_s\,dZ_s + \int_0^t Z_s\,dF_s + \int_0^t D_s\,ds$$

$$= \int_0^t F_s\,dN_s + \int_0^t F_s h_s\,ds + \int_0^t Z_s\,dM_s + \int_0^t Z_s A_s\,ds + \int_0^t D_s\,ds$$

$$= \mathcal{F}^X\text{-martingale} + \int_0^t [F_s h_s + Z_s A_s + D_s]\,ds.$$

By Lemma 29.4 and the obvious fact that Z is adapted to $\{\mathcal{F}_t^Z\}$,

$$\widehat{F_t Z_t} = \widehat{F}_t Z_t = \mathcal{F}^Z\text{-martingale} + \int_0^t (\widehat{F_s h_s} + Z_s\widehat{A}_s + \widehat{D}_s)\,ds.$$

Again using the product formula,

$$\widehat{F}_t Z_t = \int_0^t \widehat{F}_s\,dZ_s + \int_0^t Z_s\,d\widehat{F}_s + \int_0^t H_s\,ds$$

$$= \mathcal{F}^Z\text{-martingale} + \int_0^t [\widehat{F}_s h_s + Z_s\widehat{A}_s + H_s]\,ds.$$

Therefore

$$\int_0^t (\widehat{F_s h_s} + Z_s \widehat{A}_s + \widehat{D}_s - \widehat{F_s} \widehat{h}_s - Z_s \widehat{A}_s - H_s)\, ds$$

is a continuous $\mathcal{F}^Z$-martingale that has paths that are locally of bounded variation and which is zero at time zero, hence is identically zero by Theorem 9.7. Hence with probability one, $H_s = \widehat{F_s h_s} - \widehat{F_s} \widehat{h}_s + \widehat{D}_s$ for almost every s. Substituting this in (29.8) and combining with (29.7) gives our result.   □

29.5 Linear models

The filtering equation (29.6) is difficult to apply in most cases. However, in the linear model, we can get a much simpler representation. To define the *linear model* in d dimensions, let X_t solve

$$dX_t = A(t)\, d\overline{W}_t + B(t)X_t\, dt, \tag{29.9}$$

where $\overline{W}_t$ is a d-dimensional Brownian motion and $A(t)$ and $B(t)$ are deterministic $d \times d$ matrices that are continuous in t. Let

$$dZ_t = dW_t + C(t)X_t\, dt, \tag{29.10}$$

where C is a deterministic $d \times d$ matrix-valued function that is continuous in t and W_t is a d-dimensional Brownian motion independent of $\overline{W}$ and X.

Why is this model useful? Suppose X_t is two-dimensional with $X_t^{(1)}$ being the position of a particle and $X_t^{(2)}$ its velocity. Suppose the position and the velocity have some randomness and that our observations of the position and velocity are noisy. This fits into the model (29.9)–(29.10) if we take

$$A(t) = \begin{pmatrix} 1 & 0 \\ 0 & 1 \end{pmatrix}, \quad B(t) = \begin{pmatrix} 0 & 1 \\ 0 & 0 \end{pmatrix}, \quad C(t) = \begin{pmatrix} c_1 & 0 \\ 0 & c_2 \end{pmatrix}.$$

For another example, suppose a particle has a fixed unknown velocity and the position is observed, but obscured by noise. Let $X_t^{(1)}$ and $X_t^{(2)}$ be the position and velocity and let $A(t)$ be the zero matrix,

$$B(t) = \begin{pmatrix} 0 & 1 \\ 0 & 0 \end{pmatrix}, \quad C(t) = \begin{pmatrix} 1 & 0 \\ 0 & 0 \end{pmatrix}.$$

The solution of the filtering problem modeled by (29.9)–(29.10) is known as the *Kalman–Bucy filter*. For simplicity we will consider the special case where the dimension d is 1 and A, B, C are constant in t; the general case is done in exactly the same way, but the notation becomes much more complicated (see Kallianpur, 1980). We will further assume $\mathbb{E}\, X_0$ and $\mathrm{Var}\, X_0$ are known.

29.6 Kalman–Bucy filter

Let

$$V_t = \widehat{X_t^2} - (\widehat{X}_t)^2,$$

the conditional variance of X_t given $\mathcal{F}_t^Z$.

Theorem 29.6 V_t *solves the deterministic ordinary differential equation*

$$\frac{dV_t}{dt} = 1 + 2BV_t - C^2V_t^2, \qquad V_0 = \operatorname{Var} X_0 \qquad (29.11)$$

In particular, V_t is deterministic. $\widehat{X}_t$ solves

$$d\widehat{X}_t = CV_t\,dZ_t + (B - CV_t)\widehat{X}_t\,dt, \qquad \widehat{X}_0 = \mathbb{E}X_0. \qquad (29.12)$$

The equation (29.11) is an example of what is known as a Riccati equation. We get a similar equation when $d > 1$ or when A, B, and C depend on t, but in general one cannot solve the Riccati equation explicitly. However, when $d = 1$ and A, B, C do not depend on t, one can solve (29.11) by separation of variables. Write

$$\frac{dV}{1 + 2BV - C^2V^2} = dt,$$

and integrate both sides.

When $d = 1$ (and even if A, B, and C depend on time), we can solve (29.12). Let $G_t = B - CV_t$ so that we have

$$d\widehat{X}_t = CV_t\,dZ_t + G_t\widehat{X}_t\,dt,$$

or by the product formula

$$d\left[e^{-\int_0^t G_r\,dr}\widehat{X}_t\right] = e^{-\int_0^t G_r\,dr}CV_t\,dZ_t,$$

and hence

$$\widehat{X}_t = \mathbb{E}X_0 + \int_0^t e^{\int_s^t G_r\,dr}CV_s\,dZ_s.$$

(Cf. the solution of (24.15).)

Proof of Theorem 29.6 By Itô's formula, if $f \in C^2$,

$$f(X_t) - f(X_0) = \mathcal{F}^X\text{-martingale} + \int_0^t \left[\tfrac{1}{2}f''(X_s) + BX_sf'(X_s)\right]ds.$$

By the filtering equation applied with $f(x) = x$,

$$\widehat{X}_t = \mathbb{E}X_0 + B\int_0^t \widehat{X}_s\,ds + C\int_0^t V_s\,dN_s. \qquad (29.13)$$

By Exercises 29.4(2) and 29.5(3),

$$\widehat{X_t^3} - \widehat{X}_t\widehat{X_t^2} = 2\widehat{X}_tV_t. \qquad (29.14)$$

With the filtering equation applied with $f(x) = x^2$ and (29.14),

$$\widehat{X_t^2} = \mathbb{E}X_0^2 + C\int_0^t (1 + 2B\widehat{X_s^2})\,ds + C\int_0^t (\widehat{X_s^3} - \widehat{X}_s\widehat{X_s^2})\,dN_s$$

$$= \mathbb{E}X_0^2 + C\int_0^t (1 + 2B\widehat{X_s^2})\,ds + 2C\int_0^t V_s\widehat{X}_s\,dN_s.$$

Therefore

$$dV_t = d(\widehat{X_t^2} - (\widehat{X_t})^2) \tag{29.15}$$

$$= 2CV_t\widehat{X_t}\,dN_t + (1 + 2B\widehat{X_t^2}\,dt) - 2\widehat{X_t}(CV_t\,dN_t + B\widehat{X_t}\,dt) - C^2V_t^2\,dt$$

$$= (1 + 2BV_t - C^2V_t^2)\,dt.$$

This shows that V_t solves the deterministic ordinary differential equation (29.15). This equation has a unique solution (cf. Theorem 15.1), so V_t is deterministic. We obtain (29.12) from (29.2), (29.10), and (29.13). □

Exercises

29.1 Justify the first equality in (29.4).

29.2 Show that if M_t is a martingale with respect to $\{\mathcal{F}_t^X\}$, then $\widehat{M}_t$ is a martingale with respect to $\{\mathcal{F}_t^Z\}$.

29.3 Suppose W is a Brownian motion and $\{\mathcal{F}_t\}$ is its minimal augmented filtration. Let T be a bounded stopping time with respect to $\{\mathcal{F}_t\}$. Suppose Y is a $\mathcal{F}_T$ measurable random variable with $\mathbb{E}\,Y^2 < \infty$. Show that there exists a predictable process H_s with $\mathbb{E}\int_0^T H_s^2\,ds < \infty$ such that $Y = \mathbb{E}\,Y + \int_0^T H_s\,dW_s$, a.s.

29.4 (1) Show that the solution to (29.9) is a Gaussian process.

(2) Show that the solutions (X_t, Z_t) to (29.9)–(29.10) form a Gaussian process.

29.5 (1) Show that if X is a normal random variable with mean μ and variance σ^2, then

$$\mathbb{E}\,X^3 = \mu(\mu^2 + 3\sigma^2).$$

(2) Show that if $X, Y_1, \ldots, Y_n$ are jointly normal random variables, then

$$\mathbb{E}\,[X^3 \mid Y_1, \ldots, Y_n] = \mathbb{E}\,[X \mid Y_1, \ldots, Y_n](\mathbb{E}\,[X \mid Y_1, \ldots, Y_n]^2$$
$$+ 3\mathrm{Var}\,[X \mid Y_1, \ldots, Y_n]),$$

where

$$\mathrm{Var}\,[X \mid Y_1, \ldots, Y_n] = \mathbb{E}\,[(X - \mathbb{E}\,[X \mid Y_1, \ldots, Y_n])^2 \mid Y_1, \ldots, Y_n].$$

(3) Show that

$$\widehat{X_t^3} = \widehat{X_t}((\widehat{X_t})^2 + 3\mathrm{Var}\,(X_t \mid \mathcal{F}_t^Z)),$$

where

$$\mathrm{Var}\,(X_t \mid \mathcal{F}_t^Z) = \mathbb{E}\,[(X_t - \widehat{X_t})^2 \mid \mathcal{F}_t^Z] = \widehat{X_t^2} - (\widehat{X_t})^2.$$

Notes

For more on filtering, see Kallianpur (1980) and Øksendal (2003).

30

Convergence of probability measures

Suppose we have a sequence of probabilities on a metric space S and we want to define what it means for the sequence to converge weakly. Alternately, we may have a sequence of random variables and want to say what it means for the random variables to converge weakly. We will apply the results we obtain here in later chapters to the case where S is a function space such as $C[0, 1]$ and obtain theorems on the convergence of stochastic processes.

For now our state space is assumed to be an arbitrary metric space, although we will soon add additional assumptions on S. We use the Borel σ-field on S, which is the σ-field generated by the open sets in S. We write $A^0, \overline{A}$, and ∂A for the interior, closure, and boundary of A, respectively.

30.1 The portmanteau theorem

Clearly the definition of weak convergence of real-valued random variables in terms of distribution functions (see Section A.12) has no obvious analog. The appropriate generalization is the following; cf. Proposition A.41.

Definition 30.1 A sequence of probabilities $\{\mathbb{P}_n\}$ on a metric space S furnished with the Borel σ-field is said to *converge weakly* to $\mathbb{P}$ if $\int f \, d\mathbb{P}_n \to \int f \, d\mathbb{P}$ for every bounded and continuous function f on S. A sequence of random variables $\{X_n\}$ taking values in S *converges weakly* to a random variable X taking values in S if $\mathbb{E} f(X_n) \to \mathbb{E} f(X)$ whenever f is a bounded and continuous function.

Saying X_n converges weakly to X is the same as saying that the laws of X_n converge weakly to the law of X. To see this, if $\mathbb{P}_n$ is the law of X_n, that is, $\mathbb{P}_n(A) = \mathbb{P}(X_n \in A)$ for each Borel subset A of S, then $\mathbb{E} f(X_n) = \int f \, d\mathbb{P}_n$ and $\mathbb{E} f(X) = \int f \, d\mathbb{P}$. (This holds when f is an indicator by the definition of the law of X_n and X, then for simple functions by linearity, then for non-negative measurable functions by monotone convergence, and then for arbitrary bounded and Borel measurable f by linearity.)

What might cause a bit of confusion is that weak convergence in probability is not the same as weak convergence in functional analysis, but rather is equivalent to what is known as weak-$*$ convergence in functional analysis. Feel free to skip the remainder of this paragraph where we explain this. Recall that if B is a Banach space and B^* is its dual, then $x_n \in B$ converges weakly to $x \in B$ if $f(x_n) \to f(x)$ for all $f \in B^*$. $f_n \in B^*$ converges with respect to the weak-$*$ topology to $f \in B^*$ if $f_n(x) \to f(x)$ for all $x \in B$. By the Riesz representation theorem, there is a one-to-one correspondence between positive bounded linear functionals on $B = C(X)$, the continuous functions on X, where X is compact, and the set $\mathcal{M}$ of finite

237

measures on X. When $B = C(X)$, B^* can be identified with $\mathcal{M}$, and measures $\mathbb{P}_n$ with mass 1 in $\mathcal{M}$ converge to $\mathbb{P} \in \mathcal{M}$ with respect to the weak-$*$ topology if $\mathbb{P}_n(g) \to \mathbb{P}(g)$ for every $g \in B = C(X)$. Interpreting $\mathbb{P}_n(g)$ as $\int g \, d\mathbb{P}_n$ shows the connection.

Returning to weak convergence in the probability sense, the following theorem, known as the portmanteau theorem, gives some other characterizations. For this chapter we let

$$F_\delta = \{x : d(x, F) < \delta\} \tag{30.1}$$

for closed sets F, the set of points within δ of F, where $d(x, F) = \inf\{d(x, y) : y \in F\}$.

Theorem 30.2 *Suppose* $\{\mathbb{P}_n, n = 1, 2, \ldots\}$ *and* $\mathbb{P}$ *are probabilities on a metric space. The following are equivalent.*

(1) $\mathbb{P}_n$ *converges weakly to* $\mathbb{P}$.
(2) $\limsup_n \mathbb{P}_n(F) \leq \mathbb{P}(F)$ *for all closed sets* F.
(3) $\liminf_n \mathbb{P}_n(G) \geq \mathbb{P}(G)$ *for all open sets* G.
(4) $\lim_n \mathbb{P}_n(A) = \mathbb{P}(A)$ *for all Borel sets* A *such that* $\mathbb{P}(\partial A) = 0$.

Proof   The equivalence of (2) and (3) is easy because if F is closed, then $G = F^c$ is open and $\mathbb{P}_n(G) = 1 - \mathbb{P}_n(F)$.

To see that (2) and (3) imply (4), suppose $\mathbb{P}(\partial A) = 0$. Then

$$\limsup_n \mathbb{P}_n(A) \leq \limsup_n \mathbb{P}_n(\overline{A}) \leq \mathbb{P}(\overline{A})$$
$$= \mathbb{P}(A^0) \leq \liminf_n \mathbb{P}_n(A^0) \leq \liminf_n \mathbb{P}_n(A).$$

Next, let us show (4) implies (2). Let F be closed. If $y \in \partial F_\delta$, then $d(y, F) = \delta$. The sets ∂F_δ are disjoint for different δ. At most countably many of them can have positive $\mathbb{P}$-measure, hence there exists a sequence $\delta_k \downarrow 0$ such that $\mathbb{P}(\partial F_{\delta_k}) = 0$ for each k. Then

$$\limsup_n \mathbb{P}_n(F) \leq \limsup_n \mathbb{P}_n(\overline{F_{\delta_k}}) = \mathbb{P}(\overline{F_{\delta_k}}) = \mathbb{P}(F_{\delta_k})$$

for each k. Since $\mathbb{P}(F_{\delta_k}) \downarrow \mathbb{P}(F)$ as $\delta_k \to 0$, this gives (2).

We show now that (1) implies (2). Suppose F is closed. Let $\varepsilon > 0$. Take $\delta > 0$ small enough so that $\mathbb{P}(\overline{F_\delta}) - \mathbb{P}(F) < \varepsilon$. Then take f continuous, to be equal to 1 on F, to have support in $\overline{F_\delta}$, and to be bounded between 0 and 1. For example, $f(x) = 1 - (1 \wedge \delta^{-1} d(x, F))$ would do. Then

$$\limsup_n \mathbb{P}_n(F) \leq \limsup_n \int f \, d\mathbb{P}_n = \int f \, d\mathbb{P}$$
$$\leq \mathbb{P}(\overline{F_\delta}) \leq \mathbb{P}(F) + \varepsilon.$$

Since this is true for all ε, (2) follows.

Finally, let us show (2) implies (1). Let f be bounded and continuous. If we show

$$\limsup_n \int f \, d\mathbb{P}_n \leq \int f \, d\mathbb{P}, \tag{30.2}$$

for every such f, then applying this inequality to both f and $-f$ will give (1). By adding a sufficiently large positive constant to f and then multiplying by a suitable constant, without loss of generality we may assume f is bounded and takes values in $(0, 1)$. We define $F_i = \{x : f(x) \geq i/k\}$, which is closed.

$$\int f d\mathbb{P}_n \leq \sum_{i=1}^{k} \frac{i}{k} \mathbb{P}_n\left(\frac{i-1}{k} \leq f(x) < \frac{i}{k}\right)$$

$$= \sum_{i=1}^{k} \frac{i}{k} [\mathbb{P}_n(F_{i-1}) - \mathbb{P}_n(F_i)]$$

$$= \sum_{i=0}^{k-1} \frac{i+1}{k} \mathbb{P}_n(F_i) - \sum_{i=1}^{k} \frac{i}{n} \mathbb{P}_n(F_i)$$

$$\leq \frac{1}{k} + \frac{1}{k} \sum_{i=1}^{k} \mathbb{P}_n(F_i).$$

Similarly,

$$\int f d\mathbb{P} \geq \frac{1}{k} \sum_{i=1}^{k} \mathbb{P}(F_i).$$

Then

$$\limsup_n \int f d\mathbb{P}_n \leq \frac{1}{k} + \frac{1}{k} \sum_{i=1}^{k} \limsup_n \mathbb{P}_n(F_i)$$

$$\leq \frac{1}{k} + \frac{1}{k} \sum_{i=1}^{k} \mathbb{P}(F_i) \leq \frac{1}{k} + \int f d\mathbb{P}.$$

Since k is arbitrary, this gives (30.2).   □

If $x_n \to x$, $\mathbb{P}_n = \delta_{x_n}$, and $\mathbb{P} = \delta_x$, it is easy to see $\mathbb{P}_n$ converges weakly to $\mathbb{P}$. Letting $A = \{x\}$ shows that one cannot, in general, have $\lim_n \mathbb{P}_n(F) = \mathbb{P}(F)$ for all closed sets F.

30.2 The Prohorov theorem

It turns out there is a simple condition that ensures that a sequence of probability measures has a weakly convergent subsequence.

Definition 30.3 A sequence of probabilities $\mathbb{P}_n$ on a metric space $\mathcal{S}$ is *tight* if for every ε there exists a compact set K (depending on ε) such that $\sup_n \mathbb{P}_n(K^c) \leq \varepsilon$.

The important result here is Prohorov's theorem.

Theorem 30.4 *If a sequence of probability measures on a metric space $\mathcal{S}$ is tight, there is a subsequence that converges weakly to a probability measure on $\mathcal{S}$.*

Proof Suppose first that the metric space $\mathcal{S}$ is compact. Then $C(\mathcal{S})$, the collection of continuous functions on $\mathcal{S}$, is a separable metric space when furnished with the supremum norm; this is Exercise 30.1. Let $\{f_i\}$ be a countable collection of non-negative elements of $C(\mathcal{S})$ whose linear span is dense in $C(\mathcal{S})$. For each i, $\int f_i d\mathbb{P}_n$ is a bounded sequence, so we have a convergent subsequence. By a diagonalization procedure, we can find a subsequence n' such that $\int f_i \, d\mathbb{P}_{n'}$ converges for all i. By the term "diagonalization procedure" we are referring to the well-known method of proof of the Ascoli–Arzelà theorem; see any book on real analysis for a detailed explanation. Call the limit Lf_i. Clearly $0 \leq Lf_i \leq \|f_i\|_\infty$, L is linear, and so we can extend L to a bounded linear functional on $\mathcal{S}$. By the Riesz representation theorem (Rudin, 1987), there exists a measure $\mathbb{P}$ such that $Lf = \int f \, d\mathbb{P}$. Since $\int f_i \, d\mathbb{P}_{n'} \to \int f_i \, d\mathbb{P}$ for all f_i, it is not hard to see, since each $\mathbb{P}_{n'}$ has total mass 1, that $\int f \, d\mathbb{P}_{n'} \to \int f \, d\mathbb{P}$ for all $f \in C(\mathcal{S})$. Therefore $\mathbb{P}_{n'}$ converges weakly to $\mathbb{P}$. Since $Lf \geq 0$ if $f \geq 0$, then $\mathbb{P}$ is a positive measure. The function that is identically equal to 1 is bounded and continuous, so $1 = \mathbb{P}_{n'}(\mathcal{S}) = \int 1 \, d\mathbb{P}_{n'} \to \int 1 \, d\mathbb{P}$, or $\mathbb{P}(\mathcal{S}) = 1$.

Next suppose that $\mathcal{S}$ is a Borel subset of a compact metric space $\mathcal{S}'$. Extend each $\mathbb{P}_n$, initially defined on $\mathcal{S}$, to $\mathcal{S}'$ by setting $\mathbb{P}_n(\mathcal{S}' \setminus \mathcal{S}) = 0$. By the first paragraph of the proof, there is a subsequence $\mathbb{P}_{n'}$ that converges weakly to a probability $\mathbb{P}$ on $\mathcal{S}'$ (the definition of weak convergence here is relative to the topology on $\mathcal{S}'$). Since the $\mathbb{P}_n$ are tight, there exist compact subsets K_m of $\mathcal{S}$ such that $\mathbb{P}_n(K_m) \geq 1 - 1/m$ for all n. The K_m will also be compact relative to the topology on $\mathcal{S}'$, so by Theorem 30.2,

$$\mathbb{P}(K_m) \geq \limsup_{n'} \mathbb{P}_{n'}(K_m) \geq 1 - 1/m.$$

Since $\cup_m K_m \subset \mathcal{S}$, we conclude $\mathbb{P}(\mathcal{S}) = 1$.

If G is open in $\mathcal{S}$, then $G = H \cap \mathcal{S}$ for some H open in $\mathcal{S}'$. Then

$$\liminf_{n'} \mathbb{P}_{n'}(G) = \liminf_{n'} \mathbb{P}_{n'}(H) \geq \mathbb{P}(H) = \mathbb{P}(H \cap \mathcal{S}) = \mathbb{P}(G).$$

Thus by Theorem 30.2, $\mathbb{P}_{n'}$ converges weakly to $\mathbb{P}$ relative to the topology on $\mathcal{S}$.

Now let $\mathcal{S}$ be an arbitrary metric space. Since all the $\mathbb{P}_n$'s are supported on $\cup_m K_m$, we can replace $\mathcal{S}$ by $\cup_m K_m$, or we may as well assume that $\mathcal{S}$ is σ-compact, and hence separable. It remains to embed the separable metric space $\mathcal{S}$ into a compact metric space $\mathcal{S}'$. If d is the metric on $\mathcal{S}$, $d \wedge 1$ will also be an equivalent metric, that is, one that generates the same collection of open sets, so we may assume d is bounded by 1. Now $\mathcal{S}$ can be embedded in $\mathcal{S}' = [0,1]^{\mathbb{N}}$ as follows. We define a metric on $\mathcal{S}'$ by

$$d'(a, b) = \sum_{i=1}^{\infty} 2^{-i}(|a^i - b^i| \wedge 1), \qquad a = (a^1, a^2, \ldots), b = (b^1, b^2, \ldots). \quad (30.3)$$

Being the product of compact spaces, $\mathcal{S}'$ is itself compact. If $\{z_j\}$ is a countable dense subset of $\mathcal{S}$, let $I : \mathcal{S} \to [0,1]^{\mathbb{N}}$ be defined by

$$I(x) = (d(x, z_1), d(x, z_2), \ldots).$$

We leave it to the reader to check that I is a one-to-one continuous open map of $\mathcal{S}$ to a subset of $\mathcal{S}'$. Since $\mathcal{S}$ is σ-compact, and the continuous image of compact sets is compact, then $I(\mathcal{S})$ is a Borel set. □

Clearly, Prohorov's theorem is easily modified to handle the case of finite measures on $\mathcal{S}$.

30.3 Metrics for weak convergence

Since we have defined a notion of convergence of probability measures, one might wonder if one can make the set of probability measures $\mathcal{M}$ on $\mathcal{S}$ into a metric space so that weak convergence is equivalent to convergence in $\mathcal{M}$. This is indeed possible and in fact there are a number of metrics on the space of probability measures that work. We will focus on the *Prohorov metric*.

Definition 30.5 If $\mathbb{P}$ and $\mathbb{Q}$ are probability measures on a separable metric space $\mathcal{S}$, define

$$d_{\mathcal{M}}(\mathbb{P}, \mathbb{Q}) = \inf\{\varepsilon : \mathbb{P}(F) \leq \mathbb{Q}(F_{\varepsilon}) + \varepsilon \text{ for all } F \text{ closed}\}. \tag{30.4}$$

It is not immediately obvious that $d_{\mathcal{M}}$ is even a metric, so the first task is to show that it is.

Proposition 30.6 $d_{\mathcal{M}}$ *is a metric on* $\mathcal{M}$.

Proof We start with symmetry, that is, that $d_{\mathcal{M}}(\mathbb{Q}, \mathbb{P}) = d_{\mathcal{M}}(\mathbb{P}, \mathbb{Q})$. Let α be any real number larger than $d_{\mathcal{M}}(P, Q)$. If H is closed, then $H_{\alpha} = \{x : d(x, H) < \alpha\}$ is open and $K = \mathcal{S} \setminus H_{\alpha}$ is closed. Note that $H \subset \mathcal{S} - K_{\alpha}$, where $K_{\alpha} = \{x : d(x, K) < \alpha\}$, because if $x \in H$, then $d(x, K) \geq \alpha$, so $x \notin K_{\alpha}$ and hence $x \in \mathcal{S} \setminus K_{\alpha}$. Since K is closed, by the definition of $d_{\mathcal{M}}(P, Q)$,

$$\mathbb{P}(H_{\alpha}) = 1 - \mathbb{P}(K) \geq 1 - \mathbb{Q}(K_{\alpha}) - \alpha = \mathbb{Q}(\mathcal{S} \setminus K_{\alpha}) - \alpha \geq \mathbb{Q}(H) - \alpha,$$

or $\mathbb{Q}(H) \leq \mathbb{P}(H_{\alpha}) + \alpha$. Since H was an arbitrary closed set, $d_{\mathcal{M}}(\mathbb{Q}, \mathbb{P}) \leq \alpha$, and it follows that $d_{\mathcal{M}}(\mathbb{Q}, \mathbb{P}) \leq d_{\mathcal{M}}(\mathbb{P}, \mathbb{Q})$. Reversing the roles of $\mathbb{P}$ and $\mathbb{Q}$ shows symmetry.

Clearly $d_{\mathcal{M}}(\mathbb{P}, \mathbb{Q}) \geq 0$. If $d_{\mathcal{M}}(\mathbb{P}, \mathbb{Q}) = 0$, then $\mathbb{P}(F) = \mathbb{Q}(F) = 0$ for all closed sets F. Since the collection of closed sets generates the Borel σ-field, it is not hard to see that $\mathbb{P}(A) = \mathbb{Q}(A)$ for all Borel subsets A, and hence $\mathbb{P} = \mathbb{Q}$.

Finally we prove the triangle inequality. Suppose $\mathbb{P}, \mathbb{Q}, \mathbb{R} \in \mathcal{M}$. If α is any real larger than $d_{\mathcal{M}}(\mathbb{P}, \mathbb{Q})$ and β any real larger than $d_{\mathcal{M}}(\mathbb{Q}, \mathbb{R})$, then for any $\varepsilon > 0$ and any closed set F

$$
\begin{aligned}
\mathbb{P}(F) &\leq \mathbb{Q}(F_{\alpha}) + \alpha \leq \mathbb{Q}(\overline{F_{\alpha}}) + \alpha \\
&\leq \mathbb{R}((\overline{F_{\alpha}})_{\beta}) + \alpha + \beta \\
&\leq \mathbb{R}(F_{\alpha+\beta+\varepsilon}) + (\alpha + \beta + \varepsilon).
\end{aligned}
$$

Therefore $d_{\mathcal{M}}(\mathbb{P}, \mathbb{R}) \leq \alpha + \beta + \varepsilon$, and since ε is arbitrary, the triangle inequality follows. $\square$

Now we show that weak convergence is equivalent to convergence in the topology generated by $d_{\mathcal{M}}$, at least if $\mathcal{S}$ is separable. ($L^{\infty}[0, 1]$ is an example of a nonseparable metric space.)

Proposition 30.7 *Suppose $\mathcal{S}$ is a separable metric space. A sequence of probability measures $\mathbb{P}_n$ on $\mathcal{S}$ converges weakly to a probability $\mathbb{P}$ if and only if $d_{\mathcal{M}}(\mathbb{P}_n, \mathbb{P}) \to 0$.*

Proof We first suppose $d_{\mathcal{M}}(\mathbb{P}_n, \mathbb{P}) \to 0$ and show that $\mathbb{P}_n$ converges weakly to $\mathbb{P}$. Separability is not used in this part of the proof. Suppose F is closed and set $\varepsilon_n = d_{\mathcal{M}}(\mathbb{P}_n, \mathbb{P}) + 1/n$. Since $\mathbb{P}_n(F) \leq \mathbb{P}(F_{\varepsilon_n}) + \varepsilon_n$, then

$$\limsup_n \mathbb{P}_n(F) \leq \limsup_n \mathbb{P}(F_{\varepsilon_n}) = \mathbb{P}(F),$$

and we now apply Theorem 30.2(2).

We now suppose $\mathbb{P}_n$ converges weakly to $\mathbb{P}$. Let $\varepsilon > 0$. Cover $\mathcal{S}$ with countably many balls $\{B_i\}$ of diameter less than $\varepsilon/2$ (separability is used here) and let $A_1 = B_1, A_2 = B_2 \setminus B_1$, $A_3 = B_3 \setminus (B_1 \cup B_2), A_4 = B_4 \setminus (B_1 \cup B_2 \cup B_3)$, and so on. Hence the A_n form a collection of disjoint sets which cover $\mathcal{S}$ and each A_n has diameter less than $\varepsilon/2$. Choose N large enough so that $\mathbb{P}(\cup_{i=1}^N A_i) > 1 - \varepsilon/2$. Let $\mathcal{G}$ be the collection of open sets of the form $(A_{i_1} \cup \cdots \cup A_{i_j})_{\varepsilon/2}$ such that $i_1, \ldots, i_j \leq N$. That is, we look at all finite unions of $A_1, \ldots, A_N$, and then take the $(\varepsilon/2)$-enlargements. The collection $\mathcal{G}$ is finite. This fact and Theorem 30.2(3) imply that we can find n_0 such that $\mathbb{P}(G) \leq \mathbb{P}_n(G) + \varepsilon/2$ if $n \geq n_0$ and $G \in \mathcal{G}$.

Suppose F is closed. Let $G = (\cup\{A_i : i \leq N, A_i \cap F \neq \emptyset\})_{\varepsilon/2}$. Then $G \in \mathcal{G}$ and if $n \geq n_0$

$$\mathbb{P}(F) \leq \mathbb{P}(G) + \mathbb{P}(\cup_{i=N+1}^\infty A_i) \leq \mathbb{P}(G) + \varepsilon/2$$
$$\leq \mathbb{P}_n(G) + \varepsilon \leq \mathbb{P}_n(F_\varepsilon) + \varepsilon.$$

In the last inequality we used the definition of G and the fact that the A_i have diameters less than $\varepsilon/2$. This shows $d_{\mathcal{M}}(\mathbb{P}, \mathbb{P}_n) \leq \varepsilon$ if $n \geq n_0$, which in turn implies $d_{\mathcal{M}}(\mathbb{P}, \mathbb{P}_n) \to 0$. $\qquad\square$

Exercises

30.1 If $\mathcal{S}$ is a metric space, then it is well known that $C(\mathcal{S})$, the collection of continuous functions with the metric

$$d(f, g) = \sup_{x \in \mathcal{S}} |f(x) - g(x)|$$

is a metric space. Show that if $\mathcal{S}$ is compact, then $C(\mathcal{S})$ is separable.

30.2 Suppose X_n converges weakly to X and the random variables Z_n are such that $d(X_n, Z_n)$ converges to 0 in probability. Prove that Z_n converges weakly to X. This is known as *Slutsky's theorem*.

   Hint: Start with $\mathbb{P}(Z_n \in F) \leq \mathbb{P}(X_n \in \overline{F_\delta}) + \mathbb{P}(d(X_n, Z_n) \geq \delta)$.

30.3 Suppose X_n take values in a normed linear space and converge weakly to X. Suppose c_n are scalars converging to c. Show $c_n X_n$ converges weakly to cX.

30.4 Give an example of a sequence $\mathbb{P}_n$ converging weakly to $\mathbb{P}$ and a function f that is continuous but not bounded such that $\int f \, d\mathbb{P}_n$ does not converge to $\int f \, d\mathbb{P}$.

30.5 Give an example of a sequence $\mathbb{P}_n$ converging weakly to $\mathbb{P}$ and a function f that is bounded but not continuous such that $\int f \, d\mathbb{P}_n$ does not converge to $\int f \, d\mathbb{P}$.

30.6 Show that if X_n converges weakly to X and Y_n converges in probability to 0, then $X_n Y_n$ converges in probability to 0.

30.7 This exercise considers a sequence of probability measures that have densities. Suppose $\mathcal{S}$ is furnished with the Borel σ-field and μ is a measure on $\mathcal{S}$. Suppose that $f_n : \mathcal{S} \to [0, \infty)$ and $f : \mathcal{S} \to [0, \infty)$ are measurable functions, each of whose integral over $\mathcal{S}$ is one, and define $\mathbb{P}_n(A) = \int_A f_n(x) \, \mu(dx)$ for each n and $\mathbb{P}(A) = \int_A f(x) \, \mu(dx)$.

   (1) Show that if $f_n \to f$, μ-a.e., then $\mathbb{P}_n$ converges weakly to $\mathbb{P}$.

   (2) Give an example where $\mathbb{P}_n$ and $\mathbb{P}$ are as above, $\mathbb{P}_n$ converges weakly to $\mathbb{P}$, but f_n does not converge almost everywhere to f.

30.8 Give an example of continuous processes X_n and X such that all the finite-dimensional distributions of X_n converge weakly to the corresponding finite-dimensional distributions of X, but where X_n does not converge weakly to X with respect to the topology of $C[0, 1]$.

30.9 Suppose X is a random variable taking values in a complete separable metric space. If $\varepsilon > 0$, show there exists a compact set K such that $\mathbb{P}(X \notin K) < \varepsilon$.

Hint: For each n choose closed balls $\{B_{ni}, i = 1, \ldots, N_n\}$ such that

$$\mathbb{P}(X \notin \cup_{i=1}^{N_n} B_{ni}) < \varepsilon/2^{n+1}.$$

Then $K = \cap_{n=1}^{\infty} \cup_{i=1}^{N_n} B_{ni}$ is totally bounded, hence compact.

30.10 Suppose X_n converges weakly to X and the metric space S is complete and separable. Prove that the sequence $\{X_n\}$ is tight.

30.11 Let $\mathcal{L}$ be the collection of continuous functions on S such that
  (1) $\sup_{x \in S} |f(x)| \le 1$.
  (2) $|f(x) - f(y)| \le d(x, y)$ for all $x, y \in S$.
Define

$$d_{\mathcal{L}}(\mathbb{P}, \mathbb{Q}) = \sup_{f \in \mathcal{L}} \left| \int f \, d\mathbb{P} - \int f \, d\mathbb{Q} \right|.$$

Show that $d_{\mathcal{L}}$ is a metric on the collection of probability measures on the Borel σ-field of S. Prove that a sequence of probability measures $\mathbb{P}_n$ converges weakly to $\mathbb{P}$ if and only if $d_{\mathcal{L}}(\mathbb{P}_n, \mathbb{P}) \to 0$.

30.12 Suppose S is a separable metric space. Show that $\mathcal{M}$ is separable.

Notes

For more information, see Billingsley (1968) and Ethier and Kurtz (1986).

31

Skorokhod representation

Suppose S is a complete separable metric space furnished with the Borel σ-field. We are going to show that if X_n are random variables taking values in S converging weakly to a random variable X, then we can find another probability space and other random variables X_n', X' such that the law of X_n' equals the law of X_n for each n, the law of X' equals the law of X, and X_n' converges to X' almost surely.

Let $\Omega' = [0, 1]$, $\mathcal{F}'$ the Borel σ-field on $[0, 1]$, and $\mathbb{P}'$ Lebesgue measure. We first prove

Theorem 31.1 *Let $\mathbb{P}$ be a probability measure on S. Then there exists a random variable X mapping Ω' to S such that the law of X' under $\mathbb{P}'$ is equal to $\mathbb{P}$.*

Proof For each $k \geq 1$, let $\{A_{ki}\}$ be a countable disjoint covering of S by Borel sets of diameter less than $1/k$, such that $\mathbb{P}(\partial A_{ki}) = 0$, and $\{A_{ki}\}$ is a refinement of $\{A_{k-1,i}\}$. We can construct these families inductively. To start, cover S with countably many balls of radius less than 1. Since for each x_0, $\mathbb{P}(\{x : |x - x_0| = r\})$ can be nonzero for at most countably many values of r, we can arrange matters so that the $\mathbb{P}$-measure of the boundary of these balls is 0. We order the balls $B_1, B_2, \ldots$, and then let $A_{11} = B_1$, $A_{12} = B_2 \setminus B_1$, $A_{13} = B_3 \setminus (B_1 \cup B_2)$, and so on. To construct $\{A_{2i}\}$, we first find a similar covering of S by sets $\{A_{2i}'\}$ of diameter less than $1/2$, and then take all intersections of sets in $\{A_{2i}'\}$ with sets in $\{A_{1j}\}$.

We inductively define closed subintervals of $[0, 1]$ by choosing I_{11} to have left endpoint at 0 and length equal to $\mathbb{P}(A_{11})$, then I_{12} to have left endpoint equal to the right endpoint of I_{11} and length equal to $\mathbb{P}(A_{12})$, and so forth. We then decompose I_{11} into subintervals $\{I_{21}\}$ in an analogous way so that the lengths of the subintervals match the probabilities of the A_{2i}'s contained in A_{11}. We then subdivide I_{12}, and so on. We observe that $\{I_{ki}\}$ is a refinement of $\{I_{k-1,i}\}$ for all $k \geq 2$ and $\mathbb{P}'(I_{ki}) = \mathbb{P}(A_{ki})$ for all k and i.

Pick a point $x_{ki} \in A_{ki}$ for each k and i. We define X^k by setting $X^k(\omega')$ equal to x_{ki} if $\omega' \in I_{ki}$. (The set of endpoints of the I_{ki} is countable, hence has Lebesgue measure 0, and it doesn't matter how we define X^k at those points.) For each ω' except those that are endpoints of some I_{ki}, if $n \geq m$, then $X^n(\omega')$ and $X^m(\omega')$ are in the same A_{mi} for some i. Since the diameter of A_{mi} is less than $1/m$, we see that $d(X^n(\omega'), X^m(\omega')) \leq 1/m$. That is, $X^n(\omega')$ is a Cauchy sequence. The space S is complete, so we can define $X(\omega')$ to be the limit of the $X^n(\omega')$. The collection of endpoints of the I_{mi} is countable, so the limit exists for almost every ω'.

244

It remains to show that the law of X under $\mathbb{P}'$ is $\mathbb{P}$. Let F be a closed set, let $F_k = \{x : d(x, F) < 1/k\}$, and let $J_k = \{i : A_{ki} \cap F \neq \emptyset\}$. We have

$$\mathbb{P}'(X^k \in F) \leq \mathbb{P}'(X^k \in \cup_{i \in J_k} A_{ki}) \leq \sum_{i \in J_k} \mathbb{P}'(X^k \in A_{ki})$$

$$= \sum_{i \in J_k} \mathbb{P}'(I_{ki}) = \sum_{i \in J_k} \mathbb{P}(A_{ki}) \leq \mathbb{P}(\overline{F_k}).$$

We used the fact that each A_{ki} has diameter less than $1/k$. Hence

$$\limsup_k \mathbb{P}'(X^k \in F) \leq \mathbb{P}(F).$$

Therefore the laws of X^k under $\mathbb{P}'$ converge weakly to $\mathbb{P}$. But we know $d(X^k(\omega'), X(\omega')) \leq 1/k$, so X^k converges to X, a.s., with respect to $\mathbb{P}'$. If f is continuous and bounded, $\mathbb{E}'f(X^k) \to \mathbb{E}'f(X)$ by dominated convergence, so $X^k \to X$ weakly. Therefore the law of X under $\mathbb{P}'$ is equal to $\mathbb{P}$. $\qquad\square$

We did not need the fact that the A_{ki} were continuity sets, i.e., that the probability of the boundary of A_{ki} is zero, but this will be used in the next theorem, which is known as the *Skorokhod representation*.

Theorem 31.2 *Suppose $\mathbb{P}_n$ are probability measures on S converging weakly to $\mathbb{P}$. Then there exist random variables X_n mapping Ω' to S with laws $\mathbb{P}_n$ and a random variable X mapping Ω' to S with law $\mathbb{P}$ such that $X_n \to X$, a.s.*

Equivalently, if X'_n converges to X' weakly, there exist random variables X_n and X mapping Ω' to S with laws equal to X'_n and X, respectively, such that $X_n \to X$, a.s.

Proof Let the A_{ki} be as in the proof of the previous theorem, and for each $\mathbb{P}_n$ define intervals I_{ki}^n and random variables X_n^k as was done above, and let X_n be the limit of the X_n^k's. Let $K_{kn} = \{i : \mathbb{P}(A_{ki}) > \mathbb{P}_n(A_{ki})\}$ and $K_{kn}^c = \{i : \mathbb{P}(A_{ki}) \leq \mathbb{P}_n(A_{ki})\}$. Since

$$\sum_i [\mathbb{P}(A_{ki}) - \mathbb{P}_n(A_{ki})] = 1 - 1 = 0,$$

we have

$$\sum_{K_{kn}^c} [\mathbb{P}(A_{ki}) - \mathbb{P}_n(A_{ki})] = -\sum_{K_{kn}} [\mathbb{P}(A_{ki}) - \mathbb{P}_n(A_{ki})].$$

Hence

$$\sum_i |\mathbb{P}'(I_{ki}) - \mathbb{P}'(I_{ki}^n)| = \sum_i |\mathbb{P}(A_{ki}) - \mathbb{P}_n(A_{ki})| \qquad (31.1)$$

$$= \sum_{K_{kn}} [\mathbb{P}(A_{ki}) - \mathbb{P}_n(A_{ki})] - \sum_{K_{kn}^c} [\mathbb{P}(A_{ki}) - \mathbb{P}_n(A_{ki})]$$

$$= 2 \sum_{K_{kn}} [\mathbb{P}(A_{ki}) - \mathbb{P}_n(A_{ki})]$$

$$= 2 \sum_i [\mathbb{P}(A_{ki}) - \mathbb{P}_n(A_{ki})]^+.$$

Each term in the sum on the last line goes to 0 as $n \to \infty$ by Theorem 30.2 because the A_{ki} are $\mathbb{P}$-continuity sets, that is, $\mathbb{P}(\partial A_{ki}) = 0$; also each term is dominated by $\mathbb{P}(A_{ki})$, and

$\sum_i \mathbb{P}(A_{ki}) = 1$. Therefore by dominated convergence the sum on the last line of (31.1) goes to 0.

Fix k and j and let α, α_n be the left-hand endpoints of I_{kj}, I_{kj}^n, respectively. Then (31.1) allows us to use dominated convergence to conclude that

$$\alpha = \sum_{i \in J} \mathbb{P}'(I_{ki}) = \lim_{n \to \infty} \sum_{i \in J} \mathbb{P}'(I_{ki}^n) = \lim_{n \to \infty} \alpha_n,$$

where J consists of those i such that I_{ki} is to the left of I_{kj}; note that for $i \in J$ we have that I_{ki}^n is to the left of I_{kj}^n and conversely, if I_{ki}^n is to the left of I_{kj}^n, then $i \in J$. Similarly the right-hand endpoint of I_{kj}^n converges to the right-hand endpoint of I_{kj}.

If ω' is in the interior of I_{kj}, then it will be in the interior of I_{kj}^n for all sufficiently large n. This means that for n sufficiently large,

$$d(X(\omega'), X_n(\omega') \le 2/k.$$

This implies our result. $\square$

Exercises

31.1 Suppose f is bounded, X_n converges to X weakly, and also that $\mathbb{P}(X \in D_f) = 0$, where $D_f = \{x : f \text{ is not continuous at } x\}$. Show that $f(X_n)$ converges weakly to $f(X)$.

31.2 Suppose a sequence $\{X_n\}$ is uniformly integrable and X_n converges to X weakly. Show $\mathbb{E} X_n \to \mathbb{E} X$.

31.3 Give an example of a sequence of random variables X_n converging weakly to X and where each X_n is integrable, but X is not integrable.

31.4 Suppose X_n converges weakly to X and each X_n is non-negative. Prove that

$$\mathbb{E} X \le \liminf_{n \to \infty} \mathbb{E} X_n.$$

31.5 Suppose X_n converges weakly to X and each X_n has the property that with probability one,

$$|X_n(t) - X_n(s)| \le |t - s|, \qquad s, t \le 1.$$

(This might arise, for example, if each X_n is of the form $X_n(t) = \int_0^t Y_n(s)\, ds$ and each Y_n is bounded by 1.) Prove that X has this same property, that is, with probability one,

$$|X(t) - X(s)| \le |t - s|, \qquad s, t \le 1.$$

31.6 Here is a way to prove one direction of Lebesgue's theorem on Riemann integrable functions.

(1) For each $n \ge 1$ and each $i \le n$, let x_{in} be a point in $[(i-1)/n, i/n)$. Let $\mathbb{P}_n$ be the probability measure that assigns mass $1/n$ to each point x_{in}, $i = 1, 2, \ldots, n$. Show that $\mathbb{P}_n$ converges weakly to $\mathbb{P}$, where $\mathbb{P}$ is a Lebesgue measure on $[0, 1]$.

(2) Suppose f is a bounded function which is continuous at almost every point of $[0, 1]$. Show that $\int f\, d\mathbb{P}_n \to \int f\, d\mathbb{P}$. Note that $\int f\, d\mathbb{P}_n$ is a Riemann sum approximation to $\int_0^1 f(x)\, dx$.

32

The space $C[0, 1]$

We examine weak convergence for the space $C[0, 1]$, the set of continuous real-valued functions on $[0, 1]$. We give a criterion for the laws of a sequence of continuous stochastic processes to be tight. We apply these results to show that a simple symmetric random walk converges weakly to a Brownian motion, which in particular gives another construction of Brownian motion.

32.1 Tightness

Let $C[0, 1]$ be the collection of continuous real-valued functions from $[0, 1]$ into $\mathbb{R}$. We make $C[0, 1]$ into a metric space by defining

$$d(f, g) = \sup_{t \in [0,1]} |f(t) - g(t)|,$$

and it is well known that $C[0, 1]$ is separable and complete. We recall the Ascoli–Arzelà theorem: if a family $\mathcal{F}$ of functions on a compact set is equicontinuous and uniformly bounded at one point, then every subsequence in $\mathcal{F}$ has a further subsequence in $\mathcal{F}$ that converges. Rephrased another way, if the family $\mathcal{F}$ is equicontinuous and uniformly bounded at one point, then the closure of $\mathcal{F}$ is compact. We furnish $C[0, 1]$ with the Borel σ-field.

Given a continuous function f on $[0, 1]$, we define ω_f, the *modulus of continuity* of f, by

$$\omega_f(\delta) = \sup_{s,t \in [0,1], |t-s| < \delta} |f(t) - f(s)|.$$

We have the following criterion for a sequence of continuous processes to be tight.

Theorem 32.1 *Suppose the X_n are continuous real-valued processes. Suppose for each ε and $\eta > 0$ there exist $n_0, A,$ and δ (depending on ε and η) such that if $n \geq n_0$, then*

$$\mathbb{P}(\omega_{X_n}(\delta) \geq \varepsilon) \leq \eta \tag{32.1}$$

and

$$\mathbb{P}(|X_n(0)| \geq A) \leq \eta. \tag{32.2}$$

Then the X_n are tight.

Proof　Since each X_i is a continuous process, then for each i, $\mathbb{P}(\omega_{X_i}(\delta) \geq \varepsilon) \to 0$ as $\delta \to 0$ by dominated convergence. Hence, given ε and η we can, by taking δ smaller if necessary, assume that (32.1) holds for all n.

247

Choose $\varepsilon_m = \eta_m = 2^{-m}$ and consider the δ_m and A_m so that

$$\sup_n \mathbb{P}(\omega_{X_n}(\delta_m) \geq 2^{-m}) \leq 2^{-m}$$

and

$$\sup_n \mathbb{P}(|X_n(0)| \geq A_m) \leq 2^{-m}.$$

Let

$$K_{m_0} = \{f \in C[0, 1] : \sup_{s,t \in [0,1], |t-s| \leq \delta_m} |f(t) - f(s)| \leq 2^{-m} \text{ for all } m \geq m_0,$$

$$|f(0)| \leq A_{m_0}\}.$$

Each K_{m_0} is an equicontinuous family, and by the Ascoli–Arzelá theorem, each K_{m_0} is a compact subset of $C[0, 1]$. We have

$$\mathbb{P}(X_n \notin K_{m_0}) \leq \mathbb{P}(|X_n(0)| \geq A_{m_0}) + \sum_{m=m_0}^{\infty} \mathbb{P}(\omega_{X_n}(\delta_m) \geq \varepsilon_m)$$

$$\leq 2^{-m_0} + \sum_{m=m_0}^{\infty} 2^{-m} = 3 \cdot 2^{-m_0}.$$

This proves tightness.   □

We have given one criterion for a process to have continuous paths, namely, Theorem 8.1. In the case of Markov processes, we have given another: Theorem 21.5.

32.2 A construction of Brownian motion

We will now use the results of Section 32.1 to give a construction of Brownian motion, quite different from that of Chapter 6.

Let Y_i be i.i.d. random variables with $\mathbb{P}(Y_i = 1) = \mathbb{P}(Y_i = -1) = \frac{1}{2}$. Then $S_n = \sum_{i=1}^{n} Y_i$ is a *simple symmetric random walk*. Let $Z_n(t) = S_{nt}/\sqrt{n}$ for t a multiple of $1/n$ and define Z_t^n by linear interpolation for other t. That is, if $k/n \leq t \leq (k+1)/n$, then

$$Z_t^n = \frac{(k+1) - nt}{\sqrt{n}} S_k + \frac{nt - k}{\sqrt{n}} S_{k+1}. \tag{32.3}$$

The Z_n are continuous processes. Let $\mathbb{P}_n$ be the law of Z_n, which will be a probability measure on $C[0, 1]$.

Theorem 32.2 *The sequence $\mathbb{P}_n$ converges weakly to a probability measure $\mathbb{P}_\infty$ on $C[0, 1]$, and $\mathbb{P}_\infty$ is the law of a Brownian motion.*

Proof   The main step is to prove that the $\mathbb{P}_n$ are tight. We then show that any subsequential limit point is a Wiener measure, that is, the law of a Brownian motion. We can then appeal to Theorem 31.1 to obtain the process X, which will be a Brownian motion.

A computation shows that

$$\mathbb{E} S_n^4 = \sum_{i=1}^{n} \mathbb{E} Y_i^4 + \sum_{i \neq j} (\mathbb{E} Y_i^2)(\mathbb{E} Y_j^2) \leq cn^2, \tag{32.4}$$

since $\mathbb{E}\,Y_i$ and $\mathbb{E}\,Y_i^3$ are both 0, the Y_i's are independent, and the second sum has $n(n-1) \le n^2$ terms.

If s and t are multiples of $1/n$, then

$$\mathbb{E}\,|Z_t - Z_s|^4 = \frac{1}{n^2}\mathbb{E}\left(\sum_{i=ns+1}^{nt} Y_i\right)^4 = \frac{1}{n^2}\mathbb{E}\left(\sum_{i=1}^{nt-ns} Y_i\right)^4 \tag{32.5}$$

$$\le \frac{c}{n^2}n^2|t-s|^2 \le c|t-s|^2.$$

If we tried to get by with only the second moment, we would only end up with $c|t-s|$, which is not good enough for Theorem 8.1.

At this point we would like to apply Theorem 32.1, but we have the technical nuisance that s and t might not be multiples of $1/n$. If $|t-s| \le 2/n$, then by the construction of Z_n using linear interpolation and the fact that the Y_i's are bounded by one in absolute value, we have $|Z_n(t) - Z_n(s)| \le c|t-s|\sqrt{n}$ and then

$$\mathbb{E}\,|Z_n(t) - Z_n(s)|^4 \le c|t-s|^4 n^2 \le c|t-s|^2. \tag{32.6}$$

Suppose $|t-s| > 2/n$. Let s' be the largest multiple of $1/n$ less than or equal to s and t' the largest multiple of $1/n$ larger than or equal to t. Using (32.5) and (32.6),

$$\mathbb{E}\,|Z_n(t) - Z_n(s)|^4 \le c\mathbb{E}\,|Z_n(t) - Z_n(t')|^4 + c\mathbb{E}\,|Z_n(t') - Z_n(s')|^4 + \mathbb{E}\,|Z_n(s') - Z_n(s)|^4$$

$$\le c|t-t'|^2 + c|t'-s'|^2 + c|s'-s|^2$$

$$\le c|t-s|^2,$$

since $|t-t'|$, $|t'-s'|$, and $|s-s'|$ are all less than $c|t-s|$. Note $Z_n(0) = 0$ for all n. We now apply Theorems 8.1 and 32.1 to obtain the tightness.

Any subsequential limit point is a probability measure on $C[0,1]$, so to show that the limit is a Brownian motion, it is enough by Theorem 2.6 to show that the finite-dimensional distributions under the limit law $\mathbb{P}_\infty$ agree with those of Brownian motion. Fix t. Then $Z_n(t)$ differs from $S_{[nt]}/\sqrt{n}$ by at most $1/\sqrt{n}$, where $[nt]$ is the largest integer less than or equal to nt. By the central limit theorem (Theorem A.51), $S_{[nt]}/\sqrt{[nt]}$ converges weakly (with respect to the topology of $\mathbb{R}$) to a mean zero normal random variable with variance one. By Exercise 30.3, $S_{[nt]}/\sqrt{n}$ converges weakly to a mean zero normal random variable with variance t, and by Exercise 30.2, $Z_n(t)$ converges weakly to a mean zero normal random variable with variance t. This shows that the one-dimensional distributions of Z_n converge weakly to the one-dimensional distributions of a Brownian motion. We leave the analogous argument for the higher-dimensional distributions to the reader. $\square$

One can also use Doob's inequalities to obtain the necessary tightness estimate. If s and t are multiples of $1/n$, we have

$$\mathbb{P}(\max_{ns \le k \le nt} |S_k - S_{ns}| > \lambda\sqrt{n}) \le c\frac{\mathbb{E}\,|S_{nt} - S_{ns}|^4}{\lambda^4 n^2} \tag{32.7}$$

$$\le c\frac{|t-s|^2}{\lambda^4}.$$

Exercises

32.1 The support of a measure λ is the smallest closed set F such that $\lambda(F^c) = 0$. Let $\mathbb{P}$ be a Wiener measure on $C[0, 1]$, i.e., the law of a Brownian motion on $[0, 1]$. Use Exercise 13.4 to prove that the support of $\mathbb{P}$ is all of $C[0, 1]$.

32.2 Let $(\mathcal{S}, d)$ be a complete separable metric space and let $\mathcal{R}$ be a subset of $\mathcal{S}$. Then $(\mathcal{R}, d)$ is also a metric space. If X_n converges weakly to X with respect to the topology of $(\mathcal{S}, d)$ and each X_n and X take values in $\mathcal{R}$, does X_n converge weakly to X with respect to the topology of $(\mathcal{R}, d)$? Does the answer change if $\mathcal{R}$ is a closed subset of $\mathcal{S}$?

If X_n and X take values in $\mathcal{R}$ and X_n converges weakly to X with respect to the topology of $(\mathcal{R}, d)$, does X_n converge weakly to X with respect to the topology of $(\mathcal{S}, d)$? What if $\mathcal{R}$ is a closed subset of $\mathcal{S}$?

32.3 Give a proof of Theorem 32.2 using (32.7) in place of Theorem 8.1.

32.4 Suppose $(X, W, \mathbb{P})$ is a weak solution to

$$dX_t = \sigma(X_t)\, dW_t + b(X_t)\, dt, \qquad X_0 = x, \tag{32.8}$$

where W is a one-dimensional Brownian motion and σ and b are bounded and continuous, but we do not assume that σ is bounded below by a positive constant. Suppose the solution to (32.8) is unique in law.

Suppose σ_n and b_n are Lipschitz functions which are uniformly bounded and which converge uniformly to σ and b, respectively. Let $X_t(n)$ be the unique pathwise solution to

$$dY_t = \sigma_n(Y_t)\, dW_t + b_n(Y_t)\, dt, \qquad Y_0 = x;$$

the probability measure here is $\mathbb{P}$. Prove that $X(n)$ converges weakly to X with respect to $C[0, 1]$.

32.5 Let W be a d-dimensional Brownian motion and let $\{X_t, t \in [0, 1]\}$ be the solution to (24.22). If $x \in \mathbb{R}^d$, prove that the support of $\mathbb{P}^x$ is all of $C[0, 1]$.

33

Gaussian processes

A Gaussian process is a stochastic process where each of the finite-dimensional distributions is jointly normal. We will primarily, but not exclusively, be concerned with Gaussian processes that have continuous paths. For much of what we consider, it is not essential that the index set of times be $[0, \infty)$, and can in fact be almost any set. We will thus consider $\{X_t : t \in T\}$ for some index set T, and where for every finite subset S of T, the collection $\{X_s : s \in S\}$ is jointly normal.

33.1 Reproducing kernel Hilbert spaces

We define the covariance function Γ by

$$\Gamma(s, t) = \mathbb{E}\left[(X_s - \mathbb{E}X_s)(X_t - \mathbb{E}X_t)\right], \qquad s, t \in T. \tag{33.1}$$

For our purposes, having a non-zero mean just complicates formulas without adding anything interesting, so in this chapter we will assume $\mathbb{E}X_t = 0$ for all $t \in T$, and (33.1) becomes

$$\Gamma(s, t) = \mathbb{E}[X_s X_t], \qquad s, t \in T. \tag{33.2}$$

We first show how Γ can be used to construct a Hilbert space called the reproducing kernel Hilbert space (RKHS).

When we write $\Gamma(s, \cdot)$, we mean that we fix an element $s \in T$ and then consider the function $g : T \to \mathbb{R}$ defined by $g(t) = \Gamma(s, t)$ for $t \in T$. Let $\mathcal{K}$ be the collection of finite linear combinations of the functions $\Gamma(s, \cdot)$, $s \in T$. Thus each element of $\mathcal{K}$ has the form

$$\sum_{j=1}^{m} a_j \Gamma(s_j, \cdot),$$

where $m \geq 1$, the a_j's are real, and each s_j, $j = 1, \ldots, m$, is an element of T. If $f = \sum_{j=1}^{m} a_j \Gamma(s_j, \cdot)$ and $g = \sum_{k=1}^{n} b_k \Gamma(t_k, \cdot)$, define

$$\langle f, g \rangle_{RKHS} = \sum_{j=1}^{m} \sum_{k=1}^{n} a_j b_k \Gamma(s_j, t_k).$$

We define $\mathcal{H}$ to be the closure of $\mathcal{K}$ with respect to the norm induced by the inner product $\langle \cdot, \cdot \rangle_{RKHS}$.

We need to show that this bilinear form is indeed an inner product, that what is known as the reproducing property holds, and that $\mathcal{H}$ is a Hilbert space.

We start with the reproducing property. If $f = \sum_{j=1}^{m} a_j \Gamma(s_j, \cdot)$, then the *reproducing property* applied to f is the formula

$$\langle f, \Gamma(t, \cdot) \rangle_{RKHS} = f(t). \tag{33.3}$$

This follows from

$$\langle f, \Gamma(t, \cdot) \rangle_{RKHS} = \sum_{j=1}^{m} a_j \Gamma(s_j, t) = f(t).$$

By taking limits, (33.3) holds for all $f \in \mathcal{H}$.

To show that $\langle \cdot, \cdot \rangle_{RKHS}$ is an inner product, notice that when

$$f = \sum a_j \Gamma(s_j, \cdot) \in \mathcal{K},$$

then

$$\langle f, f \rangle_{RKHS} = \sum_{j=1}^{m} \sum_{k=1}^{m} a_j a_k \Gamma(s_j, s_k) = \sum_{j,k=1}^{m} a_j a_k \mathbb{E}\left[X_{s_j} X_{s_k}\right]$$

$$= \mathbb{E}\left(\sum_{j=1}^{m} a_j X_{s_j}\right)^2 \geq 0.$$

The Cauchy–Schwarz inequality holds for $\langle \cdot, \cdot \rangle_{RKHS}$ (the standard proof of the Cauchy–Schwarz inequality applies), and so if $\langle f, f \rangle_{RKHS} = 0$, then

$$|f(t)|^2 = \langle f, \Gamma(t, \cdot) \rangle_{RKHS}^2 \leq \langle f, f \rangle_{RKHS} \langle \Gamma(t, \cdot), \Gamma(t, \cdot) \rangle_{RKHS} = 0,$$

and thus f is zero.

If f_n is a Cauchy sequence with respect to the norm

$$\|g\|_{RKHS} = \langle g, g \rangle_{RKHS}^{1/2},$$

then

$$|f_n(t) - f_m(t)|^2 = \langle f_n - f_m, \Gamma(t, \cdot) \rangle_{RKHS}^2$$

$$\leq \langle f_n - f_m, f_n - f_m \rangle_{RKHS} \langle \Gamma(t, \cdot), \Gamma(t, \cdot) \rangle_{RKHS},$$

which tends to 0 as $n, m \to \infty$. Thus f_n converges pointwise. This is enough to prove $\mathcal{H}$ is complete; this is Exercise 33.1.

We summarize.

Proposition 33.1 $\mathcal{H}$ *with the inner product* $\langle \cdot, \cdot \rangle_{RKHS}$ *is a Hilbert space. Moreover, if* $f \in \mathcal{H}$ *and* $t \in T$, *then*

$$\langle f, \Gamma(t, \cdot) \rangle_{RKHS} = f(t).$$

We consider another Hilbert space $\mathcal{M}$, the closure of the linear span of $\{X_t : t \in T\}$ with respect to $L^2(\mathbb{P})$. We define

$$\langle Y, Z \rangle_{\mathcal{M}} = \mathbb{E}[YZ]$$

if Y and Z are both finite linear combinations of the X_t's. Thus if $m, n \geq 1, a_j, b_k \in \mathbb{R}$, we set

$$\left\langle \sum_{j=1}^{m} a_j X_{s_j}, \sum_{k=1}^{n} b_k X_{t_k} \right\rangle_{\mathcal{M}} = \sum_{j=1}^{m} \sum_{k=1}^{n} a_j b_k \mathbb{E}\left[X_{s_j} X_{t_k}\right], \qquad (33.4)$$

and we let $\mathcal{M}$ be the closure of the collection of random variables of the form $\sum_{j=1}^{m} a_j X_{s_j}$ with respect to $\langle \cdot, \cdot \rangle_{\mathcal{M}}$. Since $\Gamma(s_j, t_k) = \mathbb{E}\left[X_{s_j} X_{t_k}\right]$, from (33.4) we see that $\mathcal{H}$ and $\mathcal{M}$ are isomorphic, where we have a one-to-one correspondence between $\sum_{j=1}^{m} a_j \Gamma(s_j, \cdot)$ and $\sum_{j=1}^{m} a_j X_{s_j}$.

Let $\{e_n\}$ be a complete orthonormal system for $\mathcal{H}$. Let Y_n be the element of $\mathcal{M}$ corresponding to e_n. Then

$$\mathbb{E}\left[Y_n Y_m\right] = \langle Y_n, Y_m \rangle_{\mathcal{M}} = \langle e_n, e_m \rangle_{RKHS} = \delta_{nm},$$

where δ_{nm} is 0 if $n \neq m$ and 1 if $n = m$. This implies that the Y_n are independent normal random variables with mean zero and variance one; see Proposition A.55. (Recall that we are assuming that all the X_t's have mean zero.)

Since $\Gamma(s, \cdot)$ is an element of $\mathcal{H}$, we can write

$$\Gamma(s, \cdot) = \sum_{n=1}^{\infty} \langle \Gamma(s, \cdot), e_n \rangle_{RKHS} e_n(\cdot) = \sum_{n=1}^{\infty} e_n(s) e_n(\cdot).$$

Using the correspondence between $\mathcal{H}$ and $\mathcal{M}$, we have

$$X_s = \sum_{n=1}^{\infty} e_n(s) Y_n,$$

where the Y_n are i.i.d. standard normal variables. This is known as the *Karhunen–Loève expansion* of a Gaussian process.

Example 33.2 Let's see what this expansion is in the case of Brownian motion. If we define

$$\langle f, g \rangle_{CM} = \int_0^1 f'(r) g'(r) \, dr \qquad (33.5)$$

for f and g whose first derivatives are in $L^2([0, 1])$ and such that $f(0) = g(0) = 0$, then because $\Gamma(s, t) = s \wedge t$,

$$\langle \Gamma(s, \cdot), \Gamma(t, \cdot) \rangle_{CM} = \int_0^1 1_{[0,s)}(r) 1_{[0,t)}(r) \, dr = s \wedge t$$
$$= \Gamma(s, t),$$

and we see that we have identified the reproducing kernel Hilbert space for Brownian motion on $[0, 1]$. The notation $\langle \cdot, \cdot \rangle_{CM}$ is used because the Hilbert space with this inner product is called the *Cameron–Martin space*, a space that has many connections with Brownian motion.

If $e_n(s) = \sqrt{2} \sin(n\pi s)/n\pi$, then the sequence $\{e_n\}$ is a complete orthonormal sequence for the Cameron–Martin space. The Karhunen–Loève expansion is equivalent to the formula (6.2) that we used in our first construction of Brownian motion.

33.2 Continuous Gaussian processes

We now turn to the construction of Gaussian processes with continuous paths. Suppose we have an index set T and a non-negative definite kernel $\Gamma(\cdot, \cdot)$. Saying Γ is *non-negative definite* means that for each n and each $t_1, \ldots, t_n \in T$, the matrix whose (i, j) entry is $\Gamma(t_i, t_j)$ is a non-negative definite matrix. We define a metric on T by defining

$$d(s, t) = (\text{Var}\,(X_t - X_s))^{1/2}.$$

Actually, d is a pseudo-metric because $d(s, t) = 0$ does not necessarily imply $t = s$. An ε-ball is a set of the form $\{t \in T : d(t, t_0) < \varepsilon\}$ for some t_0. Let $N(\varepsilon)$ be the minimum number of ε-balls needed to cover T.

Theorem 33.3 *Let $\Gamma : T \times T \to \mathbb{R}$ be continuous with respect to the pseudo-metric d, symmetric, and non-negative definite. If for some $\beta < 1$ and some constant c we have*

$$\log N(\varepsilon) \le c\varepsilon^{-\beta}, \qquad \varepsilon \in (0, 1), \tag{33.6}$$

then there exists a continuous Gaussian process $\{X_t : t \in T\}$ with covariance kernel Γ.

One can in fact be more precise than (33.6) and give an integral condition that $N(x)$ must satisfy for x small.

Before proving Theorem 33.3, let us look at a number of examples.

Example 33.4 In the case of Brownian motion, $\text{Var}\,(X_t - X_s) = |t - s|$, so that $d(s, t) = |s - t|^{1/2}$. If T is the interval $[0, 1]$, then the set of intervals of length ε^2 and centers $k\varepsilon^2/4$, $k = 0, 1, \ldots, 4/\varepsilon^2$, is a collection of ε-balls covering $[0, 1]$. Therefore $N(\varepsilon) \le c/\varepsilon^2$, implying $\log N(\varepsilon) \le c \log(1/\varepsilon)$, which satisfies (33.6). This and Theorem 2.4 gives a construction of Brownian motion.

Example 33.5 We look at *fractional Brownian motion*. Let $H \in (0, 2)$. H is known as the *Hurst index*, where $H = 1$ corresponds to Brownian motion. Define

$$\Gamma(s, t) = |s|^H + |t|^H - |s - t|^H.$$

This leads to $d(s, t) = c|t - s|^{H/2}$. Open intervals of length $\varepsilon^{2/H}$ are ε-balls, and it takes $c\varepsilon^{-2/H}$ of them to cover $[0, 1]$. Therefore again $N(\varepsilon) \le c \log(1/\varepsilon)$, and (33.6) applies. One use of fractional Brownian motion is to model stock prices where there is more or less memory of the past than a Brownian motion has.

Example 33.6 Here is our first example of a Gaussian process where T is not a subset of $[0, \infty)$. We construct a *Brownian sheet*, $X(t_1, t_2)$, where the points $(t_1, t_2) \in [0, 1]^2$. More generally we can consider $X(t)$, where $t \in [0, 1]^d$. This is no harder, but for simplicity of notation we consider only the case $d = 2$. If $s = (s_1, s_2)$ and $t = (t_1, t_2)$, define

$$\Gamma(s, t) = (s_1 \wedge t_1)(s_2 \wedge t_2).$$

One motivation for this formula is to identify the point (t_1, t_2) with the rectangle R_t whose lower left corner is at the origin and whose upper right corner is at (t_1, t_2). Then the covariance of X_s and X_t is the area of $R_s \cap R_t$.

Some simple geometry shows that if we put ε-balls centered at the points $(c_1 j\varepsilon^2, c_1 k\varepsilon^2)$ for an appropriate c_1 and with $j, k \leq c_2 \varepsilon^{-2}$, we cover T. Therefore $N(\varepsilon) \leq c\varepsilon^{-4}$, and so $\log N(\varepsilon) \leq c \log(1/\varepsilon)$.

Example 33.7 We can generalize the last example. For every Borel subset A of $[0,1]^d$, let X_A be a Gaussian random variable. We want the covariance of X_A and X_B to be the Lebesgue measure of $A \cap B$. This is known as a *set-indexed process*. If we let T be the collection of all Borel subsets of $[0,1]^d$, one cannot get a continuous Gaussian process. In order to get a continuous process X one must restrict T to be a subcollection of sets whose boundaries are sufficiently smooth; see Dudley (1973).

Example 33.8 Our last example has a more complicated index set. Let W be a one-dimensional Brownian motion. If $f \in L^2[0,1]$, define

$$X_f = \int_0^1 f(s)\, dW_s.$$

By Exercise 24.6, X_f is a Gaussian random variable with mean 0 and variance $\int_0^1 f(s)^2\, ds$ and the covariance of X_f and X_g is $\int_0^1 f(s)g(s)\, ds$. It follows that

$$d(f,g)^2 = \int_0^1 (f(s) - g(s))^2\, ds.$$

The process X_f is known as a *Gaussian field*.

For what subsets T of $L^2([0,1])$ can one define a process X_f that has continuous paths with respect to d? This means that the map $f \to X_f(\omega)$ is continuous for almost all ω, where we use the pseudo-metric d to define open sets in T. It turns out $T = \{f \in L^2([0,1]) : \|f\|_2 \leq 1\}$ is too large to obtain a continuous Gaussian process, but, for example, $T = \{f \in C^2([0,1]) : \|f\|_\infty \leq 1, \|f'\|_\infty \leq 1, \|f''\|_\infty \leq 1\}$ is small enough to apply Theorem 33.3.

We now proceed to the proof of Theorem 33.3.

Proof of Theorem 33.3 Since T can be covered by finitely many ε-balls for each ε, it follows that if $\mathcal{A}(\varepsilon)$ is the collection of centers for the cover by ε-balls, then $\mathcal{A} = \cup_{n=1}^\infty \mathcal{A}(2^{-n})$ is a countable dense subset of T. We first label the elements of $\mathcal{A}$ by $t_1, t_2, \ldots$ For each n, we construct the law of $(X_{t_1}, \ldots, X_{t_n})$. We then use the Kolmogorov extension theorem to construct the law of $\{X_t : t \in \mathcal{A}\}$. Next we prove that $t \to X_t$ is uniformly continuous on $\mathcal{A}$, almost surely. Finally we define X_t for all $t \in T$ by continuity.

Step 1. We construct the law of $(X_{t_1}, \ldots, X_{t_n})$. Let n be fixed, and let B be an $n \times n$ matrix whose (i,j) entry is $\Gamma(t_i, t_j)$. The matrix B is symmetric, and non-negative definite by hypothesis. Let $Y_1, \ldots, Y_n$ be independent normal random variables with mean zero and variance one. If we let C be the non-negative definite square root of B and

$$X = CY$$

(viewed as vectors), or equivalently,

$$X_{t_i} = \sum_{j=1}^n C_{ij} Y_j,$$

a simple calculation shows that $\mathbb{E}\,[X_{t_k}X_{t_m}] = B_{km} = \Gamma(t_k, t_m)$. The X_{t_i}'s are jointly normal and this gives the first step of the construction.

Step 2. We apply the Kolmogorov extension theorem. Let $\mathbb{P}_n$ be the law of $(X_{t_1}, \ldots, X_{t_n})$. It is easy to see the consistency property holds for the $\mathbb{P}_n$, so by the Kolmogorov extension theorem, there exists a probability $\mathbb{P}$ on $\mathbb{R}^\mathbb{N}$ such that if we define $X_t(\omega)$ by $\omega(t)$ for $t \in \mathcal{A}$, the law of $(X_{t_1}, \ldots, X_{t_n})$ is $\mathbb{P}_n$ for each n.

Step 3. We show that except for a null set of probability zero, the map $t \to X_t(\omega)$ is uniformly continuous on $\mathcal{A}$.

To prove the uniform continuity, we proceed similarly to Theorem 8.1. For each point $t \in \mathcal{A}$, let t_j be the element of $\mathcal{A}(2^{-j})$ closest to t, with some convention for breaking ties. We will fix J in a moment, and write

$$X_t = X_{t_J} + (X_{t_{J+1}} - X_{t_J}) + (X_{t_{J+2}} - X_{t_{J+1}}) + \cdots,$$

where the sum is finite because $t \in \mathcal{A}$. Let $\lambda > 0$. If $|X_t - X_s| > \lambda$ for some $s, t \in \mathcal{A}$ with $d(s, t) < 2^-$, then ω is in one or more of the following events:

(a) the event

$$E_J = \{|X_{t_J} - X_{s_J}| > \lambda/2 \text{ for some } s_J, t_J \in \mathcal{A}(2^{-J}) \text{ with } d(s_J, t_J) \le 3 \cdot 2^{-J}\};$$

(b) the event

$$F_j = \left\{|X_{t_{j+1}} - X_{t_j}| > \frac{\lambda}{8j^2} \text{ for some } t_j \in \mathcal{A}(2^{-j}), t_{j+1} \in \mathcal{A}(2^{-(j+1)})\right.$$
$$\left. \text{with } d(t_j, t_{j+1}) < 3 \cdot 2^{-j+1}\right\}$$

for some $j \ge J$;

(c) the event

$$G_j = \left\{|X_{s_{j+1}} - X_{s_j}| > \frac{\lambda}{8j^2} \text{ for some } s_j \in \mathcal{A}(2^{-j}), s_{j+1} \in \mathcal{A}(2^{-(j+1)})\right.$$
$$\left. \text{with } d(s_j, s_{j+1}) < 3 \cdot 2^{-j+1}\right\}$$

for some $j \ge J$.

First we bound the probability of E_J. There are $N(2^{-J})$ elements of $\mathcal{A}(2^{-J})$, so there are at most $\exp(2c(2^J)^\beta)$ pairs (s_J, t_J). If $d(t_J, s_J) < 3 \cdot 2^{-J}$, then

$$\mathbb{P}(|X_{s_J} - X_{t_J}| > \lambda/2) \le 2 \exp\left(-\frac{(\lambda/2)^2}{2 \cdot 3 \cdot 2^{-J}}\right).$$

Therefore the probability of E_J is bounded by

$$\mathbb{P}(E_j) \le e^{c2^{\beta J}} e^{-c\lambda^2 2^J}.$$

Since $\beta < 1$, this can be made as small as we like by taking J large enough.

For any t_j and t_{j+1} with $d(t_j, t_{j+1}) < 3 \cdot 2^{-j+1}$,

$$\mathbb{P}(|X_{t_j} - X_{t_{j+1}}| > \lambda/(8j^2)) \le 2 \exp\left(\frac{\lambda^2/64j^4}{6 \cdot 2^{-j+1}}\right).$$

There are less than $e^{c2^{\beta j}}$ points in $\mathcal{A}(2^{-j})$ and $e^{c2^{\beta(j+1)}}$ points in $\mathcal{A}(2^{-(j+1)})$, so less than $e^{c2^{\beta j}}$ pairs. Thus the probability of F_j is bounded by

$$\mathbb{P}(F_j) \leq ce^{c2^{\beta j}}e^{-c\lambda^2 2^j/j^4}.$$

Since $\beta < 1$, this is summable in j, and $\sum_{j=J}^{\infty}\mathbb{P}(F_j)$ can be made as small as we like if we take J large enough. We handle the bound for G_j similarly.

Thus, given ε, we have

$$\mathbb{P}(\sup_{s,t\in\mathcal{A},d(s,t)<2^{-J}}|X_t - X_s| > \lambda) \leq \varepsilon$$

if we take J large enough, where J depends on ε and λ. This suffices to prove the uniform continuity.

Step 4. We use continuity to complete the proof. Define $X_t = \lim_{s\in\mathcal{A},s\to t}X_s$. The limit exists and will be a continuous function of t by virtue of the uniform continuity. By Remark A.56, X_t will have the desired covariance function. □

We have been considering Gaussian processes taking values in $\mathbb{R}$, but it is also of interest to look at Brownian motion taking values in a Hilbert space or a Banach space. There are three steps to constructing such a process:

(1) constructing Gaussian measures on Banach (or Hilbert) spaces;
(2) getting a suitable estimate on $\|X_t - X_s\|$;
(3) constructing a Brownian motion.

Of these three steps, the third follows along the lines we used for real-valued processes. Steps (1) and (2) require considerable work, and we refer the reader to Bogachev (1998) or Kuo (1975). A measure μ on a Banach space is called Gaussian if $\mu \circ L^{-1}$ is a Gaussian measure on $\mathbb{R}$ for every linear functional L on the Banach space.

Exercises

33.1 Finish the proof that $\mathcal{H}$ as defined in Section 33.1 is complete.

33.2 Show that if in Example 33.8 we let

$$T = \{f \in C^1([0, 1]); \|f\|_\infty \leq 1, \|f'\|_\infty \leq 1\},$$

then $N(\varepsilon)$ is bounded above by $c_1\varepsilon^{-1}$ and bounded below by $c_2\varepsilon^{-1}$.

33.3 Suppose X^i and Y^i are two sequences of Brownian motions with all of the Brownian motions independent of each other. Let

$$Z^n_{(s,t)} = \frac{1}{\sqrt{n}}\sum_{i=1}^{n}X^i_s Y^i_t.$$

Prove that Z^n converges weakly with respect to the topology of $C([0, 1]^2)$ as $n \to \infty$ to a Brownian sheet.

33.4 Let X be a Brownian bridge. (This will be studied further in Section 35.2.) This means that X is a mean zero Gaussian process with

$$\text{Cov}\,(X_s, X_t) = s \wedge t - st, \qquad 0 \le s, t \le 1.$$

Identify the reproducing kernel Hilbert space for X.

33.5 Let X be the Ornstein–Uhlenbeck process started at 0. This was defined in Exercise 19.5. Identify the reproducing kernel Hilbert space for X.

34

The space $D[0, 1]$

We define the space $D[0, 1]$ to be the collection of real-valued functions on $[0, 1]$ which are right continuous with left limits. We will introduce a topology on $D = D[0, 1]$, the Skorokhod topology, which makes D into a complete separable metric space. We will give a criterion for a subset of D to be compact, which will lead to some criteria for a family of probability measures on D to be tight.

34.1 Metrics for $D[0, 1]$

We write $f(t-)$ for $\lim_{s<t,s\to t} f(s)$. We will need the following observation. If f is in D and $\varepsilon > 0$, let $t_0 = 0$, and for $i > 0$ let $t_{i+1} = \inf\{t > t_i : |f(t) - f(t_i)| > \varepsilon\} \wedge 1$. Because f is right continuous with left limits, then from some i on, t_i must be equal to 1.

Our first try at a metric, ρ, makes D into a separable metric space, but one that is not complete. Let's start with ρ anyway, since we need it on the way to the metric d we end up with.

Let Λ be the set of functions λ from $[0, 1]$ to $[0, 1]$ that are continuous, strictly increasing, and such that $\lambda(0) = 0, \lambda(1) = 1$. Define

$$\rho(f, g) = \inf\{\varepsilon > 0 : \exists \lambda \in \Lambda \text{ such that } \sup_{t \in [0,1]} |\lambda(t) - t| < \varepsilon,$$

$$\sup_{t \in [0,1]} |f(t) - g(\lambda(t))| < \varepsilon\}.$$

Since the function $\lambda(t) = t$ is in Λ, then $\rho(f, g)$ is finite if $f, g \in D$. Clearly $\rho(f, g) \geq 0$. If $\rho(f, g) = 0$, then either $f(t) = g(t)$ or else $f(t) = g(t-)$ for each t; since elements of D are right continuous with left limits, it follows that $f = g$. If $\lambda \in \Lambda$, then so is λ^{-1} and we have, setting $s = \lambda^{-1}(t)$ and noting both s and t range over $[0, 1]$,

$$\sup_{t \in [0,1]} |\lambda^{-1}(t) - t| = \sup_{s \in [0,1]} |s - \lambda(s)|$$

and

$$\sup_{t \in [0,1]} |f(\lambda^{-1}(t)) - g(t)| = \sup_{s \in [0,1]} |f(s) - g(\lambda(s))|,$$

and we conclude $\rho(f, g) = \rho(g, f)$. The triangle inequality follows from

$$\sup_{t \in [0,1]} |\lambda_2 \circ \lambda_1(t) - t| \leq \sup_{t \in [0,1]} |\lambda_1(t) - t| + \sup_{s \in [0,1]} |\lambda_2(s) - s|$$

259

and

$$\sup_{t \in [0,1]} |f(t) - h(\lambda_2 \circ \lambda_1(t))| \leq \sup_{t \in [0,1]} |f(t) - g(\lambda_1(t))|$$
$$+ \sup_{s \in [0,1]} |g(s) - h(\lambda_2(s))|.$$

Look at the set of f in D for which there exists an integer k such that f is constant and equal to a rational on each interval $[(i-1)/k, i/k)$. It is not hard to check (Exercise 34.1) that the collection of such f's is dense in D with respect to ρ, which shows (D, ρ) is separable.

The space D with the metric ρ is not, however, complete; see Exercise 34.2. We therefore introduce a slightly different metric d. Define

$$\|\lambda\| = \sup_{s \neq t, s, t \in [0,1]} \left| \log \frac{\lambda(t) - \lambda(s)}{t - s} \right|$$

and let

$$d(f, g) = \inf\{\varepsilon > 0 : \exists \lambda \in \Lambda \text{ such that } \|\lambda\| \leq \varepsilon, \sup_{t \in [0,1]} |f(t) - g(\lambda(t))| \leq \varepsilon\}.$$

Note $\|\lambda^{-1}\| = \|\lambda\|$ and $\|\lambda_2 \circ \lambda_1\| \leq \|\lambda_1\| + \|\lambda_2\|$. The symmetry of d and the triangle inequality follow easily from this, and we conclude d is a metric.

Lemma 34.1 *There exists ε_0 such that*

$$\rho(f, g) \leq 2d(f, g)$$

if $d(f, g) < \varepsilon_0$.

(It turns out $\varepsilon_0 = 1/4$ will do.)

Proof   Since $\log(1 + 2x)/(2x) \to 1$ as $x \to 0$, we have

$$\log(1 - 2\varepsilon) < -\varepsilon < \varepsilon < \log(1 + 2\varepsilon)$$

if ε is small enough. Suppose $d(f, g) < \varepsilon$ and λ is the element of Λ such that $d(f, g) < \|\lambda\| < \varepsilon$ and $\sup_{t \in [0,1]} |f(t) - g(\lambda(t))| < \varepsilon$. Since $\lambda(0) = 0$, we have

$$\log(1 - 2\varepsilon) < -\varepsilon < \log \frac{\lambda(t)}{t} < \varepsilon < \log(1 + 2\varepsilon), \tag{34.1}$$

or

$$1 - 2\varepsilon < \frac{\lambda(t)}{t} < 1 + 2\varepsilon, \tag{34.2}$$

which implies $|\lambda(t) - t| < 2\varepsilon$, and hence $\rho(f, g) \leq 2d(f, g)$. $\square$

We define the analog ξ_f of the modulus of continuity for a function in D as follows. Define $\theta_f[a, b) = \sup_{s, t \in [a,b)} |f(t) - f(s)|$ and

$$\xi_f(\delta) = \inf\{ \max_{1 \leq i \leq n} \theta_f[t_{i-1}, t_i) : \exists n \geq 1, 0 = t_0 < t_1 < \cdots < t_n = 1$$

$$\text{such that } t_i - t_{i-1} > \delta \text{ for all } i \leq n\}.$$

Observe that if $f \in D$, then $\xi_f(\delta) \downarrow 0$ as $\delta \downarrow 0$.

Lemma 34.2 *Suppose* $\delta < 1/4$. *Let* $f \in D$. *If* $\rho(f, g) \leq \delta^2$, *then* $d(f, g) \leq 4\delta + \xi_f(\delta)$.

Proof Choose t_i's such that $t_i - t_{i-1} > \delta$ and $\theta_f[t_{i-1}, t_i) < \xi_f(\delta) + \delta$ for each i. Pick $\mu \in \Lambda$ such that $\sup_t |f(t) - g(\mu(t))| < \delta^2$ and $\sup_t |\mu(t) - t| < \delta^2$. Then $\sup_t |f(\mu^{-1}(t)) - g(t)| < \delta^2$. Set $\lambda(t_i) = \mu(t_i)$ and let λ be linear in between. Since $\mu^{-1}(\lambda(t_i)) = t_i$ for all i, then t and $\mu^{-1} \circ \lambda(t)$ always lie in the same subinterval $[t_{i-1}, t_i)$. Consequently

$$
\begin{aligned}
|f(t) - g(\lambda(t))| &\leq |f(t) - f(\mu^{-1}(\lambda(t)))| + |f(\mu^{-1}(\lambda(t))) - g(\lambda(t))| \\
&\leq \theta_f(\delta) + \delta^2 \\
&\leq \xi_f(\delta) + \delta + \delta^2 < \xi_f(\delta) + 4\delta.
\end{aligned}
$$

We have

$$
\begin{aligned}
|\lambda(t_i) - \lambda(t_{i-1}) - (t_i - t_{i-1})| &= |\mu(t_i) - \mu(t_{i-1}) - (t_i - t_{i-1})| \\
&\leq 2\delta^2 < 2\delta(t_i - t_{i-1}).
\end{aligned}
$$

Since λ is defined by linear interpolation,

$$
|\lambda(t) - \lambda(s)) - (t - s)| \leq 2\delta |t - s|, \qquad s, t \in [0, 1],
$$

which leads to

$$
\left| \frac{\lambda(t) - \lambda(s)}{t - s} - 1 \right| \leq 2\delta,
$$

or

$$
\log(1 - 2\delta) \leq \log\left(\frac{\lambda(t) - \lambda(s)}{t - s} \right) \leq \log(1 + 2\delta).
$$

Since $\delta < \frac{1}{4}$, we have $\|\lambda\| \leq 4\delta$. □

Proposition 34.3 *The metrics d and ρ are equivalent, i.e., they generate the same topology.*

In particular, (D, d) is separable.

Proof Let $B_\rho(f, r)$ denote the ball with center f and radius r with respect to the metric ρ and define $B_d(f, r)$ analogously. Let $\varepsilon > 0$ and let $f \in D$. If $d(f, g) < \varepsilon/2$ and ε is small enough, then $\rho(f, g) \leq 2d(f, g) < \varepsilon$, and so $B_d(f, \varepsilon/2) \subset B_\rho(f, \varepsilon)$.

To go the other direction, what we must show is that given f and ε, there exists δ such that $B_\rho(f, \delta) \subset B_d(f, \varepsilon)$. δ may depend on f; in fact, it has to in general, for otherwise a Cauchy sequence with respect to d would be a Cauchy sequence with respect to ρ, and vice versa. Choose δ small enough that $4\delta^{1/2} + \xi_f(\delta^{1/2}) < \varepsilon$. By Lemma 34.2, if $\rho(f, g) < \delta$, then $d(f, g) < \varepsilon$, which is what we want.

Finally, suppose G is open with respect to the topology generated by ρ. For each $f \in G$, let r_f be chosen so that $B_\rho(f, r_f) \subset G$. Hence $G = \cup_{f \in G} B_\rho(f, r_f)$. Let s_f be chosen so that $B_d(f, s_f) \subset B_\rho(f, r_f)$. Then $\cup_{f \in G} B_d(f, s_f) \subset G$, and in fact the sets are equal because if $f \in G$, then $f \in B_d(f, s_f)$. Since G can be written as the union of balls which are open with respect to d, then G is open with respect to d. The same argument with d and ρ interchanged shows that a set that is open with respect to d is open with respect to ρ. □

34.2 Compactness and completeness

We now show completeness for (D, d).

Theorem 34.4 *The space D with the metric d is complete.*

Proof Let f_n be a Cauchy sequence with respect to the metric d. If we can find a subsequence n_j such that f_{n_j} converges, say, to f, then it is standard that the whole sequence converges to f. Choose n_j such that $d(f_{n_j}, f_{n_{j+1}}) < 2^{-j}$. For each j there exists λ_j such that

$$\sup_t |f_{n_j}(t) - f_{n_{j+1}}(\lambda_j(t))| \le 2^{-j}, \qquad \|\lambda_j\| \le 2^{-j}.$$

As in (34.1) and (34.2),

$$|\lambda_j(t) - t| \le 2^{-j+1}.$$

Then

$$\sup_t |\lambda_{n+m+1} \circ \lambda_{m+n} \circ \cdots \circ \lambda_n(t) - \lambda_{n+m} \circ \cdots \circ \lambda_n(t)|$$

$$= \sup_s |\lambda_{n+m+1}(s) - s|$$

$$\le 2^{-(n+m)}$$

for each n. Hence for each n, the sequence $\lambda_{m+n} \circ \cdots \circ \lambda_n$ (indexed by m) is a Cauchy sequence of functions on $[0, 1]$ with respect to the supremum norm on $[0, 1]$. Let v_n be the limit. Clearly $v_n(0) = 0$, $v_n(1) = 1$, v_n is continuous, and nondecreasing. We also have

$$\left| \log \frac{\lambda_{n+m} \circ \cdots \circ \lambda_n(t) - \lambda_{n+m} \circ \cdots \circ \lambda_n(s)}{t - s} \right|$$

$$\le \|\lambda_{n+m} \circ \cdots \circ \lambda_n\|$$

$$\le \|\lambda_{n+m}\| + \cdots + \|\lambda_n\|$$

$$\le \frac{1}{2^{n-1}}.$$

If we then let $m \to \infty$, we obtain

$$\left| \log \frac{v_n(t) - v_n(s)}{t - s} \right| \le \frac{1}{2^{n-1}},$$

which implies $v_n \in \Lambda$ with $\|v_n\| \le 2^{1-n}$.

We see that $v_n = v_{n+1} \circ \lambda_n$. Consequently

$$\sup_t |f_{n_j}(v_j^{-1}(t)) - f_{n_{j+1}}(v_{j+1}^{-1}(t))| = \sup_s |f_{n_j}(s) - f_{n_{j+1}}(\lambda_j(s))| \le 2^{-j}.$$

Therefore $f_{n_j} \circ v_j^{-1}$ is a Cauchy sequence on $[0, 1]$ with respect to the supremum norm. Let f be the limit. Since

$$\sup_t |f_{n_j}(v_j^{-1}(t)) - f(t)| \to 0$$

and $\|v_j\| \to 0$ as $j \to \infty$, then $d(f_{n_j}, f) \to 0$. $\square$

We next show that if $f_n \to f$ with respect to d and $f \in C[0, 1]$, the convergence is in fact uniform.

Proposition 34.5 *Suppose $f_n \to f$ in the topology of $D[0, 1]$ with respect to d and $f \in C[0, 1]$. Then $\sup_{t \in [0,1]} |f_n(t) - f(t)| \to 0$.*

Proof Let $\varepsilon > 0$. Since f is uniformly continuous on $[0, 1]$, there exists δ such that $|f(t) - f(s)| < \varepsilon/2$ if $|t - s| < \delta$. For n sufficiently large there exists $\lambda_n \in \Lambda$ such that $\sup_t |f_n(t) - f(\lambda_n(t))| < \varepsilon/2$ and $\sup_t |\lambda_n(t) - t| < \delta$. Therefore $|f(\lambda_n(t)) - f(t)| < \varepsilon/2$, and so $|f_n(t) - f(t)| < \varepsilon$. □

We turn to compactness.

Theorem 34.6 *A set A has compact closure in $D[0, 1]$ if*

$$\sup_{f \in A} \sup_t |f(t)| < \infty$$

and

$$\lim_{\delta \to 0} \sup_{f \in A} \xi_f(\delta) = 0.$$

The converse of this theorem is also true, but we won't need this. See Billingsley (1968) or Exercise 34.9.

Proof A complete and totally bounded set in a metric space is compact, and $D[0, 1]$ is a complete metric space. Hence it suffices to show that A is totally bounded: for each $\varepsilon > 0$ there exist finitely many balls of radius ε that cover A.

Let $\eta > 0$ and choose k large such that $1/k < \eta$ and $\xi_f(1/k) < \eta$ for each $f \in A$. Let $M = \sup_{f \in A} \sup_t |f(t)|$ and let $H = \{-M + j/k : j \leq 2kM\}$, so that H is an η-net for $[-M, M]$. Let B be the set of functions $f \in D[0, 1]$ that are constant on each interval $[(i-1)/k, i/k)$ and that take values only in the set H. In particular, $f(1) \in H$.

We first prove that B is a 2η-net for A with respect to ρ. If $f \in A$, there exist $t_0, \ldots, t_n$ such that $t_0 = 0$, $t_n = 1$, $t_i - t_{i-1} > 1/k$ for each i, and $\theta_f[t_{i-1}, t_i) < \eta$ for each i. Note we must have $n \leq k$. For each i choose integers j_i such that $j_i/k \leq t_i < (j_i + 1)/k$. The j_i are distinct since the t_i are at least $1/k$ apart. Define λ so that $\lambda(j_i/k) = t_i$ and λ is linear on each interval $[j_i/k, j_{i+1}/k]$. Choose $g \in B$ such that $|g(m/k) - f(\lambda(m/k))| < \eta$ for each $m \leq k$. Observe that each $[m/k, (m+1)/k)$ lies inside some interval of the form $[j_i/k, j_{i+1}/k)$. Since λ is increasing, $[\lambda(m/k), \lambda((m+1)/k))$ is contained in $[\lambda(j_i/k), \lambda(j_{i+1}/k)) = [t_i, t_{i+1})$. The function f does not vary more than η over each interval $[t_i, t_{i+1})$, so $f(\lambda(t))$ does not vary more than η over each interval $[m/k, (m+1)/k)$. g is constant on each such interval, and hence

$$\sup_t |g(t) - f(\lambda(t))| < 2\eta.$$

We have

$$|\lambda(j_i/k) - j_i/k| = |t_i - j_i/k| < 1/k < \eta$$

for each i. By the piecewise linearity of λ, $\sup_t |\lambda(t) - t| < \eta$. Thus $\rho(f, g) < 2\eta$. We have proved that given $f \in A$, there exists $g \in B$ such that $\rho(f, g) < 2\eta$, or B is a 2η-net for A with respect to ρ.

Now let $\varepsilon > 0$ and choose $\delta > 0$ small so that $4\delta + \xi_f(\delta) < \varepsilon$ for each $f \in A$. Set $\eta = \delta^2/4$. Choose B as above to be a 2η-net for A with respect to ρ. By Lemma 34.2, if $\rho(f, g) < 2\eta < \delta^2$, then $d(f, g) \leq 4\delta + \xi_f(\delta) < \varepsilon$. Therefore B is an ε-net for A with respect to d. $\qquad\square$

The following corollary is proved exactly similarly to Theorem 32.1.

Corollary 34.7 *Suppose X_n are processes whose paths are right continuous with left limits. Suppose for each ε and η there exists n_0, R, and δ such that*

$$\mathbb{P}(\xi_{X_n}(\delta) \geq \varepsilon) \leq \eta$$

and

$$\mathbb{P}(\sup_{t \in [0,1]} |X_n(t)| \geq R) \leq \eta.$$

Then the X_n are tight with respect to the topology of $D[0, 1]$.

34.3 The Aldous criterion

A very useful criterion for tightness is the following one due to Aldous (1978).

Theorem 34.8 *Let $\{X_n\}$ be a sequence in $D[0, 1]$. Suppose*

$$\lim_{R \to \infty} \sup_n \mathbb{P}(|X_n(t)| \geq R) = 0 \tag{34.3}$$

for each $t \in [0, 1]$ and that whenever τ_n are stopping times for X_n and $\delta_n \to 0$ are reals,

$$|X_n(\tau_n + \delta_n) - X_n(\tau_n)| \tag{34.4}$$

converges to 0 in probability as $n \to \infty$.

Proof We will set $X_n(t) = X_n(1)$ for $t \in [1, 2]$ to simplify notation. The proof of this theorem comprises four steps.

Step 1. We claim that (34.4) implies the following: given ε there exist n_0 and δ such that

$$\mathbb{P}(|X_n(\tau_n + s) - X_n(\tau_n)| \geq \varepsilon) \leq \varepsilon \tag{34.5}$$

for each $n \geq n_0$, $s \leq 2\delta$, and τ_n a stopping time for X_n. For if not, we choose an increasing subsequence n_k, stopping times τ_{n_k}, and $s_{n_k} \leq 1/k$ for which (34.5) does not hold. Taking $\delta_{n_k} = s_{n_k}$ gives a contradiction to (34.4).

Step 2. Let $\varepsilon > 0$, fix $n \geq n_0$, and let $T \leq U \leq 1$ be two stopping times for X_n. We will prove

$$\mathbb{P}(U \leq T + \delta, |X_n(U) - X_n(T)| \geq 2\varepsilon) \leq 16\varepsilon. \tag{34.6}$$

To prove this, we start by letting λ be Lebesgue measure. If

$$A_T = \{(\omega, s) \in \Omega \times [0, 2\delta] : |X_n(T + s) - X_n(T)| \geq \varepsilon\},$$

then for each $s \leq 2\delta$ we have $\mathbb{P}(\omega : (\omega, s) \in A_T) \leq \varepsilon$ by (34.5) with τ_n replaced by T. Writing $\mathbb{P} \times \lambda$ for the product measure, we then have

$$\mathbb{P} \times \lambda(A_T) \leq 2\delta\varepsilon. \tag{34.7}$$

Set $B_T(\omega) = \{s : (\omega, s) \in A_T\}$ and $C_T = \{\omega : \lambda(B_T(\omega)) \geq \frac{1}{4}\delta\}$. From (34.7) and the Fubini theorem,

$$\int \lambda(B_T(\omega))\,\mathbb{P}(d\omega) \leq 2\delta\varepsilon,$$

so

$$\mathbb{P}(C_T) \leq 8\varepsilon.$$

We similarly define B_U and C_U, and obtain $\mathbb{P}(C_T \cup C_U) \leq 16\varepsilon$.

If $\omega \notin C_T \cup C_U$, then $\lambda(B_T(\omega)) \leq \frac{1}{4}\delta$ and $\lambda(B_U(\omega)) \leq \frac{1}{4}\delta$. Suppose $U \leq T + \delta$. Then

$$\lambda\{t \in [T, T + 2\delta] : |X_n(t) - X_n(T)| \geq \varepsilon\} \leq \tfrac{1}{4}\delta,$$

and

$$\lambda\{t \in [U, U + \delta] : |X_n(t) - X_n(U)| \geq \varepsilon\} \leq \tfrac{1}{4}\delta.$$

Hence there exists $t \in [T, T + 2\delta] \cap [U, U + \delta]$ such that $|X_n(t) - X_n(T)| < \varepsilon$ and $|X_n(t) - X_n(U)| < \varepsilon$; this implies $|X_n(U) - X_n(T)| < 2\varepsilon$, which proves (34.6).

Step 3. We obtain a bound on ξ_{X_n}. Let $T_{n0} = 0$ and

$$T_{n,i+1} = \inf\{t > T_{ni} : |X_n(t) - X_n(T_{ni})| \geq 2\varepsilon\} \wedge 2.$$

Note we have $|X_n(T_{n,i+1}) - X_n(T_{ni})| \geq 2\varepsilon$ if $T_{ni} < 2$. We choose n_0, δ as in Step 1. By Step 2 with $T = T_{ni}$ and $U = T_{n,i+1}$,

$$\mathbb{P}(T_{n,i+1} - T_{ni} < \delta, T_{ni} < 2) \leq 16\varepsilon. \tag{34.8}$$

Let $K = [2/\delta] + 1$ and apply (34.5) with ε replaced by ε/K to see that there exist $n_1 \geq n_0$ and $\zeta \leq \delta \wedge \varepsilon$ such that if $n \geq n_1$, $s \leq 2\zeta$, and τ_n is a stopping time, then

$$\mathbb{P}(|X_n(\tau_n + s) - X_n(\tau_n)| > \varepsilon/K) \leq \varepsilon/K. \tag{34.9}$$

By (34.6) with $T = T_{ni}$ and $U = T_{n,i+1}$ and δ replaced by ζ,

$$\mathbb{P}(T_{n,i+1} \leq T_{ni} + \zeta) \leq 16\varepsilon/K \tag{34.10}$$

for each i and hence

$$\mathbb{P}(\exists i \leq K : T_{n,i+1} \leq T_{ni} + \zeta) \leq 16\varepsilon. \tag{34.11}$$

We have

$$\begin{aligned}
\mathbb{E}\,[T_{ni} - T_{n,i-1}; T_{nK} < 1] &\geq \delta\mathbb{P}(T_{ni} - T_{n,i-1} \geq \delta, T_{nK} < 1) \\
&\geq \delta[\mathbb{P}(T_{nK} < 1) - \mathbb{P}(T_{ni} - T_{n,i-1} < \delta, T_{nK} < 1)] \\
&\geq \delta[\mathbb{P}(T_{nK} < 1) - 16\varepsilon],
\end{aligned}$$

where we used (34.8) in the last step. Summing over i from 1 to K,

$$\begin{aligned}
\mathbb{P}(T_{nK} < 1) \geq \mathbb{E}\,[T_{nK}; T_{nK} < 1] &= \sum_{i=1}^{K} \mathbb{E}\,[T_{ni} - T_{n,i-1}; T_{nK} < 1] \\
&\geq K\delta[\mathbb{P}(T_{nK} < 1) - 16\varepsilon] \geq 2[\mathbb{P}(T_{nK} < 1) - 16\varepsilon],
\end{aligned}$$

or $\mathbb{P}(T_{nK} < 1) \leq 32\varepsilon$. Hence except for an event of probability at most 32ε, we have $\xi_{X_n}(\zeta) \leq 4\varepsilon$.

Step 4. The last step is to obtain a bound on $\sup_t |X_n(t)|$. Let $\varepsilon > 0$ and choose δ and n_0 as in Step 1. Define

$$D_{Rn} = \{(\omega, s) \in \Omega \times [0, 1] : |X_n(s)(\omega)| > R\}$$

for $R > 0$. The measurability of D_{Rn} with respect to the product σ-field $\mathcal{F} \times \mathcal{B}[0, 1]$, where $\mathcal{B}[0, 1]$ is the Borel σ-field on $[0, 1]$, follows by the fact that X_n is right continuous with left limits. Let

$$G(R, s) = \sup_n \mathbb{P}(|X_n(s)| > R).$$

By (34.3), $G(R, s) \to 0$ as $R \to \infty$ for each s. Pick R large so that

$$\lambda(\{s : G(R, s) > \varepsilon\delta\}) < \varepsilon\delta.$$

Then

$$\int 1_{D_{Rn}}(\omega, s)\, \mathbb{P}(d\omega) = \mathbb{P}(|X_n(s)| > R) \leq \begin{cases} 1, & G(r, s) > \varepsilon\delta, \\ \varepsilon\delta, & \text{otherwise.} \end{cases}$$

Integrating over $s \in [0, 1]$,

$$\mathbb{P} \times \lambda(D_{Rn}) < 2\varepsilon\delta.$$

If $E_{Rn}(\omega) = \{s : (\omega, s) \in D_{Rn}\}$ and $F_{Rn} = \{\omega : \lambda(E_{Rn}) > \delta/4\}$, we have

$$\tfrac{1}{4}\delta\mathbb{P}(F_{Rn}) = \int_{F_{Rn}} \tfrac{1}{4}\delta\, \mathbb{P}(d\omega) \leq \int \int_0^1 1_{D_{Rn}}(\omega, s)\, \lambda(ds)\, \mathbb{P}(d\omega) \leq 2\varepsilon\delta,$$

so $\mathbb{P}(F_{Rn}) \leq 8\varepsilon$.

Define $T = \inf\{t : |X_n(t)| \geq R + 2\varepsilon\} \wedge 2$ and define A_T, B_T, and C_T as in Step 2. We have

$$\mathbb{P}(C_T \cup F_{Rn}) \leq 16\varepsilon.$$

If $\omega \notin C_T \cup F_{Rn}$ and $T < 2$, then $\lambda(E_{Rn}(\omega)) \leq \delta/4$. Hence there exists $t \in [T, T + 2\delta]$ such that $|X_n(t)| \leq R$ and $|X_n(t) - X_n(T)| \leq \varepsilon$. Therefore $|X_n(T)| \leq R + \varepsilon$, which contradicts the definition of T. We conclude that T must equal 2 on the complement of $C_T \cup F_{Rn}$, or in other words, except for an event of probability at most 16ε, we have $\sup_t |X_n(t)| \leq R + 2\varepsilon$, provided, of course, that $n \geq n_0$.

An application of Corollary 34.7 completes the proof. $\qquad\qquad\square$

Aldous's criterion is particularly well suited for strong Markov processes.

Proposition 34.9 *Suppose X_n is a sequence of real-valued strong Markov processes and there exists c, p, and $\gamma > 0$ such that*

$$\mathbb{E}^x |X_n(t) - X_n(0)|^p \leq ct^\gamma, \qquad x \in \mathbb{R}, \quad t \in [0, 1]. \tag{34.12}$$

Then for each $x \in \mathbb{R}$, the sequence of $\mathbb{P}^x$-laws of $\{X_n\}$ is tight with respect to the space $D[0, 1]$.

Unlike the Kolmogorov continuity criterion, we do not require $\gamma > 1$.

Proof Fix x. For each t,

$$\mathbb{P}^x(|X_n(t)| \geq R + |x|) \leq \mathbb{P}^x(|X_n(t) - X_n(0)| \geq R)$$
$$\leq \frac{\mathbb{E}^x|X_n(t) - X_n(0)|^p}{R^p}$$
$$\leq \frac{ct^\gamma}{R^p},$$

which tends to 0 as $R \to \infty$. We used Chebyshev's inequality here.

Suppose τ_n are stopping times for X_n and $\delta_n \to 0$. By the strong Markov property, for each $\varepsilon > 0$

$$\mathbb{P}^x(|X_n(\tau_n + \delta_n) - X_n(\tau_n)| > \varepsilon) \leq \frac{\mathbb{E}^x|X_n(\tau_n + \delta_n) - X_n(\tau_n)|^p}{\varepsilon^p}$$
$$= \varepsilon^{-p}\mathbb{E}^x\big[\mathbb{E}^{X_n(\tau_n)}|X_n(\delta_n) - X_n(0)|^\gamma\big]$$
$$\leq c\varepsilon^{-p}\delta_n^\gamma,$$

which tends to 0 as $n \to \infty$. Now apply Theorem 34.8. $\qquad\square$

Exercises

34.1 Show that the space D with the metric ρ is separable.

34.2 Let $f_n = 1_{[1/2, 1/2+1/n)}$. Show that this is a Cauchy sequence with respect to ρ, but does not converge to an element of D. Show $\{f_n\}$ is not a Cauchy sequence with respect to d.

34.3 Show that (with respect to the topology on D) the subset $C[0, 1]$ of D is nowhere dense.

34.4 Consider D with the metric $d_{\sup}(f, g) = \sup_{t \in [0,1]} |f(t) - g(t)|$. Show that D is not separable with respect to the metric $d_{\sup}$.

34.5 Suppose $\mathbb{P}$ and $\mathbb{P}'$ are measures supported on $D[0, 1]$ that agree on all cylindrical subsets of $D[0, 1]$. In other words, all the finite-dimensional distributions agree. Prove that $\mathbb{P} = \mathbb{P}'$ on $D[0, 1]$.

34.6 Show that the following are continuous functions on the space $D[0, 1]$.
   (1) $f(x) = \sup_{t \leq 1} x(t)$.
   (2) $f(x) = \int_0^1 x(t)\, dt$.
   (3) $f(x) = \sup_{t \leq 1} (x(t) - x(t-))$.

34.7 Let P be a Poisson process with parameter λ. Prove that

$$\frac{P_{nt} - n\lambda t}{\sqrt{n\lambda}}$$

converges weakly with respect to the topology of $D[0, 1]$ as $n \to \infty$ to a Brownian motion.

34.8 Suppose X_n converges weakly to X with respect to the topology of $C[0, 1]$. Prove that X_n converges weakly to X with respect to the topology of $D[0, 1]$.

34.9 This is the converse to Theorem 34.6. Let A be an index set, and suppose the collection of functions $\{f_\alpha, f \in A\}$ is precompact in $D[0, 1]$, i.e., its closure is compact.

(1) Prove $\sup_{\alpha \in A} \sup_{0 \le t \le 1} |f(t)| < \infty$.

(2) Prove

$$\lim_{\delta \to 0} \sup_{\alpha \in A} \xi_{f_\alpha}(\delta) = 0.$$

Notes

See Billingsley (1968) for more information.

35

Applications of weak convergence

In Chapter 32 we showed how weak convergence of stochastic processes could be used to give another construction of Brownian motion by showing that a simple symmetric random walk converges to a Brownian motion. In the first section of this chapter, we show that the sum of independent, identically distributed mean zero random variables with variance one also converges to a Brownian motion, which is known as the Donsker invariance principle.

We then consider a Brownian bridge, which is a Brownian motion conditioned to return to zero at time one. We prove in Section 35.3 that a Brownian bridge is the limit process for a sequence of normalized empirical processes.

35.1 Donsker invariance principle

Suppose the Y_i are i.i.d. real-valued random variables with mean zero and variance one, $S_n = \sum_{i=1}^{n} Y_i$, and $Z_n(t)$ is defined to be equal to $S_{nt}/\sqrt{n}$ if nt is an integer and defined by linear interpolation for other values of t. The Donsker invariance principle says that the Z_n converge weakly with respect to the space $C[0, 1]$ to a Brownian motion. This is a bit more delicate than in Section 32.2 because here our Y_i only have second moments.

The statement of the *Donsker invariance principle* is the following.

Theorem 35.1 *Let the Y_i and Z_n be as above. Then Z_n converges weakly to the law of Brownian motion on $[0, 1]$ with respect to the metric of $C[0, 1]$.*

Before we prove this, we give an application and explain the name "invariance principle." An example of how the Donsker invariance principle can be used is the following.

Corollary 35.2 *Let $M = \sup_{s \leq 1} W_s$ and $M_n = \sup_{s \leq 1} Z_n(s)$, where W is a Brownian motion. Then M_n converges weakly to M.*

Proof Let g be a bounded and continuous function on the reals and define a function F on $C[0, 1]$ by

$$F(f) = g(\sup_{s \leq 1} f(s)).$$

Notice $|\sup_{s \leq 1} f_2(s) - \sup_{s \leq 1} f_1(s)| \leq \sup_{s \leq 1} |f_2(s) - f_1(s)|$ and therefore $F : C[0, 1] \to \mathbb{R}$ is bounded and continuous. Since Z_n converges weakly to W with respect to the topology on $C[0, 1]$, then $\mathbb{E} F(Z_n) \to \mathbb{E} F(W)$. This is equivalent to $\mathbb{E} g(M_n) \to \mathbb{E} g(M)$. Because g is an arbitrary bounded continuous function on the reals, we conclude $M_n \to M$ weakly. $\quad\square$

This corollary says that the distribution of $\max_{i \leq n} S_i/\sqrt{n}$ converges to the supremum of a Brownian motion. We can actually use this to derive the distribution of the maximum of a Brownian motion: first determine the distribution of the maximum of S_n when the Y_i's are particularly simple, such as when they are a simple symmetric random walk. (That is, $\mathbb{P}(Y_i = 1) = \mathbb{P}(Y_i = -1) = \frac{1}{2}$.) Then take the limit as $n \to \infty$. In the case of a simple symmetric random walk, we can find the distribution of the maximum using the reflection principle, and there are no technical difficulties with the proof, unlike using the reflection principle with Brownian motion.

Another useful example is where $I_n = \int_0^1 |Z_n(t)|^2 \, dt$ and $I = \int_0^1 |W_t|^2 \, dt$. Here the distribution of I can be found by an eigenvalue argument (Kuo, 1975), and this is then an approximation to the distribution of I_n.

If f is a continuous function from $C[0, 1]$ to $\mathbb{R}$, an argument similar to the proof of Corollary 35.2 shows that $f(Z_n)$ converges weakly to $f(W)$. We get the same limit process, regardless of the distribution of the Y_i's, provided only that they are i.i.d. with mean zero and variance one. This is where the name "invariance principle" comes from – the limit is invariant with respect to changing the distribution of the Y_i's.

Lemma 35.3 *Suppose we have a sequence Y_i of i.i.d. random variables with mean zero and variance one and $S_n = \sum_{i=1}^n Y_i$. Suppose $\lambda > 4$. Then*

$$\mathbb{P}(\max_{i \leq n} |S_i| \geq \lambda\sqrt{n}) \leq \tfrac{4}{3}\mathbb{P}(|S_n| \geq \lambda\sqrt{n}/2).$$

Proof Let $N = \min\{i : |S_i| \geq \lambda\sqrt{n}\}$, the first time S_i is bigger than $\lambda\sqrt{n}$. N is a stopping time and $(N = i)$ is in the σ-field generated by $Y_1, \ldots, Y_i$. We have

$$\mathbb{P}(\max_{i \leq n} |S_i| \geq \lambda\sqrt{n}) \leq \mathbb{P}(|S_n| \geq \lambda\sqrt{n}/2) + \mathbb{P}(N < n, |S_n| < \lambda\sqrt{n}/2) \qquad (35.1)$$

$$\leq \mathbb{P}(|S_n| \geq \lambda\sqrt{n}/2)$$

$$+ \sum_{i=1}^{n-1} \mathbb{P}(N = i, |S_n| < \lambda\sqrt{n}/2).$$

If $N = i$ with $i < n$ and $|S_n| < \lambda\sqrt{n}/2$, then $|S_n - S_i| \geq \lambda\sqrt{n}/2$, and moreover the event $\{|S_n - S_i| \geq \lambda\sqrt{n}/2\}$ is in the σ-field generated by $Y_{i+1}, \ldots, Y_n$, and hence is independent of the event $\{N = i\}$. Using Chebyshev's inequality, the sum on the last line of (35.1) is bounded by

$$\sum_{i=1}^{n-1} \mathbb{P}(N = i)\mathbb{P}(|S_n - S_i| \geq \lambda\sqrt{n}/2) \leq \sum_{i=1}^{n-1} \mathbb{P}(N = i)\frac{\mathbb{E}|S_n - S_i|^2}{\lambda^2 n/4}$$

$$= \sum_{i=1}^{n-1} \mathbb{P}(N = i)\frac{n-i}{\lambda^2 n/4}$$

$$\leq \tfrac{1}{4}\mathbb{P}(N < i)$$

$$\leq \tfrac{1}{4}\mathbb{P}(\max_{i \leq n} |S_i| \geq \lambda\sqrt{n}),$$

since $\lambda > 4$. Therefore

$$\mathbb{P}(\max_{i \leq n} \mathbb{P}(|S_i| \geq \lambda\sqrt{n}) \leq \mathbb{P}(|S_n| \geq \lambda\sqrt{n}/2) + \tfrac{1}{4}\mathbb{P}(\max_{i \leq n} |S_i| \geq \lambda\sqrt{n}).$$

Subtracting the second term on the right from both sides and multiplying by 4/3 proves the lemma. □

Note that the central limit theorem tells us that for any $\beta > 0$

$$\mathbb{P}(|S_n| \geq \beta\sqrt{n}) \to \mathbb{P}(|Z| \geq \beta) \leq e^{-\beta^2/2},$$

where Z is a mean zero normal random variable with variance one, and hence for n large (depending on β),

$$\mathbb{P}(|S_n| \geq \beta\sqrt{n}) \leq 2e^{-\beta^2/2}. \tag{35.2}$$

Lemma 35.4 *For each $\varepsilon, \eta > 0$, there exist n_0 and δ such that if $n \geq n_0$ and $s \in [0, 1 - \delta]$, then*

$$\mathbb{P}(\sup_{s \leq t \leq s+\delta} |Z_n(t) - Z_n(s)| > \varepsilon) \leq \eta\delta.$$

Proof Let $\varepsilon, \eta > 0$, and choose δ small enough that $2e^{-\varepsilon^2/128\delta} \leq \delta\eta/2$. Then choose j_0 large enough so that, using (35.2),

$$\mathbb{P}\left(|S_j| > \frac{\varepsilon\sqrt{j}}{8\sqrt{\delta}}\right) \leq 2e^{-\varepsilon^2/128\delta} \leq \delta\eta/2$$

if $j \geq j_0$. Finally, choose $n_0 \geq j_0/\delta + 2$, so that if $n \geq n_0$, then $[n\delta] + 2 \geq j_0$ and $n\delta \geq ([n\delta] + 2)/2$, where $[x]$ is the largest integer less than or equal to x.

Let $n \geq n_0$ and set $J = [n\delta] + 2$. Suppose there exists s such that for some $t \in [s, s + \delta]$ we have $|Z_n(t) - Z_n(s)| > \varepsilon$. Then there exists $j \leq n$ such that for some i between j and $j + J$ we have $|S_i - S_j| \geq \varepsilon\sqrt{n}/2$. Therefore $n \geq J/2\delta$ and by Lemma 35.3

$$\mathbb{P}(\sup_{s \leq t \leq s+\delta} |Z_n(t) - Z_n(s)| > \varepsilon) \leq \mathbb{P}(\max_{j \leq i \leq j+J} |S_i - S_j| > \sqrt{n}\varepsilon/2)$$

$$\leq \mathbb{P}\left(\max_{j \leq i \leq j+J} |S_i - S_j| > \frac{\sqrt{J}\varepsilon}{4\sqrt{\delta}}\right)$$

$$\leq \tfrac{4}{3}\mathbb{P}\left(|S_{j+J} - S_j| > \frac{\sqrt{J}\varepsilon}{8\sqrt{\delta}}\right)$$

$$\leq \tfrac{4}{3}\mathbb{P}\left(|S_J| > \frac{\sqrt{J}\varepsilon}{8\sqrt{\delta}}\right)$$

$$\leq \delta\eta.$$

The proof is complete. □

Lemma 35.5 *For each $\varepsilon, \eta > 0$ there exist n_0 and δ such that if $n \geq n_0$,*

$$\mathbb{P}(\omega_{Z_n}(\delta) \geq \varepsilon) \leq 2\eta.$$

Proof We will take $\delta = 1/K$ for some large K. If $|t - s| \leq 1/K$, then either both s, t are in the same interval $[(i - 1)/K, i/K]$ or they are in adjoining intervals. Thus they both lie in some interval of the form $[(i - 2)/K, i/K]$. Since

$$|Z_n(t) - Z_n(s)| \leq |Z_n(t) - Z_n((i - 2)/K)| + |Z_n(s) - Z_n((i - 2)/K)|,$$

then using Lemma 35.4 with $\delta = 2/K$

$$\mathbb{P}(\exists s, t \in [0, 1] : |Z_n(t) - Z_n(s)| \geq \varepsilon, |t - s| < \delta)$$
$$\leq \mathbb{P}(\exists i \leq K : \sup_{(i-2)/K \leq s \leq i/K} |Z_n(s) - Z_{(i-2)/K}| \geq \varepsilon/2)$$
$$\leq K \sup_i \mathbb{P}(\sup_{(i-2)/K \leq s \leq i/K} |Z_n(s) - Z_{(i-2)/K}| \geq \varepsilon/2)$$
$$\leq K\eta(2/K) = 2\eta,$$

which proves the lemma. $\qquad\qquad\qquad\qquad\qquad\qquad\qquad\qquad\qquad\qquad\qquad\qquad\square$

We can now prove the Donsker invariance principle.

Proof of Theorem 35.1 By Lemma 35.5, Theorem 32.1, and the fact that $Z_n(0) = 0$ for all n, the laws of the Z_n are tight. Therefore by Prohorov's theorem (Theorem 30.4), every subsequence has a further subsequence which converges weakly with respect to the topology on $C[0, 1]$. We therefore only need to show that every subsequential limit point of the Z_n with respect to weak convergence is a Brownian motion. Since our processes lie in $C[0, 1]$, the paths of any subsequential limit point are continuous, so it suffices by Theorem 2.6 to show that the finite-dimensional distributions of Z_n converge weakly to the corresponding finite-dimensional distributions of a Brownian motion W. We will show the one-dimensional distributions converge, and leave the analogous argument for the higher-dimensional distributions to the reader.

We have

$$\mathbb{P}(\max_{i \leq n} |Y_i|/\sqrt{n} \geq \varepsilon) \leq n\mathbb{P}(|Y_1| \geq \sqrt{n}\varepsilon) \leq n\mathbb{P}(|Y_1|^2/\varepsilon^2 \geq n). \qquad (35.3)$$

For any integrable non-negative random variable X,

$$n\mathbb{P}(X \geq n) = \mathbb{E}[n; X \geq n] \leq \mathbb{E}[X; X \geq n],$$

which tends to zero by dominated convergence. Therefore

$$\mathbb{P}(\max_{i \leq n} |Y_i|/\sqrt{n} \geq \varepsilon) \to 0. \qquad (35.4)$$

Fix $t \in [0, 1]$. By the central limit theorem, $S_{[nt]}/\sqrt{[nt]}$ converges weakly on $\mathbb{R}$ to a mean zero normal random variable with variance one, and by Exercise 30.3, we see that $S_{[nt]}/\sqrt{n}$ converges weakly to a mean zero normal random variable with variance t. From the preceding paragraph we conclude that for each t, $|Z_n(t) - S_{[nt]}/\sqrt{n}|$ converges to zero in probability. By Exercise 30.2, $Z_n(t)$ has the same weak limit as $S_{[nt]}/\sqrt{n}$, namely, a mean zero normal random variable with variance t, which is the distribution of W_t. $\qquad\square$

There is an elegant proof of the Donsker invariance principle using Skorokhod embedding. Unlike the proof above, however, this second proof does not extend to random variables taking values in $\mathbb{R}^d$.

By Theorem 15.6 we can find a Brownian motion W and a random walk S_n such that

$$\sup_{i \leq n} \frac{|S_i - W_i|}{\sqrt{n}} \to 0$$

in probability. By the continuity of paths of W,

$$\mathbb{P}(\sup_{|t-s|\leq 1/n, s,t\leq 1} |W_t - W_s| > \varepsilon) \to 0.$$

If we let $W^n(t) = W_{nt}/\sqrt{n}$, we then have that $\sup_{i\leq n} |Z_n(i/n) - W_n(i/n)|$ tends to zero in probability as $n \to \infty$ and also, because W_n is again a Brownian motion,

$$\mathbb{P}(\sup_{|t-s|\leq 1/n, s,t\leq 1} |W_n(t) - W_n(s)| > \varepsilon) \to 0.$$

We conclude that

$$\sup_{t\leq 1} |Z_n(t) - W_n(t)| \to 0.$$

The law of W_n is that of a Brownian motion and does not depend on n. By Exercise 30.2 we obtain that Z_n converges weakly to the law of a Brownian motion.

If the above proof seems too simple, remember that we used Theorem 15.6, which in turn relies on Skorokhod embedding.

One might ask about the weak convergence of $\widetilde{Z}_n(t) = S_{[nt]}/\sqrt{n}$; these are the normalized partial sums without the linear interpolation. Rather than being continuous and piecewise linear like the $Z_n(t)$, the $\widetilde{Z}_n(t)$ are piecewise constant and have jumps.

Proposition 35.6 *Suppose the Y_i are independent with mean zero and variance one. The $\widetilde{Z}_n$ converge weakly with respect to the topology of $D[0, 1]$ to Brownian motion.*

Proof The Z_n converge weakly with respect to the topology of $C[0, 1]$ to a Brownian motion. By the Skorokhod representation (Theorem 31.2), we can find a probability space and random variables Z'_n having the same law as Z_n that converge almost surely with respect to the supremum norm. Therefore the Z'_n converge almost surely with respect to the metric of $D[0, 1]$, and hence the Z_n converge weakly to a Brownian motion with respect to the topology of $D[0, 1]$. If we show that $\sup_{t\leq 1} |Z_n(t) - \widetilde{Z}_n(t)|$ converges to zero in probability, then our result will follow by Exercise 30.2.

Now $Z_n(t)$ and $\widetilde{Z}_n(t)$ will differ by more than ε for some t only if some Y_i is larger than $\sqrt{n}\varepsilon$ in absolute value. But by (35.4), the probability of this tends to zero as $n \to \infty$. $\qquad\square$

35.2 Brownian bridge

A *Brownian bridge* W_t^0 is the process defined by

$$W_t^0 = W_t - tW_1, \qquad 0 \leq t \leq 1,$$

where W is a Brownian motion. W^0 has continuous paths, is jointly normal, is zero at time 0 and at time 1, has mean zero, and we calculate its covariance by

$$\begin{aligned}\text{Cov}(W_s^0, W_t^0) &= \text{Cov}(W_s, W_t) - s\,\text{Cov}(W_1, W_t) - t\,\text{Cov}(W_s, W_1) + st\,\text{Var}(W_1) \\ &= s \wedge t - st,\end{aligned}$$

recalling (2.1).

A Brownian bridge can be characterized as a Brownian motion conditioned to be zero at time 1. To make this precise, let W be a Brownian motion started at zero under $\mathbb{P}$, and for A a Borel subset of $C[0, 1]$, define

$$\mathbb{P}_\varepsilon(A) = \mathbb{P}(W \in A \mid |W_1| \leq \varepsilon);$$

cf. (A.13). Set $\mathbb{P}_0(A) = \mathbb{P}(W^0 \in A)$, the law of W^0.

Proposition 35.7 $\mathbb{P}_\varepsilon$ *converges weakly to* $\mathbb{P}_0$ *with respect to the topology of* $C[0, 1]$ *as* $\varepsilon \to 0$.

Proof Since W is a jointly normal process and

$$\operatorname{Cov}(W_t - tW_1, W_1) = \operatorname{Cov}(W_t, W_1) - t\operatorname{Var}(W_1) = 0,$$

then the process $W_t^0 = W_t - tW_1$ and the random variable W_1 are independent by Proposition A.55. Let F be any closed subset of $C[0, 1]$ and let $F_\delta = \{g \in C[0, 1] : d(g, F) < \delta\}$, where $d(g, F) = \inf\{d(g, f) : f \in F\}$ and d here is the supremum norm. Note $\sup_{t \leq 1} |W_t - W_t^0| \leq \varepsilon$ on the event $\{|W_1| \leq \varepsilon\}$. If $\delta > \varepsilon$,

$$\mathbb{P}_\varepsilon(F) = \mathbb{P}(W \in F \mid |W_1| \leq \varepsilon) \leq \mathbb{P}(W^0 \in F_\delta \mid |W_1| \leq \varepsilon)$$
$$= \mathbb{P}(W^0 \in F_\delta) = \mathbb{P}_0(F_\delta).$$

Thus $\limsup_{\varepsilon \to 0} \mathbb{P}_\varepsilon(F) \leq \mathbb{P}_0(F_\delta)$. Since F is closed, $\mathbb{P}_0(F_\delta) \to \mathbb{P}_0(F)$ as $\delta \to 0$, so $\limsup \mathbb{P}_\varepsilon(F) \leq \mathbb{P}_0(F)$. An application of Theorem 30.2 completes the proof. $\qquad\square$

We show that a Brownian bridge can also be represented as the solution X of the stochastic differential equation

$$dX_t = dW_t - \frac{X_t}{1-t}dt, \qquad X_0 = 0, \tag{35.5}$$

where W is a Brownian motion. This is plausible: X behaves much like a Brownian motion until t is close to 1, when there is a strong push toward the origin. The existence and uniqueness theory of Chapter 24 shows uniqueness and existence for the solution of (35.5) for $s \leq t$ for any $t < 1$; see Exercise 24.4. We can solve (35.5) explicitly. We have

$$dW_t = dX_t + \frac{X_t}{1-t}dt = (1-t)\,d\left[\frac{X_t}{1-t}\right],$$

or

$$X_t = (1-t)\int_0^t \frac{dW_s}{1-s}.$$

Thus X_t is a continuous Gaussian process with mean zero. The variance of X_t is

$$(1-t)^2 \int_0^t (1-s)^{-2}\,ds = t - t^2,$$

the same as the variance of a Brownian bridge. A similar calculation shows that the covariance of X_t and X_s is the same as the covariance of $W_t - tW_1$ and $W_s - sW_1$; see Exercise 24.6. Hence the finite-dimensional distributions of X_t and a Brownian bridge are the same. We now appeal to Theorem 2.6.

35.3 Empirical processes

In this section we will consider empirical processes, which are useful in statistics in estimating distribution functions. Let X_i, $i = 1, \ldots, n$, be i.i.d. random variables that are uniformly distributed on the interval $[0, 1]$. Define the *empirical process*

$$F_n(t) = \frac{1}{n} \sum_{i=1}^{n} 1_{[0,t]}(X_i). \tag{35.6}$$

The Glivenko–Cantelli theorem (Theorem A.40) says that

$$\sup_{t \in [0,1]} |F_n(t) - t| \to 0, \qquad \text{a.s.}$$

Our goal in this section is to obtain the corresponding weak limit theorem. Let

$$Z_n(t) = \sqrt{n}(F_n(t) - t) = \frac{1}{\sqrt{n}} \sum_{=1}^{n} (1_{[0,t]}(X_i) - t). \tag{35.7}$$

We will show that Z_n converges weakly with respect to $D[0, 1]$ to a Brownian bridge.
   Let

$$\omega_{Z_n}(\delta) = \sup_{s,t \in [0,1], |t-s| < \delta} |Z_n(t) - Z_n(s)|.$$

The paths of Z_n are not continuous: they have a jump of size $1/n$ at every time X_i. Thus $\omega_{Z_n}(\delta)$ does not tend to zero as $\delta \to 0$. Nevertheless we can get reasonable estimates on $\omega_{Z_n}(\delta)$.
   We need an elementary lemma on binomial random variables, the proof of which is Exercise 35.1.

Lemma 35.8 *Suppose S_n is a binomial random variable with parameters n and p. Then there exists a constant c not depending on n or p such that*

$$\mathbb{E} |S_n - \mathbb{E} S_n|^4 \leq cnp + cn^2 p^2 \tag{35.8}$$

and

$$\mathbb{E} |S_n|^4 \leq cnp + cn^4 p^4. \tag{35.9}$$

Proposition 35.9 *Let $\varepsilon, \eta > 0$. There exists δ and n_0 such that if $n \geq n_0$, then*

$$\mathbb{P}(\omega_{Z_n}(\delta) > \varepsilon) \leq \eta.$$

The idea of the proof is to use Corollary 8.4 to estimate $Z_n(t) - Z_n(s)$ when $|t - s|$ is small and use estimates on binomials when $|t - s|$ is large.

Proof Let $\varepsilon, \eta > 0$. We will choose n_0, δ later. Assuming that they have been chosen, suppose $n \geq n_0$ and choose k such that $n \leq 2^k < 2n$. If $t \in [0, 1]$, let $t(k)$ be the largest multiple of 2^{-k} less than or equal to t and similarly define $s(k)$. Let $\mathcal{D}_k = \{i/2^k : 0 \leq i \leq 2^k\}$. We will show there exists $\delta > 0$ such that

$$\mathbb{P}(\sup_{s,t \in \mathcal{D}_k, |t-s| < 2\delta} |Z_n(t) - Z_n(s)| > \varepsilon/3) < \eta/3 \tag{35.10}$$

and

$$\mathbb{P}(\sup_{s\in[0,1]} |Z_n(s) - Z_n(s(k))| > \varepsilon/3) < \eta/3. \tag{35.11}$$

Step 1. We first prove (35.10) by using Corollary 8.4. Suppose $s, t \in \mathcal{D}_k$ with $|t - s| < 2\delta$. Then either $s = t$, in which case $Z_n(t) - Z_n(s) = 0$, or else $|t - s| \geq 2^{-k} \geq 1/(2n)$. Take $p = t - s$ and note that $1_{(s,t]}(X_i)$ is a Bernoulli random variable with parameter p. Using (35.7) and Lemma 35.8,

$$\mathbb{E}\,|Z_n(t) - Z_n(s)|^4 \leq \frac{c}{n^2}(np + n^2 p^2)$$

$$= c\left(\frac{p}{n} + p^2\right) \leq c|t - s|^2,$$

where in the last line we used $1/n \leq 2|t - s|$. By Corollary 8.4,

$$\mathbb{P}(\sup_{s,t\in\mathcal{D}_k, |t-s|<2\delta} |Z_n(t) - Z_n(s)| > \varepsilon/3) \leq \mathbb{P}\left(\sup_{s,t\in\mathcal{D}_k, |t-s|<2\delta} \frac{|Z_n(t) - Z_n(s)|}{|t - s|^{1/8}} > c\frac{\varepsilon}{\delta^{1/8}}\right)$$

$$\leq c(\varepsilon/\delta^{1/8})^{-4} = c\delta^{1/2}/\varepsilon^4.$$

We choose δ small enough so that the last term is less than $\eta/3$.

Step 2. We now prove (35.11). Let

$$T_n(t) = \sum_{i=1}^{n} 1_{[0,t]}(X_i).$$

Observe that $T_n(t)$ is nondecreasing in t. If there exists $s \in [0, 1]$ such that $T_n(s) - T_n(s(k)) > \varepsilon\sqrt{n}/3$, then there exists $j \leq 2^k - 1$ such that $T_n((j+1)/2^k) - T_n(j/2^k) > \varepsilon\sqrt{n}/3$. Therefore, using (35.9),

$$\mathbb{P}\left(\sup_{s\in[0,1]} \frac{T_n(s) - T_n(s(k))}{\sqrt{n}} > \varepsilon/3\right)$$

$$\leq \mathbb{P}(\exists j \leq 2^k - 1 : T_n((j+1)/2^k) - T_n(j/2^k) > \varepsilon\sqrt{n}/3)$$

$$\leq 2^k \sup_{j\leq 2^k-1} \mathbb{P}(T_n((j+1)/2^k) - T_n(j/2^k) > \varepsilon\sqrt{n}/3)$$

$$\leq c2^k \frac{\sup_j \mathbb{E}\,|T_n((j+1)/2^k) - T_n(j/2^k)|^4}{\varepsilon^4 n^2}$$

$$\leq c2^k \frac{n2^{-k} + (n2^{-k})^4}{\varepsilon^4 n^2}.$$

Since $n2^{-k} \leq 2$, the last line is less than or equal to

$$c2^k n2^{-k}/\varepsilon^4 n^2 = c_1/\varepsilon^4 n.$$

We choose $n_0 > 1/\delta$ large enough so that if $n \geq n_0$, then $c_1/\varepsilon^4 n$ is less than $\eta/3$.

   Also,

$$\mathbb{E}\,[T_n(s) - T_n(s(k)] \leq n(s - s(k)) \leq n2^{-k} \leq 2$$

will be less than $\varepsilon \sqrt{n}/3$ if $n \geq 36/\varepsilon^2$ and we choose n_0 larger if necessary so that $n_0 > 36/\varepsilon^2$. Since

$$Z_n(t) - Z_n(s) = \frac{T_n(t) - T_n(s)}{\sqrt{n}} - \frac{\mathbb{E}[T_n(t) - T_n(s)]}{\sqrt{n}},$$

(35.11) follows.

Step 3. Now that we have (35.10) and (35.11), we write

$$|Z_n(t) - Z_n(s)| \leq |Z_n(t) - Z_n(t(k))| + |Z_n(t(k)) - Z_n(s(k))| + |Z_n(s(k)) - Z_n(s)|.$$

If $|t - s| < \delta$, then $|t(k) - s(k)| \leq \delta + 2^{-k} \leq \delta + 1/n$. Provided $n \geq n_0 > 1/\delta$, combining (35.10) and (35.11) gives

$$\mathbb{P}(\sup_{s,t \in [0,1], |t-s| < \delta} |Z_n(t) - Z_n(s)| > \varepsilon) < \eta$$

as required. $\qquad\square$

Theorem 35.10 *The Z_n converge weakly to a Brownian bridge with respect to the topology of $D[0, 1]$.*

Proof We smooth Z_n to get a continuous process V_n. Set $Z_n(t) = Z_n(1)$ for $t \in [1, 2]$ and set

$$V_n(t) = n \int_0^{n^{-1}} Z_n(u + t) \, du.$$

We have

$$|V_n(t_2) - V_n(t_1)| \leq n \int_0^{n^{-1}} |Z_n(t_2 + u) - Z_n(t_1 + u)| \, du$$

$$\leq n \int_0^{n^{-1}} \omega_{Z_n}(|t_2 - t_1|) \, du = \omega_{Z_n}(|t_2 - t_1|).$$

Note also that by (35.8) with $p = t - s$ and using Jensen's inequality with the measure $n 1_{[0,n^{-1}]}(u) \, du$,

$$\mathbb{E}|V_n(0)|^4 \leq n \int_0^{n^{-1}} \mathbb{E}|Z_n(u)|^4 \, du \leq c.$$

Hence

$$\mathbb{P}(|V_n(0)| \geq A) \leq \frac{\mathbb{E}|V_n(0)|^4}{A^4} \leq \frac{c}{A^4}.$$

Therefore by Theorem 8.1, the V_n are tight with respect to weak convergence on $C[0, 1]$. If the V_{n_j} converges weakly (with respect to $C[0, 1]$), by the Skorokhod representation we may find V'_{n_j} with the same law as V_{n_j} that converge almost surely. Then the V'_{n_j} will also converge almost surely in the space $D[0, 1]$. This proves that the V_n are tight in $D[0, 1]$ by Exercise 30.10.

Given ε and η, choose δ and n_0 such that $\mathbb{P}(\omega_{Z_n}(\delta) > \varepsilon) < \eta$ if $n \geq n_0$. We have

$$|V_n(t) - Z_n(t)| \leq n \int_0^{n^{-1}} |Z_n(u + t) - Z_n(t)| \, du \leq \omega_{Z_n}(n^{-1}).$$

If n is large enough so that $n^{-1} < \delta$, then

$$\mathbb{P}(\sup_t |V_n(t) - Z_n(t)| > \varepsilon) \leq \mathbb{P}(\omega_{Z_n}(n^{-1}) > \varepsilon) \leq \mathbb{P}(\omega_{Z_n}(\delta) > \varepsilon) < \eta.$$

Therefore $V_n - Z_n$ converges to 0 in probability, and by Exercise 30.2 the subsequential limit points of V_n are the same as those of Z_n.

It remains to show that any subsequential limit point of the Z_n is a Brownian bridge. This follows from the multidimensional central limit theorem for multinomials (see Remark A.57) and is left as Exercise 35.2. $\square$

Exercises

35.1 Prove Lemma 35.8.

35.2 Prove that the finite-dimensional distributions of Z_n in Theorem 35.10 converge to those of a Brownian bridge.

35.3 If W_t^0 is a Brownian bridge, prove that $Y_t = W_{1-t}^0$ is also a Brownian bridge.

35.4 Let $t_0 < 1$. The SDE (35.5) has a unique solution when $X_0 = 0$ is replaced by $X_0 = x$. Let $\mathbb{P}^x$ be the law of the solution when $X_0 = x$ and let Z_t be the canonical process. Show that $(Z_t, \mathbb{P}^x)$ is not a Markov process.

35.5 Let $N_t(A)$ be a Poisson point process with respect to the measure space (S, m) and let $A_s, s > 0$, be an increasing sequence of subsets of S with $m(A_s) \to \infty$ as $s \to \infty$. Does

$$\frac{N_t(A_s) - m(A_s)}{\sqrt{m(A_s)}}$$

converge weakly with respect to $D[0, 1]$ as $s \to \infty$? What is the limit?

This can be applied to get central limit theorems for the number of downcrossings of a Brownian motion, for example.

35.6 This exercise asks you to prove that the Poisson process conditioned to be equal to n at time 1 has the same law as n times the empirical process. Here is the precise statement. Suppose P_t is a Poisson process with parameter $\lambda > 0$. Let $\mathbb{Q}$ be the law of $\{P_t, t \in [0, 1]\}$ conditioned so that $P_1 = n$. Thus $\mathbb{Q}$ is a probability on $D[0, 1]$ with

$$\mathbb{Q}(P \in A) = \mathbb{P}(P \in A \mid P_1 = n).$$

Since $(P_1 = n)$ is an event with positive probability, there is no difficulty defining these conditional probabilities. Prove that $\mathbb{Q}$ is also the law of $\{nF_n(t), t \in [0, 1]\}$, where F_n is defined in Section 35.3.

36

Semigroups

In this chapter we suppose we have a semigroup of positive contraction operators $\{P_t\}$, and we show how to construct a Markov process X corresponding to this semigroup. In Chapters 37 and 38, we will show how such semigroups might arise.

We suppose that we have a state space S that is a separable locally compact metric space S. Let C_0 be the set of continuous functions on S that vanish at infinity. Recall that $f \in C_0$ if f is continuous, and given ε, there exists a compact set K depending on ε and f such that

$$|f(x)| < \varepsilon, \qquad x \notin K.$$

We use the usual supremum norm on C_0. We assume we have a semigroup $\{P_t\}$ of positive contractions mapping C_0 to C_0. More precisely, we assume

Assumption 36.1 There exists a family $\{P_t\}$, $t \geq 0$, of operators on C_0 such that
(1) If $f \in C_0$, then

$$P_t(P_s f)(x) = P_{t+s} f(x), \qquad x \in S, \quad s, t \geq 0.$$

(2) If $f(x) \geq 0$ for all x and if $t \geq 0$, then $P_t f(x) \geq 0$ for all x.
(3) For all t, $\|P_t f\| \leq \|f\|$.
(4) If $f \in C_0$, then $P_t f \to f$ uniformly as $t \to 0$.

Our goal in this section is to construct a process X corresponding to the semigroup P_t. The steps we use are the following.
(1) We temporarily assume each P_t maps the function 1 into itself. We define X_t for t in the dyadic rationals and define $\mathbb{P}^x$ using the Kolmogorov extension theorem.
(2) We verify a preliminary version of the Markov property.
(3) We use the regularity theorem for supermartingales to show that X has left and right limits along the dyadic rationals, and then define X_t for all t.
(4) We verify that our process $(X_t, \mathbb{P}^x)$ corresponds to the semigroup P_t.
(5) We remove the assumption that $P_t 1 = 1$.

36.1 Constructing the process

Let us assume the following for now. We will remove this assumption at the end of this section.

Assumption 36.2 $P_t 1(x) = 1$ for all x and all $t \geq 0$.

We now begin the construction of $(X_t, \mathbb{P}^x)$.

Step 1. Let $\mathcal{D}_n = \{k/2^n : k \geq 0\}$ and let $\mathcal{D} = \cup_n \mathcal{D}_n$, the dyadic rationals. Let Ω be the set of functions from $\mathcal{D}$ to $\mathcal{S}$. Define

$$X_t(\omega) = \omega(t), \qquad t \in \mathcal{D}, \quad \omega \in \Omega.$$

We let $\mathcal{F}$ be the σ-field on Ω generated by the collection of cylindrical subsets of Ω.

By the Riesz representation theorem (see Rudin (1987)), for each $t > 0$ there exists a measure $P_t(x, dy)$ such that

$$P_t f(x) = \int f(y) P_t(x, dy), \qquad f \in C_0. \tag{36.1}$$

(The Riesz representation theorem is most often phrased for continuous functions on compact spaces; since we are working with C_0, we can let the state space satisfy slightly weaker hypotheses; see Folland (1999), p. 223.) We can use (36.1) to define $P_t f$ for all bounded Borel measurable functions f. Since P_t maps C_0 to C_0, and continuous functions are Borel measurable, a limit argument shows that $P_t f$ is Borel measurable whenever f is bounded and Borel measurable.

Our main task in this step is to define $\mathbb{P}^x$. $\mathcal{D}$ is countable and we fix a labeling $\mathcal{D} = \{t_1, t_2, \ldots\}$. Let $E_n = \{t_1, \ldots, t_n\}$. Let $s_1 \leq \cdots \leq s_n$ be the ordering of E_n according to the usual ordering of the reals, so that s_1 is the smallest element of the set $\{t_1, \ldots, t_n\}$, s_2 is the next smallest, and so on. Define

$$\mathbb{P}_n^x(X_{s_1} \in A_1, \ldots, X_{s_n} \in A_n) \tag{36.2}$$

$$= \int_{A_n} \cdots \int_{A_1} P_{s_1}(x, dx_1) P_{s_2 - s_1}(x_1, dx_2) \cdots P_{s_n - s_{n-1}}(x_{n-1}, dx_n)$$

for $A_1, \ldots, A_n$ Borel subsets of $\mathcal{S}$. The $\mathbb{P}_n^x$ are consistent in the sense of Appendix D. The key to checking this is to observe that if $s_1, \ldots, s_n$ is the ordering of E_n and we temporarily write $s_1, \ldots, s_i, \bar{s}, s_{i+1}, \ldots, s_n$ for the ordering of E_{n+1}, then

$$\int_{\mathcal{S}} P_{\bar{s} - s_i}(x_{i-1}, d\bar{x}) P_{s_{i+1} - \bar{s}}(\bar{x}, dx_i) = P_{s_{i+1} - s_i}(x_{i-1}, dx_i)$$

by the semigroup property; cf. (19.10).

By the Kolmogorov extension theorem (Theorem D.1), for each x there exists a probability $\mathbb{P}^x$ such that

$$\mathbb{P}^x(X_{t_1} \in A_1, \ldots, X_{t_n} \in A_n) = \mathbb{P}_n^x(X_{t_1} \in A_1, \ldots, X_{t_n} \in A_n)$$

for each n whenever $A_1, \ldots, A_n$ are Borel subsets of $\mathcal{S}$.

If $\mathbb{E}^x$ is the expectation corresponding to $\mathbb{P}^x$, (36.2) can be rewritten as

$$\mathbb{E}^x[f_1(X_{s_1}) \cdots f_n(X_{s_n})] \tag{36.3}$$

$$= \int \cdots \int f_1(x_1) \cdots f_n(x_n) P_{s_1}(x, dx_1) P_{s_2 - s_1}(x_1, dx_2) \cdots$$

$$\times P_{s_n - s_{n-1}}(x_{n-1}, dx_n)$$

when $f_i = 1_{A_i}$ for each i. To see this, by linearity we have (36.3) when the functions f_i are simple functions; by a limit argument we have (36.3) when the f_i are Borel measurable and non-negative, and by linearity, (36.3) holds when the f_i are bounded and Borel measurable.

By (36.2) we have

$$\mathbb{P}^x(X_t \in A) = \mathbb{E} \, 1_A(X_t) = \int_A P_t(x, dy) = P_t 1_A(x).$$

Using linearity and a limit argument, we have $\mathbb{E}^x f(X_t) = P_t f(x)$ when f is bounded and Borel measurable.

Proposition 36.3 *If f is bounded and Borel measurable, $s, t > 0$, and $x \in S$, then*

$$\mathbb{E}^x\left[\mathbb{E}^{X_t} f(X_s)\right] = \mathbb{E}^x f(X_{s+t}). \tag{36.4}$$

Proof The proof of (36.4) is mainly a matter of sorting out notation. Let $\varphi(x) = \mathbb{E}^x f(X_s) = P_s f(x)$. Hence $\mathbb{E}^{X_t} f(X_s) = \varphi(X_t) = P_s f(X_t)$. Then the left-hand side is $\mathbb{E}^x(P_s f)(X_t) = P_t(P_s f)(x)$. The right-hand side of (36.4) is $P_{s+t} f(x)$, and so the two sides agree by the semigroup property. $\square$

Step 2. We so far only have X_t constructed for $t \in \mathcal{D}$. To extend the definition to all t, we want to let $X_t = \lim_{u>t, u\in\mathcal{D}, u\to t} X_u$. But before we can make that definition, we need to know that the limits exist. We will use the regularity of supermartingales to show this, so we need to look at conditional expectations. Let

$$\mathcal{F}'_s = \sigma(X_r; r \le s, r \in \mathcal{D}).$$

Proposition 36.4 *If $s < t$ with $s, t \in \mathcal{D}$ and f is bounded and Borel measurable, then*

$$\mathbb{E}^x[f(X_t) \mid \mathcal{F}'_s] = \mathbb{E}^{X_s} f(X_{t-s}), \qquad \mathbb{P}^x\text{-a.s.} \tag{36.5}$$

Proof Take $n \ge 1$, $r_1 \le r_2 \le \cdots \le r_n \le s$ with each r_j in $\mathcal{D}$, and $A_1, \ldots, A_n$ Borel subsets of S. It suffices to show that

$$\mathbb{E}^x[f(X_t)1_{A_1}(X_{r_1}) \cdots 1_{A_n}(X_{r_n})] = \mathbb{E}^x[(\mathbb{E}^{X_s} f(X_{t-s}))1_{A_1}(X_{r_1}) \cdots 1_{A_n}(X_{r_n})], \tag{36.6}$$

since the events $(X_{r_1} \in A_1, \ldots, X_{r_n} \in A_n)$ generate $\mathcal{F}'_s$. The right-hand side of (36.6) is equal to

$$\mathbb{E}^x[P_{t-s}f(X_s)1_{A_1}(X_{r_1}) \cdots 1_{A_n}(X_{r_n})]. \tag{36.7}$$

From (36.3)

$$\mathbb{E}^x[P_{t-s}f(X_s)1_{A_1}(X_{r_1}) \cdots 1_{A_n}(X_{r_n})] = \int \cdots \int P_{t-s}f(y)1_{A_1}(x_1) \cdots 1_{A_n}(x_n) \tag{36.8}$$

$$\times P_{r_1}(x, dx_1) \cdots P_{r_n-r_{n-1}}(x_{n-1}, x_n)P_{s-r_n}(x_n, dy).$$

But $P_{t-s}f(y) = \int f(z)P_{t-s}(y, dz)$. Substituting this in (36.8) and using (36.3) again, we obtain the left-hand side of (36.6). $\square$

Step 3. We define R_λ, the resolvent or λ-resolvent of P_t, by

$$R_\lambda f(x) = \int_0^\infty e^{-\lambda t} P_t f(x) \, dt. \tag{36.9}$$

Lemma 36.5 *If $f \geq 0$ is bounded and Borel measurable and $x \in \mathcal{S}$, then $M_t = e^{-\lambda t} R_\lambda f(X_t)$, $t \in \mathcal{D}$, is a supermartingale with respect to the filtration $\{\mathcal{F}'_t; t \in \mathcal{D}\}$ and the probability measure $\mathbb{P}^x$.*

Proof   What we need to show is that if $s < t \in \mathcal{D}$, then

$$\mathbb{E}^x[e^{-\lambda t} R_\lambda f(X_t) \mid \mathcal{F}'_s] \leq e^{-\lambda s} R_\lambda f(X_s), \qquad \mathbb{P}^x\text{-a.s.}$$

By Proposition 36.3 the left-hand side is

$$e^{-\lambda t} \mathbb{E}^{X_s} R_\lambda f(X_{t-s}),$$

so what we need to show is that

$$\mathbb{E}^y R_\lambda f(X_{t-s}) \leq e^{\lambda(t-s)} R_\lambda f(y) \tag{36.10}$$

for all y. The left-hand side of (36.10) is

$$
\begin{aligned}
P_{t-s} R_\lambda f(y) &= \int_0^\infty e^{-\lambda r} P_{t-s} P_r f(y) \, dr \\
&= \int_0^\infty e^{-\lambda r} P_{r+t-s} f(y) \, dr \\
&= e^{\lambda(t-s)} \int_{t-s}^\infty e^{-\lambda r} P_r f(y) \, dr \\
&\leq e^{\lambda(t-s)} \int_0^\infty e^{-\lambda r} P_r f(y) \, dr \\
&= e^{\lambda(t-s)} R_\lambda f(y).
\end{aligned}
$$

The first equality is the Fubini theorem, the second the semigroup property, and the third equality comes from a change of variables. $\square$

Next, if f is non-negative and bounded, by Theorem 3.12 with $\mathbb{P}$ replaced by $\mathbb{P}^x$, we see that $e^{-\lambda t} R_\lambda f(X_t)$ has left and right limits along $t \in \mathcal{D}$. Therefore the same is true for $R_\lambda f(X_t)$.

By Assumption 36.1 and dominated convergence, we have that if $f \in C_0$,

$$
\begin{aligned}
\lambda R_\lambda f(x) - f(x) &= \int_0^\infty e^{-\lambda t} (P_t f(x) - f(x)) \, dt \\
&= \int_0^\infty e^{-t} (P_{t/\lambda} f(x) - f(x)) \, dt
\end{aligned}
$$

tends to zero uniformly in x as $\lambda \to 0$. Take a countable dense subset $\{f_i\}$ of C_0 and look at $j R_j f_i(X_t)$ for all positive integers j. Since $j R_j f_i(X_t)$ has left and right limits along $\mathcal{D}$, a.s., letting $j \to \infty$, we see that $f_i(X_t)$ does also. We conclude that X_t has left and right limits along $\mathcal{D}$.

Now define $X_t = \lim_{u > t, u \in \mathcal{D}, u \to t} X_u$. Then X_u is right continuous with left limits. We check that

$$\mathbb{P}^x(X_{t_1} \in A_1, \ldots, X_{t_n} \in A_n) = \int_{A_1} \cdots \int_{A_n} P_{t_1}(x, dx_1) \cdots P_{t_n - t_{n-1}}(x_{n-1}, dx_n).$$

To see this, we know this holds when the t_i are in $\mathcal{D}$. By linearity and a limit argument, we conclude

$$\mathbb{E}^x[f_1(X_{t_1}) \cdots f_n(X_{t_n})] = \int \cdots \int f(x_1) \cdots f(x_n) P_{t_1}(x, dx_1) \cdots P_{t_n - t_{n-1}}(x_{n-1}, dx_n)$$

(36.11)

when the f_i are bounded and continuous. Using a limit argument, we know (36.11) holds when the t_i are arbitrary non-negative real numbers. Using a limit argument again, (36.11) holds for all bounded and measurable f, in particular, when $f_i = 1_{A_i}$.

Step 4. It remains to show that $(X_t, \mathbb{P}^x)$ satisfies Definition 19.1 and that P_t is the semigroup of this process. Let $\mathcal{F}_t^{00} = \sigma(X_s; s \leq t)$. Then we have already shown that $(X_t, \mathbb{P}^x)$ is a Markov process with respect to the filtration $\{\mathcal{F}_t^{00}\}$, except for showing that

$$\mathbb{P}^x(X_{s+t} \in A \mid \mathcal{F}_s^{00}) = \mathbb{P}^{X_s}(X_t \in A).$$

However, this can be proved almost identically to the way we proved Proposition 36.4.

Step 5. Sometimes the semigroup is a contraction semigroup and satisfies Assumption 36.1 but not Assumption 36.2. In this case the $P_t(x, A)$ are called *sub-Markov transition probability kernels*. The missing probability is due to the process being killed, and we can handle this situation as follows. Let $\mathcal{S}_\Delta = \mathcal{S} \cup \{\Delta\}$, where we introduce an isolated point $\{\Delta\}$. The topology on $\mathcal{S}_\Delta$ is the one generated by the open sets on $\mathcal{S}$ together with the set $\{\Delta\}$. Given a function f on $\mathcal{S}$, we extend it to $\mathcal{S}_\Delta$ by setting $f(\Delta) = 0$. We replace $P_t(x, A)$ by $\overline{P}_t(x, A)$, where

$$\begin{cases} \overline{P}_t(x, A) = P_t(x, A), & x \in \mathcal{S}, \quad A \subset \mathcal{S}, \\ \overline{P}_t(x, \{\Delta\}) = 1 - P_t(x, \mathcal{S}), & x \in \mathcal{S}, \\ \overline{P}_t(\Delta, \{\Delta\}) = 1. \end{cases}$$

(36.12)

One can go through the above construction with $\overline{P}_t$ and obtain a strong Markov process X_t whose state space is $\mathcal{S}_\Delta$. It is not hard to show that starting at Δ, the process stays at Δ forever; see Exercise 36.1.

We remark that by the results of Chapter 20 and also Exercise 20.1, we can expand the filtration from $\{\mathcal{F}_t^{00}\}$ to $\{\mathcal{F}_t\}$, where $\{\mathcal{F}_t\}$ is right continuous and each $\mathcal{F}_t$ contains all the sets that are null with respect to each $\mathbb{P}^x$. In addition, the strong Markov property will hold for $(X_t, \mathbb{P}^x)$.

36.2 Examples

Example 36.6 Our first example is a Brownian motion. Let

$$p(t, x, y) = (2\pi t)^{d/2} e^{-|x-y|^2/2t},$$

and set

$$P_t(x, A) = \int_A p(t, x, y) \, dy.$$

We know

$$\int p(t, x, z) p(s, z, y) \, dz = p(t + s, x, y)$$

by Proposition 19.5, and so P_t satisfies the semigroup property. We showed in Section 19.4 that Assumption 36.1 is satisfied, except for the fact that P_t maps C_0 to C_0; this is Exercise 36.2. Therefore we have a strong Markov process associated with P_t. By Proposition 21.5, the paths of the strong Markov process can be taken to be continuous. This gives yet another construction of a Brownian motion.

Example 36.7 We now use the machinery we have developed in this chapter to construct the Poisson process. Define transition probabilities by

$$P_t(x, A) = e^{-\lambda t} \sum_{k=0}^{\infty} \frac{(\lambda t)^k}{k!} 1_A(x + k),$$

where λ is some fixed parameter. If $p(t, k) = e^{-\lambda t} (\lambda t)^k / k!$, then

$$P_t f(x) = \sum_{k=0}^{\infty} f(x + k) p(t, k). \tag{36.13}$$

Thus

$$P_s(P_t f)(x) = \sum_{j=0}^{\infty} P_t f(x + j) p(s, j) = \sum_{j=0}^{\infty} \sum_{k=0}^{\infty} f(x + j + k) p(t, k) p(s, j).$$

This is equal to

$$\sum_{m=0}^{\infty} f(x + m) \sum_{k=0}^{m} p(t, m - k) p(s, k), \tag{36.14}$$

which by Exercise 36.3 is equal to

$$\sum_{m=0}^{\infty} f(x + m) p(s + t, m) = P_{s+t} f(x). \tag{36.15}$$

Therefore the semigroup property holds.

We therefore have a strong Markov process X whose paths are right continuous with left limits. We want to show that the process X_t under the probability measure $\mathbb{P}^0$ is a Poisson process. That $\mathbb{P}^0(X_0 = 0) = 1$ is obvious. We need to show that Definition 5.1(3) and (4) hold. For the former,

$$\mathbb{P}^0(X_t - X_s = k) = \sum_{j=0}^{\infty} \mathbb{P}(X_t = j + k, X_s = j) \tag{36.16}$$

$$= \sum_{j=0}^{\infty} p(s, j) p(t - s, k) = p(t - s, k), \tag{36.17}$$

as desired. For Definition 5.1(4), suppose $r_1 \leq r_2 \leq \cdots \leq r_n \leq s < t$, $a_1, \ldots, a_n$ are integers, and let $A = (X_{r_1} = a_1, \ldots, X_{r_n} = a_n)$. We will be done if we show

$$\mathbb{P}^0(X_t - X_s = k, A) = \mathbb{P}^0(X_t - X_s = k)\mathbb{P}^0(A). \tag{36.18}$$

The left-hand side of (36.18) is equal to

$$\sum_{j=0}^{\infty} \mathbb{P}^0(X_t = j + k, X_s = j, A) = \sum_{j=0}^{\infty} \mathbb{E}^0[\mathbb{P}^0(X_t = j + k \mid \mathcal{F}_s); X_s = j, A]$$

$$= \sum_{j=0}^{\infty} \mathbb{E}^0\left[\mathbb{P}^{X_s}(X_{t-s} = j + k); X_s = j, A\right]$$

$$= \sum_{j=0}^{\infty} \mathbb{E}^0\left[\mathbb{P}^j(X_{t-s} = j + k); X_s = j, A\right]$$

$$= \sum_{j=0}^{\infty} \mathbb{E}^0[p(t - s, k); X_s = j, A]$$

$$= p(t - s, k)\mathbb{P}^0(A).$$

Together with (36.16) this proves (36.18).

Exercises

36.1 Suppose P_t is a family of sub-Markov transition probabilities and we define $\overline{P}_t$ by (36.12). Show that $\overline{P}_t$ is a family of Markov transition probabilities. Show that $\mathbb{P}^{\Delta}(X_t \neq \Delta$ for some $t > 0) = 0$, i.e., starting at Δ, the process stays there forever.

36.2 Show that if $P_t(x, A)$ is defined by (19.17), and $P_t f(x) = \int f(y) P_t(x, dy)$, then P_t maps C_0 into C_0.

36.3 Show that (36.14) equals (36.15).

36.4 Show that P_t defined by (36.13) satisfies all the parts of Assumption 36.1.

36.5 Suppose $\{\mu_t, t \geq 0\}$ is a tight family of probability measures on the real line. Suppose there exists a function $\psi : \mathbb{R} \to \mathbb{C}$ such that the Fourier transforms of the μ_t have the following form:

$$\int e^{iux} \mu_t(dx) = e^{t\psi(u)}, \qquad t \geq 0, u \in \mathbb{R}.$$

(1) Prove that μ_t converges weakly to μ_0 as $t \to 0$. Note that μ_0 is the same as point mass at 0.

(2) Define the operators P_t by

$$P_t f(x) = \int f(x - y) \mu_t(dy).$$

Prove that the P_t form a strongly continuous semigroup of contraction operators mapping C_0 into C_0. Conclude that there exists a strong Markov process whose semigroup is given by the P_t.

This semigroup is called a *convolution semigroup* because $\mu_{t+s} = \mu_t * \mu_s$, in the sense of convolution of measures. We will see later that these are associated with Lévy processes.

Notes

See Blumenthal and Getoor (1968) for further information.

Infinitesimal generators

Often a Markov process is specified in terms of its behavior at each point, and one wants to form a global picture of the process. This means one is given the infinitesimal generator, which is a linear operator that is an unbounded operator in general, and one wants to come up with the semigroup for the Markov process.

We will begin by looking further at semigroups and resolvents, and then define the infinitesimal generator of a semigroup. We will prove the Hille–Yosida theorem, which is the primary tool for constructing semigroups from infinitesimal generators. Then we will look at two important examples: elliptic operators in nondivergence form and Lévy processes.

37.1 Semigroup properties

Let S be a locally compact separable metric space. We will take $\mathcal{B}$ to be a separable Banach space of real-valued functions on S. For the most part, we will take $\mathcal{B}$ to be the continuous functions on S that vanish at infinity (with the supremum norm), although another common example is to let $\mathcal{B}$ be the set of functions on S that are in L^2 with respect to some measure. We use $\| \cdot \|$ for the norm on $\mathcal{B}$.

For the duration of this chapter we will make the following assumption.

Assumption 37.1 *Suppose that P_t, $t \geq 0$, are operators acting on $\mathcal{B}$ such that*
(1) the P_t are contractions: $\|P_t f\| \leq \|f\|$ for all $t \geq 0$ and all $f \in \mathcal{B}$,
(2) the P_t form a semigroup: $P_s P_t = P_{t+s}$ for all $s, t \geq 0$, and
(3) the P_t are strongly continuous: if $f \in \mathcal{B}$, then $P_t f \to f$ as $t \to 0$.

Note that the semigroup property implies in particular that P_s and P_t commute. For a bounded operator A on $\mathcal{B}$, $\|A\| = \sup\{\|Af\| : \|f\| \leq 1\}$, so saying P_t is a contraction is the same as saying $\|P_t\| \leq 1$.

Define the *resolvent* or *λ-resolvent* operator of a semigroup P_t by

$$R_\lambda f(x) = \int_0^\infty e^{-\lambda t} P_t f(x) \, dt. \tag{37.1}$$

The resolvent equation is

$$R_\lambda - R_\mu = (\mu - \lambda) R_\lambda R_\mu. \tag{37.2}$$

We show that the semigroup property implies the resolvent equation.

Proposition 37.2 *The resolvent equation (37.2) holds.*

Proof We write

$$
\begin{aligned}
R_\lambda(R_\mu f)(x) &= \int_0^\infty e^{-\lambda t} P_t(R_\mu f)(x)\, dt \\
&= \int_0^\infty e^{-\lambda t} \int_0^\infty e^{-\mu s} P_t(P_s f)(x)\, ds\, dt \\
&= \int_0^\infty e^{-\lambda t} \int_0^\infty e^{-\mu s} P_{t+s} f(x)\, ds\, dt \\
&= \int_0^\infty e^{-\lambda t} e^{\mu t} \int_t^\infty e^{-\mu s} P_s f(x)\, ds\, dt \\
&= \int_0^\infty \int_0^s e^{-(\lambda-\mu)t} e^{-\mu s} P_s f(x)\, dt\, ds \\
&= \int_0^\infty \frac{1 - e^{-(\lambda-\mu)s}}{\lambda - \mu} e^{-\mu s} P_s f(x)\, ds \\
&= \frac{1}{\mu - \lambda}\Big[\int_0^\infty e^{-\lambda s} P_s f(x)\, ds - \int_0^\infty e^{-\mu s} P_s f(x)\, ds\Big] \\
&= \frac{1}{\mu - \lambda}[R_\lambda f(x) - R_\mu f(x)].
\end{aligned}
$$

The second equality uses Exercise 37.2, the fourth a change of variables, and the fifth the Fubini theorem. □

We have the following corollary to Proposition 37.2.

Corollary 37.3 *If $\mu, \lambda > 0$ and $|\mu - \lambda| < \lambda$, then*

$$
R_\mu f = R_\lambda f + \sum_{i=1}^\infty (\lambda - \mu)^i R_\lambda^{i+1} f. \tag{37.3}
$$

Here $R_\lambda^2 f = R_\lambda(R_\lambda f)$, and similarly for $R_\lambda^i f$.

Proof By Proposition 37.2, we have

$$
R_\mu f = R_\lambda f + (\lambda - \mu) R_\lambda R_\mu f. \tag{37.4}
$$

If we substitute for $R_\mu f$ in the last term on the right-hand side of (37.4), we have

$$
R_\mu f = R_\lambda f + (\lambda - \mu) R_\lambda R_\lambda f + (\lambda - \mu)^2 R_\lambda R_\lambda R_\mu f.
$$

We again substitute for $R_\mu f$, and repeat. Since

$$
\|(\lambda - \mu) R_\lambda\| \le \frac{|\lambda - \mu|}{\lambda},
$$

which is less than one in absolute value, $(\lambda - \mu)^i R_\lambda^{i+1} R_\mu f$ converges to zero as $i \to \infty$ and the series converges. □

Remark 37.4 In particular, if R_λ and S_λ are two resolvents that agree at one value of λ, say λ_0, then Corollary 37.3 applied once with R_λ and once with S_λ implies that if $\lambda < 2\lambda_0$, then

$$R_\lambda f = R_{\lambda_0} f + \sum_{i=1}^\infty (\lambda_0 - \lambda)^i (R_{\lambda_0})^{i+1} f$$

$$= S_{\lambda_0} f + \sum_{i=1}^\infty (\lambda_0 - \lambda)^i (S_{\lambda_0})^{i+1} f = S_\lambda f,$$

or R_λ and S_λ agree for $\lambda < 2\lambda_0$. Applying this observation again with λ_0 replaced by $3\lambda_0/2$, then R_λ and S_λ agree for $\lambda < 3\lambda_0$. Continuing this argument, we see that R_λ and S_λ must agree for each positive value of λ.

If for some $f \in \mathcal{B}$,

$$\left\| \frac{P_h f - f}{h} - g \right\| \to 0$$

as $h \to 0$, we say that f is in the domain of the *infinitesimal generator* of the semigroup, we write $g = \mathcal{L}f$ and write $f \in \mathcal{D} = \mathcal{D}(\mathcal{L})$. Generally $\mathcal{D}(\mathcal{L})$ is a proper subset of $\mathcal{B}$. If $f \in \mathcal{D}$ and $t > 0$, then

$$\frac{P_h P_t f - P_t f}{h} = \frac{P_t P_h f - P_t f}{h} = P_t \left(\frac{P_h f - f}{h} \right) \to P_t \mathcal{L}f, \qquad (37.5)$$

since P_t is a contraction. Therefore $P_t f \in \mathcal{D}$ when $f \in \mathcal{D}$ and $\mathcal{L}(P_t f) = P_t(\mathcal{L}f)$.

Proposition 37.5 *Fix $\lambda > 0$ and let $C = \{R_\lambda f : f \in \mathcal{B}\}$. Then $C = \mathcal{D}(\mathcal{L})$ and for $f \in \mathcal{B}$,*

$$\mathcal{L}R_\lambda f = \lambda R_\lambda f - f.$$

Proof Suppose that $g \in C$, so that $g = R_\lambda f$ for some $f \in \mathcal{B}$. Then

$$P_h R_\lambda f = \int_0^\infty e^{-\lambda t} P_{h+t} f\, dt = e^{\lambda h} \int_h^\infty e^{-\lambda t} P_t f\, dt, \qquad (37.6)$$

and so

$$P_h g - g = P_h R_\lambda f - R_\lambda f = (e^{\lambda h} - 1) \int_h^\infty e^{-\lambda t} P_t f\, dt - \int_0^h e^{-\lambda t} P_t f\, dt. \qquad (37.7)$$

Dividing by h and letting $h \to 0$, the first term on the right of (37.7) converges (use Exercise 37.2) to

$$\lambda \int_0^\infty e^{-\lambda t} P_t f\, dt = R_\lambda f.$$

Since $f \in \mathcal{B}$, then $P_t f \to f$ as $t \to 0$. After dividing by h, the second term on the right-hand side of (37.7) converges to f. Thus

$$\mathcal{L}(R_\lambda f) = \lambda R_\lambda f - f, \qquad (37.8)$$

as required.

We have shown that $C \subset \mathcal{D}(\mathcal{L})$, and we now show the opposite inclusion. Suppose $f \in \mathcal{D}(\mathcal{L})$. Let $g = \lambda f - \mathcal{L}f$, which is in $\mathcal{B}$. Since P_t and $\mathcal{L}$ commute, then R_λ and $\mathcal{L}$ commute, and by (37.8),

$$f = \lambda R_\lambda f - (\lambda R_\lambda f - f) = \lambda R_\lambda f - R_\lambda \mathcal{L}f$$
$$= R_\lambda g,$$

which is in C. $\qquad\square$

Example 37.6 Let us compute the infinitesimal generator when $(X_t, \mathbb{P}^x)$ is a one-dimensional Brownian motion. For our space $\mathcal{B}$ we take the continuous functions on $\mathbb{R}$ that vanish at infinity. Suppose $f \in C^2$ with compact support. By a Taylor series expansion,

$$P_h f(x) = \mathbb{E}^x f(X_h) = f(x) + f'(x)\mathbb{E}^x(X_h - x) + \tfrac{1}{2}f''(x)\mathbb{E}^x(X_h - x)^2 + R_h,$$

where R_h is the remainder term. We know R_h is bounded by

$$\|f''\|_\infty \mathbb{E}^x[\varphi(X_h - x)],$$

where φ is bounded and $|\varphi(y)/y^2| \to 0$ as $y \to 0$. Since W_h started at x has mean x and variance h, we have

$$P_h f(x) = f(x) + \tfrac{1}{2}f''(x)h + R_h,$$

where $|R_h/h|$ tends to zero as $h \to 0$. Therefore

$$\frac{P_h f - f}{h} \to \tfrac{1}{2}f'',$$

the convergence being with respect to the supremum norm. Exactly the same argument holds in higher dimensions to show that $\mathcal{L}f = \tfrac{1}{2}\Delta f$. We have shown that $\mathcal{D}(\mathcal{L})$ contains the C^2 functions with compact support, but have not actually identified the domain of the infinitesimal generator. We refer the reader to Knight (1981) for a detailed discussion.

The domain of an infinitesimal generator is nearly as important as the operator itself. We will briefly discuss aspects of the domains of the infinitesimal generator for absorbing Brownian motion and for reflecting Brownian motion on $[0, \infty)$. Both have the same operator $\mathcal{L}f = \tfrac{1}{2}f''$ but different domains.

Absorbing Brownian motion on $[0, \infty)$ is Brownian motion killed on first hitting $(-\infty, 0)$. Let W_t be standard Brownian motion on $\mathbb{R}$ and let X_t be W_t killed on first hitting $(-\infty, 0)$. If $f \in C^2[0, \infty)$ with f and its first and second derivatives being bounded and uniformly continuous and $x \neq 0$, $(\mathbb{E}^x f(X_t) - f(x))/t$ differs from $(\mathbb{E}^x f(W_t) - f(x))/t$ by at most

$$\|f\|_\infty \mathbb{P}^x(T_0 < t)/t,$$

where T_0 is the first time W_t hits $(-\infty, 0)$. If $x \neq 0$,

$$\frac{\mathbb{P}^x(T_0 < t)}{t} \leq \frac{\mathbb{P}^x(\sup_{s \leq t} |W_s - W_0| \geq x)}{t} \leq \frac{2}{t}e^{-x^2/2t} \to 0$$

as $t \to 0$. Therefore for $x \neq 0$, the infinitesimal generator of absorbing Brownian motion is the same as the infinitesimal generator of standard Brownian motion, namely, $\tfrac{1}{2}f''(x)$.

If $f = R_\lambda g$ for g bounded and continuous, we have

$$f(0) = R_\lambda g(0) = \mathbb{E}^0 \int_0^{T_0} e^{-\lambda t} g(X_t) \, dt = 0.$$

We use the fact that starting at 0, $T_0 = 0$, a.s., by Theorem 7.2. Using Proposition 37.5, every function in the domain of the infinitesimal generator of absorbing Brownian motion must satisfy $f(0) = 0$.

We can define reflecting Brownian motion on $[0, \infty)$ by $X_t = |W_t|$, where W is a one-dimensional Brownian motion on $\mathbb{R}$. As in the preceding paragraph, the infinitesimal generator for X agrees with $\frac{1}{2} f''(x)$ if $x \neq 0$. For $x = 0$, an application of Taylor's theorem gives

$$\mathbb{E}^0 f(|W_t|) = f(0) + f'(0) \mathbb{E}^0 |W_t| + \tfrac{1}{2} f''(0) \mathbb{E}^0 |W_t|^2 + \mathbb{E}^0 R_t,$$

where R_t is a remainder term. Subtracting $f(0)$ from both sides and dividing by t, and noting $\mathbb{E}^0 |W_t| / t = c_1 \sqrt{t} / t \to \infty$ as $t \to 0$, the only way we can get convergence is if $f'(0) = 0$. Thus every function in the domain of the infinitesimal generator of reflecting Brownian motion must satisfy $f'(0) = 0$.

In higher dimensions, the analogous restriction for reflecting Brownian motion is that the normal derivative $\partial f / \partial n$ must equal zero on the boundary of the domain, where n is the inward-pointing unit normal vector. In the partial differential equations literature, this is known as the *Neumann boundary condition*, and models situations where there is no heat flow across the boundary. For absorbing Brownian motion the analogous restriction is that $f = 0$ on the boundary of the domain, and this is called the *Dirichlet boundary condition*.

Example 37.7 Next we compute the generator for a Poisson process with parameter λ. We can let $\mathcal{B}$ be as in Example 37.6. We have

$$P_h f(x) = \sum_{i=0}^{\infty} e^{-\lambda h} \frac{(\lambda h)^i}{i!} f(x+i)$$

$$= e^{-\lambda h} f(x) + e^{-\lambda h} \lambda h f(x+1) + \sum_{i=2}^{\infty} e^{-\lambda h} \frac{(\lambda h)^i}{i!} f(x+i).$$

Subtracting $f(x)$ from both sides, dividing by h, and letting $h \to 0$, we obtain

$$\mathcal{L} f(x) = -\lambda f(x) + \lambda f(x+1) = \lambda [f(x+1) - f(x)].$$

In this case the domain of $\mathcal{L}$ is all of $\mathcal{B}$.

A very useful result is Dynkin's formula.

Theorem 37.8 *Suppose P_t operating on the space $\mathcal{B}$ of continuous functions vanishing at infinity is the semigroup of a Markov process $(X_t, \mathbb{P}^x)$, $f \in \mathcal{D}(\mathcal{L})$, and f and $\mathcal{L} f$ are bounded. If $x \in S$ and T is a stopping time with $\mathbb{E}^x T < \infty$, then*

$$\mathbb{E}^x f(X_T) - f(x) = \mathbb{E}^x \int_0^T \mathcal{L} f(X_r) \, dr.$$

Proof If $f \in \mathcal{D}(\mathcal{L})$, then $\mathcal{L}f \in \mathcal{B}$, and so $P_t \mathcal{L}f$ is continuous in t. Moreover, as we saw in (37.5),

$$\frac{\partial}{\partial t} P_t f(y) = P_t \mathcal{L}f(y).$$

By the fundamental theorem of calculus,

$$P_t f(y) - f(y) = \int_0^t P_r \mathcal{L}f(y)\, dr,$$

which can be rewritten as

$$\mathbb{E}^y f(X_t) - f(y) = \mathbb{E}^y \int_0^t \mathcal{L}f(X_r)\, dr; \tag{37.9}$$

we used the Fubini theorem here as well. This holds for each $y \in S$ and each $t > 0$.

Set $M_t = f(X_t) - f(X_0) - \int_0^t \mathcal{L}f(X_r)\, dr$. What (37.9) says is that $\mathbb{E}^y M_t = 0$ for all y and all t. By the Markov property,

$$\mathbb{E}^x[M_t - M_s \mid \mathcal{F}_s] = \mathbb{E}^x\Big[f(X_t) - f(X_s) - \int_s^t \mathcal{L}f(X_r)\, dr \mid \mathcal{F}_s \Big]$$

$$= \mathbb{E}^x\Big[\Big(f(X_{t-s}) - f(X_0) - \int_0^{t-s} \mathcal{L}f(X_r)\, dr\Big) \circ \theta_s \mid \mathcal{F}_s \Big]$$

$$= \mathbb{E}^{X_s} M_{t-s} = 0.$$

Therefore M_t is a martingale with respect to $\mathbb{P}^x$ for each x. If T is a bounded stopping time, then by optional stopping, $\mathbb{E}^x M_T = 0$. If T is instead only integrable with respect to $\mathbb{P}^x$, we have $\mathbb{E}^x M_{T \wedge n} = 0$ for each n. We then let $n \to \infty$ and use the fact that f and $\mathcal{L}f$ are bounded to conclude $\mathbb{E}^x M_T = 0$, which is what we want. $\qquad\qquad\square$

We say a few words about the Kolmogorov backward and forward equations. Suppose the semigroup P_t can be written

$$P_t f(x) = \int f(y) p(t, x, y)\, dy,$$

for functions $p(t, x, y)$, which are called *transition densities*. Provided there are no difficulties interchanging integration and differentiation, the equation

$$\frac{\partial}{\partial t} P_t f(x) = \mathcal{L} P_t f(x)$$

can be rewritten as

$$\int f(y) \frac{\partial}{\partial t} p(t, x, y)\, dy = \int f(y) \mathcal{L} p(t, x, y)\, dy,$$

which leads to the *Kolmogorov backward equation*

$$\frac{\partial}{\partial t} p(t, x, y) = \mathcal{L} p(t, x, y),$$

where $\mathcal{L}$ operates on the x variable and y is held fixed.

If $\mathcal{L}$ has an adjoint operator $\mathcal{L}^*$, which means $\int f(\mathcal{L}g) = \int (\mathcal{L}^* f)g$ for f and g in the domains of $\mathcal{L}^*$ and $\mathcal{L}$, respectively, the equation

$$\frac{\partial}{\partial t} P_t f(x) = P_t \mathcal{L} f(x)$$

can be rewritten as

$$\int f(y) \frac{\partial}{\partial t} p(t, x, y) \, dy = \int \mathcal{L} f(y) p(t, x, y) \, dy = \int f(y) \mathcal{L}^* p(t, x, y) \, dy,$$

which leads to the *Kolmogorov forward equation*

$$\frac{\partial}{\partial t} p(t, x, y) = \mathcal{L}^* p(t, x, y),$$

where $\mathcal{L}^*$ operates on the y variable and x is held fixed.

37.2 The Hille–Yosida theorem

We now show how to construct a semigroup given the infinitesimal generator. We start with a few preliminary observations. If A is a bounded operator, we can define

$$e^A = I + A + A^2/2! + \cdots = \sum_{i=0}^{\infty} A^i/i!$$

To see that the series converges, note that

$$\left\| \sum_{i=m}^{n} A^i/i! \right\| \leq \sum_{i=m}^{n} \|A^i\|/i! \leq \sum_{i=m}^{\infty} \|A\|^i/i!,$$

which will be small if m is large since $\|A\|$ is a finite number. Similarly,

$$\|e^A\| \leq \sum_{i=0}^{\infty} \|A\|^i/i! = e^{\|A\|}.$$

Proposition 37.9 *Suppose $\{R_\lambda\}$ is a family of bounded operators defined on $\mathcal{B}$ such that*
(1) the resolvent equation holds,
(2) $\|R_\lambda\| \leq 1/\lambda$ for each $\lambda > 0$, and
(3) $\|\lambda R_\lambda f - f\| \to 0$ as $\lambda \to \infty$ for each $f \in \mathcal{B}$.
Then there exists a strongly continuous semigroup P_t whose resolvent is R_λ.

Proof Let $D_\lambda = \lambda(\lambda R_\lambda - I)$ and $Q_t^\lambda = e^{tD_\lambda}$. Note that the resolvent equation implies that D_λ and D_μ commute and therefore all the operators D_λ, Q_t^λ, D_μ, and Q_t^μ commute. Since $\|\lambda R_\lambda\| \leq 1$, then

$$\|Q_t^\lambda\| = e^{-\lambda t} \|e^{t\lambda^2 R_\lambda}\| \leq e^{-\lambda t} e^{\|t\lambda^2 R_\lambda\|} \leq e^{-\lambda t} e^{\lambda t} = 1.$$

We first show that the set of f such that $D_\lambda f$ converges as $\lambda \to \infty$ is a dense subset of $\mathcal{B}$. If $f = R_a g$ for some $a > 0$ and some $g \in \mathcal{B}$, then by the resolvent equation

$$D_\lambda f = \lambda(\lambda R_\lambda - I)(R_a g) = \lambda^2 R_\lambda R_a g - \lambda R_a g$$

$$= \frac{\lambda^2}{\lambda - a}(R_a g - R_\lambda g) - \lambda R_a g.$$

We have

$$\frac{\lambda^2}{\lambda - a}R_\lambda g = \frac{\lambda}{\lambda - a}\lambda R_\lambda g \to g$$

as $\lambda \to \infty$ by hypothesis (3) and

$$\frac{\lambda^2}{\lambda - a}R_a g - \lambda R_a g = \frac{\lambda}{\lambda - a}a R_a g \to a R_a g$$

as $\lambda \to \infty$. Therefore

$$D_\lambda R_a g \to a R_a g - g. \tag{37.10}$$

Thus D_λ converges on $E = \cup_{a>0}\{R_a f : f \in \mathcal{B}\}$. But for any $f \in \mathcal{B}$, $a R_a f \to f$ as $a \to \infty$ and $a R_a f = R_a(af) \in E$, which proves that E is a dense subset of $\mathcal{B}$.

Next we show that if $D_\lambda f$ converges, then $Q_t^\lambda f$ converges. Suppose $D_\lambda f$ converges and $\varepsilon > 0$. Choose M such that if $\lambda, \mu \geq M$, then $\|D_\lambda f - D_\mu f\| < \varepsilon$. Since $\partial Q_t^\lambda f / dt = D_\lambda Q_t^\lambda f$ and Q_0^λ, Q_0^μ are both the identity operator, we have

$$Q_t^\lambda f - Q_t^\mu f = \int_0^t \frac{\partial}{\partial s}(Q_s^\lambda Q_{t-s}^\mu f)\,ds$$

$$= \int_0^t [Q_s^\lambda D_\lambda Q_{t-s}^\mu f - Q_s^\lambda D_\mu Q_{t-s}^\mu f]\,ds$$

$$= \int_0^t [Q_s^\lambda Q_{t-s}^\mu (D_\lambda f - D_\mu f)]\,ds,$$

so

$$\|Q_t^\lambda f - Q_t^\mu f\| \leq t\|D_\lambda f - D_\mu f\| < \varepsilon t,$$

using that Q_s^λ and Q_{t-s}^μ are contractions.

Since ε is arbitrary, this proves that $Q_t^\lambda f$ is a Cauchy sequence in $\mathcal{B}$ and hence converges. Call the limit $P_t f$. We can easily check that Q_t^λ is a semigroup for each $\lambda > 0$ and we saw that Q_t^λ is a contraction for each t and λ. It follows that P_t is a semigroup and that the norm of each P_t is bounded by 1. Each Q_t^λ is strongly continuous, and by the uniform convergence, it follows that $P_t f \to f$ as $t \to 0$ for $f \in E$. Since each P_t is a contraction and E is dense in $\mathcal{B}$, we can extend each P_t so as to have domain $\mathcal{B}$ and so that the P_t will be a strongly continuous semigroup on $\mathcal{B}$.

Let S_λ be the resolvent for P_t. It remains to prove that $S_\lambda = R_\lambda$. Fix a and let $f = R_a g$. We saw in (37.10) that $D_\lambda R_a g \to a R_a g - g$. Now Q_t^λ is a semigroup for each λ and by Exercise 37.4, the infinitesimal generator of Q_t^λ is D_λ. By the fundamental theorem of calculus,

$$Q_t^\lambda(R_a g) - R_a g = \int_0^t \frac{\partial}{\partial s}(Q_s^\lambda R_a g)\,ds = \int_0^t Q_s^\lambda(D_\lambda R_a g)\,ds.$$

Letting $\lambda \to \infty$,

$$P_t(R_a g) - R_a g = \int_0^t P_s(a R_a g - g)\,ds.$$

Let $b < a$. Multiply the above equation by e^{-bt} and integrate over t from 0 to ∞. Then

$$S_b(R_a g) - \frac{1}{b} R_a g = \int_0^\infty e^{-bt} \int_0^t P_s(aR_a g - g)\, ds\, dt$$

$$= \int_0^\infty \int_s^\infty e^{-bt} P_s(aR_a g - g)\, dt\, ds$$

$$= \int_0^\infty \frac{1}{b} e^{-bs} P_s(aR_a g - g)\, ds$$

$$= \frac{1}{b} S_b(aR_a g - g).$$

Therefore

$$S_b g = R_a g + (a - b) S_b R_a g.$$

Applying this with g replaced by $R_a g$, iterating, and using Corollary 37.3, we obtain

$$S_b g = R_a g + (a - b) R_a^2 g + (a - b)^3 R_a^3 g + \cdots = R_b g.$$

By Remark 37.4, this proves $S_b = R_b$ for all b.   □

We now show that under appropriate hypotheses on $\mathcal{L}$, there exists a semigroup whose infinitesimal generator is $\mathcal{L}$. This is known as the Hille–Yosida theorem. We say that an operator $\mathcal{L}$ is *dissipative* if

$$\|(\lambda - \mathcal{L})f\| \geq \lambda \|f\|, \qquad f \in \mathcal{D}(\mathcal{L}). \tag{37.11}$$

Theorem 37.10 *Suppose $\mathcal{L}$ is an operator such that*
  (1) the domain of $\mathcal{L}$ is a dense subset of $\mathcal{B}$,
  (2) the range of $\lambda - \mathcal{L}$ is $\mathcal{B}$ for each λ, and
  (3) $\mathcal{L}$ is dissipative.
Then there exists a semigroup on $\mathcal{B}$ which has $\mathcal{L}$ as its infinitesimal generator.

Proof   If $(\lambda - \mathcal{L})f = (\lambda - \mathcal{L})g$, then

$$\lambda \|f - g\| \leq \|(\lambda - \mathcal{L})(f - g)\| = 0,$$

or $f = g$. Thus $\lambda - \mathcal{L}$ is a one-to-one map, hence is invertible because the range of $\lambda - \mathcal{L}$ is $\mathcal{B}$. We let R_λ be the inverse, and thus the domain of R_λ is all of $\mathcal{B}$.

We first show that the resolvent equation holds. We observe

$$(\mu - \mathcal{L}) \frac{1}{\lambda - \mu} R_\mu f = \frac{1}{\lambda - \mu} f$$

and

$$(\mu - \mathcal{L}) \frac{1}{\lambda - \mu} R_\lambda f = (\mu - \lambda) \frac{1}{\lambda - \mu} R_\lambda f + (\lambda - \mathcal{L}) \frac{1}{\lambda - \mu} R_\lambda f$$

$$= -R_\lambda f + \frac{1}{\lambda - \mu} f.$$

Combining,

$$(\mu - \mathcal{L}) R_\mu R_\lambda f = R_\lambda f = (\mu - \mathcal{L}) \frac{1}{\lambda - \mu} (R_\mu - R_\lambda) f.$$

Applying R_μ to both sides yields the resolvent equation.

The hypothesis that $\|(\lambda - \mathcal{L})f\| \geq \lambda\|f\|$ immediately implies $\|R_\lambda f\| \leq \|f\|/\lambda$.

We next show $\lambda R_\lambda f \to f$ as $\lambda \to \infty$. If $f \in \mathcal{D}$, then

$$R_\lambda \mathcal{L}f = \mathcal{L}R_\lambda f = \lambda R_\lambda f - f,$$

and so

$$\|\lambda R_\lambda f - f\| \leq \frac{1}{\lambda}\|\mathcal{L}f\| \to 0$$

as $\lambda \to \infty$. Since $\|\lambda R_\lambda\| \leq 1$ and the domain of $\mathcal{L}$ is dense in $\mathcal{B}$, we conclude $\lambda R_\lambda f \to f$ for all $f \in \mathcal{B}$.

We use Proposition 37.9 to construct P_t. By Proposition 37.9, R_λ is the resolvent for P_t. If $\mathcal{M}$ is the infinitesimal generator for P_t, then by Proposition 37.5, the domain of $\mathcal{M}$ is $\{R_\lambda f : f \in \mathcal{B}\}$. Since we know $\mathcal{L}(R_\lambda f) = \lambda R_\lambda f - f \in \mathcal{B}$, then the domain of $\mathcal{L}$ contains $\{R_\lambda f : f \in \mathcal{B}\}$. Since $\mathcal{M}$ is the infinitesimal generator of P_t, by Proposition 37.5, $\mathcal{M}(R_\lambda f) = \lambda R_\lambda f - f$. Therefore $\mathcal{L}$ is an extension of $\mathcal{M}$.

If $f \in \mathcal{D}(\mathcal{L})$, then $g = (\lambda - \mathcal{L})f \in \mathcal{B}$, and thus

$$(\lambda - \mathcal{M})^{-1}g \in \mathcal{D}(\mathcal{M}) \subset \mathcal{D}(\mathcal{L}).$$

Hence

$$(\lambda - \mathcal{L})f = g = (\lambda - \mathcal{M})(\lambda - \mathcal{M})^{-1}g = (\lambda - \mathcal{L})(\lambda - \mathcal{M})^{-1}g.$$

Since $\lambda - \mathcal{L}$ is one-to-one, then $f = (\lambda - \mathcal{M})^{-1}g$, which implies $f \in \mathcal{D}(\mathcal{M})$. Therefore $\mathcal{M} = \mathcal{L}$ and so $\mathcal{L}$ is the generator of P_t. $\square$

When applying the Hille–Yosida theorem, it is quite often the case that it is easier to show that the range of $\lambda - \mathcal{L}$ is only dense in $\mathcal{B}$, rather than being all of $\mathcal{B}$. When that occurs, one needs to look at a closed extension $\overline{\mathcal{L}}$ of $\mathcal{L}$. An operator $\overline{\mathcal{L}}$ is *closed* if whenever $f_n \to f$ and $\overline{\mathcal{L}}f_n \to g$, then $f \in \mathcal{D}(\overline{\mathcal{L}})$ and $\overline{\mathcal{L}}f = g$. To construct the closed extension of $\mathcal{L}$, where we assume that $\mathcal{L}$ is dissipative (defined by (37.11)), let $R_\lambda g = f$ if $(\lambda - \mathcal{L})f = g$. $\mathcal{L}$ being dissipative is equivalent to the norm of R_λ being bounded by $1/\lambda$ on the range of $\lambda - \mathcal{L}$, and so we can extend the domain of R_λ uniquely to all of $\mathcal{B}$. Now define $\mathcal{D}(\overline{\mathcal{L}})$ to be the range of R_λ and set

$$\overline{\mathcal{L}}R_\lambda g = \lambda R_\lambda g - g. \tag{37.12}$$

We will soon give two examples where infinitesimal generators can be used to construct very useful processes. The first is where the infinitesimal generator is an elliptic operator of second order in non-divergence form. The second case studies the infinitesimal generators of Lévy processes.

We should mention that there is another important example where infinitesimal generators are useful in constructing a process, that of *infinite particle systems*. The name "infinite particle systems" refers to a class of models with discrete space and continuous time that are useful in mathematical biology and in statistical mechanics. One of the simplest examples is the voter model. Suppose at every point in $\mathbb{Z}^2$, the integer lattice in the plane, there is a voter, who is leaning either toward the Democrat candidate or the Republican candidate. At each point, the voter waits a length of time that is exponential with parameter one, chooses one of his four nearest neighbors at random, and then changes his view to agree with that neighbor. Other infinite particle systems include the contact process (modeling the spread of infection), Ising model (modeling ferromagnetism), and the exclusion model (used in solid state physics). See Liggett (2010) for how to construct these processes using infinitesimal generators, and for much more.

37.3 Nondivergence form elliptic operators

Let us consider the operator $\mathcal{L}$ defined on C^2 functions on $\mathbb{R}^d$ by

$$\mathcal{L}f(x) = \sum_{i,j=1}^{d} a_{ij}(x)\frac{\partial^2 f}{\partial x_i \partial x_j}(x) + \sum_{i=1}^{d} b_i(x)\frac{\partial f}{\partial x_i}(x).$$

We suppose $a_{ij}(x) = a_{ji}(x)$ for all x. We assume the a_{ij} and b_i are bounded and Hölder continuous of order $\alpha \in (0, 1)$: there exists c such that

$$|a_{ij}(x) - a_{ij}(y)| \leq c|x - y|^{\alpha}, \qquad |b_i(x) - b_i(y)| \leq c|x - y|^{\alpha},$$

for $i, j = 1, \ldots, d$. We also assume a *uniform ellipticity* condition on the a_{ij}: there exists $\Lambda > 0$ such that

$$\sum_{i,j=1}^{d} a_{ij}(x)y_i y_j \geq \Lambda \sum_{i=1}^{d} y_i^2, \qquad (y_1, \ldots, y_d) \in \mathbb{R}^d.$$

Uniform ellipticity says that the matrix whose (i, j)th element is $a_{ij}(x)$ is positive definite, uniformly in x.

If the a_{ij} and b_i were Lipschitz continuous, we can construct the Markov process with infinitesimal generator $\mathcal{L}$ using stochastic differential equations (see Chapter 39), which is a more probabilistic way of doing it. Even when the a_{ij} are continuous and the b_i only measurable, it is possible to construct the Markov process via SDEs, although this is much harder. Here we illustrate how the Hille–Yosida theorem can be used in constructing these processes.

Let $\mathcal{B}$ be the space of continuous functions that vanish at infinity. We will want the domain of $\mathcal{L}$ to include the class $\mathcal{C}$ of functions f such that f and its first and second partial derivatives are continuous and vanish at infinity. Then $\mathcal{C}$ is dense in $\mathcal{B}$ and $\mathcal{L}$ maps $\mathcal{C}$ into $\mathcal{B}$.

We show that $\mathcal{L}$ is dissipative. Let $f \in \mathcal{C}$ and let x_0 be a point where $|f(x_0)| = \|f\|$. There is nothing to prove if f is identically zero. If $f(x_0) < 0$, we can look at $-f$, so let us suppose $f(x_0) > 0$. Such a point x_0 exists because f is continuous and vanishes at infinity. It suffices to show that $\mathcal{L}f(x_0) \leq 0$, since then

$$\lambda\|f\| = \lambda f(x_0) \leq (\lambda - \mathcal{L})f(x_0) \leq \|(\lambda - \mathcal{L})f\|.$$

Let A be the matrix whose (i, j) element is $a_{ij}(x_0)$ and let H be the Hessian at x_0 so that

$$H_{ij} = \frac{\partial^2 f}{\partial x_i \partial x_j}(x_0).$$

Let $y \in \mathbb{R}^d$ and consider the function $f(x_0 + ty)$, $t \in \mathbb{R}$. Since x_0 is the location of a local maximum for this function, its second derivative, which is $\sum_{i,j=1}^{d} y_i y_j H_{ij}$, will be less than or equal to 0. The first derivative of this function will be zero at x_0.

Since A is positive definite, there exists an orthogonal matrix P and a diagonal matrix D with positive entries such that $A = P^T D P$. Recall the trace of a square matrix is defined by Trace $(C) = \sum_{i=1}^{d} C_{ii}$ and Trace $(AB) = $ Trace (BA). Note

$$\sum_{i,j=1}^{d} a_{ij}(x_0) \frac{\partial^2 f}{\partial x_i \partial x_j}(x_0) = \text{Trace } (AH).$$

We have

$$\text{Trace } (AH) = \text{Trace } (P^T DPH) = \text{Trace } (PHP^T D) = \sum_{i=1}^{d} (PHP^T)_{ii} D_{ii},$$

since D is a diagonal matrix. Thus to show that Trace $(AH) \leq 0$, it suffices to show that $(PHP^T)_{ii} \leq 0$ for each i. If we let e_i be the unit vector in the x_i direction and $y = P^T e_i$, we have

$$(PHP^T)_{ii} = e_i^t PHP^T e_i = y^t Hy = \sum_{i,j=1}^{d} y_i y_j H_{ij} \leq 0.$$

Since x_0 is the location of a local maximum, then $\frac{\partial f}{\partial x_i}(x_0) = 0$, and we conclude $\mathcal{L}f(x_0) \leq 0$.

Since $\mathcal{L}1 = 0$, then $P_t 1 = 1$ for all t. This and Exercise 37.1 imply that the P_t are non-negative operators.

To apply the Hille–Yosida theorem, it remains to show that the range of $\lambda - \mathcal{L}$ is dense in $\mathcal{B}$. For this we refer the reader to the PDE literature, e.g., Bass (1997), Chapter 3 or Gilbarg and Trudinger (1983), Chapters 5,6.

37.4 Generators of Lévy processes

Let n be a measure on $\mathbb{R} \setminus \{0\}$ satisfying

$$\int (h^2 \wedge 1) \, n(dh) < \infty.$$

Consider the operator $\mathcal{L}$ defined on C^2 functions by

$$\mathcal{L}f(x) = \int [f(x+h) - f(x) - 1_{(|h| \leq 1)} f'(x)h] \, n(dh).$$

We will show that $\mathcal{L}$ is the infinitesimal generator of a Markov semigroup. We construct these processes, the Lévy processes, probabilistically in Chapter 42. We confine ourselves to the one-dimensional case, although the argument for higher dimensions is completely analogous.

We let $\mathcal{B}$ be the continuous functions vanishing at infinity. We let $\mathcal{C}$ be the class of Schwartz functions, which is the class of C^∞ functions, all of whose kth partial derivatives go to zero faster than $|x|^{-m}$ as $|x| \to \infty$ for every $k = 0, 1, \ldots$ and every $m = 1, 2, \ldots$; see Section B.2.

First we show that $\mathcal{L}$ maps $\mathcal{C}$ into $\mathcal{B}$, so that the domain of $\mathcal{L}$ contains $\mathcal{C}$, and hence is dense in $\mathcal{B}$. Given $M > 1$ and $f \in \mathcal{C}$, by Taylor's theorem

$$|\mathcal{L}f(x)| \leq \int |f(x+h) - f(x) - 1_{(|h|\leq 1)}f'(x)h| \, n(dh) \tag{37.13}$$

$$\leq \sup_{|y-x|\leq 1} (|f''(y)|) \int_{0<|h|\leq 1} h^2 \, n(dh) + 2(\sup_{|y-x|\leq M} |f(y)|) \int_{1<|h|\leq M} n(dh)$$

$$+ 2 \int_{|h|>M} \|f\|_\infty \, n(dh).$$

This shows $|\mathcal{L}f(x)|$ is finite. Given $\varepsilon > 0$ and $f \in \mathcal{C}$, choose M large so that

$$\int_{|h|>M} n(dh) < \varepsilon / \|f\|_\infty.$$

Since the first two terms on the right-hand side of (37.13) tend to zero as $|x| \to \infty$, then $\mathcal{L} : \mathcal{C} \to \mathcal{B}$.

To show $\mathcal{L}$ is dissipative, let $f \in \mathcal{C}$ and choose x_0 such that $|f(x_0)| = \|f\|$. There is nothing to prove if $\|f\| = 0$, so assume $\|f\| > 0$. Because f is in the Schwartz class, it takes on its maximum and its minimum. By looking at $-f$ if necessary, we may suppose $f(x_0) > 0$. Since x_0 is the location of a local maximum, $f'(x_0) = 0$ and $f(x_0 + h) - f(x_0) \leq 0$ for each h, hence $\mathcal{L}f(x_0) \leq 0$. Then

$$\lambda \|f\| = \lambda f(x_0) \leq (\lambda - \mathcal{L})f(x_0) \leq \|(\lambda - \mathcal{L})f\|.$$

Taking limits, this holds for every f in the domain of $\mathcal{L}$.

Finally we need to show that the range of $\lambda - \mathcal{L}$ is dense in $\mathcal{B}$. This is the most complicated part and we break the argument into steps.

Step 1. We start by computing the Fourier transform of $\mathcal{L}f$ if $f \in \mathcal{C}$. Let $n_\delta(dh) = 1_{(|h|\geq\delta)}n(dh)$ and let

$$\mathcal{L}_\delta f(x) = \int [f(x+h) - f(x) - 1_{(|h|\leq 1)}f'(x)h] \, n_\delta(dh).$$

Then n_δ is a finite measure. Using the Fubini theorem and the fact that the Fourier transform of the function $x \to f(x+h)$ is $e^{iuh}\widehat{f}(u)$ and the Fourier transform of $f'(x)$ is $-iu\widehat{f}(u)$,

$$\widehat{\mathcal{L}_\delta f}(u) = \int \int [e^{iux}f(x+h) - e^{iux}f(x) - 1_{(|h|\leq 1)}e^{iux}f'(x)h] \, dx \, n_\delta(dh)$$

$$= \widehat{f}(u) \int [e^{-iuh} - 1 + 1_{(|h|\leq 1)}iuh] \, n_\delta(dh)$$

$$= \widehat{f}(u) \int [e^{-iuh} - 1 + 1_{(|h|\leq 1)}iuh]1_{(|h|\geq\delta)} \, n(dh). \tag{37.14}$$

The expression in brackets on the last line is bounded by $c(h^2 \wedge 1)$ and by dominated convergence the last line converges to $\widehat{f}(u)\psi(u)$ as $\delta \to 0$, where

$$\psi(u) = \int [e^{-iuh} - 1 + 1_{(|h|\leq 1)}iuh] \, n(dh). \tag{37.15}$$

Since

$$|\widehat{\mathcal{L}f}(u) - \widehat{\mathcal{L}_\delta f}(u)|$$

$$= \left| \int e^{iux} \int_{|h|<\delta} [f(x+h) - f(x) - 1_{(|h|\leq1)} f'(x)h] \, n(dh) \, dx \right|$$

$$\leq \int \left(\sup_{|y-x|<\delta} |f''(y)| \right) \int_{|h|<\delta} h^2 \, n(dh) \, dx,$$

which tends to zero as $\delta \to 0$ because $f \in \mathcal{C}$, we conclude

$$\widehat{\mathcal{L}f}(u) = \widehat{f}(u)\psi(u). \tag{37.16}$$

Step 2. Now let $g \in \mathcal{C}$, let $\varepsilon > 0$, choose $K > 1$ such that $\int_{|h|\geq K} n(dh) < \varepsilon$, let $m_K(dh) = 1_{(|h|\geq K)} n(dh)$, and define $\mathcal{L}_K$ and ψ_K in terms of m_K. We show there exists $f \in \mathcal{C}$ such that $g = (\lambda - \mathcal{L}_K) f = g$.

We have

$$\psi_K(u) = \int_{|h|\leq K} [e^{-iuh} - 1 + iuh 1_{(|h|\leq1)}] \, n(dh),$$

so using dominated convergence,

$$\psi'_K(u) = \int_{|h|\leq K} [-ihe^{-iuh} + ih 1_{(|h|\leq1)}] \, n(dh),$$

$$\psi''_K(u) = \int_{|h|\leq K} [-h^2 e^{-iuh}] \, n(dh),$$

with similar formulas for the higher derivatives. Thus all the derivatives of ψ_K are bounded. Moreover the real part of $\psi_K(u)$ is $\int_{|h|\leq K} [\cos(uh) - 1] \, n(dh)$, which is less than or equal to 0. Since $g \in \mathcal{C}$, by Section B.2, $\widehat{g} \in \mathcal{C}$. If we define f by

$$\widehat{f}(u) = \frac{1}{\lambda - \psi_K(u)} \widehat{g}(u), \tag{37.17}$$

we see that $\widehat{f}$ and all its derivatives are continuous and tend to zero faster than $|u|^{-m}$ for every m. Hence $\widehat{f} \in \mathcal{C}$, which implies $f \in \mathcal{C}$ by Section B.2.

Notice $(\lambda - \mathcal{L}_K) f = g$ because

$$\lambda \widehat{f}(u) - \widehat{\mathcal{L}_K f}(u) = \frac{\lambda - \psi_K(u)}{\lambda - \psi_K(u)} \widehat{g}(u) = \widehat{g}(u).$$

Step 3. We prove that $\|\mathcal{L}f - \mathcal{L}_K f\| \leq c\varepsilon \|g\|$.

Since $g \in \mathcal{C}$, then $\widehat{g} \in L^1$. From (37.17) we have $|\widehat{f}(u)| \leq |\widehat{g}(u)|/\lambda$. Then

$$\|f\|_\infty \leq c\|\widehat{f}\|_{L^1} \leq c\|\widehat{g}\|_{L^1}$$

and

$$|\mathcal{L}f(x) - \mathcal{L}_K f(x)| \leq \int_{|h| \geq K} |f(x+h) - f(x)|\, n(dh)$$

$$\leq 2\|f\|_\infty \int_{|h| \geq K} n(dh)$$

$$\leq c\varepsilon \|\widehat{g}\|_{L^1}.$$

Step 4. We complete the proof that the range of $\lambda - \mathcal{L}$ is dense in $\mathcal{B}$. Since $\|\mathcal{L}f - \mathcal{L}_K f\| \leq c\varepsilon \|g\|$ by Step 3 and $(\lambda - \mathcal{L}_K)f = g$, then

$$\|(\lambda - \mathcal{L})f - g\| \leq c\varepsilon \|g\|.$$

Because $f \in \mathcal{C} \subset \mathcal{D}(\mathcal{L})$ and ε is arbitrary, this proves the range of $\lambda - \mathcal{L}$ is dense in $\mathcal{C}$, hence in $\mathcal{B}$.

We thus have $\mathcal{L}$ satisfying all the hypotheses of the Hille–Yosida theorem, and hence there exists a semigroup P_t mapping $\mathcal{B}$ into $\mathcal{B}$. We again note that $\mathcal{L}1 = 0$, hence $P_t = 1$ for all t, and so by Exercise 37.1, the P_t are non-negative operators.

Exercises

37.1 Let $\mathcal{B}$ be either the space L^2 with respect to a finite measure or else the continuous functions vanishing at infinity for some locally compact separable metric space S. In the former case, we say $f \geq 0$ if $f(x) \geq 0$ for almost every x, in the latter case if $f(x) \geq 0$ for all x. A semigroup is non-negative if $f \geq 0$ implies $P_t f \geq 0$ for all $t \geq 0$. Suppose that P_t is a semigroup, the space $\mathcal{B}$ contains the constant functions, and $P_t 1 = 1$ for all t. Show that P_t is a contraction if and only if P_t is non-negative.

37.2 Show that P_t and R_λ commute and that

$$P_t R_\lambda f = \int_0^\infty e^{-\lambda s} P_{s+t} f\, ds.$$

Show that for any $a < b$ we have

$$\left\| \int_a^b e^{\lambda t} P_t f\, dt \right\| \leq \int_a^b e^{-\lambda t} \|P_t f\|\, dt.$$

Hint: Approximate $R_\lambda f$ by a Riemann sum.

37.3 Show that if P_t is a contraction semigroup and R_λ is the resolvent, then

$$\|R_\lambda\| \leq 1/\lambda. \tag{37.18}$$

37.4 Show that if A is a bounded operator and $T_t = e^{tA}$, then T_t is a strongly continuous semigroup of operators with infinitesimal generator A. (We cannot assert that the T_t are contractions.)

37.5 Prove that if $\mathcal{L}$ is dissipative, the domain of $\mathcal{L}$ is dense in $\mathcal{B}$, and the range of $\lambda - \mathcal{L}$ is dense in $\mathcal{B}$, then $\overline{\mathcal{L}}$ defined in (37.12) is a closed extension of $\mathcal{L}$ that is dissipative and the range of $\lambda - \overline{\mathcal{L}}$ is equal to $\mathcal{B}$. Show there is only one such closed extension of $\mathcal{L}$.

37.6 If the range of $\lambda - \mathcal{L}$ equals $\mathcal{B}$ for a single value of λ, then the range of $\lambda - \mathcal{L}$ equals $\mathcal{B}$ for every value of λ.
    Hint: Define R_λ as the inverse of $\lambda - \mathcal{L}$, then use (37.3) to define R_a for other values of a.

37.7 Let $(X_t, \mathbb{P}^x)$ be a Markov process with transition probabilities given by $P_t f(x) = f(x + t)$. Determine $\mathcal{L}$ and $\mathcal{D}(\mathcal{L})$.

37.8 Let P_t be a strongly continuous semigroup of contraction operators and let $\mathcal{L}$ be the infinitesimal generator. Show that $\mathcal{D}(\mathcal{L}^n)$ is dense in $\mathcal{B}$ for every positive integer n.

37.9 This is a continuation of Exercise 36.5. Prove that if $f \in C^2$ with compact support, P_t is the semigroup given in Exercise 36.5, and $\mathcal{L}$ is the infinitesimal generator, then the Fourier transform of $\mathcal{L}f$ is $\widehat{f}(u)\psi(u)$.

37.10 Suppose that P_t is a strongly continuous semigroup, but not necessarily of contractions. Thus $P_{t+s} = P_t P_s$ and $P_t f \to f$ in norm if $f \in \mathcal{B}$, but we do not assume $\|P_t\| \leq 1$. Prove that there exist constants $K, b > 0$ such that $\|P_t\| \leq K e^{bt}$ for all $t \geq 0$.

   Hint: Use the uniform boundedness principle from functional analysis to prove there exists c, t_0 such that $\|P_t\| \leq c$ if $t \leq t_0$. Then use the semigroup property.

38

Dirichlet forms

When constructing semigroups, it is sometimes easier to start with a bilinear form, called the Dirichlet form, than to work with the infinitesimal generator, and to construct the semigroup from the form. For example, let Δ be the Laplacian. If $f, g \in C^2$ with compact support, then integration by parts shows

$$\int_{\mathbb{R}^d} f(x) (\tfrac{1}{2} \Delta g)(x)\, dx = \tfrac{1}{2} \int_{\mathbb{R}^d} f(x) \sum_{i=1}^{d} \frac{\partial^2 g}{\partial x_i^2}(x)\, dx$$

$$= -\tfrac{1}{2} \int_{\mathbb{R}^d} \sum_{i=1}^{d} \frac{\partial f}{\partial x_i}(x) \frac{\partial g}{\partial x_i}(x)\, dx.$$

If we write

$$\mathcal{E}(f, g) = \tfrac{1}{2} \int \sum_{i=1}^{d} \frac{\partial f}{\partial x_i}(x) \frac{\partial g}{\partial x_i}(x)\, dx,$$

we thus have

$$\int_{\mathbb{R}^d} f(\tfrac{1}{2} \Delta g) = -\mathcal{E}(f, g). \tag{38.1}$$

Clearly $\mathcal{E}(f, g)$ is symmetric in f and g, so

$$\int_{\mathbb{R}^d} f(\tfrac{1}{2} \Delta g) = -\mathcal{E}(f, g) = -\mathcal{E}(g, f) = \int_{\mathbb{R}^d} g(\tfrac{1}{2} \Delta f)\, dx.$$

If R_λ is the resolvent for Brownian motion, (38.1) and the fact that $\tfrac{1}{2} \Delta R_\lambda f = \lambda R_\lambda f - f$ tells us that

$$\mathcal{E}(R_\lambda f, g) + \lambda \int (R_\lambda f)g = -\int (\tfrac{1}{2} \Delta R_\lambda f)g + \lambda \int (R_\lambda f)g \tag{38.2}$$

$$= -\int (\lambda R_\lambda f - f)g + \lambda \int (R_\lambda f)g$$

$$= \int fg.$$

The bilinear form $\mathcal{E}(f, g)$ makes sense even if f, g are only in C^1 with compact support, which is one major advantage of the Dirichlet form. Since $\mathcal{E}$ is clearly linear in each variable, we have

$$\mathcal{E}(f, g) = \tfrac{1}{2} [\mathcal{E}(f + g, f + g) - \mathcal{E}(f, f) - \mathcal{E}(g, g)],$$

302

so to specify the Dirichlet form, it is only necessary to know $\mathcal{E}(f, f)$, a number, rather than $\mathcal{L}f$, a function. One disadvantage of Dirichlet forms is that one needs a self-adjoint operator, and not every infinitesimal generator is self-adjoint. Another disadvantage is that when working with Dirichlet forms, L^2 is the natural space to work with, which means there are null sets one has to worry about. In particular, the construction of Chapter 36 is not directly applicable, because there we required our Banach space to be the set of continuous functions vanishing at infinity. (Modifications of the methods in Chapter 36 do work, however.)

38.1 Framework

Let us now suppose $\mathcal{S}$ is a locally compact separable metric space together with a σ-finite measure m defined on the Borel subsets of $\mathcal{S}$. We want to give a definition of the Dirichlet form in this more general context. We suppose there exists a dense subset $\mathcal{D} = \mathcal{D}(\mathcal{E})$ of $L^2(S, m)$ and a non-negative bilinear symmetric form $\mathcal{E}$ defined on $\mathcal{D} \times \mathcal{D}$, which means

$$\mathcal{E}(f, g) = \mathcal{E}(g, f), \quad \mathcal{E}(f + g, h) = \mathcal{E}(f, h) + \mathcal{E}(g, h)$$
$$\mathcal{E}(af, g) = a\mathcal{E}(f, g), \quad \mathcal{E}(f, f) \geq 0$$

for $f, g, h \in \mathcal{D}, a \in \mathbb{R}$.

We will frequently write $\langle f, g \rangle$ for $\int f(x)g(x) \, m(dx)$. For $a > 0$ define

$$\mathcal{E}_a(f, f) = \mathcal{E}(f, f) + a\langle f, f \rangle.$$

We can define a norm on $\mathcal{D}$ using the inner product $\mathcal{E}_a$: the norm of f equals $(\mathcal{E}_a(f, f))^{1/2}$; we call this the norm induced by $\mathcal{E}_a$. Since $a\langle f, f \rangle \leq \mathcal{E}_a(f, f)$, then

$$\mathcal{E}_a(f, f) \leq \mathcal{E}_b(f, f) = \mathcal{E}_a(f, f) + (b - a)\langle f, f \rangle$$
$$\leq \left(1 + \frac{b - a}{a}\right)\mathcal{E}_a(f, f)$$

if $a < b$, so the norms induced by different a's are all equivalent. We say $\mathcal{E}$ is *closed* if $\mathcal{D}$ is complete with respect to the norm induced by $\mathcal{E}_a$ for some a. Equivalently, $\mathcal{E}$ is closed if whenever $u_n \in \mathcal{D}$ satisfies $\mathcal{E}_1(u_n - u_m, u_n - u_m) \to 0$ as $n, m \to \infty$, then there exists $u \in \mathcal{D}$ such that $\mathcal{E}(u_n - u, u_n - u) \to 0$ as $n \to \infty$.

We say $\mathcal{E}$ is *Markovian* if whenever $u \in \mathcal{D}$, then $v = 0 \vee (u \wedge 1) \in \mathcal{D}$ and $\mathcal{E}(v, v) \leq \mathcal{E}(u, u)$. (A slightly weaker definition of Markovian is sometimes used.) A *Dirichlet form* is a non-negative bilinear symmetric form that is closed and Markovian.

Absorbing Brownian motion on $[0, \infty)$ is a symmetric process. The corresponding Dirichlet form is

$$\mathcal{E}(f, f) = \tfrac{1}{2} \int_0^\infty |f'(x)|^2 \, dx,$$

and the appropriate domain turns out to be the completion of the set of C^1 functions with compact support contained in $(0, \infty)$ with respect to the norm induced by $\mathcal{E}_1$. In particular, any function with compact support contained in $(0, \infty)$ will be zero in a neighborhood of 0. In a domain D in higher dimensions, the Dirichlet form for absorbing Brownian motion becomes

$$\mathcal{E}(f, f) = \tfrac{1}{2} \int |\nabla f(x)|^2 \, dx, \tag{38.3}$$

with the domain of $\mathcal{E}$ being the completion with respect to $\mathcal{E}_1$ of the C^1 functions whose support is contained in the interior of D.

Reflecting Brownian motion is also a symmetric process. For a domain D, the Dirichlet form is given by (38.3) and the domain $\mathcal{D}(\mathcal{E})$ of the form is given by the completion with respect to the norm induced by $\mathcal{E}_1$ of the C^1 functions on $\overline{D}$ with compact support, where $\overline{D}$ is the closure of D. One might expect there to be some restriction on the normal derivative $\partial f/\partial n$ on the boundary of D, but in fact there is no such restriction. To examine this further, consider the case of $D = (0, \infty)$. If one takes the class of functions f which are C^1 with compact support and with $f'(0) = 0$ and takes the closure with respect to the norm induced by $\mathcal{E}_1$, one gets the same class as $\mathcal{D}(\mathcal{E})$; this is Exercise 38.1.

One nice consequence of the fact that we don't need to impose a restriction on the normal derivative in the domain of $\mathcal{E}$ for reflecting Brownian motion is that this allows us to define reflecting Brownian motion in any domain, even when the boundary is not smooth enough for the notion of a normal derivative to be defined.

38.2 Construction of the semigroup

We now want to construct the resolvent corresponding to a Dirichlet form. The motivation given in (38.2) shows we should expect

$$\mathcal{E}_a(R_a f, g) = \langle f, g \rangle \tag{38.4}$$

for all $a > 0$ and all f, g such that $R_a f, g \in \mathcal{D}$. Our Banach space $\mathcal{B}$ will be $L^2(\mathcal{S}, m)$.

Theorem 38.1 *If $\mathcal{E}$ is a Dirichlet form, there exists a family of resolvent operators $\{R_\lambda\}$ such that*

(1) the R_λ satisfy the resolvent equation,
(2) $\|\lambda R_\lambda\| \leq 1$ for all $\lambda > 0$,
(3) $\lambda R_\lambda f \to f$ as $\lambda \to \infty$ for $f \in \mathcal{B}$, and
(4) $\mathcal{E}_a(R_a f, g) = \langle f, g \rangle$ if $a > 0$, $R_a f, g \in \mathcal{D}$.
Moreover, if $f \in \mathcal{B}$ satisfies $0 \leq f(x) \leq 1$, m-a.e., then for all $a > 0$

$$0 \leq a R_a f \leq 1, \qquad \text{m-a.e.} \tag{38.5}$$

Proof Fix $f \in \mathcal{B}$ and define a linear functional on $\mathcal{B}$ by $I(g) = \langle f, g \rangle$. This functional is also a bounded linear functional on $\mathcal{D}$ with respect to the norm induced by $\mathcal{E}_a$, that is, there exists c such that $|I(g)| \leq c\mathcal{E}_a(g, g)^{1/2}$. This follows because

$$|I(g)| = \left| \int fg \right| \leq \langle f, f \rangle^{1/2} \langle g, g \rangle^{1/2} \leq \langle f, f \rangle^{1/2} (\tfrac{1}{a} \mathcal{E}_a(g, g))^{1/2}$$

by the Cauchy–Schwarz inequality. Since $\mathcal{E}$ is closed, $\mathcal{D}$ is a Hilbert space with respect to the norm induced by $\mathcal{E}_a$. By the Riesz representation theorem for Hilbert spaces (see, e.g., Folland (1999), Theorem 5.25), there exists a unique element $u \in \mathcal{D}$ such that $I(g) = \mathcal{E}_a(u, g)$ for all $g \in \mathcal{D}$. We set $R_a f = u$. In particular, (38.4) holds, and $R_a f \in \mathcal{D}$.

We show the resolvent equation holds. If $g \in \mathcal{D}$,

$$
\begin{aligned}
\mathcal{E}_a(R_a f - R_b f, g) &= \mathcal{E}_a(R_a f, g) - \mathcal{E}(R_b f, g) - a\langle R_b f, g\rangle \\
&= \langle f, g\rangle - \mathcal{E}(R_b f, g) - b\langle R_b f, g\rangle + (b - a)\langle R_b f, g\rangle \\
&= \langle f, g\rangle - \mathcal{E}_b(R_b f, g) + (b - a)\langle R_b f, g\rangle \\
&= (b - a)\langle R_b f, g\rangle \\
&= \mathcal{E}_a((b - a)R_a R_b f, g).
\end{aligned}
$$

Since this holds for all $g \in \mathcal{D}$ and $\mathcal{D}$ is dense in $\mathcal{B}$, then $R_a f - R_b f = (b - a)R_a R_b f$.
Next we show that $\|a R_a f\| \le \|f\|$, or equivalently,

$$
\langle a R_a f, a R_a f\rangle \le \langle f, f\rangle. \tag{38.6}
$$

If $\langle R_a f, R_a f\rangle$ is zero, then (38.6) trivially holds, so suppose it is positive. We have

$$
a\langle R_a f, R_a f\rangle \le \mathcal{E}_a(R_a f, R_a f) = \langle f, R_a f\rangle \le \langle f, f\rangle^{1/2}\langle R_a f, R_a f\rangle^{1/2}
$$

by (38.4) and the Cauchy–Schwarz inequality. If we now divide both sides by $\langle R_a f, R_a f\rangle^{1/2}$ and then square both sides, we obtain (38.6).

We show that $b R_b f \to f$ as $b \to \infty$ when $f \in \mathcal{B}$. If $f \in \mathcal{D}$, then by the Cauchy–Schwarz inequality and (38.6)

$$
\begin{aligned}
\langle b R_b f, f\rangle &\le \langle b R_b f, b R_b f\rangle^{1/2}\langle f, f\rangle^{1/2} \\
&\le \langle f, f\rangle.
\end{aligned}
$$

Using this,

$$
\begin{aligned}
b\langle b R_b f - f, b R_b f - f\rangle &\le \mathcal{E}_b(b R_b f - f, b R_b f - f) \\
&= b^2 \mathcal{E}_b(R_b f, R_b f) - 2b \mathcal{E}_b(R_b f, f) + \mathcal{E}_b(f, f) \\
&= b^2 \langle R_b f, f\rangle - 2b\langle f, f\rangle + \mathcal{E}(f, f) + b\langle f, f\rangle \\
&\le \mathcal{E}(f, f).
\end{aligned}
$$

Now divide both sides by b to get $\|b R_b f - f\|^2 \le \mathcal{E}(f, f)/b \to 0$ as $b \to \infty$. Since $\mathcal{D}$ is dense in $\mathcal{B}$ and $\|b R_b\| \le 1$ for all b, we conclude $b R_b f \to f$ for all $f \in \mathcal{B}$.

It remains to show $0 \le b R_b f \le 1$, m-a.e., if $0 \le f \le 1$, m-a.e. Fix $f \in \mathcal{B}$ with $0 \le f \le 1$, m-a.e., and let $a > 0$. Define a functional ψ on $\mathcal{D}$ by

$$
\psi(v) = \mathcal{E}(v, v) + a\Big\langle v - \frac{f}{a}, v - \frac{f}{a}\Big\rangle.
$$

We claim

$$
\psi(R_a f) + \mathcal{E}_a(R_a f - v, R_a f - v) = \psi(v), \qquad v \in \mathcal{D}. \tag{38.7}
$$

To see this, start with the left-hand side, which is equal to

$$\mathcal{E}(R_af, R_af) + a\left\langle R_af - \frac{1}{a}f, R_af - \frac{1}{a}f\right\rangle + \mathcal{E}_a(R_af - v, R_af - v)$$

$$= \mathcal{E}_a(R_af, R_af) - 2\langle R_af, f\rangle + \frac{1}{a}\langle f, f\rangle + \mathcal{E}_a(R_af, R_af) - 2\mathcal{E}_a(R_af, v) + \mathcal{E}_a(v, v)$$

$$= \frac{1}{a}\langle f, f\rangle - 2\langle f, v\rangle + \mathcal{E}(v, v) + a\langle v, v\rangle$$

$$= \psi(v).$$

If follows from (38.7) and the fact that $\mathcal{E}_a(g, g)$ is non-negative for any $g \in \mathcal{D}$ that R_af is the function that minimizes ψ.

Set $\phi(x) = 0 \vee (x \wedge (1/a))$ and let $w = \phi(R_af)$. Observe that $|\phi(t) - s| \leq |t - s|$ for $t \in \mathbb{R}$ and $s \in [0, 1/a]$, so

$$\left|w(x) - \frac{f(x)}{a}\right| \leq \left|R_af(x) - \frac{f(x)}{a}\right|,$$

and therefore

$$\left\langle w - \frac{f}{a}, w - \frac{f}{a}\right\rangle \leq \left\langle R_af - \frac{f}{a}, R_af - \frac{f}{a}\right\rangle. \tag{38.8}$$

Since $\mathcal{E}$ is Markovian, then $aw = 0 \vee ((aR_af) \wedge 1)$, which leads to

$$\mathcal{E}(w, w) \leq \frac{1}{a^2}\mathcal{E}(aR_af, aR_af) = \mathcal{E}(R_af, R_af). \tag{38.9}$$

Adding (38.8) and (38.9), we conclude $\psi(w) \leq \psi(R_af)$. Since R_af is the minimizer for ψ, then $w = R_af$, m-a.e. But $0 \leq w \leq 1/a$, and hence aR_af takes values in $[0, 1]$, m-a.e.   □

If we combine Proposition 37.9 and Theorem 38.1, we obtain a semigroup P_t whose resolvent satisfies (38.4). We would like to know that the analog of (38.5) holds for P_t.

Corollary 38.2 *If* $0 \leq f \leq 1$, *m-a.e., then* $0 \leq P_tf \leq 1$, *m-a.e.*

Proof   If $0 \leq f \leq 1$, m-a.e., then $0 \leq bR_bf \leq 1$, m-a.e, by Theorem 38.1, and iterating, $0 \leq (bR_b)^if \leq 1$, m-a.e., for every i. Using the notation of the proof of Proposition 37.9,

$$Q_t^bf(x) = e^{-bt}\sum_{i=0}^{\infty}(bt)^i(bR_b)^if(x)/i!,$$

which will be non-negative, m-a.e., and bounded by $e^{-bt}\sum_{i=0}^{\infty}(bt)^i/i!$, m-a.e. Passing to the limit as $b \to \infty$, we see that P_tf takes values in $[0, 1]$, m-a.e.   □

When it comes to using the semigroup P_t derived from a Dirichlet form to construct a Markov process X, there is a difficulty that we did not have before. Since P_t is constructed using an L^2 procedure, P_tf is defined only up to almost everywhere equivalence. Without some continuity properties of P_tf for enough f's, we must neglect some null sets. If the only null sets we could work with were sets of m-measure 0, we would be in trouble. For example, when S is the plane and m is a two-dimensional Lebesgue measure, the x axis has measure zero, but a continuous process will (in general) hit the x axis. Fortunately there is a notion of sets of capacity zero, which are null sets that are smaller than sets of measure zero. It is possible to construct a process X starting from all points x in $\mathcal{S}$ except for those in a set $\mathcal{N}$ of capacity zero and to show that starting from any point not in $\mathcal{N}$, the process never hits $\mathcal{N}$.

There is another difficulty when working with Dirichlet forms. In general, one must look at $\widetilde{\mathcal{S}}$, a certain compactification of $\mathcal{S}$, which is a compact set containing $\mathcal{S}$. Even when our state space is a domain in $\mathbb{R}^d$, $\widetilde{\mathcal{S}}$ is not necessarily equal to $\overline{\mathcal{S}}$, the Euclidean closure of $\mathcal{S}$, and one must work with $\widetilde{\mathcal{S}}$ instead of $\overline{\mathcal{S}}$. It can be shown that this problem will not occur if the Dirichlet form is regular. Let C_K be the set of continuous functions with compact support. A Dirichlet form $\mathcal{E}$ is *regular* if $\mathcal{D} \cap C_K$ is dense in $\mathcal{D}$ with respect to the norm induced by $\mathcal{E}_1$ and $\mathcal{D} \cap C_K$ also is dense in C_K with respect to the supremum norm.

38.3 Divergence form elliptic operators

We want to show how to construct the Markov process corresponding to the operator

$$\mathcal{L}f(x) = \sum_{i,j=1}^{d} \frac{\partial}{\partial x_i}\left(a_{ij}(\cdot)\frac{\partial f}{\partial x_j}(\cdot)\right)(x). \tag{38.10}$$

If the a_{ij}'s are smooth in x, this can be interpreted as first calculating the partial derivative of f with respect to x_j, multiplying the result by $a_{ij}(x)$, taking the partial derivative of the product with respect to x_i, and then summing over i and j. If, however, the a_{ij}'s are only bounded and measurable, one cannot even in general give any nontrivial examples of functions in the domain of $\mathcal{L}$. Here is where Dirichlet forms are the perfect tool. Operators of the form (38.10) are known as elliptic operators in divergence form or in variational form, and the study of their properties has a long history in PDE.

We assume $a_{ij}(x) = a_{ji}(x)$ for each i and j and each x. We suppose the $a_{ij}(x)$ are measurable functions and are uniformly bounded in x for each i and j. We also require *uniform ellipticity*: there exists Λ such that

$$\sum_{i,j=1}^{d} a_{ij}(x)y_i y_j \geq \Lambda \sum_{i=1}^{d} y_i^2, \qquad (y_1, \ldots, y_d) \in \mathbb{R}^d.$$

Just as in the nondivergence elliptic operator case, the matrix whose (i, j)th element is $a_{ij}(x)$ is positive definite, uniformly in x.

We will shortly define a Dirichlet form, but let us first specify a domain. Let C_K^1 be the collection of C^1 functions with compact support, and define H^1 to be the completion of C_K^1 with respect to the norm

$$\|f\|_{H^1} = \left(\int (|f(x)|^2 + |\nabla f(x)|^2)\,dx\right)^{1/2}. \tag{38.11}$$

One can show that H^1 with this norm is a Banach space; this is Exercise 38.2.

Now for $f \in C_K^1$ define

$$\mathcal{E}(f, f) = \int_{\mathbb{R}^d} \sum_{i,j=1}^{d} a_{ij}(x)\frac{\partial f}{\partial x_i}(x)\frac{\partial f}{\partial x_j}(x)\,dx. \tag{38.12}$$

We can use the fact that C_K^1 is dense in H^1 to extend the definition of $\mathcal{E}$ to all of $H^1 \times H^1$. The connection with the operator $\mathcal{L}$ is that when the a_{ij} are smooth, integration by parts yields

$$\int (\mathcal{L}f)g\,dx = -\mathcal{E}(f,g)$$

if g is C^1 with compact support; cf. (38.1).

Because of the boundedness and uniform ellipticity, there exist positive constants c_1 and c_2 not depending on f such that

$$c_1 \int |\nabla f(x)|^2\,dx \leq \mathcal{E}(f,f) \leq c_2 \int |\nabla f(x)|^2\,dx.$$

Therefore the norm induced by $\mathcal{E}_1$ and the norm in H^1 are equivalent. This implies $\mathcal{E}$ is closed. By the definition of H^1, $\mathcal{E}$ is regular, and clearly $\mathcal{E}$ is symmetric. Thus we need only to show that $\mathcal{E}$ is Markovian.

Let $\phi(x) = (0 \vee x) \wedge 1$. For each $\varepsilon > 0$ let ϕ_ε be C^∞, bounded, agreeing with ϕ on $[0, 1]$, with $\|\phi_\varepsilon'\|_\infty \leq 1$, and such that $\phi_\varepsilon(x) \to \phi(x)$ uniformly in x as $\varepsilon \to 0$ and $\phi_\varepsilon'(x) \to 1_{[0,1]}(x)$ pointwise as $\varepsilon \to 0$. Note $\nabla \phi_\varepsilon(f) = \phi_\varepsilon'(f)\nabla f$, so if $f \in C_K^1$,

$$\mathcal{E}(\phi_\varepsilon(f), \phi_\varepsilon(f)) = \sum_{i,j=1}^{d} \int (\phi_\varepsilon'(f)(x))^2 a_{ij}(x) \frac{\partial f}{\partial x_i}(x) \frac{\partial f}{\partial x_j}(x)\,dx. \tag{38.13}$$

Since

$$\sum_{i,j=1}^{d} a_{ij}(x) \frac{\partial f}{\partial x_i}(x) \frac{\partial f}{\partial x_j}(x) \geq \Lambda |\nabla f(x)|^2 \geq 0$$

and $|\phi_\varepsilon'(f)(x)| \leq 1$, we see that

$$\mathcal{E}(\phi_\varepsilon(f), \phi_\varepsilon(f)) \leq \mathcal{E}(f,f).$$

Taking the limit as $\varepsilon \to 0$ in (38.13) we obtain

$$\mathcal{E}(\phi(f), \phi(f)) \leq \mathcal{E}(f,f) < \infty. \tag{38.14}$$

In particular, $\phi(f) \in H^1 = \mathcal{D}(\mathcal{E})$. We now pass to the limit to show that (38.14) holds for all $f \in H^1$, which says that $\mathcal{E}$ is Markovian.

We can therefore apply Theorem 38.1 to obtain a semigroup corresponding to the Dirichlet form $\mathcal{E}$. As mentioned earlier, there is potentially a problem in that the semigroup is only defined for points not in a certain null set. However, a famous result of Nash and of DeGiorgi shows that the semigroup P_t can be written as $P_t f(x) = \int f(y)p(t,x,y)\,dy$ with $p(t,x,y)$ Hölder continuous in x and y; see Bass (1997), Chapter VII for a presentation of this result. This allows us to take the null set to be empty and to see that our semigroup satisfies the assumptions of Chapter 36. Therefore there exists a strong Markov process having P_t as its semigroup.

Exercises

38.1 Let $F_1 = \{f \in C^1[0, \infty) : f$ has compact support$\}$ and $F_2 = F_1 \cap \{f \in C^1[0, \infty) :$ f has compact support, $f'(0) = 0\}$. Show that the closures of F_1 and F_2 with respect to the norm $(\int (|f(x)|^2 + |f'(x)|^2) \, dx)^{1/2}$ are the same.

38.2 If H^1 is the completion of C_K^1, the C^1 functions on $\mathbb{R}^d$ with compact support, relative to the norm given by (38.11), show H^1 is a Hilbert space.

38.3 Show that the resolvent operator R_λ defined in Theorem 38.1 is a symmetric operator, that is, if $f, g \in \mathcal{B}$, then $\langle R_\lambda f, g \rangle = \langle f, R_\lambda g \rangle$.

38.4 Show that if the resolvent operator R_λ is a symmetric operator, then the transition operators P_t are also symmetric: if $f, g \in \mathcal{B}$, then $\langle P_t f, g \rangle = \langle f, P_t g \rangle$.

38.5 To do the next few exercises, you will have to know some functional analysis, specifically, the spectral theorem for self-adjoint operators. See Lax (2002).

Let $\mathcal{E}$ be a Dirichlet form with domain $\mathcal{D}(\mathcal{E})$ and let $\mathcal{L}$ be the infinitesimal generator of the semigroup P_t that corresponds to $\mathcal{L}$. Let $E(d\lambda)$ be a spectral resolution of the identity for $-\mathcal{L}$. (The operator $\mathcal{L}$ is a negative operator, so $-\mathcal{L}$ is a positive one.) Then a consequence of the spectral theorem is that

$$P_t f = \int_0^\infty e^{-\lambda t} E(d\lambda) f$$

and

$$R_a f = \int_0^\infty \frac{1}{a + \lambda} E(d\lambda) f.$$

Also

$$\langle f, g \rangle = \int_0^\infty \langle E(d\lambda) f, g \rangle.$$

Show that if $f, g \in \mathcal{D}$, then

$$\mathcal{E}(f, g) = \int_0^\infty \lambda \langle E(d\lambda) f, g \rangle.$$

Hint: First prove it for $f = R_a h$. Write

$$\mathcal{E}(R_a h, g) = \langle h, g \rangle - a \langle R_a h, g \rangle = \int_0^\infty \left(1 - \frac{a}{a + \lambda}\right) \langle E(d\lambda) h, g \rangle$$

$$= \int_0^\infty \frac{\lambda}{a + \lambda} \langle E(d\lambda) h, g \rangle = \int_0^\infty \lambda \langle E(d\lambda)(R_a h), g \rangle.$$

To extend this to all f in the domain of $\mathcal{E}$, use the fact that $\mathcal{E}$ is closed.

38.6 If $\mathcal{L}$ is the infinitesimal generator of the semigroup associated with the Dirichlet form $\mathcal{E}$, show that $\mathcal{D}(\sqrt{-\mathcal{L}}) = \mathcal{D}(\mathcal{E})$.

38.7 Show that if $f \in \mathcal{D}(\mathcal{E})$, then $a R_a f$ converges to f with respect to the norm induced by $\mathcal{E}_1$.

38.8 Show that if $b > 0$, then $\{R_b f : f \in L^2\}$ is a dense subset of $\mathcal{D}(\mathcal{E})$ with respect to the norm induced by $\mathcal{E}_1$.

38.9 Show that $\{P_t f : f \in L^2, t > 0\}$ is a dense subset of $\mathcal{D}(\mathcal{E})$ with respect to the norm induced by $\mathcal{E}_1$.

38.10 This exercise shows how to approximate $\mathcal{E}$ by forms whose domain is all of $\mathcal{B}$. Let

$$\mathcal{E}^{(t)}(f, g) = \frac{1}{t}\langle f - P_t f, g\rangle.$$

Show that if $f \in \mathcal{D}(\mathcal{E})$, then $\mathcal{E}^{(t)}(f, f)$ increases to $\mathcal{E}(f, f)$. Show that if $f, g \in \mathcal{D}(\mathcal{E})$, then $\mathcal{E}^{(t)}(f, g)$ converges to $\mathcal{E}(f, g)$.

38.11 Show that if $u \in \mathcal{D}(\mathcal{E})$, then $|u| \in \mathcal{D}(\mathcal{E})$ and $\mathcal{E}(|u|, |u|) \leq \mathcal{E}(u, u)$.
     Hint: Use Exercise 38.10.

38.12 Use Exericse 38.11 to show that if $u \in \mathcal{D}(\mathcal{E})$, then $\mathcal{E}(u^+, u^-) \leq 0$.

38.13 Suppose $\{P_t\}$ are the transition probabilities corresponding to a Dirichlet form $\mathcal{E}$. Suppose there exist functions $p_t(x, y)$ such that for each t,

$$P_t f(x) = \int p_t(x, y)\, m(dy)$$

for almost every x. Prove that for almost every pair (x, y) with respect to the product measure $m \times m$, $p_t(x, y) = p_t(y, x)$.

38.14 Let $f \in L^2(m)$ and define the functional

$$\psi(u) = \mathcal{E}(u, u) + \lambda\langle u, u\rangle - 2\langle f, u\rangle$$

for u in the domain of $\mathcal{E}$. Prove that ψ is minimized by $u = R_\lambda f$, and that this function is the unique minimizer.

38.15 Let P_t be the semigroup associated with a Dirichlet form and define

$$J(dx, dy) = P_t(x, dy)\, m(dx).$$

(1) Prove that if f, g are continuous with compact support, then

$$\int \int f(x)g(y)\, J(dx, dy) = \int \int g(x)f(y)\, J(dx, dy).$$

(2) With f and g continuous with compact support, prove that

$$\int f(x)g(y)\, J(dx, dy) = \langle f, P_t g\rangle$$

and

$$\int \int f(x)g(x)\, J(dx, dy) = \langle fg, P_t 1\rangle.$$

(3) Let $k(x) = 1 - P_t 1(x)$. Prove that if $\mathcal{E}^{(t)}$ is defined as in Exercise 38.10, then

$$2t\mathcal{E}^{(t)}(f, g) = \int \int (f(x) - f(y))(g(x) - g(y))\, J(dx, dy) + \int f(x)g(x)k(x)\, m(dx).$$

(4) Is $\mathcal{E}^{(t)}$ a Dirichlet form? A regular Dirichlet form?

38.16 This is a continuation of the previous exercise. If f is a function on the state space, we say that g is a normal contraction of f if $|g(x)| \leq |f(x)|$ for all x and $|g(x) - g(y)| \leq |f(x) - f(y)|$ for all x and y. As an example, note that if $g(x) = -1 \vee (f(x) \wedge 1)$, then g is a normal contraction of f. Prove that if $f \in \mathcal{D}(\mathcal{E})$, where $\mathcal{E}$ is a Dirichlet form and g is a normal contraction of f, then for each $t > 0$,

$$\mathcal{E}^{(t)}(g, g) \leq \mathcal{E}^{(t)}(f, f) \leq \mathcal{E}(f, f).$$

Notes

See Fukushima *et al.* (1994) for further information.

Markov processes and SDEs

One common way of constructing Markov processes is via stochastic differential equations. Roughly speaking, if there is uniqueness for every starting point, then one can create a strong Markov process. After proving this, we establish a connection between stochastic differential equations and partial differential equations, and then we describe what is known as the martingale problem.

39.1 Markov properties

Let $\mathbb{P}$ be a probability and suppose W is a d-dimensional Brownian motion with respect to $\mathbb{P}$. Consider the SDE

$$dX_t = \sigma(X_t)\,dW_t + b(X_t)\,dt. \tag{39.1}$$

Here σ is a $d \times d$ matrix-valued function and b is a vector-valued function, both Borel measurable and bounded. This can be written in terms of components as

$$dX_t^i = \sum_{j=1}^{d} \sigma_{ij}(X_t)\,dW_t^j + b_i(X_t)\,dt, \qquad i = 1, \dots, d,$$

where $W = (W^1, \dots, W^d)$. Let X_t^x be the solution to (39.1) when $X_0 = x$. Let $\mathbb{P}^x$ be the law of X_t^x.

Let $\Omega = C[0, \infty)$, let $\mathcal{F}$ be the cylindrical subsets of Ω, and define $Z_t(\omega) = \omega(t)$. The main result of this section is that if weak existence and weak uniqueness hold for (39.1) for every starting point x, then the solutions $(Z_t, \mathbb{P}^x)$ form a strong Markov process.

We begin by considering regular conditional probabilities.

Definition 39.1 Let $(\Omega, \mathcal{F}, \mathbb{P})$ be a probability space, and let $\mathcal{E}$ be a σ-field contained in $\mathcal{F}$. A *regular conditional probability* for $\mathbb{E}[\cdot \mid \mathcal{E}]$ is a kernel $Q(\omega, d\omega')$ such that

(1) $Q(\omega, \cdot)$ is a probability measure on $(\Omega, \mathcal{E})$ for each ω;
(2) for each $A \in \mathcal{F}$, $Q(\cdot, A)$ is a random variable that is measurable with respect to $\mathcal{F}$;
(3) for each $A \in \mathcal{F}$ and each $B \in \mathcal{E}$,

$$\int_B Q(\omega, A)\,\mathbb{P}(d\omega) = \mathbb{P}(A \cap B).$$

Regular conditional probabilities need not always exist, but if the probability space has sufficient structure, then they do. We provide a proof in the appendix; see Theorem C.1. $Q(\omega, A)$ can be thought of as $\mathbb{P}(A \mid \mathcal{E})(\omega)$, regularized so as to have some joint measurability.

Recall that the definition of minimal augmented filtration for a Markov process was given in Section 20.1.

Theorem 39.2 *Suppose weak existence and weak uniqueness hold for the SDE* (39.1) *whenever X_0 is a random variable that is in L^2 and is measurable with respect to $\mathcal{F}_0$. Suppose the matrix $\sigma(y)$ is invertible for each y. Let $(\Omega, \mathcal{F}, \mathbb{P})$ be defined as above. Let $\mathbb{P}^x$ be the law of the weak solution when X_0 is identically equal to x. Let $\{\mathcal{F}_t\}$ be the minimal augmented filtration generated by Z. Then $(\mathbb{P}^x, Z_t)$ is a strong Markov process.*

Proof We will prove that if T is a bounded stopping time and f is a bounded and Borel measurable function on $\mathbb{R}^d$, then

$$\mathbb{E}^x[f(Z_{T+t}) \mid \mathcal{F}_T] = \mathbb{E}^{Z_T} f(Z_t), \qquad \text{a.s.} \tag{39.2}$$

As in Section 20.3, this is sufficient to get the strong Markov property.

Fix x. Let

$$Y_t = Z_t - \int_0^t b(Z_r)\,dr \tag{39.3}$$

and

$$W_t' = \int_0^t \sigma^{-1}(Z_r)\,dY_r. \tag{39.4}$$

Since the $\mathbb{P}^x$ law of Z_t is the same as the $\mathbb{P}$ law of X_t^x, then the $\mathbb{P}^x$ law of W' is the same as the $\mathbb{P}$ law of W, or in other words, W' is a Brownian motion under $\mathbb{P}^x$. Rearranging (39.3) and (39.4), we have the equation

$$Z_t = Z_0 + \int_0^t \sigma(Z_r)\,dW_r' + \int_0^t b(Z_r)\,dr. \tag{39.5}$$

Let Q be a regular conditional probability for $\mathbb{E}^x[\,\cdot \mid \mathcal{F}_T]$. Let $\widetilde{Z}_t = Z_{T+t}$ and $\widetilde{W}_t = W_{T+t}' - W_T'$. Using (39.5) with t replaced by $T+t$ and then with t replaced by T, and taking the difference, we obtain

$$Z_{T+t} - Z_T = \int_T^{T+t} \sigma(Z_r)\,dW_r + \int_T^{T+t} b(Z_r)\,dr,$$

and hence

$$\widetilde{Z}_t = \widetilde{Z}_0 + \int_0^t \sigma(\widetilde{Z}_r)\,\widetilde{W}_r + \int_0^t b(\widetilde{Z}_r)\,dr. \tag{39.6}$$

We will show in a moment that $\widetilde{W}$ is a Brownian motion with respect to $Q(\omega, \cdot)$ for $\mathbb{P}^x$-almost all ω. Thus except for ω in a $\mathbb{P}^x$-null set, (39.6) implies that under $Q(\omega, \cdot)$, $\widetilde{Z}$ is a solution to (39.1) with starting point $\widetilde{Z}_0 = Z_T(\omega)$. If $\mathbb{E}_Q$ denotes the expectation with respect to Q, the weak uniqueness tells us that

$$\mathbb{E}_Q f(\widetilde{Z}_t) = \mathbb{E}^{Z_T} f(Z_t), \qquad \mathbb{P}^x(d\omega)\text{-a.s.} \tag{39.7}$$

On the other hand,

$$\mathbb{E}_Q f(\widetilde{Z}_t) = \mathbb{E}_Q f(Z_{T+t}) = \mathbb{E}^x[f(Z_{T+t}) \mid \mathcal{F}_T], \qquad \mathbb{P}^x(d\omega)\text{-a.s.} \tag{39.8}$$

Combining (39.7) and (39.8) proves (39.2).

It remains to prove that under Q the process $\widetilde{W}$ is a Brownian motion. $Q(\omega, \cdot)$ is a probability measure on Ω', so $t \to \widetilde{W}_t$ is continuous for every ω'. Let $t_1 < \cdots < t_n$ and

$$N(u_2, \ldots, u_n, t_1, \ldots, t_n) = \left\{ \omega : \mathbb{E}_Q \exp \left(i \sum_{j=2}^{n} u_j (W'_{T+t_j} - W'_{T+t_{j-1}}) \right) \right.$$

$$\left. \neq \exp \left(- \sum_{j=2}^{n} |u_j|^2 (t_j - t_{j-1})/2 \right) \right\}.$$

By the strong Markov property of the Brownian motion W' and the definition of Q,

$$\mathbb{E}_Q \exp \left(i \sum_{j=2}^{n} u_j (W'_{T+t_j} - W'_{T+t_{j-1}}) \right) = \mathbb{E} \left[\exp \left(i \sum_{j=2}^{n} u_j (W'_{T+t_j} - W'_{T+t_{j-1}}) \right) \mid \mathcal{F}_T \right]$$

$$= \mathbb{E}^{W'_T} \exp \left(i \sum_{j=2}^{n} u_j (W'_{T+t_j} - W'_{T+t_{j-1}}) \right)$$

$$= \exp \left(- \sum_{j=2}^{n} |u_j|^2 (t_j - t_{j-1})/2 \right),$$

where the second equality holds almost surely, that is, except for a $\mathbb{P}^x$-null set of ω's. This shows that $N(u_2, \ldots, u_n, t_1, \ldots, t_n)$ is a null set with respect to $\mathbb{P}^x$.

Let N be the union of all such $N(u_1, \ldots, u_n, t_1, \ldots, t_n)$ for $n \geq 1$, $u_1, \ldots, u_n$ rational, and $t_1 < \ldots < t_n$ rational. Therefore N is a $\mathbb{P}^x$-null set.

Suppose $\omega \notin N$. By the continuity of the paths of W',

$$\mathbb{E}_Q \exp \left(i \sum_{j=2}^{n} u_j (W'_{T+t_j} - W'_{T+t_{j-1}}) \right) = \exp \left(- \sum_{j=2}^{n} |u_j|^2 (t_j - t_{j-1})/2 \right)$$

for all $t, \ldots, t_n \in [0, \infty)$ and $u_2, \ldots, u_n \in \mathbb{R}$. Thus the finite-dimensional distributions of $\widetilde{W}$ under $Q_T(\omega, \cdot)$ are those of a Brownian motion. By the continuity of $\widetilde{W}$ and Theorem 2.6, under Q_T, $\widetilde{W}$ is a Brownian motion, except for a null set of ω's. $\qquad\square$

By a slight abuse of notation, we will say $(X_t, \mathbb{P}^x)$ is a strong Markov family when $(Z_t, \mathbb{P}^x)$ is a strong Markov family.

39.2 SDEs and PDEs

The connection between stochastic differential equations and partial differential equations comes about through the following theorem, which is simply an application of Itô's formula. Let $\mathcal{L}$ be the operator on functions in C^2 defined by

$$\mathcal{L}f(x) = \frac{1}{2} \sum_{i,j=1}^{d} a_{ij}(x) \frac{\partial^2 f}{\partial x_i \partial x_j}(x) + \sum_{i=1}^{d} b_i(x) \frac{\partial f}{\partial x_i}(x). \tag{39.9}$$

Theorem 39.3 *Suppose X_t is a solution to (39.1), σ and b are bounded and Borel measurable, and $a = \sigma\sigma^T$. Suppose $f \in C^2$. Then*

$$f(X_t) = f(X_0) + M_t + \int_0^t \mathcal{L}f(X_s)\, ds, \tag{39.10}$$

where

$$M_t = \int_0^t \sum_{i,j=1}^d \frac{\partial f}{\partial x_i}(X_s)\sigma_{ij}(X_s)\, dW_s^j \tag{39.11}$$

is a local martingale.

Proof Since the components of the Brownian motion W_t are independent, we have $d\langle W^k, W^\ell\rangle_t = 0$ if $k \neq \ell$; see Exercise 9.4. Therefore

$$d\langle X^i, X^j\rangle_t = \sum_k \sum_\ell \sigma_{ik}(X_t)\sigma_{jl}(X_t)\, d\langle W^k, W^\ell\rangle_t$$

$$= \sum_k \sigma_{ik}(X_t)\sigma_{kj}^T(X_t)\, dt = a_{ij}(X_t)\, dt.$$

We now apply Itô's formula:

$$f(X_t) = f(X_0) + \sum_i \int_0^t \frac{\partial f}{\partial x_i}(X_s)\, dX_s^i + \tfrac{1}{2}\int_0^t \sum_{i,j} \frac{\partial^2 f}{\partial x_i \partial x_j}(X_s)\, d\langle X^i, X^j\rangle_s$$

$$= f(X_0) + M_t + \sum_i \int_0^t \frac{\partial f}{\partial x_i}(X_s)b_i(X_s)\, ds + \tfrac{1}{2}\int_0^t \sum_{i,j} \frac{\partial^2 f}{\partial x_i \partial x_j}(X_s)a_{ij}(X_s)\, ds$$

$$= f(X_0) + M_t + \int_0^t \mathcal{L}f(X_s)\, ds,$$

and we are finished. $\qquad\square$

39.3 Martingale problems

In this section we consider operators in *nondivergence form*, that is, operators of the form given by (39.9). We assume throughout this section that the coefficients a_{ij} and b_i are bounded and measurable and that $a_{ij}(x) = a_{ji}(x)$ for all $i, j = 1, \ldots, d$ and all $x \in \mathbb{R}^d$. The coefficients a_{ij} are called the *diffusion coefficients* and the b_i are called the *drift coefficients*. We also assume that the operator $\mathcal{L}$ is *uniformly elliptic*, which means that there exists $\Lambda > 0$ such that

$$\sum_{i,j=1}^d y_i a_{ij}(x)y_j \geq \Lambda |y|^2, \qquad y \in \mathbb{R}^d, x \in \mathbb{R}^d. \tag{39.12}$$

This says that the matrix $a_{ij}(x)$ is positive definite, uniformly in x.

We saw in the previous section that if X_t is the solution to (39.1), $a = \sigma\sigma^T$, and $f \in C^2$, then

$$f(X_t) - f(X_0) - \int_0^t \mathcal{L}f(X_s)\, ds \tag{39.13}$$

is a local martingale under $\mathbb{P}$. A very fruitful idea of Stroock and Varadhan is to phrase the association of X_t with $\mathcal{L}$ in terms which use (39.13) as a key element. Let Ω consist of all continuous functions ω mapping $[0, \infty)$ to $\mathbb{R}^d$. Let $X_t(\omega) = \omega(t)$ and given a probability $\mathbb{P}$, let $\{\mathcal{F}_t\}$ be the minimal augmented filtration generated by X. A probability measure $\mathbb{P}$ is a solution to the *martingale problem for $\mathcal{L}$ started at x_0* if

$$\mathbb{P}(X_0 = x_0) = 1 \tag{39.14}$$

and

$$f(X_t) - f(X_0) - \int_0^t \mathcal{L}f(X_s)\, ds \tag{39.15}$$

is a local martingale under $\mathbb{P}$ whenever $f \in C^2(\mathbb{R}^d)$. The martingale problem is *well posed* if there exists a solution $\mathbb{P}$ and this solution is unique.

Uniqueness of the martingale problem for $\mathcal{L}$ is closely connected to weak uniqueness or, equivalently, uniqueness in law of (39.1).

Theorem 39.4 *Suppose $a = \sigma\sigma^T$ and suppose the matrix $\sigma(x)$ is invertible for each x. Weak uniqueness for (39.1) holds if and only if the solution for the martingale problem for $\mathcal{L}$ started at x is unique. Weak existence for (39.1) holds if and only if there exists a solution to the martingale problem for $\mathcal{L}$ started at x.*

Proof We prove the uniqueness assertion. Let Ω be the continuous functions on $[0, \infty)$ and Z_t the coordinate process: $Z_t(\omega) = \omega(t)$. First suppose the solution to the martingale problem is unique. If $(X_t^1, W_t^1, \mathbb{P}_1)$ and $(X_t^2, W_t^2, \mathbb{P}_2)$ are two weak solutions to (39.1), define $\mathbb{P}_i^x$ on Ω to be the law of X^i under $\mathbb{P}_i$, $i = 1, 2$. Clearly $\mathbb{P}_i^x(Z_0 = x) = \mathbb{P}_i(X_0^i = x) = 1$. The expression in (39.13) is a local martingale under $\mathbb{P}_i^x$ for each i and each $f \in C^2$. By the uniqueness for the solution of the martingale problem, $\mathbb{P}_1^x = \mathbb{P}_2^x$. This implies that the laws of X_t^1 and X_t^2 are the same, or weak uniqueness holds.

Now suppose weak uniqueness holds for (39.1). Let

$$Y_t = Z_t - \int_0^t b(Z_s)\, ds.$$

Let $\mathbb{P}_1^x$ and $\mathbb{P}_2^x$ be solutions to the martingale problem. If $f(x) = x_k$, the kth coordinate of x, then $\partial f/\partial x_i(x) = \delta_{ik}$ and $\partial^2 f/\partial x_i \partial x_j(x) = 0$, where δ_{ik} is 1 if $i = k$ and 0 otherwise, and so $\mathcal{L}f(Z_s) = b_k(Z_s)$. We see from (39.13) that the kth coordinate of Y_t is a local martingale under $\mathbb{P}_i^x$.

Now let $f(x) = x_k x_m$. A simple computation shows that $\mathcal{L}f(x) = a_{km}(x)$, hence $Y_t^k Y_t^m - \int_0^t a_{km}(Z_s)\, ds$ is a local martingale. We set

$$W_t = \int_0^t \sigma^{-1}(Z_s)\, dY_s.$$

The stochastic integral is finite since

$$\mathbb{E} \int_0^t \sum_{j=1}^d (\sigma^{-1})_{ij}(Z_s) \sum_{k=1}^d (\sigma^{-1})_{ik}(Z_s)\, d\langle Y^j, Y^k \rangle_s \tag{39.16}$$

$$= \mathbb{E} \int_0^t \sum_{i,k=1}^d (a^{-1})_{ik}(Z_s) a_{ik}(Z_s)\, ds = t < \infty.$$

Since Y_t is a local martingale, it follows that W_t is a local martingale, and a calculation similar to (39.16) shows that $W_t^k W_t^m - \delta_{km} t$ is also a martingale under $\mathbb{P}_i^x$. By Lévy's theorem (Exercise 12.4), W_t is a Brownian motion under both $\mathbb{P}_1^x$ and $\mathbb{P}_2^x$, and $(Z_t, W_t, \mathbb{P}_i^x)$ is a weak solution to (39.1). By the weak uniqueness hypothesis, the laws of Z_t under $\mathbb{P}_1^x$ and $\mathbb{P}_2^x$ agree, which is what we wanted to prove.

Exercise 39.1 asks you to prove that the existence of a weak solution to (39.1) is equivalent to the existence of a solution to the martingale problem. $\qquad\square$

If the σ_{ij} and b_i are Lipschitz functions, the solution to (39.1) is pathwise unique; see Exercise 24.5. By Proposition 25.2, weak existence and uniqueness hold, and then the martingale problem for $\mathcal{L}$ is well posed for every starting point.

A process that can be described in terms of a martingale problem (as well as other ways) is *super-Brownian motion*. Super-Brownian motion, also known as a *measure-valued branching diffusion process*, is a process whose state space is the set $\mathcal{M}$ of finite positive measures on $\mathbb{R}^d$. The intuitive picture is as follows. Given an initial finite measure μ as a starting point, let X_t^n be the process that starts with $[n\mu(R^d)]$ particles, each with mass $1/n$, each distributed according to $\mu(dx)/\mu(\mathbb{R}^d)$, where $[\cdot]$ denotes the integer part. Each particle moves as an independent Brownian motion for a time $1/n$, at which time each particle splits into two or dies, independently of the other particles. The particles that are now alive move as independent Brownian motions for time $1/n$, at which time each particle splits into two or dies, and so on. X_t^n is the measure that assigns mass $1/n$ at each point at which there is a particle alive at time t. We take the right-continuous version of X_t^n. It turns out that the sequence converges weakly with respect to the topology of $D[0, 1]$, but where the state space is the set of right-continuous functions with left limits taking values in $\mathcal{M}$ (rather than the set of real-valued functions) and the limit law can be characterized as the unique solution to a martingale problem. A solution to this martingale problem started at $\mu \in \mathcal{M}$ is a probability measure on the space of continuous processes taking values in $\mathcal{M}$ such that

(1) $\mathbb{P}(X_0 = \mu) = 1$;

(2) if $f \in C^\infty$ has compact support and we write $\nu(f)$ for $\int f \, d\nu$, then

$$M_t^f = X_t(f) - \int_0^t X_r(\tfrac{1}{2}\Delta f) \, dr$$

is a continuous martingale with quadratic variation process given by

$$\langle M_t^f \rangle = \int_0^t X_r(f^2) \, dr.$$

See Dawson (1993) and Perkins (2002) for more on these processes.

Exercises

39.1 Show that the existence of a weak solution to (39.1) is equivalent to the existence of a solution to the martingale problem for $\mathcal{L}$.

39.2 Suppose the a_{ij} are Lipschitz functions in x and the matrices $a(x)$ are positive definite, uniformly in x; see Exercise 25.4. Show that we can find matrices $\sigma(x)$ so that each σ_{ij} is a Lipschitz function of x and $a(x) = \sigma(x)\sigma^T(x)$ for each x.

39.3 If X is a solution to (39.1), give formulas for A_t and M_t in terms of σ and b, where M_t is a local martingale, A_t is a process whose paths are locally of bounded variation, and $|X_t| = M_t + A_t$.

39.4 Let $A \in (-1, \infty)$ and let X be a solution to (39.1), where all the b_i's are equal to 0, $a = \sigma\sigma^T$, and

$$a_{ij}(x) = \frac{\delta_{ij} + Ax_ix_j/|x|^2}{1 + A}$$

for $x \neq 0$, where δ_{ij} is equal to 1 if $i = j$ and 0 otherwise. Let $a(0)$ be the identity matrix.

  (1) Prove that the matrices $a(x)$ are uniformly elliptic.

  (2) Show that $|X_t|$ has the same law as a Bessel process of order

$$\frac{d + A}{1 + A}.$$

Conclude that if A is sufficiently close to -1, then X is transient, i.e, $\lim_{t \to \infty} |X_t| = \infty$, a.s., while if A is sufficiently large, there exist arbitrarily large times t such that $X_t = 0$.

39.5 Suppose for each $n \geq 1$, $a_{ij}^n(x)$ is symmetric in i and j, is continuous in x, and the matrix whose (i, j)th entry is $a_{ij}^n(x)$ is positive definite, uniformly in x and n. Let

$$\mathcal{L}^n f(x) = \sum_{i,j=1}^d a_{ij}^n(x) \frac{\partial^2 f}{\partial x_i \partial x_j}(x) \tag{39.17}$$

for $f \in C^2$. Suppose $a_{ij}^n(x)$ converges to $a_{ij}(x)$ uniformly in x as $n \to \infty$, and define $\mathcal{L}$ analogously to (39.17). Fix x_0 and let $\mathbb{P}_n$ be a solution to the martingale problem for $\mathcal{L}^n$ started at x_0.

  (1) Prove that $\mathbb{P}_n$ converges weakly to a solution $\mathbb{P}$ to the martingale problem for $\mathcal{L}$ started at x_0.

  (2) Prove that if the a_{ij} are continuously differentiable functions of x whose first partial derivatives are bounded, then there exists a solution to the martingale problem for $\mathcal{L}$ started at x_0.

  (3) Prove that if the a_{ij} are continuous functions of x, then there exists a solution to the martingale problem for $\mathcal{L}$ started at x_0.

39.6 Suppose X is a solution to $dX_t = \sigma(X_t)\,dW_t$, where W is a d-dimensional Brownian motion, $\sigma(x)$ is a $d \times d$ matrix-valued function that is bounded, and $\sigma^T\sigma$ is positive definite, uniformly in x. Prove the following estimate for the time to leave a ball: there exist constants c_1 and c_2 not depending on x_0 such that

$$c_1 r^2 \leq \mathbb{E}^{x_0} \tau_{B(x_0, r)} \leq c_2 r^2, \qquad r > 0,$$

where $\tau_{B(x_0, r)} = \inf\{> 0 : X_t \notin B(x_0, r)\}$.

Notes

See Bass (1997) for more information.

Solving partial differential equations

We will be concerned with giving probabilistic representations of the solutions to certain PDEs. Throughout we will be assuming that the given PDE has a solution, the solution is unique, and the solution is sufficiently smooth. We will consider Poisson's equation, the Dirichlet problem, the Cauchy problem (with an application to Brownian passage times), and Schrödinger's equation.

We let X_t be the solution to

$$dX_t = \sigma(X_t)\, dW_t + b(X_t)\, dt. \tag{40.1}$$

Here W is a d-dimensional Brownian motion, σ is a bounded Lipschitz continuous $d \times d$ matrix-valued function, b is a bounded Lipschitz continuous $d \times 1$ matrix-valued function, and X takes values in $\mathbb{R}^d$. We let $a = \sigma\sigma^T$ and we consider the operator on C^2 functions given by

$$\mathcal{L}f(x) = \tfrac{1}{2} \sum_{i,j=1}^{d} a_{ij}(x) \frac{\partial^2 f}{\partial x_i \partial x_j}(x) + \sum_{i=1}^{d} b_i(x) \frac{\partial f}{\partial x_i}(x). \tag{40.2}$$

We suppose the operator $\mathcal{L}$ is uniformly elliptic: there exists $\Lambda > 0$ such that

$$\sum_{i,j=1}^{d} a_{ij}(x) y_i y_j \geq \Lambda \sum_{i=1}^{d} y_i^2, \qquad y_1, \ldots, y_d \in \mathbb{R}^d.$$

In fact, the uniform ellipticity of $\mathcal{L}$ will be used only to guarantee that the exit times of bounded domains are finite, a.s.; see Exercise 40.1. For many non-uniformly elliptic operators, it is often the case that the finiteness of the exit times is known for other reasons, and the results then apply to equations involving these operators.

Let X_t^x be the solution to (40.1) when $X_0 = x$ and let $\mathbb{P}^x$ be the law of X_t^x. As in Chapter 39, we slightly abuse notation and say that $(X_t, \mathbb{P}^x)$ is a strong Markov process.

40.1 Poisson's equation

We consider first *Poisson's equation* in $\mathbb{R}^d$. Suppose $\lambda > 0$ and f is a C^1 function with compact support. Poisson's equation is

$$\mathcal{L}u(x) - \lambda u(x) = -f(x), \qquad x \in \mathbb{R}^d. \tag{40.3}$$

Theorem 40.1 *Suppose u is a C^2 solution to (40.3) such that u and its first and second partial derivatives are bounded. Then*

$$u(x) = \mathbb{E}^x \int_0^\infty e^{-\lambda t} f(X_t) \, dt.$$

Proof Let u be the solution to (40.3). By Theorem 39.3,

$$u(X_t) - u(X_0) = M_t + \int_0^t \mathcal{L}u(X_s) \, ds,$$

where M_t is a martingale. By the product formula,

$$e^{-\lambda t} u(X_t) - u(X_0) = \int_0^t e^{-\lambda s} dM_s + \int_0^t e^{-\lambda s} \mathcal{L}u(X_s) \, ds - \lambda \int_0^t e^{-\lambda s} u(X_s) \, ds.$$

Taking the expectation with respect to $\mathbb{P}^x$ and letting $t \to \infty$,

$$-u(x) = \mathbb{E}^x \int_0^\infty e^{-\lambda s} (\mathcal{L}u - \lambda u)(X_s) \, ds.$$

Since $\mathcal{L}u - \lambda u = -f$, the result follows.   □

Let us now let D be a nice bounded domain, e.g., a ball. Poisson's equation in D requires one to find a function u such that $\mathcal{L}u - \lambda u = -f$ in D and $u = 0$ on ∂D, where $f \in C^2(\overline{D})$ and $\lambda \geq 0$. Here we can allow λ to be equal to 0.

Theorem 40.2 *Suppose u is a solution to Poisson's equation in a bounded domain D that is C^2 in D and continuous on $\overline{D}$. Then*

$$u(x) = \mathbb{E}^x \int_0^{\tau_D} e^{-\lambda s} f(X_s) \, ds.$$

Proof The proof is nearly identical to that of the previous theorem. We already mentioned that $\tau_D < \infty$, a.s.; see Exercise 40.1. Let $S_n = \inf\{t : \text{dist}(X_t, \partial D) < 1/n\}$. By Theorem 39.3,

$$u(X_{t \wedge S_n}) - u(X_0) = \text{martingale} + \int_0^{t \wedge S_n} \mathcal{L}u(X_s) \, ds.$$

By the product formula,

$$\mathbb{E}^x e^{-\lambda(t \wedge S_n)} u(X_{t \wedge S_n}) - u(x) = \mathbb{E}^x \int_0^{t \wedge S_n} e^{-\lambda s} \mathcal{L}u(X_s) \, ds - \mathbb{E}^x \int_0^{t \wedge S_n} e^{-\lambda s} u(X_s) \, ds$$

$$= -\mathbb{E}^x \int_0^{t \wedge S_n} e^{-\lambda s} f(X_s) \, ds.$$

Now let $n \to \infty$ and then $t \to \infty$ and use the fact that u is zero on ∂D.   □

40.2 Dirichlet problem

Let D be a ball (or other nice bounded domain) and let us consider the solution to the *Dirichlet problem*: given a continuous function f on ∂D, find $u \in C(\overline{D})$ such that u is C^2 in D and

$$\mathcal{L}u = 0 \text{ in } D, \qquad u = f \text{ on } \partial D. \tag{40.4}$$

We considered the Dirichlet problem in the special case when $\mathcal{L}$ is the Laplacian in Section 21.4.

Theorem 40.3 *Suppose u is a solution to the Dirichlet problem specified by* (40.4). *Then u satisfies*

$$u(x) = \mathbb{E}^x f(X_{\tau_D}).$$

Proof As we mentioned above, $\tau_D < \infty$, a.s. Let $S_n = \inf\{t : \text{dist}\,(X_t, \partial D) < 1/n\}$. By Theorem 39.3,

$$u(X_{t \wedge S_n}) = u(X_0) + \text{martingale} + \int_0^{t \wedge S_n} \mathcal{L}u(X_s)\,ds.$$

Since $\mathcal{L}u = 0$ inside D, taking expectations shows

$$u(x) = \mathbb{E}^x u(X_{t \wedge S_n}).$$

We let $t \to \infty$ and then $n \to \infty$. By dominated convergence, we obtain $u(x) = \mathbb{E}^x u(X_{\tau_D})$. This is what we want since $u = f$ on ∂D. $\qquad\square$

If $v \in C^2$ and $\mathcal{L}v = 0$ in D, we say v is *$\mathcal{L}$-harmonic* in D.

40.3 Cauchy problem

The related parabolic partial differential equation

$$\frac{\partial u}{\partial t} = \mathcal{L}u$$

is often of interest. Here u is a function of $x \in \mathbb{R}^d$ and $t \in [0, \infty)$. When we write $\mathcal{L}u$, we mean

$$\mathcal{L}u(x, t) = \sum_{im j=1}^d a_{ij}(x) \frac{\partial^2 u}{\partial x_i \partial x_j}(x, t) + \sum_{i=1}^d b_i(x) \frac{\partial u}{\partial x_i}(x, t).$$

We will sometimes write u_t for $\partial u / \partial t$.

Suppose for simplicity that the function f is a continuous function with compact support. The *Cauchy problem* is to find u such that u is bounded, u is C^2 with bounded first and second partial derivatives in x, u is C^1 in t for $t > 0$, and

$$\begin{aligned}
u_t(x, t) &= \mathcal{L}u(x, t), & t > 0, x \in \mathbb{R}^d, \\
u(x, 0) &= f(x), & x \in \mathbb{R}^d.
\end{aligned} \tag{40.5}$$

Theorem 40.4 *Suppose there exists a solution to* (40.5) *that is C^2 in x and C^1 in t for $t > 0$. Then u satisfies*

$$u(x, t) = \mathbb{E}^x f(X_t).$$

Proof Fix t_0 and let $M_t = u(X_t, t_0 - t)$. Note

$$\frac{\partial}{\partial t} u(x, t_0 - t) = -u_t(x, t_0 - t).$$

Similarly to the proof of Theorem 39.3 (see Exercise 40.2) but using now the multivariate version of Itô's formula,

$$u(X_t, t_0 - t) = \text{ martingale } + \int_0^t \mathcal{L}u(X_s, t_0 - s)\, ds - \int_0^t u_t(X_s, t_0 - s)\, ds. \quad (40.6)$$

Since $u_t = \mathcal{L}u$, M_t is a martingale, and $\mathbb{E}^x M_0 = \mathbb{E}^x M_{t_0}$. On the one hand,

$$\mathbb{E}^x M_{t_0} = \mathbb{E}^x u(X_{t_0}, 0) = \mathbb{E}^x f(X_{t_0}),$$

while on the other hand,

$$\mathbb{E}^x M_0 = \mathbb{E}^x u(X_0, t_0) = u(x, t_0).$$

Since t_0 is arbitrary, the result follows. $\qquad\qquad\qquad\qquad\qquad\qquad\qquad\square$

A very similar proof allows one to represent the solution to the Cauchy problem in a bounded domain. Suppose $u(x, t)$ is C^2 in the x variable, C^1 in the t variable, and satisfies

$$\frac{\partial u}{\partial t}(x, t) = \mathcal{L}u(x, t)$$

for $(x, t) \in D \times (0, t_1]$, where D is a bounded domain in $\mathbb{R}^d$ and $t_1 > 0$. Suppose $u(x, 0) = f(x)$ and $u(x, t) = 0$ for all $x \in \partial D$. Exercise 40.3 asks you to show that in this case

$$u(x, t) = \mathbb{E}^x f(X_{t \wedge \tau_D}),$$

where again τ_D is the first exit time of X from the domain D.

The Cauchy problem has an application to the passage times of Brownian motion. Suppose we look at the equation

$$u_x(x, t) = \tfrac{1}{2} u_{xx}(x, t), \qquad 0 < x < b, \quad t > 0,$$

with

$$u(x, 0) = f(x) \text{ for all } x, \qquad u(0, t) = u(b, t) = 0 \text{ for all } t,$$

where f is a bounded function on $[0, b]$. This is a partial differential equation (the *heat equation*) that is sometimes solved in undergraduate classes; see, e.g., Boyce and DiPrima (2009), Section 10.5. Using a combination of the technique of separation of variables and Fourier series expansions, the solution can then be shown to be

$$u(x, t) = \int f(y) p^0(t, x, y)\, dy,$$

where

$$p^0(t, x, y) = \frac{2}{b} \sum_{n=1}^{\infty} e^{-n^2 \pi^2 t / 2b^2} \sin(n\pi x / b) \sin(n\pi y / b).$$

See also Knight (1981), p. 62. Since $u(x, t)$ is also equal to $\mathbb{E}^x f(X_{t \wedge \tau_D})$, where D is the interval $(0, b)$, then the $p^0(t, x, y)$ are the transition densities for Brownian motion killed on exiting $(0, b)$.

In particular, if we take f identically equal to 1 on $(0, b)$, we see that starting at x inside $(0, b)$, $\mathbb{P}^x(t < \tau_D)$ is asymptotically equal to $ce^{-\pi^2 t / 2b^2}$. If b is 2, this becomes $ce^{-\pi^2 t / 8}$.

Since the time for a Brownian motion started at 0 to leave $(-1, 1)$ is the same as the time for a Brownian motion started at 1 to leave $(0, 2)$, we obtain the estimate that was used in Exercise 7.2.

40.4 Schrödinger operators

Finally we look at what happens when one adds a potential term, that is, when one considers the operator

$$\mathcal{L}u(x) + q(x)u(x). \tag{40.7}$$

This is known as the *Schrödinger operator*, and $q(x)$ is known as the *potential*. Equations involving the operator in (40.7) are considerably simpler than the quantum mechanics Schrödinger equation because here all terms are real-valued.

If X_t is the diffusion corresponding to $\mathcal{L}$, then solutions to PDEs involving the operator in (40.7) can be expressed in terms of X_t by means of the *Feynman–Kac formula*. To illustrate, let D be a nice bounded domain, e.g., a ball, q a C^2 function on $\overline{D}$, and f a continuous function on ∂D; q^+ denotes the positive part of q.

Theorem 40.5 *Let D, q, f be as above. Let u be a C^2 function on $\overline{D}$ that agrees with f on ∂D and satisfies $\mathcal{L}u + qu = 0$ in D. If*

$$\mathbb{E}^x \exp \left(\int_0^{\tau_D} q^+(X_s)\, ds \right) < \infty,$$

then

$$u(x) = \mathbb{E}^x \left[f(X_{\tau_D}) e^{\int_0^{\tau_D} q(X_s)\, ds} \right]. \tag{40.8}$$

Proof Let $B_t = \int_0^{t \wedge \tau_D} q(X_s)\, ds$. By Itô's formula and the product formula,

$$e^{B(t \wedge \tau_D)} u(X_{t \wedge \tau_D}) = u(X_0) + \text{martingale} + \int_0^{t \wedge \tau_D} u(X_r) e^{B_r}\, dB_r + \int_0^{t \wedge \tau_D} e^{B_r} \mathcal{L}u(X_r)\, dr.$$

Taking the expectation with respect to $\mathbb{P}^x$ and using Proposition 39.3,

$$\mathbb{E}^x e^{B(t \wedge \tau_D)} u(X_{t \wedge \tau_D}) = u(x) + \mathbb{E}^x \int_0^{t \wedge \tau_D} e^{B_r} u(X_r) q(X_r)\, dr + \mathbb{E}^x \int_0^{t \wedge \tau_D} e^{B_r} \mathcal{L}u(X_r)\, dr.$$

Since $\mathcal{L}u + qu = 0$,

$$\mathbb{E}^x e^{B(t \wedge \tau_D)} u(X_{t \wedge \tau_D}) = u(x).$$

If we let $t \to \infty$ and use the exponential integrability of q^+, the result follows. $\square$

The existence of a solution to $\mathcal{L}u + qu = 0$ in D depends on the finiteness of $\mathbb{E}^x \exp(\int_0^{\tau_D} q^+(X_s)\, ds)$, an expression that is sometimes known as the *gauge*.

Even in one dimension with $D = (0, 1)$ and q a constant function, the gauge need not be finite. With $x = 1/2$, $\mathbb{P}^x(\tau_D > t)$ is asymptotically equal to $ce^{-\pi^2 t/2}$ as $t \to \infty$ by

Section 40.3. Hence

$$\mathbb{E}^x \exp\left(\int_0^{\tau_D} q \, ds\right) = \mathbb{E}^x e^{q\tau_D}$$

$$= \int_0^\infty q e^{qt} \mathbb{P}^x(\tau_D > t) \, dt;$$

this is infinite if $q \geq \pi^2/2$.

Exercises

40.1 This (lengthy) exercise is designed to guide you through a proof that solutions to (40.1) exit bounded sets in finite time, a.s.

(1) Suppose

$$X_t = W_t + \int_0^t a_s \, ds,$$

where W is a one-dimensional Brownian motion, and a_s is an adapted process bounded by K. Let $L > K > 0$ and $t_0 > 0$. Show that there exists $\varepsilon > 0$, depending only on L, K, and t_0 such that $\mathbb{P}(|X_{t_0}| > 3L) > \varepsilon$.

(2) Suppose $X_t = M_t + \int_0^t a_s \, ds$, where a_s is as in (1) and M is a continuous martingale with $K^{-1} \leq d\langle M \rangle_t/dt \leq K$, a.s. Use a time change argument to show that there exist $L, \varepsilon > 0$ such that

$$\mathbb{P}(\sup_{s \leq 1} |X_s| \leq L) \leq 1 - \varepsilon.$$

(3) If now X is a solution to (40.1), $a = \sigma\sigma^T$, and $\mathcal{L}$ given by (40.2) is uniformly elliptic, show by looking at the first coordinate of X that there exist L, ε such that

$$\mathbb{P}^x(\sup_{s \leq 1} |X_s| \leq L) \leq 1 - \varepsilon, \qquad x \in B(0, L).$$

(4) What you have proved in (3) can be rephrased as saying that if $(X_t, \mathbb{P}^x)$ is a strong Markov process that solves (40.1) for every starting point and $\tau = \inf\{t : X_t \notin B(0, L)\}$, then $\mathbb{P}^x(\tau > 1) \leq 1 - \varepsilon$, where ε does not depend on x. Now use the strong Markov property (cf. the proof of Proposition 21.2) to show $\mathbb{P}^x(\tau > k) \leq (1 - \varepsilon)^k$. Conclude that $\tau < \infty$, $\mathbb{P}^x$-a.s., for each starting point x.

40.2 Prove (40.6).

40.3 Let D be a ball in $\mathbb{R}^d$ and suppose u is the solution to the Cauchy problem in the domain $D \times [0, t_1]$ as described in Section 40.3. Show that $u(x, t) = \mathbb{E}^x f(X_{t \wedge \tau_D})$.

40.4 Suppose f is such that the solution u to

$$u_t(x, t) = \mathcal{L}u(x, t) + q(x), \qquad u(x, 0) = f(x),$$

is C^2 in x and t and X is the diffusion associated with $\mathcal{L}$. Prove that

$$u(x, t) = \mathbb{E}^x\left[f(X_t)e^{\int_0^t q(X_s) \, ds}\right].$$

40.5 Suppose $(X_t, \mathbb{P}^x)$ is a Brownian motion on $[0, b]$ with reflection at 0 and b. Find a series expansion for $p(t, x, y)$, the transition densities for X.

Hint: Imitate the argument for absorbing Brownian motion in Section 40.3, but now use the boundary conditions $u_x(0, t) = u_x(b, t) = 0$.

Notes

See Bass (1997) for more on the connection between probability and PDEs.

41

One-dimensional diffusions

Under very mild regularity conditions, every one-dimensional diffusion arises from first time-changing a one-dimensional Brownian motion and then making a transformation of the state space. We will prove this fact in this chapter.

41.1 Regularity

Throughout this chapter we suppose that we have a continuous process $(X_t, \mathbb{P}^x)$ defined on an interval I contained in $\mathbb{R}$. For almost all of the chapter, we suppose for simplicity that the interval is in fact all of $\mathbb{R}$. We further suppose that $(X_t, \mathbb{P}^x)$ is a strong Markov process with respect to a right-continuous filtration $\{\mathcal{F}_t\}$ such that each $\mathcal{F}_t$ contains all the sets that are $\mathbb{P}^x$-null for every x. We call such a process a *one-dimensional diffusion*.

Write

$$T_y = \inf\{t : X_t = y\}, \tag{41.1}$$

the first time the process X hits the point y. We will also assume that every point can be hit from every other point: for all x, y,

$$\mathbb{P}^x(T_y < \infty) = 1. \tag{41.2}$$

When (41.2) holds, we say the diffusion is *regular*.

For any interval J, define $\tau_J = \inf\{t : X_t \notin J\}$, the first time the process leaves J. When X_t is a Brownian motion, we know (Proposition 3.16) that the distribution of X_t upon exiting $[a, b]$ is

$$\mathbb{P}^x(X(\tau_{[a,b]}) = a) = \frac{b-x}{b-a}, \qquad \mathbb{P}^x(X(\tau_{[a,b]}) = b) = \frac{x-a}{b-a}. \tag{41.3}$$

We say that a regular diffusion X_t is on *natural scale* if (41.3) holds for every interval $[a, b]$. We also say a regular diffusion X defined on an interval I properly contained in $\mathbb{R}$ is on natural scale if (41.3) holds whenever $[a, b] \subset I$ and $x \in (a, b)$.

If X_t is regular, then the process started at x must leave x immediately. That is, if $S = \inf\{t > 0 : X_t \neq x\}$, then $\mathbb{P}^x(S = 0) = 1$. To see this, let $\varepsilon > 0$ and $U = \inf\{t : |X_t - x| \geq \varepsilon\}$. By the regularity of X, $\mathbb{E}^x e^{-U} > 0$. Observe that $U = S + U \circ \theta_S$, where θ_t is the shift operator. By the strong Markov property at time S,

$$\mathbb{E}^x e^{-U} = \mathbb{E}^x[e^{-S}\mathbb{E}^x[e^{-U} \circ \theta_S \mid \mathcal{F}_S]] = \mathbb{E}^x[e^{-S}\mathbb{E}^{X_S}[e^{-U}]] = \mathbb{E}^x[e^{-S}\mathbb{E}^x e^{-U}],$$

since $X_S = x$ by the continuity of the paths of X. The only way this can happen is if $\mathbb{E}^x e^{-S} = 1$, which implies $S = 0$, $\mathbb{P}^x$-a.s.

41.2 Scale functions

We will show that given a regular diffusion, there exists a *scale function* that is continuous, strictly increasing, and such that $s(X_t)$ is on natural scale.

We first look at a special case, when the diffusion is given as the solution to an SDE. Suppose X_t is given as the solution to

$$dX_t = \sigma(X_t)\,dW_t + b(X_t)\,dt, \qquad (41.4)$$

where we assume σ and b are real-valued, continuous and bounded above and σ is bounded below by a positive constant. Let $a(x) = \sigma^2(x)$. In this case we can give a formula for the scale function.

Theorem 41.1 *The scale function $s(x)$ is the solution to*

$$\tfrac{1}{2}a(x)s''(x) + b(x)s'(x) = 0,$$

and for some constants c_1, c_2, and x_0 is given by

$$s(x) = c_1 + c_2 \int_{x_0}^{x} \exp\left(-\int_{x_0}^{y} \frac{2b(w)}{a(w)}\,dw\right) dy. \qquad (41.5)$$

Proof To solve the differential equation, we write

$$\frac{s''(x)}{s'(x)} = -2\frac{b(x)}{a(x)},$$

or $(\log s'(x))' = -2b(x)/a(x)$, from which (41.5) follows. Since we assumed that σ and b are continuous, $s(x)$ given by (41.5) is C^2. Since σ is bounded below by a positive constant and b and σ are bounded, s given by (41.5) is strictly increasing. Applying Itô's formula,

$$s(X_t) - s(X_0) = \int_0^t s'(X_r)\sigma(X_r)\,dW_r \qquad (41.6)$$

because

$$\int_0^t \left[\tfrac{1}{2}s''(X_r)\sigma(X_r)^2 + s'(X_r)b(X_r)\right] dr = 0.$$

This implies that $s(X_t) - s(X_0)$ is a martingale, hence a time change of Brownian motion. Therefore the exit probabilities of $s(X_t)$ for an interval $[a, b]$ are the same as those of a Brownian motion, namely, those given by (41.3). □

From (41.6), if $Y_t = s(X_t)$, then

$$dY_t = (s'\sigma)(s^{-1}(Y_t))\,dW_t. \qquad (41.7)$$

Now we show there exists a scale function for general regular diffusions on $\mathbb{R}$. Let J be an interval $[a, b]$. We define

$$p(x) = p_J(x) = \mathbb{P}^x(X_{\tau_J} = b). \qquad (41.8)$$

Proposition 41.2 *Let $J = [a, b]$ be a finite interval. Then $p(X_{t \wedge \tau_J})$ is a regular diffusion on $[0, 1]$ on natural scale.*

Proof   First we show that p is increasing. To get to the point b starting from x, the process must first hit every point between x and b because X has continuous paths. If $a < x < y < b$, by the strong Markov property at time T_y, $p(x) \leq p(y)$. We claim there is a positive probability that the process starting from x hits a before y, that is,

$$\mathbb{P}^x(T_a < T_y) > 0. \tag{41.9}$$

If (41.9) did not hold, then the process started at x must hit y before hitting a, then by the continuity of paths must hit x before hitting a, and once the process is again at x, it again hits y with probability one before a and so on. Therefore the process never hits a, a contradiction to the regularity; Exercise 41.2 asks you to make this argument precise. Therefore (41.9) does hold, and by the strong Markov property at T_y,

$$p(x) = \mathbb{P}^x(T_y < T_a)p(y).$$

Since $\mathbb{P}^x(T_y < T_a) = 1 - \mathbb{P}^x(T_a < T_y)$ is strictly less than 1, p is strictly increasing.

Next we show that p is continuous. We show continuity from the right; the proof of continuity from the left is similar. Suppose $x_n \downarrow x$. The process X_t has continuous paths, so given ε we can find t small enough so that $\mathbb{P}^x(T_a < t) < \varepsilon$. By the Blumenthal 0–1 law (Proposition 20.8), $\mathbb{P}^x(T_{(x,b]} = 0)$ is zero or one, where $T_{(x,b]}$ is the first time the process hits the interval $(x, b]$. If it is zero, the process immediately moves to the left from x, a.s., and by the strong Markov property at T_x, it never hits b, a contradiction. The probability must therefore be one. Thus by the continuity of paths, for n large enough, $\mathbb{P}^x(T_{x_n} < t) \geq 1 - \varepsilon$. Hence with probability at least $1 - 2\varepsilon$, X_t hits x_n before a. Since

$$p(x) = \mathbb{P}^x(T_{x_n} < T_a)\, p(x_n) \geq (1 - 2\varepsilon)p(x_n)$$

and ε is arbitrary, we see that $p(x) \geq \liminf_{n\to\infty} p(x_n)$. Since p is strictly increasing, $p(x_n)$ decreases, and therefore $p(x) = \lim p(x_n)$.

Finally, we show $p(X_t)$ is on natural scale. Let $[e, f] \subset (0, 1)$ and let

$$r(y) = \mathbb{P}^y(X_t \text{ hits } p^{-1}(f) \text{ before hitting } p^{-1}(e)).$$

Note that

$$\mathbb{P}^x(p(X_t) \text{ hits } f \text{ before } e) = \mathbb{P}^{p^{-1}(x)}(X_t \text{ hits } p^{-1}(f) \text{ before } p^{-1}(e))$$
$$= r(p^{-1}(x)). \tag{41.10}$$

For $y \in [p^{-1}(a), p^{-1}(b)]$, the strong Markov property tells us that

$$p(y) = \mathbb{P}^y\big(X_t \text{ hits } p^{-1}(f) \text{ before } p^{-1}(e)\big)p\big(p^{-1}(f)\big) \tag{41.11}$$
$$+ \mathbb{P}^y\big(X_t \text{ hits } p^{-1}(e) \text{ before } p^{-1}(f)\big)p\big(p^{-1}(e)\big)$$
$$= r(y)f + (1 - r(y))e.$$

Solving for $r(y)$, we obtain $r(y) = (p(y) - e)/(f - e)$. Substituting in (41.10),

$$\mathbb{P}^x(p(X_t) \text{ hits } f \text{ before } e) = r(p^{-1}(x)) = (p(p^{-1}(x)) - e)/(f - e)$$
$$= (x - e)/(f - e),$$

which is the formula we wanted. $\square$

Note that if X_t is on natural scale, then so is $c_1 X_t + c_2$ for any constants $c_1 > 0, c_2 \in \mathbb{R}$.

Theorem 41.3 *There exists a continuous strictly increasing function s such that $s(X_t)$ is on natural scale on $s(\mathbb{R})$.*

Proof Let J_n be closed intervals increasing up to $\mathbb{R}$. Pick two points in J_1; label them a and b with $a < b$. Choose A_n and B_n so that if $s_n(x) = A_n p_{J_n}(x) + B_n$, then $s_n(a) = 0$ and $s_n(b) = 1$.

We will show that if $n \geq m$, then $s_n = s_m$ on J_m. Once we have that, we can set $s(x) = s_n(x)$ on J_n, and the theorem will be proved.

Suppose $J_m = [e, f]$. By Proposition 41.2, both $s_m(X_t)$ and $s_n(X_t)$ are on natural scale. For all $x \in J_m$,

$$\frac{s_m(x) - s_m(e)}{s_m(f) - s_m(e)} = \mathbb{P}^{s_m(x)}\big(s_m(X_t) \text{ hits } s_m(f) \text{ before } s_m(e)\big)$$
$$= \mathbb{P}^x(X_t \text{ hits } f \text{ before } e).$$

We have a similar equation with s_m replaced everywhere by s_n. It follows that

$$\frac{s_m(x) - s_m(e)}{s_m(f) - s_m(e)} = \frac{s_n(x) - s_n(e)}{s_n(f) - s_n(e)}$$

for all x, which implies that $s_n(x) = C s_m(x) + D$ for some constants C and D. Since s_n and s_m are equal at both $x = a$ and $x = b$, then C must be 1 and D must be 0. $\qquad\square$

41.3 Speed measures

Suppose that $(\mathbb{P}^x, X_t)$ is a regular diffusion on $\mathbb{R}$ on natural scale. For each finite interval (a, b), define

$$G_{ab}(x, y) = \begin{cases} \frac{2(x-a)(b-y)}{b-a}, & a < x \leq y < b, \\ \frac{2(y-a)(b-x)}{b-a}, & a < y \leq x < b, \end{cases} \tag{41.12}$$

and set $G_{ab}(x, y) = 0$ if x or y is not in (a, b). A measure $m(dx)$ is the *speed measure* for the diffusion $(X_t, \mathbb{P}^x)$ if

$$\mathbb{E}^x \tau_{(a,b)} = \int G_{ab}(x, y) \, m(dy) \tag{41.13}$$

for each finite interval (a, b) and each $x \in (a, b)$. As (41.13) indicates, the speed measure governs how quickly the diffusion moves through intervals.

As an example, let us argue that the speed measure for Brownian motion is a Lebesgue measure. By Proposition 3.16, if $(X_t, \mathbb{P}^x)$ is a Brownian motion,

$$\mathbb{E}^x \tau_{(a,b)} = (x - a)(b - x).$$

On the other hand, a calculation shows that

$$\int G_{ab}(x, y) \, dy = (x - a)(b - x).$$

Since

$$\mathbb{E}^x \tau_{(a,b)} = \int G_{ab}(x,y)\, dy$$

and Brownian motion is on natural scale, we see that the speed measure $m(dy)$ of Brownian motion is equal to a Lebesgue measure.

We will show that

(1) a regular diffusion on natural scale has one and only one speed measure,

(2) the law of the diffusion is determined by the speed measure, and

(3) there exists a diffusion with a given speed measure.

We first want to show that any speed measure must satisfy $0 < m(a,b) < \infty$ for any finite interval $[a,b]$. To start we have the following lemma.

Lemma 41.4 *If $[a,b]$ is a finite interval, then $\sup_x \mathbb{E}^x \tau_{(a,b)}^k < \infty$ for each positive integer k.*

Proof Pick $y \in (a,b)$. Since X_t is a regular diffusion, $\mathbb{P}^y(T_a < \infty) = 1$, and hence there exists t_0 such that $\mathbb{P}^y(T_a > t_0) < 1/2$. Similarly, taking t_0 larger if necessary, $\mathbb{P}^y(T_b > t_0) \le 1/2$. If $a < x \le y$, then

$$\mathbb{P}^x(\tau_{(a,b)} > t_0) \le \mathbb{P}^x(T_a > t_0) \le \mathbb{P}^y(T_a > t_0) \le 1/2,$$

and similarly, $\mathbb{P}^x(\tau_{(a,b)} > t_0) \le 1/2$ if $y \le x < b$. By the Markov property,

$$\mathbb{P}^x(\tau_{(a,b)} > (n+1)t_0) = \mathbb{E}^x[\mathbb{P}^{X(nt_0)}(\tau_{(a,b)} > t_0); \tau_{(a,b)} > nt_0]$$
$$\le \tfrac{1}{2}\mathbb{P}^x(\tau_{(a,b)} > nt_0),$$

and by induction, $\mathbb{P}^x(\tau_{(a,b)} > nt_0) \le 2^{-n}$. The lemma is now immediate. □

Lemma 41.5 *If $(X_t, \mathbb{P}^x)$ has a speed measure m and $[a,b]$ is a non-empty finite interval, then $0 < m(a,b) < \infty$.*

Proof If $m(a,b) = 0$, then for $x \in (a,b)$, we have

$$\mathbb{E}^x \tau_{(a,b)} = \int G_{ab}(x,y)\, m(dy) = 0,$$

which implies $\tau_{(a,b)} = 0$, $\mathbb{P}^x$-a.s., a contradiction to the continuity of the paths of X_t.

Next we show the finiteness of $m(a,b)$. Pick (e,f) such that $[a,b] \subset (e,f)$. There exists a constant c such that for $x, y \in (a,b)$, $G_{ef}(x,y)$ is bounded below by c, so

$$m(a,b) \le c^{-1} \int_e^f G_{ef}(x,y)\, m(dy) = c^{-1}\mathbb{E}^x \tau_{(e,f)} < \infty.$$

This completes the proof. □

Theorem 41.6 *A regular diffusion on natural scale on $\mathbb{R}$ has one and only one speed measure.*

Proof First let $I = (e,f)$ be a finite open interval. For $n > 1$ let $x_i = e + i(f-e)/2^n$, $i = 0, 1, 2, \ldots, 2^n$. Let $\mathcal{D}_n = \{x_i : 0 \le i \le 2^n\}$. Let

$$m_n(dx) = 2^n \sum_{i=1}^{2^n-1} B(x_i)\delta_{x_i}, \tag{41.14}$$

where $B(x_i) = \mathbb{E}^{x_i} \tau_{(x_{i-1}, x_{i+1})}$. We first want to show that if $[a, b]$ is a subinterval of I with a, b each in $\mathcal{D}_n$ and x is also in $\mathcal{D}_n$, then

$$\mathbb{E}^x \tau_{(a,b)} = \int G_{ab}(x, y) \, m_n(dy). \tag{41.15}$$

To see this, let $S_0 = 0$ and $S_{j+1} = \inf\{t > S_j : |X_t - X_{S_j}| = 2^{-n}\} \wedge \tau_{(a,b)}$. The S_j's are the successive times that X moves 2^{-n}, up until the time of leaving (a, b). Because X is on natural scale, $X_{S_{j+1}}$ is equal to $X_{S_j} + 2^{-n}$ with probability $\frac{1}{2}$ and equal to $X_{S_j} - 2^{-n}$ with probability $\frac{1}{2}$, until leaving (a, b). Therefore X_{S_j} is a simple symmetric random walk on the lattice with step size 2^{-n}, stopped on leaving (a, b).

Let $J(x_i) = (x_i - 2^{-n}, x_i + 2^{-n})$ for $x_i \neq a, b$. Let $J(a) = J(b) = \emptyset$. By repeated use of the strong Markov property,

$$\mathbb{E}^x \tau_{(a,b)} = \sum_{j=0}^{\infty} \mathbb{E}^x (S_{j+1} - S_j)$$

$$= \mathbb{E}^x \sum_{j=0}^{\infty} \mathbb{E}^{X(S_j)}[\tau_{J(X_0)}] = \mathbb{E}^x \sum_{j=0}^{\infty} B(X_{S_j}) 1_{(a,b)}(X_{S_j}).$$

Let $N_i = \sum_{j=0}^{\infty} 1_{\{x_i\}}(X_{S_j})$, the number of visits to x_i before exiting (a, b). Then

$$\mathbb{E}^x \tau_{(a,b)} = \mathbb{E}^x \sum_{j=0}^{\infty} B(X_{S_j}) 1_{(a,b)}(X_{S_j}) \tag{41.16}$$

$$= \mathbb{E}^x \sum_{j=0}^{\infty} \sum_{i=1}^{2^n-1} B(X_{S_j}) 1_{\{x_i\}}(X_{S_j})$$

$$= \mathbb{E}^x \sum_{i=1}^{2^n-1} B(x_i) N_i.$$

$\mathbb{E}^x N_i$ must equal 0 when $x = a$ or $x = b$ and satisfies the equation

$$\mathbb{E}^{x_j} N_i = \delta_{ij} + \tfrac{1}{2}(\mathbb{E}^{x_{j+1}} N_i + \mathbb{E}^{x_{j-1}} N_i), \tag{41.17}$$

where δ_{ij} is 1 if $i = j$ and 0 otherwise. This holds because for $j \neq i$, the process goes left or right, each with probability $1/2$, while if $j = 1$, we add one to N_i before going left or right. The function $x \to \mathbb{E}^x N_i$ is hence piecewise linear on (a, x_i) and on (x_i, b). Some algebra shows that we must have

$$\mathbb{E}^x N_i = 2^n G_{ab}(x, x_i). \tag{41.18}$$

Combining (41.16) and (41.18),

$$\mathbb{E}^x \tau_{(a,b)} = \sum_{i=1}^{2^n-1} B(x_i) 2^n G_{ab}(x, x_i)$$

$$= \int G_{ab}(x, y) \, m_n(dy),$$

which is (41.15).

Using (41.15) and the same proof as that of Lemma 41.5, $m_n(a, b)$ is bounded above by a constant independent of n. By a diagonalization procedure, there exists a subsequence n_k such that m_{n_k} converges weakly to m, where m is a measure that is finite on every subinterval (a, b) such that $[a, b] \subset I$. By the continuity of G_{ab},

$$\mathbb{E}^x \tau_{(a,b)} = \int G_{ab}(x, y) \, m(dy) \tag{41.19}$$

whenever a, b, and x are in $\mathcal{D}_n$ for some n.

We now remove this last restriction. If a, b are not of this form, take a_r, b_r to be in $\cup_n \mathcal{D}_n$ such that $(a_r, b_r) \uparrow (a, b)$. Then $\tau_{(a_r,b_r)} \uparrow \tau_{(a,b)}$, and by the continuity of G_{ab} in a, b, x, and y, we have (41.19) for all a and b. Take $y_r \uparrow x$, $z_r \downarrow x$ such that y_r and z_r are in $\mathcal{D}_n$ for some n. By the strong Markov property,

$$\mathbb{E}^x \tau_{(a,b)} = \mathbb{E}^x \tau_{(y_r,z_r)} + \mathbb{E}^{y_r} \tau_{(a,b)} \mathbb{P}^x (X_{\tau_{(y_r,z_r)}} = y_r)$$
$$+ \mathbb{E}^{z_r} \tau_{(a,b)} \mathbb{P}^x (X_{\tau_{(y_r,z_r)}} = z_r).$$

By the continuity of G_{ab} in x, and the fact that $\mathbb{E}^x \tau_{(y'_r,z_r)} \to 0$ as $r \to \infty$, we obtain (41.19) for all x.

We leave the uniqueness as Exercise 41.3.

Finally, let I_k be finite subintervals increasing up to $\mathbb{R}$. Let m_k be the speed measure for X_t on the interval I_k. By the uniqueness result, m_k agrees with m_ℓ on I_ℓ if $I_\ell \subset I_k$. Setting m to be the measure whose restriction to I_k is m_k gives us the speed measure. $\square$

The speed measure completely characterizes occupation times.

Corollary 41.7 *Suppose X_t is a diffusion on natural scale on $\mathbb{R}$. If f is bounded and measurable, for each $a < b$,*

$$\mathbb{E}^x \int_0^{\tau_{(a,b)}} f(X_s) \, ds = \int G_{ab}(x, y) f(y) \, m(dy). \tag{41.20}$$

Proof   Suppose that f is continuous and bounded on $[a, b]$. Let x_i, S_j, $B(x_i)$, N_i, and m_n be as in the proof of Theorem 41.6. Let

$$\varepsilon_n = \sup\{|f(x) - f(y)| : |x - y| \le 2^{-n}\}.$$

Note that if $(x - a)/(b - a)$ is a multiple of 2^{-n},

$$\mathbb{E}^x \int_0^{\tau_{(a,b)}} f(X_s) \, ds = \sum_{j=0}^{\infty} \mathbb{E}^x \int_{S_j}^{S_{j+1}} f(X_s) \, ds \tag{41.21}$$

and

$$\mathbb{E}^x \sum_{j=0}^{\infty} f(X_{S_j})(S_{j+1} - S_j) = \mathbb{E}^x \sum_{j=0}^{\infty} f(X_{S_j}) 1_{(a,b)}(X_{S_j}) \mathbb{E}^{X_{S_j}} S_1 \tag{41.22}$$

$$= \sum_{i=1}^{2^n - 1} f(x_i) B(x_i) 1_{(a,b)}(x_i) \mathbb{E}^x N_i.$$

Moreover, the right-hand side of (41.21) differs from the left-hand side of (41.22) by at most $\varepsilon_n \mathbb{E}^x \tau_{(a,b)}$. By (41.18) the right-hand side of (41.22) is equal to

$$\sum_{i=1}^{2^n-1} 2^n f(x_i) B(x_i) 1_{(a,b)}(x_i) G_{ab}(x, x_i) = \int G_{ab}(x, x_i) f(x_i) m_n(dx).$$

By weak convergence along an appropriate subsequence, the left-hand side and the right-hand side of (41.20) differ by at most $\limsup_n \varepsilon_n \mathbb{E}^x \tau_{(a,b)}$, which is zero. A limit argument then shows that (41.20) holds for all $x \in [a, b]$, and another limit argument shows that (41.20) holds for all bounded f. $\square$

41.4 The uniqueness theorem

We next turn to showing that the speed measure characterizes the law of a diffusion.

Theorem 41.8 *If* $(X_t, \mathbb{P}_i^x)$, $i = 1, 2$, *are two diffusions on natural scale with the same speed measure* m, *then* $\mathbb{P}_1^x = \mathbb{P}_2^x$.

Proof We start by letting $(a, b) \subset \mathbb{R}$ and considering the operator

$$R_\lambda^i f(x) = \mathbb{E}_i^x \int_0^{\tau_{(a,b)}} e^{-\lambda t} f(X_t) \, dt, \qquad \lambda \ge 0, \tag{41.23}$$

for $i = 1, 2$. We show first that $R_0^1 = R_0^2$, that is, that

$$\mathbb{E}_1^x \int_0^{\tau_{(a,b)}} f(X_t) \, dt = \mathbb{E}_2^x \int_0^{\tau_{(a,b)}} f(X_t) \, dt$$

if f is bounded and Borel measurable. This is easy, because by Corollary 41.7, both sides are equal to

$$\int_a^b G_{ab}(x, y) \, m(dy).$$

Since $(\widehat{X}_t, \mathbb{P}_i^x)$ is a Markov process, where $\widehat{X}$ is the process X killed on exiting (a, b), the resolvent equation (37.2) holds. We have

$$\|R_0^i f\|_\infty \le \|f\|_\infty \sup_x \mathbb{E}^x \tau_{(a,b)}$$

$$= \|f\|_\infty \sup_x \int G_{ab}(x, y) \, m(dy)$$

$$\le c\|f\|_\infty m(a, b) < \infty.$$

Since $\|R_0^i\|_\infty < \infty$, we can let μ go to zero in (37.2). We can repeat the proof of Corollary 37.3 with $\lambda = 0$ to see that

$$R_\mu^i f = R_0^i f + \sum_{i=1}^\infty (-\mu)^j (R_0^i)^{j+1} f$$

provided $\mu < \|R_0^i\|_\infty$. We can then use Remark 37.4 to obtain that $R_\lambda^1 = R_\lambda^2$ for all $\lambda > 0$. We now take open intervals I_n increasing up to $\mathbb{R}$. Applying the above to I_n and letting $n \to \infty$,

we have

$$\mathbb{E}^x_1 \int_0^\infty e{-}\lambda t f(X_t)\, dt = \mathbb{E}^x_2 \int_0^\infty e{-}\lambda t f(X_t)\, dt$$

whenever f is bounded and Borel measurable and $x \in \mathbb{R}$.

Suppose f is continuous as well. By the uniqueness of the Laplace transform, we see that $\mathbb{E}^x_1 f(X_t) = \mathbb{E}^x_2 f(X_t)$ for almost every t, and since both terms are continuous in t, this equality holds for all t. By a limit argument, this equality holds for all bounded and Borel measurable f. Therefore the one-dimensional distributions of X under $\mathbb{P}^x_1$ and $\mathbb{P}^x_2$ agree.

If $s < t$ and f and g are bounded and Borel measurable,

$$
\begin{aligned}
\mathbb{E}^x_1[f(X_s)g(X_t)] &= \mathbb{E}^x_1[f(X_s)P^1_{t-s}g(X_s)] = \mathbb{E}^x_1[f(X_s)P^2_{t-s}g(X_s)] \\
&= \mathbb{E}^x_1[(fP^2_{t-s}g)(X_s)] = \mathbb{E}^x_2[(fP^2_{t-s}g)(X_s)] \\
&= \mathbb{E}^x_2[f(X_s)P^2_{t-s}g(X_s)] = \mathbb{E}^x_2[f(X_s)g(X_t)].
\end{aligned}
$$

Here P^i_{t-s} is the semigroup for $(X_t, \mathbb{P}^x_i)$; since the one-dimensional distributions agree, $P^1_{t-s} = P^2_{t-s}$. We have thus shown the two-dimensional distributions of X under $\mathbb{P}^x_1$ and $\mathbb{P}^x_2$ agree. Continuing, we see that all the finite-dimensional distributions under $\mathbb{P}^x_1$ and $\mathbb{P}^x_2$ agree. By the continuity of the paths of X and Theorem 2.6, that is enough to show equality of $\mathbb{P}^x_1$ and $\mathbb{P}^x_2$. $\square$

41.5 Time change

We now want to show that if m is a measure such that $0 < m(a, b) < \infty$ for all intervals $[a, b]$, then there exists a regular diffusion on natural scale on $\mathbb{R}$ having m as a speed measure. If $m(dx)$ had a density, say $m(dx) = r(x)\, dx$, we would proceed as follows. Let W_t be a one-dimensional Brownian motion and let

$$A_t = \int_0^t r(W_s)\, ds, \qquad B_t = \inf\{u : A_t > u\}, \qquad X_t = W_{B_t}.$$

In other words, we let X_t be a certain time change of Brownian motion. In general, where $m(dx)$ does not have a density, we make use of the local times L^x_t of Brownian motion; see Chapter 14.

Let

$$A_t = \int L^x_t\, m(dx), \qquad B_t = \inf\{u : A_u > t\}, \qquad X_t = W_{B_t}. \qquad (41.24)$$

Theorem 41.9 *Let $(W_t, \mathbb{P}^x)$ be a Brownian motion and m a measure on $\mathbb{R}$ such that $0 < m(a, b) < \infty$ for every finite interval (a, b). Then, under $\mathbb{P}^x$, X_t as defined by (41.24) is a regular diffusion on natural scale with speed measure m.*

Proof First we show that X_t is a continuous process. Fix ω. If we choose $a < \inf_{s \le t} W_s$ and $b > \sup_{s \le t} W_s$, then

$$A_t = \int L^x_t\, m(dx) = \int L^x_t 1_{[a,b]}(x)\, m(dx)$$

since L^x_t increases only for those times s when $W_s = x$. By the continuity of L^x_t and dominated convergence, we conclude that $A_t(\omega)$ is continuous at time t. Next we show that A_t is strictly increasing. Fix ω. If $s < u$, pick $t \in (s, u)$. Set $x = W_t$. Because the support of the measure

dL_t^x is the set $\{r : W_r = x\}$, then $L_u^x - L_s^x > 0$. By the continuity of local times, $L_u^y - L_s^y > 0$ for all y in a neighborhood of x, say $(x - \delta, x + \delta)$. Since $m(x - \delta, x + \delta) > 0$, then $A_u - A_s > 0$. Hence A_t is strictly increasing. This and the continuity of A_t imply that B_t is continuous, and therefore X_t is continuous.

Next we show that X_t is a regular diffusion on natural scale. By monotone convergence and the fact that $L_t^x \to \infty$, a.s., for each x, $A_t \uparrow \infty$, hence $B_t \uparrow \infty$, so $\tau_{(a,b)}^X < \infty$, $\mathbb{P}^x$-a.s., where $\tau_{(a,b)}^X$ denotes the exit time of (a, b) by X_t and $\tau_{(a,b)}^W$ denotes the corresponding exit time of W_t. Moreover,

$$\mathbb{P}^x(X(\tau_{(a,b)}^X) = b) = \mathbb{P}^x(W(\tau_{(a,b)}^W) = b) = \frac{x - a}{b - a},$$

since X_t is a time change of W_t.

To verify the strong Markov property, we repeat the argument of Section 22.3. Let $\mathcal{F}_t' = \mathcal{F}_{B_t}$. Then if T is a stopping time for $\mathcal{F}_t'$, we have

$$\mathbb{E}^x[f(X_{T+t}) \mid \mathcal{F}_T'] = \mathbb{E}^x[f(W(B_{T+t})) \mid \mathcal{F}_{B_T}].$$

B_T can be seen to be a stopping time for $\mathcal{F}_t$ and $B_{T+t} = B_t \circ \theta_{B_T}$ where θ_t are the shift operators, so this is

$$\mathbb{E}^x \mathbb{E}^{W(B_T)} f(W_{B_t}) = \mathbb{E}^x \mathbb{E}^{X_T} f(X_t).$$

As in Section 20.3, this suffices to show that X_t is a strong Markov process.

It remains to determine the speed measure of X_t. Fix (a, b) and write τ_X for $\tau_{(a,b)}^X$ and τ_W for $\tau_{(a,b)}^W$. We have

$$\begin{aligned}
\mathbb{E}^x \tau_X &= \mathbb{E}^x \int_0^\infty 1_{(a,b)}(X_{s \wedge \tau_X}) \, ds \\
&= \mathbb{E}^x \int_0^\infty 1_{(a,b)}(W_{B_{s \wedge \tau_X}}) \, ds \\
&= \mathbb{E}^x \int_0^\infty 1_{(a,b)}(W_{t \wedge \tau_W}) \, dA_t \\
&= \mathbb{E}^x \int \int_0^\infty 1_{(a,b)}(W_{t \wedge \tau_W}) L_t^y \, m(dy) \\
&= \mathbb{E}^x \int \int_0^{\tau_W} L_t^y \, m(dy) = \int \mathbb{E}^x L_{\tau_W}^y \, m(dy).
\end{aligned}$$

We also have

$$\mathbb{E}^x L_{\tau_W}^y = \mathbb{E}^x |W_{\tau_W} - y| - |x - y|$$

by (14.5). This is equal to

$$\begin{aligned}
|a - y| \mathbb{P}^x(W_{\tau_W} = a) &+ |b - y| \mathbb{P}^x(W_{\tau_W} = b) - |x - y| \\
&= |a - y| \frac{b - x}{b - a} + |b - y| \frac{x - a}{b - a} - |x - y| = G_{ab}(x, y).
\end{aligned}$$

We thus have

$$\mathbb{E}^x \tau_X = \int G_{ab}(x, y) \, m(dy),$$

as required. $\qquad\square$

As a corollary to the proof, we see that a regular diffusion on natural scale is a local martingale, since it is a time change of Brownian motion.

41.6 Examples

Let us calculate the scale function and the speed measure for some examples of diffusions. First we need to connect the speed measure with the coefficients of an SDE.

Let us look at the solutions to the SDE (41.4), but now suppose b is identically zero, or $dX_t = \sigma(X_t)\,dW_t$. We again set $a(x) = \sigma(x)^2$.

Theorem 41.10 *Suppose $c_1 < \sigma(x) < c_2$ for all x and σ is continuous. The speed measure of X_t is given by*

$$m(dx) = \frac{1}{a(x)}\,dx.$$

Proof Since $dX_t = \sigma(X_t)\,dW_t$, then $\langle X \rangle_t = \int_0^t a(X_s)\,ds$. To obtain a Brownian motion $\overline{W}_t$ by time-changing the martingale X_t, we must time-change by the inverse of $\langle X \rangle_t$. On the other hand, from Theorem 41.9, X_t is the time-change of a Brownian motion by B_t, where B_t is given by (41.24). Hence

$$B_t = \langle X \rangle_t = \int_0^t a(X_s)\,ds.$$

The inverse of B_t, namely, A_t, must then satisfy

$$\frac{dA_t}{dt} = \frac{1}{a(X_{A_t})} = \frac{1}{a(W_t)},$$

or

$$A_t = \int_0^t \frac{1}{a(W_s)}\,ds = \int L_t^y \frac{1}{a(y)}\,dy$$

for all t, using Theorem 14.4. However, $A_t = \int L_t^y\, m(dy)$ by (41.24). Hence

$$\int L_t^y \frac{1}{a(y)}\,dy = \int L_t^y\, m(dy).$$

We know $\mathbb{E}^x L_{\tau_{(c,d)}}^y = G_{cd}(x,y)$. Therefore

$$\int G_{cd}(x,y)\,m(dy) = \int \mathbb{E}^x L_{\tau_{(c,d)}}^y\, m(dy)$$

$$= \int \mathbb{E}^x L_{\tau_{(c,d)}}^y \frac{1}{a(y)}\,dy$$

$$= \int G_{cd}(x,y) \frac{1}{a(y)}\,dy$$

for all c, d, and x, which implies $m(dy) = (1/a(y))\,dy$. $\square$

Now we can look at some examples and do calculations.

Brownian motion with constant drift. This process is the solution to the SDE $dX_t = dW_t + b\,dt$. From Theorem 41.1, $s(x) = \exp(-2bx)$ is the scale function. If $Y_t = s(X_t)$, then

$(s'\sigma)(s^{-1}(y)) = -2by$, or Y_t corresponds to the operator $2b^2y^2 f''$, and the speed measure is $(4b^2y^2)^{-1} dx$.

Bessel processes. The process is only defined on the state space $[0, \infty)$ instead of all of $\mathbb{R}$ and there is a boundary condition at 0. We ignore this here and consider a Bessel process of order ν up until the first hit of 0. Then X solves the SDE

$$dX_t = dW_t + \frac{\nu - 1}{2X_t} dt.$$

If $\nu \neq 2$, a calculation using Theorem 41.1 shows that $s(x) = x^{2-\nu}$. Then $Y_t = s(X_t)$ satisfies

$$dY_t = (2 - \nu)Y_t^{(1-\nu)/(2-\nu)} dW_t,$$

and the speed measure is

$$m(dx) = (2 - \nu)^{-2} x^{(2\nu-2)/(2-\nu)} dx, \qquad x > 0.$$

Exercises

41.1 In the proof of Proposition 41.2 we used the strong Markov property numerous times. Write out carefully in terms of shift operators and conditional expectations how the strong Markov property is applied in each case.

41.2 Give a rigorous proof of (41.9).

41.3 Show that if

$$\int G_{ab}(x, y) m_1(dy) = \int G_{ab}(x, y) m_2(dy)$$

for all x, a, and b, then $m_1 = m_2$.

41.4 Show that if X is a Bessel process of order 2, then the scale function is given by $s(x) = \log x$, $Y_t = s(X_t)$ satisfies $dY_t = e^{-Y_t} dW_t$, and the speed measure is $m(dx) = e^{2x} dx$.

41.5 Suppose X is a regular diffusion whose state space is $\mathbb{R}$. Prove that X is on natural scale if and only if

$$\mathbb{P}^{(a+b)/2}(T_a < T_b) = \tfrac{1}{2}$$

whenever $a < b$.

41.6 Let $a > 0$ and let $m(dx) = dx + a\,\delta_0(dx)$, where δ_0 is the point mass at 0. Let $(X_t, \mathbb{P}^x)$ be the diffusion on the line on natural scale whose speed measure is given by m. Show that under $\mathbb{P}^0$,

$$\int_0^t 1_{\{0\}}(X_s)\, ds > 0$$

with probability one for each $t > 0$. Prove that for each $t > 0$, $Z_t = \{t : X_t = 0\}$ contains no intervals. Thus the zero set of the process X spends an amount of time at 0 that has positive Lebesgue measure, but the zero set contains no intervals.

41.7 Define

$$m_a(dx) = \begin{cases} dx, & x \geq 0, \\ a\,dx, & x < 0. \end{cases}$$

Let $(X_t, \mathbb{P}_a^x)$ be the diffusion on natural scale on the line whose speed measure is given by m_a. Suppose $x > 0$.

Prove that if $a \to \infty$, then $\mathbb{P}_a^x$ converges weakly to the law of Brownian motion absorbed (i.e., killed) at 0, started at x. What do you think happens when $a \to 0$?

Notes

We have considered diffusions on $\mathbb{R}$ but most of what we discussed goes through for diffusions whose state space is an interval properly contained in $\mathbb{R}$. In this case, one must specify what the process does when it hits the boundary. Being absorbed (i.e., killed) or reflected are two options, but much more complicated behavior is possible. See Itô and McKean (1965) and Knight (1981) for the complete story.

Lévy processes

A *Lévy process* is a process with stationary and independent increments whose paths are right continuous with left limits. Having *stationary increments* means that the law of $X_t - X_s$ is the same as the law of $X_{t-s} - X_0$ whenever $s < t$. Saying that X has *independent increments* means that $X_t - X_s$ is independent of $\sigma(X_r; r \leq s)$ whenever $s < t$.

We want to examine the structure of Lévy processes. We have three examples already: the Poisson process, Brownian motion, and the deterministic process $X_t = t$. It turns out that all Lévy processes can be built up out of these building blocks. We will show how to construct Lévy processes and give a representation of an arbitrary Lévy process.

Recall that we use $X_{t-} = \lim_{s<t, s\to t} X_s$ and $\Delta X_t = X_t - X_{t-}$.

42.1 Examples

Let us begin by looking at some simple Lévy processes. Let P_t^j, $j = 1, \ldots, J$, be a sequence of independent Poisson processes with parameters λ_j, respectively. Each P_t^j is a Lévy process and the formula for the characteristic function of a Poisson random variable (see Section A.13) shows that the characteristic function of P_t^j is

$$\mathbb{E}\, e^{iuP_t^j} = \exp(t\lambda_j(e^{iu} - 1)).$$

Therefore the characteristic function of $a_j P_t^j$ is

$$\mathbb{E}\, e^{iua_j P_t^j} = \exp(t\lambda_j(e^{iua_j} - 1))$$

and the characteristic function of $a_j P_t^j - a_j \lambda_j t$ is

$$\mathbb{E}\, e^{iua_j P_j^t - a)j\lambda_j t} = \exp(t\lambda_j(e^{iua_j} - 1 - iua_j)).$$

If we let m_j be the measure on $\mathbb{R}$ defined by $m_j(dx) = \lambda_j \delta_{a_j}(dx)$, where $\delta_{a_j}(dx)$ is point mass at a_j, then the characteristic function for $a_j P_t^j$ can be written as

$$\exp\left(t \int_{\mathbb{R}} [e^{iux} - 1]\, m_j(dx)\right) \tag{42.1}$$

and the one for $a_j P_t^j - a_j \lambda_j t$ as

$$\exp\left(t \int_{\mathbb{R}} [e^{iux} - 1 - iux]\, m_j(dx)\right). \tag{42.2}$$

Now let

$$X_t = \sum_{j=1}^{J} a_j P_t^j.$$

It is clear that the paths of X_t are right continuous with left limits, and the fact that X has stationary and independent increments follows from the corresponding property of the P^j's. Moreover, the characteristic function of a sum of independent random variables is the product of the characteristic functions, so the characteristic function of X_t is given by

$$\mathbb{E}\,e^{iuX_t} = \exp\left(t \int_{\mathbb{R}} [e^{iux} - 1]\,m(dx)\right) \tag{42.3}$$

with $m(dx) = \sum_{j=1}^{J} \lambda_j \delta_{a_j}(dx)$.

The process $Y_t = X_t - t\sum_{j=1}^{J} a_j \lambda_j$ is also a Lévy process and its characteristic function is

$$\mathbb{E}\,e^{iuY_t} = \exp\left(t \int_{\mathbb{R}} [e^{iux} - 1 - iux]\,m(dx)\right), \tag{42.4}$$

again with $m(dx) = \sum_{j=1}^{J} \lambda_j \delta_{a_j}(dx)$.

Remark 42.1 Recall from Proposition A.50 that if φ is the characteristic function of a random variable Z, then $\varphi'(0) = i\mathbb{E}\,Z$ and $\varphi''(0) = -\mathbb{E}\,Z^2$. If Y_t is as in the paragraph above, then clearly $\mathbb{E}\,Y_t = 0$, and calculating the second derivative of $\mathbb{E}\,e^{iuY_t}$ at 0, we obtain

$$\mathbb{E}\,Y_t^2 = t \int x^2\,m(dx).$$

The following lemma is a restatement of Corollary 4.3.

Lemma 42.2 *If X_t is a Lévy process and T is a finite stopping time, then $X_{T+t} - X_T$ is a Lévy process with the same law as $X_t - X_0$ and independent of $\mathcal{F}_T$.*

42.2 Construction of Lévy processes

A process X has *bounded jumps* if there exists a real number $K > 0$ such that $\sup_t |\Delta X_t| \le K$, a.s.

Lemma 42.3 *If X_t is a Lévy process with bounded jumps and with $X_0 = 0$, then X_t has moments of all orders, that is, $\mathbb{E}\,|X_t|^p < \infty$ for all positive integers p.*

Proof Suppose the jumps of X_t are bounded in absolute value by K. Since X_t is right continuous with left limits, there exists $M > K$ such that $\mathbb{P}(\sup_{s \le t} |X_s| \ge 2M) \le 1/2$.

Let $T_1 = \inf\{t : |X_t| \geq M\}$ and $T_{i+1} = \inf\{t > T_i : |X_t - X_{T_i}| > M\}$. For $s < T_1$, $|X_s| \leq M$, and then $|X_{T_1}| \leq |X_{T_1-}| + |\Delta X_{T_1}| \leq M + K \leq 2M$. We have

$$\mathbb{P}(\sup_{s \leq t} |X_s| \geq 2(i+1)M) \leq \mathbb{P}(T_{i+1} \leq t) \leq \mathbb{P}(T_i \leq t, T_{i+1} - T_i \leq t)$$

$$= \mathbb{P}(\sup_{s \leq t} |X_{T_i+s} - X_{T_i}| \geq 2M, T_i \leq t)$$

$$= \mathbb{P}(\sup_{s \leq t} |X_s| \geq 2M)\mathbb{P}(T_i \leq t)$$

$$\leq \tfrac{1}{2}\mathbb{P}(T_i \leq t),$$

using Lemma 42.2 in the last equality. By induction, $\mathbb{P}(\sup_{s \leq t} |X_s| \geq 2iM) \leq 2^{-i}$, and the lemma now follows immediately. □

A key lemma is the following.

Lemma 42.4 *Suppose I is a finite interval of the form (a, b), $[a, b)$, $(a, b]$, or $[a, b]$ with $a > 0$ and m is a finite measure on $\mathbb{R}$ giving no mass to I^c. Then there exists a Lévy process X_t satisfying* (42.3).

Proof First let us consider the case where $I = [a, b)$. We approximate m by a discrete measure. If $n \geq 1$, let $z_j = a + j(b-a)/n$, $j = 0, \ldots, n-1$, and let

$$m_n(dx) = \sum_{j=0}^{n-1} m([z_j, z_{j+1}))\delta_{z_j}(dx),$$

where δ_{z_j} is the point mass at z_j. The measures m_n converge weakly to m as $n \to \infty$ in the sense that

$$\int f(x) m_n(dx) \to \int f(x) dx$$

whenever f is a bounded continuous function on $\mathbb{R}$. For each n, let $P_t^{n,j}$, $j = 0, \ldots, n-1$, be independent Poisson processes with parameters $m([z_j, z_{j+1}))$ and let

$$X_t^n = \sum_{j=0}^{n-1} z_j P_t^{n,j}.$$

Then X^n is a Lévy process with jumps bounded by b.

By Lemma 42.2, if T_n is a stopping time for X^n, $\varepsilon > 0$, and $\delta > 0$, then

$$\mathbb{P}(|X_{T_n+\delta}^n - X_{T_n}^n| > \varepsilon) = \mathbb{P}(|X_\delta^n| > \varepsilon) \leq \mathbb{P}(X_\delta^n \neq 0) \tag{42.5}$$

$$\leq \mathbb{P}\left(\sum_{j=0}^{n-1} P_\delta^{n,j} \neq 0\right).$$

Since the sum of independent Poisson processes is a Poisson process, then $\sum_{j=0}^{n-1} P_t^{n,j}$ is a Poisson process with parameter

$$\sum_{j=0}^{n-1} m([z_j, z_{j+1})) = m(I).$$

The last line of (42.5) is then bounded by

$$1 - e^{-\delta m(I)} \le \delta m(I),$$

which tends to zero uniformly in n as $\delta \to 0$. Note $X_0^n = 0$, a.s. We can therefore apply the Aldous criterion (Theorem 34.8) to see that the X^n are tight with respect to weak convergence on the space $D[0, t_0)$ for any t_0.

Any subsequential weak limit X will have paths that are right continuous with left limits. For any continuous bounded function f on $\mathbb{R}$,

$$\mathbb{E} f(X_t^n - X_s^n) = \mathbb{E} f(X_{t-s}^n - X_0^n).$$

Passing to the limit along an appropriate subsequence,

$$\mathbb{E} f(X_t - X_s) = \mathbb{E} f(X_{t-s} - X_0).$$

Since f is an arbitrary bounded continuous function, we see that the laws of $X_t - X_s$ and $X_{t-s} - X_0$ are the same. Similarly we prove the increments are independent.

Since $x \to e^{iux}$ is a bounded continuous function and m_n converges weakly to m, starting with

$$\mathbb{E} \exp(iuX_t^n) = \exp \left(t \int [e^{iux} - 1] m_n(dx) \right),$$

and passing to the limit, we obtain that the characteristic function of X under $\mathbb{P}$ is given by (42.3).

If now the interval I contains the point b, we follow the above proof, except we let $P_t^{n,n-1}$ be a Poisson random variable with parameter $m([z_{n-1}, b])$. Similarly, if I does not contain the point a, we change $P_t^{n,0}$ to be a Poisson random variable with parameter $m((a, z_1))$. With these changes, the proof works for intervals I, whether or not they contain either of their endpoints. $\square$

Remark 42.5 If X is the Lévy process constructed in Lemma 42.4, then $Y_t = X_t - \mathbb{E} X_t$ will be a Lévy process satisfying (42.4).

Here is the main theorem of this section.

Theorem 42.6 *Suppose m is a measure on $\mathbb{R}$ with $m(\{0\}) = 0$ and*

$$\int (1 \wedge x^2) m(dx) < \infty.$$

Suppose $b \in \mathbb{R}$ and $\sigma \ge 0$. There exists a Lévy process X_t such that

$$\mathbb{E} e^{iuX_t} = \exp \left(t \left\{ iub - \sigma^2 u^2 / 2 + \int_{\mathbb{R}} [e^{iux} - 1 - iux 1_{(|x| \le 1)}] m(dx) \right\} \right). \tag{42.6}$$

The above equation is called the *Lévy–Khintchine formula*. The measure m is called the *Lévy measure*. If we let

$$m(dx) = \frac{1 + x^2}{x^2} m'(dx)$$

and

$$b = b' + \int_{(|x|\leq 1)} \frac{x^3}{1+x^2} m(dx) - \int_{(|x|>1)} \frac{x}{1+x^2} m(dx),$$

then we can also write

$$\mathbb{E}\, e^{iuX_t} = \exp\left(t\left\{iub' - \sigma^2 u^2/2 + \int_{\mathbb{R}} \left[e^{iux} - 1 - \frac{iux}{1+x^2}\right]\frac{1+x^2}{x^2} m'(dx)\right\}\right).$$

Both expressions for the Lévy–Khintchine formula are in common use.

Proof Let $m(dx)$ be a measure supported on $(0, 1]$ with $\int x^2 m(dx) < \infty$. Let $m_n(dx)$ be the measure m restricted to $(2^{-n}, 2^{-n+1}]$. Let Y_t^n be independent Lévy processes whose characteristic functions are given by (42.4) with m replaced by m_n; see Remark 42.5. Note $\mathbb{E}\, Y_t^n = 0$ for all n by Remark 42.1. By the independence of the Y^n's, if $M < N$,

$$\mathbb{E}\left(\sum_{n=M}^{N} Y_t^n\right)^2 = \sum_{n=M}^{N} \mathbb{E}\, (Y_t^n)^2 = \sum_{n=M}^{N} t \int x^2 m_n(dx) = t \int_{2^{-N}}^{2^{-M}} x^2 m(dx).$$

By our assumption on m, this goes to zero as $M, N \to \infty$, and we conclude that $\sum_{n=0}^{N} Y_t^n$ converges in L^2 for each t. Call the limit Y_t. It is routine to check that Y_t has independent and stationary increments. Each Y_t^n has independent increments and is mean zero, so

$$\mathbb{E}\, [Y_t^n - Y_s^n \mid \mathcal{F}_s] = \mathbb{E}\, [Y_t^n - Y_s^n] = 0,$$

or Y^n is a martingale. By Doob's inequalities and the L^2 convergence,

$$\mathbb{E} \sup_{s\leq t} \left|\sum_{n=M}^{N} Y_s^n\right|^2 \to 0$$

as $M, N \to \infty$, and hence there exists a subsequence M_k such that $\sum_{n=1}^{M_k} Y_s^n$ converges uniformly over $s \leq t$, a.s. Therefore the limit Y_t will have paths that are right continuous with left limits.

If m is a measure supported in $(1, \infty)$ with $m(\mathbb{R}) < \infty$, we do a similar procedure starting with Lévy processes whose characteristic functions are of the form (42.3). We let $m_n(dx)$ be the restriction of m to $(2^n, 2^{n+1}]$, let X_t^n be independent Lévy processes corresponding to m_n, and form $X_t = \sum_{n=0}^{\infty} X_t^n$. Since $m(\mathbb{R}) < \infty$, for each t_0, the number of times t less than t_0 at which any one of the X_t^n jumps is finite. This shows X_t has paths that are right continuous with left limits, and it is easy to then see that X_t is a Lévy process.

Finally, suppose $\int x^2 \wedge 1 \, m(dx) < \infty$. Let X_t^1, X_t^2 be Lévy processes with characteristic functions given by (42.3) with m replaced by the restriction of m to $(1, \infty)$ and $(-\infty, -1)$, respectively, let X_t^3, X_t^4 be Lévy processes with characteristic functions given by (42.4) with m replaced by the restriction of m to $(0, 1]$ and $[-1, 0)$, respectively, let $X_t^5 = bt$, and let X_t^6 be σ times a Brownian motion. Suppose the X^i's are all independent. Then their sum will be a Lévy process whose characteristic function is given by (42.6). □

A key step in the construction was the centering of the Poisson processes to get Lévy processes with characteristic functions given by (42.4). Without the centering one is forced to work only with characteristic functions given by (42.3).

42.3 Representation of Lévy processes

We now work toward showing that every Lévy process has a characteristic function of the form given by (42.6).

Lemma 42.7 *If X_t is a Lévy process and A is a Borel subset of $\mathbb{R}$ that is a positive distance from 0, then*

$$N_t(A) = \sum_{s \leq t} 1_A(\Delta X_s)$$

is a Poisson process.

Saying that A is a positive distance from 0 means that $\inf\{|x| : x \in A\} > 0$.

Proof   Since X_t has paths that are right continuous with left limits and A is a positive distance from 0, then there can only be finitely many jumps of X that lie in A in any finite time interval, and so $N_t(A)$ is finite and has paths that are right continuous with left limits. It follows from the fact that X_t has stationary and independent increments that $N_t(A)$ also has stationary and independent increments. We now apply Proposition 5.4.   □

Theorem 42.8 *Let X_t be a Lévy process with $X_0 = 0$ and let $A_1, \ldots, A_n$ be disjoint bounded Borel subsets of $(0, \infty)$, each a finite distance from 0. Set*

$$N_t(A_k) = \sum_{s \leq t} 1_{A_k}(\Delta X_s)$$

and

$$Y_t = X_t - \sum_{k=1}^{n} N_t(A_k).$$

Then the processes $N_t(A_1), \ldots, N_t(A_n)$, and Y_t are mutually independent.

Proof   Define $\lambda(A) = \mathbb{E} N_1(A)$. The previous lemma shows that if $\lambda(A) < \infty$, then $N_t(A)$ is a Poisson process, and clearly its parameter is $\lambda(A)$. The result now follows from Theorem 18.3.   □

Here is the representation theorem for Lévy processes.

Theorem 42.9 *Suppose X_t is a Lévy process with $X_0 = 0$. Then there exists a measure m on $\mathbb{R} - \{0\}$ with*

$$\int (1 \wedge x^2) \, m(dx) < \infty$$

and real numbers b and σ such that the characteristic function of X_t is given by (42.6).

Proof   Define $m(A) = \mathbb{E} N_1(A)$ if A is a bounded Borel subset of $(0, \infty)$ that is a positive distance from 0. Since $N_1(\cup_{k=1}^{\infty} A_k) = \sum_{k=1}^{\infty} N_1(A_k)$ if the A_k are pairwise disjoint and each is a positive distance from 0, we see that m is a measure on $[a, b]$ for each $0 < a < b < \infty$, and m extends uniquely to a measure on $(0, \infty)$.

First we want to show that $\sum_{s\leq t} \Delta X_s 1_{(\Delta X_s > 1)}$ is a Lévy process with characteristic function

$$\exp\left(t\int_1^\infty [e^{iux} - 1]\, m(dx)\right).$$

Since the characteristic function of the sum of independent random variables is equal to the product of the characteristic functions, it suffices to suppose $0 < a < b$ and to show that

$$\mathbb{E}\, e^{iuZ_t} = \exp\left(t\int_{(a,b]} [e^{iux} - 1]\, m(dx)\right),$$

where

$$Z_t = \sum_{s\leq t} \Delta X_s 1_{(a,b]}(\Delta X_s).$$

Let $n > 1$ and $z_j = a + j(b-a)/n$. By Lemma 42.7, $N_t((z_j, z_{j+1}])$ is a Poisson process with parameter

$$\ell_j = \mathbb{E}\, N_1((z_{j-1}, z_j]) = m((z_j, z_{j+1}]).$$

Thus $\sum_{j=0}^{n-1} z_j N_t((z_j, z_{j+1}])$ has characteristic function

$$\prod_{j=0}^{n-1} \exp(t\ell_j(e^{iuz_j} - 1)) = \exp\left(t\sum_{j=0}^{n-1}(e^{iuz_j} - 1)\ell_j\right),$$

which is equal to

$$\exp\left(t\int (e^{iux} - 1)\, m_n(dx)\right), \tag{42.7}$$

where $m_n(dx) = \sum_{j=0}^{n-1} \ell_j \delta_{z_j}(dx)$. Since Z_t^n converges to Z_t as $n \to \infty$, passing to the limit shows that Z_t has a characteristic function of the form (42.6).

Next we show that $m(1,\infty) < \infty$. (We write $m(1,\infty)$ instead of $m((1,\infty))$ for esthetic reasons.) If not, $m(1,K) \to \infty$ as $K \to \infty$. Then for each fixed L and each fixed t,

$$\limsup_{K\to\infty} \mathbb{P}(N_t(1,K) \leq L) = \limsup_{K\to\infty} \sum_{j=0}^L e^{-tm(1,K)}\frac{m(1,K)^j}{j!} = 0.$$

This implies that $N_t(1,\infty) = \infty$ for each t. However, this contradicts the fact that X_t has paths that are right continuous with left limits.

We define m on $(-\infty, 0)$ similarly.

We now look at

$$Y_t = X_t - \sum_{s\leq t} \Delta X_s 1_{(|\Delta X_s|>1)}.$$

This is again a Lévy process, and we need to examine its structure. This process has bounded jumps, hence has moments of all orders. By subtracting $c_1 t$ for an appropriate constant c_1, we may suppose Y_t has mean 0. Let $I_1, I_2, \ldots$ be an ordering of the intervals $\{[2^{-(m+1)}, 2^{-m}), (-2^{-m}, -2^{-(m+1)}] : m \geq 0\}$. Let

$$\tilde{X}_t^k = \sum_{s\leq t} \Delta X_s 1_{(\Delta X_s \in I_k)}$$

and let $X_t^k = \widetilde{X}_t^k - \mathbb{E}\,\widetilde{X}_t^k$. By Corollary 18.3 and the fact that all the X^k have mean zero,

$$\sum_{k=1}^{\infty} \mathbb{E}\,(X_t^k)^2 \leq \mathbb{E}\left[\left(Y_t - \sum_{k=1}^{\infty} X_t^k\right)^2\right] + \mathbb{E}\left[\left(\sum_{k=1}^{\infty} X_t^k\right)^2\right] = \mathbb{E}\,(Y_t)^2 < \infty.$$

Hence

$$\mathbb{E}\left[\sum_{k=M}^{N} X_t^k\right]^2 = \sum_{k=M}^{N} \mathbb{E}\,(X_t^k)^2$$

tends to zero as $M, N \to \infty$, and thus $X_t - \sum_{k=1}^{N} X_t^k$ converges in L^2. The limit, X_t^c, say, will be a Lévy process independent of all the X_t^k. Moreover, X^c has no jumps, i.e., it is continuous. Since all the X^k have mean zero, then $\mathbb{E}\,X_t^c = 0$. By the independence of the increments,

$$\mathbb{E}\,[X_t^c - X_s^c \mid \mathcal{F}_s] = \mathbb{E}\,[X_t^c - X_s^c] = 0,$$

and we see X^c is a continuous martingale. Using the stationarity and independence of the increments,

$$\mathbb{E}\,[(X_{s+t}^c)^2] = \mathbb{E}\,[(X_s^c)^2] + 2\mathbb{E}\,[X_s^c(X_{s+t}^c - X_s^c)] + \mathbb{E}\,[(X_{s+t}^c - X_s^c)^2]$$
$$= \mathbb{E}\,[(X_s^c)^2] + \mathbb{E}\,[(X_t^c)^2],$$

which implies that there exists a constant c_2 such that $\mathbb{E}\,(X_t^c)^2 = c_2 t$. We then have

$$\mathbb{E}\,[(X_t^c)^2 - c_2 t \mid \mathcal{F}_s] = (X_s^c)^2 - c_2 s + \mathbb{E}\,[(X_t^c - X_s^c)^2 \mid \mathcal{F}_s] - c_2(t-s)$$
$$= (X_s^c)^2 - c_2 s + \mathbb{E}\,[(X_t^c - X_s^c)^2] - c_2(t-s)$$
$$= (X_s^c)^2 - c_2 s.$$

The quadratic variation process of X^c is therefore $c_2 t$, and by Lévy's theorem (Theorem 12.1), $X_t^c/\sqrt{c_2}$ is a constant multiple of Brownian motion.

To complete the proof, it remains to show that $\int_{-1}^{1} x^2\, m(dx) < \infty$. But by Remark 42.1,

$$\int_{I_k} x^2\, m(dx) = \mathbb{E}\,(X_1^k)^2,$$

and we have seen that

$$\sum_k \mathbb{E}\,(X_1^k)^2 \leq \mathbb{E}\,Y_1^2 < \infty.$$

Combining gives the finiteness of $\int_{-1}^{1} x^2\, m(dx)$. $\square$

Exercises

42.1 Let $\alpha \in (0, 2)$ and let X be a Lévy process where $b = \sigma = 0$ in the Lévy–Khintchine formula and the Lévy measure is $m(dx) = c|x|^{-1-\alpha}\, dx$. Show that if $a > 0$ and $Y_t = a^{1/\alpha} X_{at}$, then Y has the same law as X. The process X is known as a *symmetric stable process of index* α.

42.2 Suppose $W_t = (W_t^1, W_t^2)$ is a two-dimensional Brownian motion started at 0. Let $\tau_s = \inf\{t > 0 : W_t^1 > s\}$. Prove that $W_{\tau_s}^2$ is a Lévy process and determine the Lévy measure.
   Hint: Use scaling to make a guess.

42.3 Let W be a one-dimensional Brownian motion and let L^0 be the local time at 0. Let T_t be the inverse of L^0, that is, $T_t = \inf\{s : L_s^0 \geq t\}$. Show T_t is a Lévy process and determine the Lévy measure.

     Hint: Use scaling to get started.

42.4 Let W_t be a one-dimensional Brownian motion, L_t^y the local time at level y, and T_t the inverse local time at 0, that is, $T_t = \inf\{s : L_s^0 \geq t\}$. Let $x > 0$ be fixed. Prove that $L_{T_t}^x$ is a Lévy process.

42.5 Let X be a Lévy process with Lévy measure m. Prove that if A and B are disjoint closed sets, then

$$\mathbb{E}^x \sum_{s \leq t} 1_A(X_{s-}) 1_B(X_s) = \mathbb{E}^x \int_0^t 1_A(X_s) m(B - X_s)\, ds$$

for each x, where $B - y = \{z - y : z \in B\}$. This is the *Lévy system formula* in the case of Lévy processes. There is an analogous formula for Hunt processes.

42.6 A *stable subordinator* X of order $\alpha \in (0, 1)$ is a Lévy process whose characteristic function is given by (42.6), where $b = \sigma^2 = 0$ and $m(dx) = c 1_{(x>0)} |x|^{-\alpha-1}\, dx$. Suppose X is a stable subordinator of index α and W is a Brownian motion. Show that, up to a deterministic time change, the process $Z_t = W_{X_t}$ is a symmetric stable process of index 2α.

     Hint: Start by using scaling.

42.7 Let Z_t be a symmetric stable process of order $\alpha \in (0, 2)$. Show that if $\varepsilon > 0$, then

$$\lim_{t \to \infty} \frac{|Z_t|}{t^{\alpha + \varepsilon}} = 0, \qquad \text{a.s.}$$

Appendix A

Basic probability

This appendix covers the facts from basic probability that we will need. The presentation here is not precisely what I use when I teach such a course. For example, in a course I prove the strong law of large numbers without using martingales, I present the inversion theorem for characteristic functions, I make use of Lévy's continuity theorem, and so on. Nevertheless, proofs of all the facts from probability needed in the main part of the text are given.

A.1 First notions

A *probability* or *probability measure* is a measure whose total mass is one. Instead of denoting a measure space by $(X, \mathcal{A}, \mu)$, probabilists use $(\Omega, \mathcal{F}, \mathbb{P})$. Here Ω is a set, $\mathcal{F}$ is called a *σ-field* (which is the same thing as a σ-algebra), and $\mathbb{P}$ is a measure with $\mathbb{P}(\Omega) = 1$. Elements of $\mathcal{F}$ are called *events*. A typical element of Ω is denoted ω.

Instead of saying a property occurs almost everywhere, we talk about properties occurring *almost surely*, written *a.s.* Real-valued measurable functions from Ω to $\mathbb{R}$ are called *random variables* and are usually denoted by X or Y or other capital letters.

Integration (in the sense of Lebesgue) with respect to $\mathbb{P}$ is called *expectation* or *expected value*, and we write $\mathbb{E} X$ for $\int X \, d\mathbb{P}$. The notation $\mathbb{E}[X; A]$ is often used for $\int_A X \, d\mathbb{P}$.

The random variable 1_A is the function that is one if $\omega \in A$ and zero otherwise. It is called the *indicator* of A (the name "characteristic function" in probability refers to the Fourier transform). Events such as $\{\omega : X(\omega) > a\}$ are almost always abbreviated by $(X > a)$ or $\{X > a\}$.

Given a random variable X, we can define a probability on the Borel σ-field of $\mathbb{R}$ by

$$\mathbb{P}_X(A) = \mathbb{P}(X \in A), \qquad A \subset \mathbb{R}. \tag{A.1}$$

The probability $\mathbb{P}_X$ is called the *law* of X or the *distribution* of X. We define $F_X : \mathbb{R} \to [0, 1]$ by

$$F_X(x) = \mathbb{P}_X((-\infty, x]) = \mathbb{P}(X \le x). \tag{A.2}$$

The function F_X is called the *distribution function* of X.

Proposition A.1 *The distribution function F_X of a random variable X satisfies:*

(1) F_X is increasing;

(2) F_X is right continuous with left limits;

(3) $\lim_{x \to \infty} F_X(x) = 1$ and $\lim_{x \to -\infty} F_X(x) = 0$.

Proof We prove the right continuity of F_X and leave the rest of the proof to the reader. If $x_n \downarrow x$, then $(X \le x_n) \downarrow (X \le x)$, and so $\mathbb{P}(X \le x_n) \downarrow \mathbb{P}(X \le x)$ since $\mathbb{P}$ is a finite measure. $\square$

Note that if $x_n \uparrow x$, then $(X \le x_n) \uparrow (X < x)$, and so $F_X(x_n) \uparrow \mathbb{P}(X < x)$.

Any function $F : \mathbb{R} \to [0, 1]$ satisfying (1)–(3) of Proposition 1.1 is called a distribution function, whether or not it comes from a random variable.

Proposition A.2 *Suppose F is a distribution function. There exists a random variable X such that $F = F_X$.*

Proof Let $\Omega = [0, 1]$, $\mathcal{F}$ the Borel σ-field, and $\mathbb{P}$ a Lebesgue measure. Define $X(\omega) = \sup\{x : F(x) < \omega\}$. It is routine to check that $F_X = F$. $\square$

In the above proof, essentially $X = F^{-1}$. However F may have jumps or be constant over some intervals, so some care is needed in defining X.

Certain distributions or laws are very common. We list some of them.

(1) *Bernoulli*. A random variable is Bernoulli if $\mathbb{P}(X = 1) = p$, $\mathbb{P}(X = 0) = 1 - p$ for some $p \in [0, 1]$.

(2) *Binomial*. This is defined by $\mathbb{P}(X = k) = \binom{n}{k} p^k (1 - p)^{n-k}$, where n is a positive integer, $0 \le k \le n$, $p \in [0, 1]$, and $\binom{n}{k} = \frac{n!}{k!(n-k)!}$.

(3) *Point mass at a*. Here $\mathbb{P}(X = a) = 1$.

(4) *Poisson*. For $\lambda > 0$ we set $\mathbb{P}(X = k) = e^{-\lambda} \lambda^k / k!$ Again k is a non-negative integer.

If F is absolutely continuous, we call $f = F'$ the *density* of F. If such an F is the distribution function of a random variable X, then

$$\mathbb{P}(X \in A) = \int_A f(x)\, dx.$$

Some examples of distributions characterized by densities are the following.

(5) *Uniform on $[a, b]$*. Define $f(x) = (b - a)^{-1} 1_{[a,b]}(x)$.

(6) *Exponential*. For $x \ge 0$ let $f(x) = \lambda e^{-\lambda x}$ and set $f(x) = 0$ for $x < 0$.

(7) *Standard normal*. Define $f(x) = \frac{1}{\sqrt{2\pi}} e^{-x^2/2}$ for $x \in \mathbb{R}$.

Let us verify that the integral of f is one. To do that, let

$$I = \int_0^\infty e^{-x^2/2}\, dx,$$

and it suffices to show $I = \sqrt{\pi/2}$. Using the Fubini theorem, the monotone convergence theorem, and a change of variables to polar coordinates, we write

$$
\begin{aligned}
I^2 &= \left(\int_0^\infty e^{-x^2/2}\, dx \right)\left(\int_0^\infty e^{-y^2/2}\, dy \right) \\
&= \int_0^\infty \int_0^\infty e^{-(x^2+y^2)/2}\, dx\, dy \\
&= \lim_{R \to \infty} \int\int_{x,y\geq 0, x^2+y^2 \leq R^2} e^{-(x^2+y^2)/2}\, dx\, dy \\
&= \lim_{R \to \infty} \int_0^{\pi/2} \int_0^R e^{-r^2/2}\, r\, dr\, d\theta \\
&= \lim_{R \to \infty} \frac{\pi}{2}(1 - e^{-R^2/2}) = \frac{\pi}{2}
\end{aligned}
$$

as desired.

We shall see later ((A.4) and (A.5)) that a standard normal random variable Z has mean zero and variance one, which means that $\mathbb{E}\, Z = 0$ and $\mathbb{E}\, Z^2 = 1$.

(8) *Normal random variables with mean μ and variance σ^2.* If Z is a standard normal random variable, then a normal random variable X with mean μ and variance σ^2 has the same distribution as $\mu + \sigma Z$. It is an exercise in calculus to check that such a random variable has density

$$
f(x) = \frac{1}{\sqrt{2\pi}\,\sigma}e^{-(x-\mu)^2/2\sigma^2}. \tag{A.3}
$$

(9) *Gamma.* A random variable X has a gamma distribution with parameters r and λ (both r and λ must be positive) if it has density

$$
f(x) = \lambda e^{-\lambda x}(\lambda x)^{r-1}/\Gamma(r)
$$

for $x \geq 0$ and $f(x) = 0$ if $x < 0$, where $\Gamma(r) = \int_0^\infty e^{-y}y^{r-1}\, dy$ is the Gamma function. Recall $\Gamma(k) = (k-1)!$ for k a non-negative integer.

We can use the law of a random variable to calculate expectations.

Proposition A.3 *Let X be a random variable. If g is bounded or non-negative, then*

$$
\mathbb{E}\, g(X) = \int g(x)\, \mathbb{P}_X(dx).
$$

Proof If g is the indicator of an event A, this is just the definition of $\mathbb{P}_X$. By linearity, the result holds for simple functions. By the monotone convergence theorem, the result holds for non-negative functions, and by linearity again, it holds for bounded g. □

If F_X has a density f, then $\mathbb{P}_X(dx) = f(x)\, dx$. In this case $\mathbb{E}\, X = \int xf(x)\, dx$ and $\mathbb{E}\, X^2 = \int x^2 f(x)\, dx$. (We need $\mathbb{E}\, |X|$ finite to justify the first equality if X is not necessarily non-negative.) We define the *mean* of a random variable to be its expectation, and the *variance* of a random variable is defined by

$$
\mathrm{Var}\, X = \mathbb{E}\, (X - \mathbb{E}\, X)^2.
$$

The *pth moment* of X is $\mathbb{E}\, X^p$ if p is a positive integer.

Note

$$\operatorname{Var} X = \mathbb{E}\left[X^2 - 2(X)(\mathbb{E}X) + (\mathbb{E}X)^2\right] = \mathbb{E}X^2 - (\mathbb{E}X)^2.$$

Let us calculate a few examples. Since $xe^{-x^2/2}$ is an odd function, if Z is a standard normal random variable, then

$$\mathbb{E}Z = \int x \frac{1}{\sqrt{2\pi}} e^{-x^2/2}\, dx = 0. \tag{A.4}$$

Using integration by parts,

$$\mathbb{E}Z^2 = \int x^2 \frac{1}{\sqrt{2\pi}} e^{-x^2/2}\, dx \tag{A.5}$$

$$= \lim_{N\to\infty} \int_{-N}^{N} x^2 \frac{1}{\sqrt{2\pi}} e^{-x^2/2}\, dx$$

$$= \lim_{N\to\infty} -2N e^{-N^2/2} + \int_{-N}^{N} \frac{1}{\sqrt{2\pi}} e^{-x^2/2}\, dx$$

$$= \frac{1}{\sqrt{2\pi}} \int e^{-x^2/2}\, dx = 1,$$

and so $\operatorname{Var} Z = 1$.

By completing the square and a change of variables, we calculate

$$\mathbb{E}\, e^{aZ} = \frac{1}{\sqrt{2\pi}} \int e^{ax} e^{-x^2/2}\, dx$$

$$= \frac{1}{\sqrt{2\pi}} e^{a^2/2} \int e^{-(x-a)^2/2}\, dx = e^{a^2/2}.$$

If X is a normal random variable with mean μ and variance σ^2, we can write $X = \mu + \sigma Z$ for Z a standard normal random variable, and obtain

$$\mathbb{E}\, e^{aX} = e^{a\mu} \mathbb{E}\, e^{a\sigma Z} = e^{a\mu + a^2\sigma^2/2}. \tag{A.6}$$

If X is a Poisson random variable with parameter λ, then

$$\mathbb{E}X = \sum_{k=0}^{\infty} k e^{-\lambda} \frac{\lambda^k}{k!} = \sum_{k=1}^{\infty} k e^{-\lambda} \frac{\lambda^k}{k!} \tag{A.7}$$

$$= \lambda e^{-\lambda} \sum_{k=1}^{\infty} \frac{\lambda^{k-1}}{(k-1)!} = \lambda.$$

A similar calculation shows that $\mathbb{E}\left[X(X-1)\right] = \lambda^2$, so

$$\operatorname{Var} X = \mathbb{E}\left[X(X-1)\right] + \mathbb{E}X - (\mathbb{E}X)^2 = \lambda. \tag{A.8}$$

A straightforward application of integration by parts shows that if X is an exponential random variable with parameter λ, then

$$\mathbb{E}X = \int_0^\infty \lambda x e^{-\lambda x}\, dx = \frac{1}{\lambda}. \tag{A.9}$$

Another equality that is useful is the following.

Proposition A.4 *If $X \geq 0$, a.s., and $p > 0$, then*

$$\mathbb{E} X^p = \int_0^\infty p\lambda^{p-1} \mathbb{P}(X > \lambda) \, d\lambda.$$

The proof will show that this equality is also valid if we replace $\mathbb{P}(X > \lambda)$ by $\mathbb{P}(X \geq \lambda)$.

Proof Using the Fubini theorem and writing

$$\int_0^\infty p\lambda^{p-1} \mathbb{P}(X > \lambda) \, d\lambda = \mathbb{E} \int_0^\infty p\lambda^{p-1} 1_{(\lambda,\infty)}(X) \, d\lambda$$

$$= \mathbb{E} \int_0^X p\lambda^{p-1} \, d\lambda = \mathbb{E} X^p$$

gives the proof.   □

We need two elementary inequalities. The first is known as Chebyshev's inequality.

Proposition A.5 *If $X \geq 0$,*

$$\mathbb{P}(X \geq a) \leq \frac{\mathbb{E} X}{a}.$$

Proof We write

$$\mathbb{P}(X \geq a) = \mathbb{E}\left[1_{[a,\infty)}(X)\right] \leq \mathbb{E}\left[\left(\frac{X}{a}\right)1_{[a,\infty)}(X)\right] \leq \mathbb{E} X/a,$$

since X/a is bigger than or equal to 1 when $X \in [a, \infty)$.   □

If we apply this to $X = (Y - \mathbb{E} Y)^2$, we obtain

$$\mathbb{P}(|Y - \mathbb{E} Y| \geq a) = \mathbb{P}((Y - \mathbb{E} Y)^2 \geq a^2) \leq \operatorname{Var} Y/a^2. \tag{A.10}$$

This special case of Chebyshev's inequality is sometimes itself referred to as Chebyshev's inequality, while Proposition A.5 is sometimes called the Markov inequality.

The second inequality we need is Jensen's inequality, not to be confused with Jensen's formula of complex analysis.

Proposition A.6 *Suppose g is convex and X and $g(X)$ are both integrable. Then*

$$g(\mathbb{E} X) \leq \mathbb{E} g(X).$$

Proof One property of convex functions is that they lie above their tangent lines, and more generally, their support lines. Thus if $x_0 \in \mathbb{R}$, we have

$$g(x) \geq g(x_0) + c(x - x_0)$$

for some constant c. Letting $x = X(\omega)$ and taking expectations, we obtain

$$\mathbb{E} g(X) \geq g(x_0) + c(\mathbb{E} X - x_0).$$

Now set x_0 equal to $\mathbb{E} X$.   □

If A_n is a sequence of sets, define $(A_n \text{ i.o.})$, read "A_n infinitely often," by

$$(A_n \text{ i.o.}) = \cap_{n=1}^\infty \cup_{i=n}^\infty A_i.$$

This set consists of those ω that are in infinitely many of the A_n.

A simple but very important proposition is the Borel–Cantelli lemma. It has two parts, and we prove the first part here, leaving the second part to the next section.

Proposition A.7 *Let $A_1, A_2, \ldots$ be a sequence of events. If $\sum_n \mathbb{P}(A_n) < \infty$, then $\mathbb{P}(A_n \text{ i.o.}) = 0$.*

Proof We write

$$\mathbb{P}(A_n \text{ i.o.}) = \lim_{n \to \infty} \mathbb{P}(\cup_{i=n}^{\infty} A_i) \leq \limsup_{n \to \infty} \sum_{i=n}^{\infty} \mathbb{P}(A_i) = 0,$$

and we are done. $\qquad\square$

A.2 Independence

We say two events A and B are *independent* if $\mathbb{P}(A \cap B) = \mathbb{P}(A)\mathbb{P}(B)$. The events $A_1, \ldots, A_n$ are independent if

$$\mathbb{P}(A_{i_1} \cap A_{i_2} \cap \cdots \cap A_{i_j}) = \mathbb{P}(A_{i_1})\mathbb{P}(A_{i_2}) \cdots \mathbb{P}(A_{i_j})$$

for each subset $\{i_1, \ldots, i_j\}$ of $\{1, \ldots, n\}$ with $1 \leq i_1 < \cdots < i_j \leq n$.

Proposition A.8 *If A and B are independent, then A^c and B are independent.*

Proof We write

$$\mathbb{P}(A^c \cap B) = \mathbb{P}(B) - \mathbb{P}(A \cap B) = \mathbb{P}(B) - \mathbb{P}(A)\mathbb{P}(B)$$
$$= \mathbb{P}(B)(1 - \mathbb{P}(A)) = \mathbb{P}(B)\mathbb{P}(A^c).$$

This is all there is to the proof. $\qquad\square$

We say two σ-fields $\mathcal{F}$ and $\mathcal{G}$ are independent if A and B are independent whenever $A \in \mathcal{F}$ and $B \in \mathcal{G}$. Two random variables X and Y are independent if $\sigma(X)$, the σ-field generated by X, and $\sigma(Y)$, the σ-field generated by Y, are independent. (Recall that the σ-field generated by a random variable X is given by $\{(X \in A) : A \text{ a Borel subset of } \mathbb{R}\}$.) We define the independence of n σ-fields or n random variables in a similar way.

Remark A.9 If f and g are Borel functions and X and Y are independent, then $f(X)$ and $g(Y)$ are independent. This follows because the σ-field generated by $f(X)$ is a sub-σ-field of the one generated by X, and similarly for $g(Y)$.

To construct independent random variables, we can use the following.

Proposition A.10 *If $F_1, \ldots, F_n$ are distribution functions, there exist independent random variables $X_1, \ldots, X_n$ such that $F_{X_i} = F_i$, $i = 1, \ldots, n$.*

Proof Let $\Omega = [0, 1]^n$, $\mathcal{F}$ the Borel σ-field on Ω, and $\mathbb{P}$ an n-dimensional Lebesgue measure on Ω. If $\omega = (\omega_1, \ldots, \omega_n)$, define $X_i(\omega) = \sup\{x : F_i(x) < \omega_i\}$. As in Proposition A.2, $F_{X_i} = F_i$. We deduce the independence from the fact that $\mathbb{P}$ is a product measure, in fact, the n-fold product of one-dimensional Lebesgue measure on $[0, 1]$. $\qquad\square$

Let $F_{X,Y}(x,y) = \mathbb{P}(X \leq x, Y \leq y)$ denote the joint distribution function of two random variables X and Y. (The comma inside the set means "and"; this is a standard convention in probability.)

Proposition A.11 $F_{X,Y}(x,y) = F_X(x)F_Y(y)$ *if and only if* X *and* Y *are independent.*

Proof If X and Y are independent, then

$$F_{X,Y}(x,y) = \mathbb{P}(X \leq x, Y \leq y) = \mathbb{P}(X \leq x)\mathbb{P}(Y \leq y) = F_X(x)F_Y(y).$$

Conversely, if the inequality holds, fix y and let $\mathcal{M}_y$ denote the collection of sets A for which $\mathbb{P}(X \in A, Y \leq y) = \mathbb{P}(X \in A)\mathbb{P}(Y \leq y)$. $\mathcal{M}_y$ contains all sets of the form $(-\infty, x]$. It follows by linearity that $\mathcal{M}_y$ contains all sets of the form $(x, z]$, and then by linearity again, all sets that are the finite union of such half-open, half-closed intervals. Note that the collection of finite unions of such intervals, $\mathcal{A}$, is an algebra generating the Borel σ-field. It is clear that $\mathcal{M}_y$ is a monotone class, so by the monotone class theorem (Theorem B.2), $\mathcal{M}_y$ contains the Borel σ-field.

For a fixed set A, let $\mathcal{M}_A$ denote the collection of sets B for which $\mathbb{P}(X \in A, Y \in B) = \mathbb{P}(X \in A)\mathbb{P}(Y \in B)$. Again, $\mathcal{M}_A$ is a monotone class and by the preceding paragraph contains the σ-field generated by the collection of finite unions of intervals of the form $(x, z]$, and hence contains the Borel sets. Therefore X and Y are independent. $\square$

The following is known as the multiplication theorem.

Proposition A.12 *If* X, Y, *and* XY *are integrable and* X *and* Y *are independent, then* $\mathbb{E}[XY] = (\mathbb{E}X)(\mathbb{E}Y)$.

Proof Consider the pairs (Z_X, Z_Y) with Z_X being $\sigma(X)$ measurable and Z_Y being $\sigma(Y)$ measurable for which the multiplication theorem is true. It holds for $Z_X = 1_A(X)$ and $Z_Y = 1_B(Y)$ with A and B Borel subsets of $\mathbb{R}$ by the definition of X and Y being independent. It holds for simple random variables (Z_X, Z_Y), that is, linear combinations of indicators, by the linearity of both sides. It holds for non-negative random variables by monotone convergence. And it holds for integrable random variables by linearity again. $\square$

If $X_1, \ldots, X_n$ are independent, then so are $X_1 - \mathbb{E}X_1, \ldots, X_n - \mathbb{E}X_n$. Assuming everything is integrable,

$$\mathbb{E}[(X_1 - \mathbb{E}X_1) + \cdots (X_n - \mathbb{E}X_n)]^2 = \mathbb{E}(X_1 - \mathbb{E}X_1)^2 + \cdots + \mathbb{E}(X_n - \mathbb{E}X_n)^2,$$

using the multiplication theorem to show that the expectations of the cross-product terms are zero. We have thus shown

$$\text{Var}(X_1 + \cdots + X_n) = \text{Var}X_1 + \cdots + \text{Var}X_n. \tag{A.11}$$

We finish up this section by proving the second half of the Borel–Cantelli lemma.

Proposition A.13 *Suppose* A_n *is a sequence of independent events. If*

$$\sum_{n=1}^{\infty} \mathbb{P}(A_n) = \infty,$$

then $\mathbb{P}(A_n \text{ i.o.}) = 1$.

Note that here the A_n are independent, while in the first half of the Borel–Cantelli lemma no such assumption was necessary.

Proof Note

$$\mathbb{P}(\cup_{i=n}^N A_i) = 1 - \mathbb{P}(\cap_{i=n}^N A_i^c) = 1 - \prod_{i=n}^N \mathbb{P}(A_i^c)$$

$$= 1 - \prod_{i=n}^N (1 - \mathbb{P}(A_i)) \geq 1 - \exp\left(-\sum_{i=n}^N \mathbb{P}(A_i)\right),$$

using the inequality $1 - x \leq e^{-x}$ for $x > 0$. As $N \to \infty$, the right-hand side tends to one, so $\mathbb{P}(\cup_{i=n}^\infty A_i) = 1$. This holds for all n, which proves the result. $\square$

A.3 Convergence

In this section we consider three ways a sequence of random variables X_n can converge.

We say X_n *converges to X almost surely* if the event $(X_n \not\to X)$ has probability zero. X_n *converges to X in probability* if for each ε, $\mathbb{P}(|X_n - X| > \varepsilon) \to 0$ as $n \to \infty$. For $p \geq 1$, X_n *converges to X in L^p* if $\mathbb{E}|X_n - X|^p \to 0$ as $n \to \infty$.

The following proposition shows some relationships among the types of convergence.

Proposition A.14 *(1) If $X_n \to X$ almost surely, then $X_n \to X$ in probability.*

(2) If $X_n \to X$ in L^p, then $X_n \to X$ in probability.

(3) If $X_n \to X$ in probability, there exists a subsequence n_j such that X_{n_j} converges to X almost surely.

Proof To prove (1), note $X_n - X$ tends to zero almost surely, so $1_{(-\varepsilon,\varepsilon)^c}(X_n - X)$ also converges to zero almost surely. Now apply the dominated convergence theorem.

(2) comes from Chebyshev's inequality:

$$\mathbb{P}(|X_n - X| > \varepsilon) = \mathbb{P}(|X_n - X|^p > \varepsilon^p) \leq \mathbb{E}|X_n - X|^p/\varepsilon^p \to 0$$

as $n \to \infty$.

To prove (3), choose n_j larger than n_{j-1} such that $\mathbb{P}(|X_n - X| > 2^{-j}) < 2^{-j}$ whenever $n \geq n_j$. Thus if we let $A_i = (|X_{n_j} - X| > 2^{-i}$ for some $j \geq i)$, then $\mathbb{P}(A_i) \leq 2^{-i+1}$. By the Borel–Cantelli lemma $\mathbb{P}(A_i \text{ i.o.}) = 0$. This implies $X_{n_j} \to X$ almost surely on the complement of $(A_i \text{ i.o.})$. $\square$

Let us give some examples to show there need not be any other implications among the three types of convergence.

Let $\Omega = [0, 1]$, $\mathcal{F}$ the Borel σ-field, and $\mathbb{P}$ a Lebesgue measure. Let $X_n = n^2 1_{(0,1/n)}$. Then clearly X_n converges to zero almost surely and in probability, but $\mathbb{E}X_n^p = n^{2p}/n \to \infty$ for any $p \geq 1$.

Let Ω be the unit circle, and let $\mathbb{P}$ be a Lebesgue measure on the circle normalized to have total mass 1. We use θ to denote the angle that the ray from 0 through a point on the circle makes with the x axis. Let $t_n = \sum_{i=1}^n i^{-1}$, and let $A_n = \{e^{i\theta} : t_{n-1} \leq \theta < t_n\}$. Let $X_n = 1_{A_n}$.

Any point on the unit circle will be in infinitely many A_n, so X_n does not converge almost surely to zero. But $\mathbb{P}(A_n) = 1/(2\pi n) \to 0$, so $X_n \to 0$ in probability and in L^p.

A.4 Uniform integrability

A sequence $\{X_i\}$ of random variables is *uniformly integrable* if

$$\sup_i \int_{(|X_i|>M)} |X_i| \, d\mathbb{P} \to 0$$

as $M \to \infty$. This can be rephrased by saying: given $\varepsilon > 0$ there exists $M > 0$ such that $\mathbb{E}\,[\,|X_i|; |X_i| > M\,] < \varepsilon$ for all i. Here M can be chosen independently of i.

Lemma A.15 *If $\{X_i\}$ is a uniformly integrable sequence of random variables, then $\sup_i \mathbb{E}\,|X_i| < \infty$.*

Proof There exists M such that $\mathbb{E}\,[\,|X_i|; |X_i| > M\,] \leq 1$. Then

$$\mathbb{E}\,|X_i| \leq \mathbb{E}\,[\,|X_i|; |X_i| \leq M\,] + \mathbb{E}\,[\,|X_i|; |X_i| > M\,] \leq M + 1,$$

and we are done. $\qquad\square$

We say a sequence of random variables $\{X_i\}$ is *uniformly absolutely continuous* if given ε there exists δ such that $\sup_i \mathbb{E}\,[\,|X_i|; A\,] \leq \varepsilon$ whenever $\mathbb{P}(A) < \delta$.

Proposition A.16 *The following are equivalent.*
(1) The sequence $\{X_i\}$ is uniformly integrable.
(2) The sequence $\{X_i\}$ is uniformly absolutely continuous and $\sup_i \mathbb{E}\,|X_i| < \infty$.

Proof If (1) holds, we showed in Lemma A.15 that the expectations are uniformly bounded. Let $\varepsilon > 0$ and choose M such that $\sup_i \mathbb{E}\,[\,|X_i| : |X_i| > M\,] < \varepsilon/2$. Then if $\delta = \varepsilon/(2M)$ and $\mathbb{P}(A) < \delta$, we have

$$\mathbb{E}\,[\,|X_i|; A\,] \leq \mathbb{E}\,[\,|X_i|; |X_i| > M\,] + \mathbb{E}\,[\,|X_i|; |X_i| \leq M, A\,] < \frac{\varepsilon}{2} + M\mathbb{P}(A) \leq \varepsilon.$$

Now suppose (2) holds. Let $\varepsilon > 0$ and choose δ such that $\mathbb{E}\,[\,|X_i|; A\,] < \varepsilon$ for all i if $\mathbb{P}(A) \leq \delta$. Let $M = \sup_i \mathbb{E}\,|X_i|/\delta$. Then by the Chebyshev inequality

$$\mathbb{P}(|X_i| > M) \leq \frac{\mathbb{E}\,|X_i|}{M} = \delta,$$

so $\mathbb{E}\,[\,|X_i|; |X_i| > M\,] < \varepsilon$. $\qquad\square$

Proposition A.17 *Suppose $\{X_i\}$ and $\{Y_i\}$ are each uniformly integrable sequences of random variables. Then $\{X_i + Y_i\}$ is also a uniformly integrable sequence.*

Proof By Proposition A.16,

$$\sup_i \mathbb{E}\,|X_i + Y_i| \leq \sup_i \mathbb{E}\,|X_i| + \sup_i \mathbb{E}\,|Y_i| < \infty.$$

Using Proposition A.16 again, given ε there exists δ such that $\mathbb{E}\,[\,|X_i|; A\,] < \varepsilon/2$ and $\mathbb{E}\,[\,|Y_i|; A\,] < \varepsilon/2$ if $\mathbb{P}(A) < \delta$. But then $\mathbb{E}\,[\,|X_i + Y_i|; A\,] < \varepsilon$ and a third use of Proposition A.16 yields our result. $\qquad\square$

Proposition A.18 *Suppose there exists* $\varphi : [0, \infty) \to [0, \infty)$ *such that* φ *is increasing,* $\varphi(x)/x \to \infty$ *as* $x \to \infty$, *and* $\sup_i \mathbb{E}\,\varphi(|X_i|) < \infty$. *Then the sequence* $\{X_i\}$ *is uniformly integrable.*

Proof Let $\varepsilon > 0$ and choose x_0 such that $x/\varphi(x) < \varepsilon$ if $x \geq x_0$. If $M \geq x_0$,

$$\int_{(|X_i|>M)} |X_i| = \int \frac{|X_i|}{\varphi(|X_i|)} \varphi(|X_i|) 1_{(|X_i|>M)} \leq \varepsilon \int \varphi(|X_i|) \leq \varepsilon \sup_i \mathbb{E}\,\varphi(|X_i|).$$

Since ε is arbitrary, we are done. $\square$

The main result we need in this section is the Vitali convergence theorem.

Theorem A.19 *If* $X_n \to X$ *almost surely and the sequence* $\{X_n\}$ *is uniformly integrable, then* $\mathbb{E}\,|X_n - X| \to 0$.

Proof By Proposition A.17 with $Y_i = -X$ for each i, the sequence $X_i - X$ is uniformly integrable. Let $\varepsilon > 0$ and choose M such that

$$\int_{(|X_i-X|>M)} |X_i - X| < \varepsilon.$$

By dominated convergence,

$$\limsup_{i\to\infty} \mathbb{E}\,|X_i - X| \leq \limsup_{i\to\infty} \mathbb{E}\,[\,|X_i - X|; |X_i - X| \leq M] + \varepsilon = \varepsilon.$$

Since ε is arbitrary, then $\mathbb{E}\,|X_i - X| \to 0$. $\square$

A.5 Conditional expectation

If $\mathcal{F} \subset \mathcal{G}$ are two σ-fields and X is an integrable $\mathcal{G}$ measurable random variable, the *conditional expectation* of X given $\mathcal{F}$, written $\mathbb{E}[X \mid \mathcal{F}]$ and read as "the expectation (or expected value) of X given $\mathcal{F}$," is any $\mathcal{F}$ measurable random variable Y such that $\mathbb{E}[Y; A] = \mathbb{E}[X; A]$ for every $A \in \mathcal{F}$. The *conditional probability* of $A \in \mathcal{G}$ given $\mathcal{F}$ is defined by $\mathbb{P}(A \mid \mathcal{F}) = \mathbb{E}[1_A \mid \mathcal{F}]$.

If Y_1, Y_2 are two $\mathcal{F}$ measurable random variables with $\mathbb{E}[Y_1; A] = \mathbb{E}[Y_2; A]$ for all $A \in \mathcal{F}$, then $Y_1 = Y_2$, a.s., and so conditional expectation is unique up to almost sure equivalence.

In the case X is already $\mathcal{F}$ measurable, $\mathbb{E}[X \mid \mathcal{F}] = X$. If X is independent from $\mathcal{F}$, $\mathbb{E}[X \mid \mathcal{F}] = \mathbb{E}X$. Both of these facts follow immediately from the definition. For another example, if $\{A_i\}$ is a finite collection of pairwise disjoint sets whose union is Ω, $\mathbb{P}(A_i) > 0$ for all i, and $\mathcal{F}$ is the σ-field generated by the A_i's, then

$$\mathbb{P}(A \mid \mathcal{F}) = \sum_i \frac{\mathbb{P}(A \cap A_i)}{\mathbb{P}(A_i)} 1_{A_i}. \tag{A.12}$$

This follows since the right-hand side is $\mathcal{F}$ measurable and its expectation over any set A_i is $\mathbb{P}(A \cap A_i)$. Equation (A.12) provides the link with the definition of conditional probability from elementary probability: if $\mathbb{P}(B) \neq 0$, then

$$\mathbb{P}(A \mid B) = \frac{\mathbb{P}(A \cap B)}{\mathbb{P}(B)}. \tag{A.13}$$

We have

$$\mathbb{E}\left[\mathbb{E}\left[X \mid \mathcal{F}\right]\right] = \mathbb{E}X \tag{A.14}$$

because $\mathbb{E}\left[\mathbb{E}\left[X \mid \mathcal{F}\right]\right] = \mathbb{E}\left[\mathbb{E}\left[X \mid \mathcal{F}\right]; \Omega\right] = \mathbb{E}\left[X; \Omega\right] = \mathbb{E}X$.

The following is easy to establish.

Proposition A.20 *(1) If $X \geq Y$ are both integrable, then*

$$\mathbb{E}\left[X \mid \mathcal{F}\right] \geq \mathbb{E}\left[Y \mid \mathcal{F}\right], \qquad \text{a.s.}$$

(2) If X and Y are integrable and $a \in \mathbb{R}$, then

$$\mathbb{E}\left[aX + Y \mid \mathcal{F}\right] = a\mathbb{E}\left[X \mid \mathcal{F}\right] + \mathbb{E}\left[Y \mid \mathcal{F}\right].$$

It is easy to check that limit theorems such as monotone convergence and dominated convergence have conditional expectation versions, as do inequalities like Jensen's and Chebyshev's inequalities. Thus, for example, we have Jensen's inequality for conditional expectations.

Proposition A.21 *If g is convex and X and $g(X)$ are integrable,*

$$\mathbb{E}\left[g(X) \mid \mathcal{F}\right] \geq g(\mathbb{E}\left[X \mid \mathcal{F}\right]), \qquad \text{a.s.}$$

A key fact is the following.

Proposition A.22 *If X and XY are integrable and Y is measurable with respect to $\mathcal{F}$, then*

$$\mathbb{E}\left[XY \mid \mathcal{F}\right] = Y\mathbb{E}\left[X \mid \mathcal{F}\right]. \tag{A.15}$$

Proof If $A \in \mathcal{F}$, then for any $B \in \mathcal{F}$,

$$\mathbb{E}\left[1_A \mathbb{E}\left[X \mid \mathcal{F}\right]; B\right] = \mathbb{E}\left[\mathbb{E}\left[X \mid \mathcal{F}\right]; A \cap B\right] = \mathbb{E}\left[X; A \cap B\right] = \mathbb{E}\left[1_A X; B\right].$$

Since $1_A \mathbb{E}\left[X \mid \mathcal{F}\right]$ is $\mathcal{F}$ measurable, this shows that (A.15) holds when $Y = 1_A$ and $A \in \mathcal{F}$. Using linearity shows that (A.15) holds whenever Y is a simple $\mathcal{F}$ measurable random variable. Taking limits, (A.15) holds whenever $Y \geq 0$ is $\mathcal{F}$ measurable and X and XY are integrable. Using linearity again completes the proof. $\qquad\square$

Two other equalities are contained in the following.

Proposition A.23 *If $\mathcal{E} \subset \mathcal{F} \subset \mathcal{G}$ are σ-fields, then*

$$\mathbb{E}\left[\mathbb{E}\left[X \mid \mathcal{F}\right] \mid \mathcal{E}\right] = \mathbb{E}\left[X \mid \mathcal{E}\right] = \mathbb{E}\left[\mathbb{E}\left[X \mid \mathcal{E}\right] \mid \mathcal{F}\right].$$

Proof The right equality holds because $\mathbb{E}\left[X \mid \mathcal{E}\right]$ is $\mathcal{E}$ measurable, hence $\mathcal{F}$ measurable. We then use the fact that if Y is $\mathcal{F}$ measurable, $\mathbb{E}\left[Y \mid \mathcal{F}\right] = Y$.

To show the left equality, let $A \in \mathcal{E}$. Then since A is also in $\mathcal{F}$,

$$\mathbb{E}\left[\mathbb{E}\left[\mathbb{E}\left[X \mid \mathcal{F}\right] \mid \mathcal{E}\right]; A\right] = \mathbb{E}\left[\mathbb{E}\left[X \mid \mathcal{F}\right]; A\right] = \mathbb{E}\left[X; A\right] = \mathbb{E}\left[\mathbb{E}\left[X \mid \mathcal{E}\right]; A\right].$$

Since both sides are $\mathcal{E}$ measurable, the equality follows. $\qquad\square$

To show the existence of $\mathbb{E}\left[X \mid \mathcal{F}\right]$, we proceed as follows.

Proposition A.24 *If X is integrable, then $\mathbb{E}\left[X \mid \mathcal{F}\right]$ exists.*

Proof Using linearity, we need only consider $X \geq 0$. Define a finite measure $\mathbb{Q}$ on $\mathcal{F}$ by $\mathbb{Q}(A) = \mathbb{E}[X; A]$ for $A \in \mathcal{F}$. This is trivially absolutely continuous with respect to $\mathbb{P}|_{\mathcal{F}}$, the restriction of $\mathbb{P}$ to $\mathcal{F}$. Let $\mathbb{E}[X \mid \mathcal{F}]$ be the Radon–Nikodym derivative of $\mathbb{Q}$ with respect to $\mathbb{P}|_{\mathcal{F}}$. Since $\mathbb{Q}$ and $\mathbb{P}|_{\mathcal{F}}$ are measures on $\mathcal{F}$, the Radon–Nikodym derivative is $\mathcal{F}$ measurable, and so provides the desired random variable. $\qquad\square$

When $\mathcal{F} = \sigma(Y)$, one usually writes $\mathbb{E}[X \mid Y]$ for $\mathbb{E}[X \mid \mathcal{F}]$. Notation that is commonly used is $\mathbb{E}[X \mid Y = y]$. The definition is as follows. If $A \in \sigma(Y)$, then $A = (Y \in B)$ for some Borel set B by the definition of $\sigma(Y)$, or $1_A = 1_B(Y)$. By linearity and taking limits, it follows that if Z is $\sigma(Y)$ measurable, then $Z = f(Y)$ for some Borel measurable function f. Set $Z = \mathbb{E}[X \mid Y]$ and choose f Borel measurable so that $Z = f(Y)$. Then $\mathbb{E}[X \mid Y = y]$ is defined to be $f(y)$.

If $X \in L^2$ and $\mathcal{M} = \{Y \in L^2 : Y \text{ is } \mathcal{F} \text{ measurable}\}$, one can show that $\mathbb{E}[X \mid \mathcal{F}]$ is equal to the projection of X onto the subspace $\mathcal{M}$.

A.6 Stopping times

We next want to talk about stopping times. Suppose we have a sequence of σ-fields $\mathcal{F}_i$ such that $\mathcal{F}_i \subset \mathcal{F}_{i+1}$ for each i. An example would be if $\mathcal{F}_i = \sigma(X_1, \ldots, X_i)$. A random mapping N from Ω to $\{0, 1, 2, \ldots\}$ is called a *stopping time* if for each n, $(N \leq n) \in \mathcal{F}_n$.

The proof of the following is immediate from the definitions.

Proposition A.25 *(1) Fixed times n are stopping times.*
(2) If N_1 and N_2 are stopping times, then so are $N_1 \wedge N_2$ and $N_1 \vee N_2$.
(3) If N_n is an increasing sequence of stopping times, then so is $N = \sup_n N_n$.
(4) If N_n is a decreasing sequence of stopping times, then so is $N = \inf_n N_n$.
(5) If N is a stopping time, then so is $N + n$.

We define

$$\mathcal{F}_N = \{A : A \cap (N \leq n) \in \mathcal{F}_n \text{ for all } n\}. \tag{A.16}$$

A.7 Martingales

In this section we consider martingales. Let $\mathcal{F}_n$ be an increasing sequence of σ-fields. A sequence of random variables M_n is *adapted* to $\mathcal{F}_n$ if for each n, M_n is $\mathcal{F}_n$ measurable.

M_n is a *martingale* if M_n is adapted to $\mathcal{F}_n$, M_n is integrable for all n, and

$$\mathbb{E}[M_n \mid \mathcal{F}_{n-1}] = M_{n-1}, \qquad \text{a.s.}, \qquad n = 2, 3, \ldots \tag{A.17}$$

If we have $\mathbb{E}[M_n \mid \mathcal{F}_{n-1}] \geq M_{n-1}$, a.s., for every n, then M_n is a *submartingale*. If we have $\mathbb{E}[M_n \mid \mathcal{F}_{n-1}] \leq M_{n-1}$, we have a *supermartingale*.

Let us look at some examples. If X_i is a sequence of mean zero independent random variables and $S_n = \sum_{i=1}^n X_i$, then $M_n = S_n$ is a martingale, since

$$\mathbb{E}[M_n \mid \mathcal{F}_{n-1}] = M_{n-1} + \mathbb{E}[M_n - M_{n-1} \mid \mathcal{F}_{n-1}]$$
$$= M_{n-1} + \mathbb{E}[M_n - M_{n-1}] = M_{n-1},$$

using independence.

Another example is the following. If the X_i's are independent and have mean zero and variance one, S_n is as in the previous example, and $M_n = S_n^2 - n$, then

$$\mathbb{E}[S_n^2 \mid \mathcal{F}_{n-1}] = \mathbb{E}[(S_n - S_{n-1})^2 \mid \mathcal{F}_{n-1}] + 2S_{n-1}\mathbb{E}[S_n \mid \mathcal{F}_{n-1}] - S_{n-1}^2 = 1 + S_{n-1}^2,$$

using independence. It follows that M_n is a martingale.

A third example is the following: if $X \in L^1$ and $M_n = \mathbb{E}[X \mid \mathcal{F}_n]$, then M_n is a martingale. The proof of this is simple:

$$\mathbb{E}[M_{n+1} \mid \mathcal{F}_n] = \mathbb{E}[\mathbb{E}[X \mid \mathcal{F}_{n+1}] \mid \mathcal{F}_n] = \mathbb{E}[X \mid \mathcal{F}_n] = M_n.$$

If M_n is a martingale, g is convex, and $g(M_n)$ is integrable for each n, then by Jensen's inequality for conditional expectations,

$$\mathbb{E}[g(M_{n+1}) \mid \mathcal{F}_n] \geq g(\mathbb{E}[M_{n+1} \mid \mathcal{F}_n]) = g(M_n), \tag{A.18}$$

or $g(M_n)$ is a submartingale. Similarly if g is convex and increasing on $[0, \infty)$ and M_n is a positive submartingale, then $g(M_n)$ is a submartingale because

$$\mathbb{E}[g(M_{n+1}) \mid \mathcal{F}_n] \geq g(\mathbb{E}[M_{n+1} \mid F_n]) \geq g(M_n).$$

A.8 Optional stopping

Note that if one takes expectations in (A.17), one has $\mathbb{E}M_n = \mathbb{E}M_{n-1}$, and by induction $\mathbb{E}M_n = \mathbb{E}M_0$. The theorem about martingales that lies at the basis of all other results is Doob's optional stopping theorem, which says that the same is true if we replace n by a stopping time N. There are various versions, depending on what conditions one puts on the stopping times.

Theorem A.26 *If N is a stopping time with respect to $\mathcal{F}_n$ that is bounded by a positive real K and M_n a martingale, then $\mathbb{E}M_N = \mathbb{E}M_0$.*

Proof   We write

$$\mathbb{E}M_N = \sum_{k=0}^{K} \mathbb{E}[M_N; N = k] = \sum_{k=0}^{K} \mathbb{E}[M_k; N = k].$$

Note $(N = k)$ is $\mathcal{F}_j$ measurable if $j \geq k$, so

$$\mathbb{E}[M_k; N = k] = \mathbb{E}[M_{k+1}; N = k] = \mathbb{E}[M_{k+2}; N = k]$$
$$= \cdots = \mathbb{E}[M_K; N = k].$$

Hence

$$\mathbb{E}M_N = \sum_{k=0}^{K} \mathbb{E}[M_K; N = k] = \mathbb{E}M_K = \mathbb{E}M_0.$$

This completes the proof.   □

The same proof as that in Theorem A.26 gives the following corollary.

Corollary A.27 *If N is a stopping time bounded by K and M_n is a submartingale, then*
$\mathbb{E} M_N \leq \mathbb{E} M_K$.

The same proof also gives

Corollary A.28 *If N is a stopping time bounded by K, $A \in \mathcal{F}_N$, and M_n is a submartingale, then* $\mathbb{E}[M_N; A] \leq \mathbb{E}[M_K; A]$.

Proposition A.29 *If $N_1 \leq N_2$ are stopping times bounded by K and M is a martingale, then* $\mathbb{E}[M_{N_2} \mid \mathcal{F}_{N_1}] = M_{N_1}$, *a.s.*

Proof Suppose $A \in \mathcal{F}_{N_1}$. We need to show $\mathbb{E}[M_{N_1}; A] = \mathbb{E}[M_{N_2}; A]$. Define a new stopping time N_3 by

$$N_3(\omega) = \begin{cases} N_1(\omega), & \omega \in A \\ N_2(\omega), & \omega \notin A. \end{cases}$$

It is easy to check that N_3 is a stopping time, so $\mathbb{E} M_{N_3} = \mathbb{E} M_K = \mathbb{E} M_{N_2}$ implies

$$\mathbb{E}[M_{N_1}; A] + \mathbb{E}[M_{N_2}; A^c] = \mathbb{E}[M_{N_2}].$$

Subtracting $\mathbb{E}[M_{N_2}; A^c]$ from each side completes the proof. $\qquad\square$

The following is known as the Doob decomposition for discrete time martingales.

Proposition A.30 *Suppose X_k is a submartingale with respect to an increasing sequence of σ-fields $\mathcal{F}_k$. Then we can write $X_k = M_k + A_k$ such that M_k is a martingale adapted to the $\mathcal{F}_k$ and A_k is a sequence of random variables with A_k being $\mathcal{F}_{k-1}$ measurable and $A_0 \leq A_1 \leq \cdots$.*

Proof Let $a_k = \mathbb{E}[X_k \mid \mathcal{F}_{k-1}] - X_{k-1}$ for $k = 1, 2, \ldots$ Since X_k is a submartingale, each $a_k \geq 0$. Let $A_k = \sum_{i=1}^{k} a_i$. The fact that the A_k are increasing and measurable with respect to $\mathcal{F}_{k-1}$ is clear. Set $M_k = X_k - A_k$. Then

$$\mathbb{E}[M_{k+1} - M_k \mid \mathcal{F}_k] = \mathbb{E}[X_{k+1} - X_k \mid F_k] - a_{k+1} = 0,$$

or M_k is a martingale. $\qquad\square$

Combining Propositions A.29 and A.30 we have

Corollary A.31 *Suppose X_k is a submartingale, and $N_1 \leq N_2$ are bounded stopping times. Then*

$$\mathbb{E}[X_{N_2} \mid \mathcal{F}_{N_1}] \geq X_{N_1}.$$

A.9 Doob's inequalities

The first interesting consequences of the optional stopping theorems are Doob's inequalities. If M_n is a martingale, set $M_n^* = \max_{i \leq n} |M_i|$.

Theorem A.32 *If M_n is a martingale or a positive submartingale,*

$$\mathbb{P}(M_n^* \geq a) \leq \frac{1}{a}\mathbb{E}[\,|M_n|; M_n^* \geq a] \leq \frac{1}{a}\mathbb{E}\,|M_n|.$$

Proof Fix n. Set $M_{n+1} = M_n$. Let $N = \min\{j : |M_j| \geq a\} \wedge (n + 1)$. Since the function $f(x) = |x|$ is convex, $|M_n|$ is a submartingale. If $A = (M_n^* \geq a)$, then $A \in \mathcal{F}_N$ and we have

$$a\mathbb{P}(M_n^* \geq a) \leq \mathbb{E}\,[\,|M_N|; A\,] \leq \mathbb{E}\,[\,|M_n|; A\,] \leq \mathbb{E}\,|M_n|,$$

the first inequality by the definition of N, the second by Corollary A.28.   □

For $p > 1$, we have the following inequality.

Theorem A.33 *If $p > 1$, M is a martingale or positive submartingale, and $\mathbb{E}\,|M_i|^p < \infty$ for $i \leq n$, then*

$$\mathbb{E}\,(M_n^*)^p \leq \Big(\frac{p}{p-1}\Big)^p \mathbb{E}\,|M_n|^p.$$

Proof Note $M_n^* \leq \sum_{i=1}^n |M_n|$, hence $M_n^* \in L^p$. We write, using Theorem A.32,

$$\mathbb{E}\,(M_n^*)^p = \int_0^\infty pa^{p-1}\mathbb{P}(M_n^* > a)\,da \leq \int_0^\infty pa^{p-1}\mathbb{E}\,[\,|M_n|1_{(M_n^* \geq a)}/a\,]\,da$$

$$= \mathbb{E}\int_0^{M_n^*} pa^{p-2}|M_n|\,da = \frac{p}{p-1}\mathbb{E}\,[(M_n^*)^{p-1}|M_n|]$$

$$\leq \frac{p}{p-1}(\mathbb{E}\,(M_n^*)^p)^{(p-1)/p}(\mathbb{E}\,|M_n|^p)^{1/p}.$$

The last inequality follows by Hölder's inequality. Now divide both sides by the quantity $(\mathbb{E}\,(M_n^*)^p)^{(p-1)/p}$.   □

A.10 Martingale convergence theorem

The martingale convergence theorem is another important consequence of optional stopping. The main step is the upcrossing lemma. The number of upcrossings of an interval $[a, b]$ is the number of times a process M crosses from below a to above b.

To be more exact, let

$$S_1 = \min\{k : M_k \leq a\}, \qquad T_1 = \min\{k > S_1 : M_k \geq b\},$$

and

$$S_{i+1} = \min\{k > T_i : M_k \leq a\}, \qquad T_{i+1} = \min\{k > S_{i+1} : M_k \geq b\}.$$

The number of upcrossings U_n before time n is $U_n = \max\{j : T_j \leq n\}$.

Theorem A.34 (Upcrossing lemma) *If M_k is a submartingale,*

$$\mathbb{E}\,U_n \leq \frac{1}{b-a}\mathbb{E}\,[(M_n - a)^+].$$

Proof The number of upcrossings of $[a, b]$ by M_k is the same as the number of upcrossings of $[0, b - a]$ by $Y_k = (M_k - a)^+$, where $x^+ = x \vee 0$. Moreover Y_k is still a submartingale. If we obtain the inequality for the number of upcrossings of the interval $[0, b - a]$ by the process Y_k, we will have the desired inequality for upcrossings of M.

Thus we may assume $a = 0$. Fix n and define $Y_{n+1} = Y_n$. This will still be a submartingale. Define S_i, T_i as above, and let $S_i' = S_i \wedge (n + 1)$, $T_i' = T_i \wedge (n + 1)$. Since $T_{i+1} > S_{i+1} > T_i$, then $T_{n+1}' = n + 1$.

We write

$$\mathbb{E}\,Y_{n+1} = \mathbb{E}\,Y_{S_1'} + \sum_{i=0}^{n+1} \mathbb{E}\,[Y_{T_i'} - Y_{S_i'}] + \sum_{i=0}^{n+1} \mathbb{E}\,[Y_{S_{i+1}'} - Y_{T_i'}].$$

All the summands in the third term on the right are non-negative since Y_k is a submartingale. The first term on the right will be non-negative since Y is non-negative. For the jth upcrossing, $Y_{T_j'} - Y_{S_j'} \geq b - a$, while $Y_{T_j'} - Y_{S_j'}$ is always greater than or equal to 0. Thus

$$\sum_{i=0}^{n+1} (Y_{T_i'} - Y_{S_i'}) \geq (b-a)U_n.$$

Hence

$$\mathbb{E}\,U_n \leq \frac{1}{b-a}\mathbb{E}\,Y_{n+1}. \tag{A.19}$$

$\square$

This leads to the martingale convergence theorem.

Theorem A.35 *If M_n is a submartingale such that $\sup_n \mathbb{E}\,M_n^+ < \infty$, then M_n converges almost surely as $n \to \infty$.*

Proof For each $a < b$, let $U_n(a,b)$ be the number of upcrossings of $[a,b]$ by M up to time n, and let $U(a,b) = \lim_{n\to\infty} U_n$. For each pair $a < b$ of rational numbers, by monotone convergence,

$$\mathbb{E}\,U(a,b) \leq \frac{1}{b-a}\sup_n \mathbb{E}\,(M_n - a)^+ < \infty.$$

Thus $U(a,b) < \infty$, a.s. If $N_{a,b}$ is the set of ω's where $U(a,b) = \infty$ and $N = \cup_{a<b,a,b\in\mathbb{Q}_+} N_{a,b}$, then $\mathbb{P}(N) = 0$. If $\omega \notin N$, we cannot have $\limsup_{n\to\infty} M_n(\omega) > \liminf_{n\to\infty} M_n(\omega)$. Therefore M_n converges almost surely, although we still have to rule out the possibility of the limit being infinite. Since M_n is a submartingale, $\mathbb{E}\,M_n \geq \mathbb{E}\,M_0$, and thus

$$\mathbb{E}\,|M_n| = \mathbb{E}\,M_n^+ + \mathbb{E}\,M_n^- = 2\mathbb{E}\,M_n^+ - \mathbb{E}\,M_n \leq 2\mathbb{E}\,M_n^+ - \mathbb{E}\,M_0.$$

By Fatou's lemma,

$$\mathbb{E}\,\lim_n |M_n| \leq \sup_n \mathbb{E}\,|M_n| \leq 2\sup_n \mathbb{E}\,M_n^+ - \mathbb{E}\,M_0 < \infty,$$

or M_n converges almost surely to a finite limit. $\square$

Corollary A.36 *If X_n is a positive supermartingale or a martingale bounded above or below, X_n converges almost surely.*

Proof If X_n is a positive supermartingale, $-X_n$ is a submartingale bounded above by 0. Now apply Theorem A.35.

If X_n is a martingale bounded above, by considering $-X_n$, we may assume X_n is bounded below. Looking at $X_n + M$ for fixed M will not affect the convergence, so we may assume X_n is bounded below by 0. Now apply the first assertion of the corollary. $\square$

M_n is a *uniformly integrable martingale* if the collection of random variables $\{M_n\}$ is uniformly integrable.

Proposition A.37 *(1) If M_n is a martingale with $\sup_n \mathbb{E}|M_n|^p < \infty$ for some $p > 1$, then the convergence is in L^p as well as almost surely. This is also true when M_n is a submartingale.*
   (2) If M_n is a uniformly integrable martingale, then the convergence is in L^1.
   (3) If $M_n \to M_\infty$ in L^1, then $M_n = \mathbb{E}[M_\infty \mid \mathcal{F}_n]$.

Proof   (1) If $\sup_n \mathbb{E}|M_n|^p < \infty$, then $\sup_n \mathbb{E}M_n^+ < \infty$ and M_n converges almost surely. Let M_∞ be the limit. Then $|M_n - M_\infty| \to 0$, a.s., and

$$\mathbb{E}\sup_n |M_n - M_\infty|^p \leq c\mathbb{E}\sup_n |M_n|^p + c\mathbb{E}|M_\infty|^p$$
$$\leq c\mathbb{E}\sup_n |M_n|^p$$
$$\leq c\sup_n \mathbb{E}|M_n|^p < \infty.$$

The second inequality is by Fatou's lemma and the last by Doob's inequalities, Theorem A.33. The L^p convergence assertion now follows by dominated convergence.
   (2) The L^1 convergence assertion follows since almost sure convergence together with uniform integrability implies L^1 convergence by the Vitali convergence theorem, Theorem A.19.
   (3) Finally, if $j < n$, we have $M_j = \mathbb{E}[M_n \mid \mathcal{F}_j]$. If $A \in \mathcal{F}_j$,

$$\mathbb{E}[M_j; A] = \mathbb{E}[M_n; A] \to \mathbb{E}[M_\infty; A]$$

by the L^1 convergence of M_n to M_∞. Since this is true for all $A \in \mathcal{F}_j$, $M_j = \mathbb{E}[M_\infty \mid \mathcal{F}_j]$. $\square$

A.11 Strong law of large numbers

Suppose we have a sequence $X_1, X_2, \ldots$ of independent and identically distributed random variables. This means that the X_i are independent and each has the same law as X_1. This situation is very common, and we abbreviate this by saying the X_i are *i.i.d.*
   Define

$$S_n = \sum_{i=1}^{n} X_i.$$

The S_n are called *partial sums*. In this section we suppose $\mathbb{E}|X_1| < \infty$. The strong law of large number is the precise version of the law of averages.

Theorem A.38 *If X_i is an i.i.d. sequence and $\mathbb{E}|X_1| < \infty$, then*

$$\frac{S_n}{n} \to \mathbb{E}X_1, \qquad \text{a.s.}$$

The proof we give is a mixture of the standard one and some martingale techniques. The standard proof (see, e.g., Chung (2001)) uses no martingale methods, while there is a proof (see Durrett (1996)) that is entirely martingale based.

Proof We may assume $\mathbb{E}\,X_i = 0$, for otherwise we replace X_i by $X_i - \mathbb{E}\,X_i$. Let $Y_n = X_n \mathbf{1}_{(|X_n| \leq n)}$, $Z_n = Y_n - \mathbb{E}\,Y_n$, and

$$M_n = \sum_{i=1}^{n} \frac{Z_i}{i}.$$

Let $\mathcal{F}_n = \sigma(X_1, \ldots, X_n)$. Note that the Z_i are independent but not identically distributed. Using the independence, M_n is a martingale:

$$\mathbb{E}\,[M_{n+1} \mid \mathcal{F}_n] = M_n + \frac{1}{n+1} \mathbb{E}\,[Z_{n+1} \mid \mathcal{F}_n] = M_n + \frac{1}{n+1} \mathbb{E}\,[Z_{n+1}] = M_n.$$

We will need the estimate

$$\sum_{i=1}^{\infty} \mathbb{P}(|X_1| \geq i) = \sum_{i=1}^{\infty} \int_{i-1}^{i} \mathbb{P}(|X_i| \geq i)\,dx \tag{A.20}$$

$$\leq \int_{0}^{\infty} \mathbb{P}(|X_1| \geq x)\,dx = \mathbb{E}\,|X_1| < \infty,$$

using Proposition A.4.

We show that $\mathbb{E}\,|M_n|$ is bounded by a constant not depending on n. In fact, again using Proposition A.4,

$$\mathbb{E}\,M_n^2 = \operatorname{Var} M_n = \sum_{i=1}^{n} \frac{\operatorname{Var} Z_i}{i^2} = \sum_{i=1}^{n} \frac{1}{i^2} \operatorname{Var} Y_i$$

$$\leq \sum_{i=1}^{n} \frac{1}{i^2} \mathbb{E}\,Y_i^2 \leq \sum_{i=1}^{\infty} \frac{1}{i^2} \int_{0}^{i} 2y \mathbb{P}(|X_i| \geq y)\,dy$$

$$= 2 \sum_{i=1}^{\infty} \frac{1}{i^2} \int_{0}^{\infty} \mathbf{1}_{(y \leq i)} y \mathbb{P}(|X_1| \geq y)\,dy$$

$$= 2 \int_{0}^{\infty} \sum_{i=1}^{\infty} \frac{1}{i^2} \mathbf{1}_{(y \leq i)} y \mathbb{P}(|X_1| \geq y)\,dy$$

$$\leq c \int_{0}^{\infty} \frac{1}{y} \cdot y \mathbb{P}(|X_1| \geq y)\,dy$$

$$= c \int_{0}^{\infty} \mathbb{P}(|X_1| \geq y)\,dy = c\mathbb{E}\,|X_1| < \infty.$$

The uniform bound on $\mathbb{E}\,|M_n|$ follows by Jensen's inequality.

By the martingale convergence theorem, M_n converges almost surely; let M_∞ be the limit. Some elementary calculus shows that $\frac{1}{n} \sum_{i=1}^{n} M_i$ also converges to M_∞, a.s. We now use summation by parts as follows. Since $i(M_i - M_{i-1}) = Z_i$ and $M_0 = 0$, then

$$\frac{1}{n} \sum_{i=1}^{n} Z_i = \frac{1}{n} \sum_{i=1}^{n} (iM_i - iM_{i-1}) = \frac{1}{n} \left(\sum_{i=1}^{n} iM_i - \sum_{i=1}^{n-1} (i+1)M_i \right)$$

$$= M_n - \frac{n-1}{n} \left(\frac{1}{n-1} \sum_{i=1}^{n-1} M_i \right) \to M_\infty - M_\infty = 0.$$

By dominated convergence and the fact that the X_i are identically distributed,

$$\mathbb{E}\, Y_n = \mathbb{E}\,[X_n 1_{(|X_n| \le i)}] = \mathbb{E}\,[X_1 1_{(|X_1| \le n)}] \to \mathbb{E}\, X_1 = 0$$

as $n \to \infty$, and this implies $\frac{1}{n} \sum_{i=1}^n \mathbb{E}\, Y_i \to 0$. Since $Y_i = Z_i + \mathbb{E}\, Y_i$, we conclude

$$\frac{1}{n} \sum_{i=1}^n Y_i \to 0, \qquad \text{a.s.}$$

Finally,

$$\sum_{i=1}^\infty \mathbb{P}(X_i \neq Y_i) = \sum_{i=1}^\infty \mathbb{P}(|X_i| \ge i) = \sum_{i=1}^\infty \mathbb{P}(|X_1| \ge i) < \infty,$$

so by the Borel–Cantelli lemma, except for a set of probability zero, $X_i = Y_i$ for all i greater than some positive integer I (I depends on ω). Hence

$$\left| \frac{1}{n} \sum_{i=1}^n X_i - \frac{1}{n} \sum_{i=1}^n Y_i \right| \le \frac{1}{n} \sum_{i=1}^I |X_i - Y_i| \to 0, \qquad \text{a.s.}$$

This completes the proof. $\square$

The following extension of the strong law will be needed when comparing a random walk and a Brownian motion.

Proposition A.39 *Suppose X_i is an i.i.d. sequence and $\mathbb{E}\,|X_1| < \infty$. Then*

$$\frac{\max_{k \le n} |S_k - \mathbb{E}\, S_k|}{n} \to 0, \qquad \text{a.s.}$$

Proof   By looking at $X_i - \mathbb{E}\, X_i$, we may assume $\mathbb{E}\, X_i = 0$. Let $j(n)$ be (one of) the value(s) of j such that $|S_j| = \max_{k \le n} |S_k|$. Suppose $S_n(\omega)/n \to 0$. It suffices to show $|S_{j(n)}(\omega)|/n \to 0$, a.s.

If not, for this ω, either

(1) there is a subsequence $n_k \to \infty$ and $\varepsilon > 0$ such that $j(n_k) \to \infty$ and $|S_{j(n_k)}|/n_k \ge \varepsilon$ for all k; or

(2) there exists a subsequence $n_k \to \infty$, $\varepsilon > 0$, and $N > 1$ such that $j(n_k) \le N$ and $|S_{j(n_k)}|/n_k \ge \varepsilon$ for all k.

In case (1), since $j(n_k) \to \infty$,

$$\frac{|S_{j(n_k)}|}{n_k} = \frac{|S_{j(n_k)}|}{j(n_k)} \frac{j(n_k)}{n_k} \le \frac{|S_{j(n_k)}|}{j(n_k)} \to 0,$$

a contradiction. In case (2),

$$\frac{|S_{j(n_k)}|}{n_k} \le \frac{\max_{m \le N} |S_m|}{n_k} \to 0,$$

also a contradiction. $\square$

Another application of the strong law of large numbers is the *Glivenko–Cantelli theorem*. Let X_i be i.i.d. random variables which have a uniform distribution on $[0, 1]$, that is, $\mathbb{P}(X_1 \leq t) = t$ if $0 \leq t \leq 1$. Let

$$F_n(t) = \frac{1}{n} \sum_{i=1}^{n} 1_{[0,t]}(X_i), \qquad 0 \leq t \leq 1.$$

By the strong law, $F_n(t) \to t$, a.s., for each t. The Glivenko–Cantelli theorem says that the convergence is uniform over t.

Theorem A.40 *With F_n as above,*

$$\sup_{0 \leq t \leq 1} |F_n(t) - t| \to 0, \qquad \text{a.s.}$$

Proof For each $t \in [0, 1]$, $1_{[0,t]}(X_i)$ is a sequence of i.i.d. random variables with expectation $\mathbb{P}(X_i \leq t) = t$. By the strong law of large numbers, for each t, $F_n(t) \to t$, a.s. Let N_t be the set of ω such that $F_n(t)(\omega)$ does not converge to t, and let $N = \cup_{\mathbb{Q}_+} N_t$. Then $\mathbb{P}(N) = 0$.

Let $\varepsilon > 0$ and take $\omega \notin N$. Take $m > 2/\varepsilon$ and choose n_0 large enough (depending on ω) such that

$$|F_n(k/m)(\omega) - (k/m)| < \varepsilon/2, \qquad k = 0, 1, 2, \ldots, m,$$

if $n \geq n_0$. Then if $n \geq n_0$ and $k/m \leq t < (k+1)/m$,

$$F_n(t) - t \leq F_n((k+1)/m) - k/m \leq F_n((k+1)/m) - (k+1)/m + \varepsilon/2 < \varepsilon,$$

and similarly $F_n(t) - t > -\varepsilon$. Hence for $n \geq n_0$,

$$\sup_{t \in [0,1]} |F_n(t) - t| \leq \varepsilon.$$

Since ε is arbitrary, this proves the uniform convergence. $\square$

A.12 Weak convergence

We will see soon that if the X_i are i.i.d. with mean zero and variance one, then $S_n/\sqrt{n}$ converges in the sense that

$$\mathbb{P}(S_n/\sqrt{n} \in [a, b]) \to \mathbb{P}(Z \in [a, b]),$$

where Z is a standard normal. We want to generalize the above type of convergence.

We say F_n *converges weakly* to F if $F_n(x) \to F(x)$ for all x at which F is continuous. Here F_n and F are distribution functions. We say X_n *converges weakly* to X if F_{X_n} converges weakly to F_X. We also say X_n *converges in distribution* or *converges in law* to X. Probabilities μ_n converge weakly if their corresponding distribution functions converge, that is, if $F_{\mu_n}(x) = \mu_n(-\infty, x]$ converges weakly.

An example that illustrates why we restrict the convergence to continuity points of F is the following. Let $X_n = 1/n$ with probability one, and $X = 0$ with probability one. $F_{X_n}(x)$ is 0 if $x < 1/n$ and 1 otherwise. Note $F_{X_n}(x)$ converges to $F_X(x)$ for all x except $x = 0$.

Proposition A.41 *X_n converges weakly to X if and only if $\mathbb{E}g(X_n) \to \mathbb{E}g(X)$ for all g bounded and continuous.*

Proof Suppose $\mathbb{E}\,g(X_n) \to \mathbb{E}\,g(X)$ whenever g is bounded and continuous. Let $\varepsilon > 0$ and suppose x is a continuity point of F_X. Choose δ such that $F_X(x) - \varepsilon < F_X(x - \delta) \leq F_X(x + \delta) < F_X(x) + \varepsilon$. Let g be a continuous function taking values in $[0, 1]$ such that g equals 1 on $(-\infty, x]$ and equals 0 on $[x + \delta, \infty)$. Then

$$\limsup_{n\to\infty} F_{X_n}(x) \leq \limsup_{n\to\infty} \mathbb{E}\,g(X_n)$$

$$= \mathbb{E}\,g(X) \leq F_X(x + \delta) < F_X(x) + \varepsilon.$$

A similar argument shows that $\liminf_{n\to\infty} F_{X_n} > F_X(x) - \varepsilon$. Since ε is arbitrary, $\lim_{n\to\infty} F_{X_n}(x) = F_X(x)$.

Now suppose $X_n \to X$ weakly. Let $\varepsilon > 0$ and choose $M > 0$ such that M and $-M$ are continuity points for F_X and also continuity points for each of the F_{X_n}, $F_X(-M) < \varepsilon$, and $F_X(M) > 1 - \varepsilon$. Suppose g is bounded and continuous on $\mathbb{R}$ and without loss of generality suppose g is bounded by 1. Then

$$\limsup_{n\to\infty} |\mathbb{E}\,[g(X_n); X_n \notin [-M, M)]| \tag{A.21}$$

$$\leq \limsup_{n\to\infty} \mathbb{P}(|X_n| \geq M)$$

$$= \limsup_{n\to\infty} F_{X_n}(-M) + \limsup_{n\to\infty}(1 - F_{X_n}(M))$$

$$\leq 2\varepsilon.$$

Similarly,

$$|\mathbb{E}\,[g(X); X \notin [-M, M)]| \leq 2\varepsilon. \tag{A.22}$$

Take f to be a step function of the form $\sum_{i=1}^{m} c_i 1_{(a_i, b_i]}$ such that $|f(x) - g(x)| < \varepsilon$ for $x \in [-M, M)$ and each a_i and b_i is a continuity point for F_X and also continuity points for each of the F_{X_n}. Then

$$\mathbb{E}\,f(X_n) = \sum_{i=1}^{m} c_i(F_{X_n}(b_i) - F_{X_n}(a_i)) \tag{A.23}$$

$$\to \sum_{i=1}^{m} c_i(F_X(b_i) - F_X(a_i)) = \mathbb{E}\,f(X).$$

Finally, since f differs from g by at most ε on $[-M, M)$, then

$$|\mathbb{E}\,f(X_n) - \mathbb{E}\,[g(X_n); X_n \in [-M, M)]| \leq \varepsilon \tag{A.24}$$

and similarly when X_n is replaced by X. Combining (A.21), (A.22), (A.23), and (A.24) and using the fact that ε is arbitrary shows that $\mathbb{E}\,g(X_n) \to \mathbb{E}\,g(X)$. □

Let us examine the relationship between weak convergence and convergence in probability. If X_i is an i.i.d. sequence, then X_i converges weakly, in fact, to X_1, since all the X_i's have the same distribution. But from the independence it is not hard to see that the sequence X_i does not converge in probability unless the X_i's are identically constant. Therefore one can have weak convergence without convergence in probability.

Proposition A.42 *(1) If X_n converges to X in probability, then it converges weakly.*

(2) If X_n converges weakly to a constant, it converges in probability.

(3) (Slutsky's theorem) *If X_n converges weakly to X and Y_n converges weakly to a constant b, then $X_n + Y_n$ converges weakly to $X + b$ and $X_n Y_n$ converges weakly to bX.*

Proof To prove (1), let g be a bounded and continuous function. If n_j is any subsequence, then there exists a further subsequence such that $X(n_{j_k})$ converges almost surely to X. Then by dominated convergence, $\mathbb{E} g(X(n_{j_k})) \to \mathbb{E} g(X)$. That suffices to show $\mathbb{E} g(X_n)$ converges to $\mathbb{E} g(X)$.

For (2), if X_n converges weakly to b,

$$\mathbb{P}(X_n - b > \varepsilon) = \mathbb{P}(X_n > b + \varepsilon) = 1 - \mathbb{P}(X_n \leq b + \varepsilon) \to 1 - \mathbb{P}(b \leq b + \varepsilon) = 0.$$

We use the fact that if Y is identically equal to b, then $b + \varepsilon$ is a point of continuity for F_Y. A similar equation shows $\mathbb{P}(X_n - b \leq -\varepsilon) \to 0$, so $\mathbb{P}(|X_n - b| > \varepsilon) \to 0$.

We now prove the first part of (3), leaving the second part for the reader. Let x be a point such that $x - b$ is a continuity point of F_X. Choose ε so that $x - b + \varepsilon$ is again a continuity point. Then

$$\mathbb{P}(X_n + Y_n \leq x) \leq \mathbb{P}(X_n + b \leq x + \varepsilon) + \mathbb{P}(|Y_n - b| > \varepsilon) \to \mathbb{P}(X \leq x - b + \varepsilon).$$

Hence $\limsup \mathbb{P}(X_n + Y_n \leq x) \leq \mathbb{P}(X + b \leq x + \varepsilon)$. Since ε can be arbitrarily small and $x - b$ is a continuity point of F_X, then $\limsup \mathbb{P}(X_n + Y_n \leq x) \leq \mathbb{P}(X + b \leq x)$. The lim inf is done similarly. $\square$

We say a sequence of distribution functions $\{F_n\}$ is *tight* if for each $\varepsilon > 0$ there exists M such that $F_n(M) \geq 1 - \varepsilon$ and $F_n(-M) \leq \varepsilon$ for all n. A sequence of random variables $\{X_n\}$ is tight if the corresponding distribution functions are tight; this is equivalent to $\mathbb{P}(|X_n| \geq M) \leq \varepsilon$.

Theorem A.43 (Helly's theorem) *Let F_n be a sequence of distribution functions that is tight. There exists a subsequence n_j and a distribution function F such that F_{n_j} converges weakly to F.*

What could conceivably happen is that X_n is identically equal to n, so that $F_{X_n} \to 0$, but the function F that is identically equal to 0 is not a distribution function; the tightness precludes this.

Proof Let q_k be an enumeration of the rationals. Since $F_n(q_k) \in [0, 1]$, any subsequence has a further subsequence that converges. Use a diagonalization argument (as in the proof of the Ascoli–Arzelà theorem; see Rudin (1976)) so that $F_{n_j}(q_k)$ converges for each q_k and call the limit $F(q_k)$. F is increasing, and define $F(x) = \inf_{q_k \geq x} F(q_k)$. Hence F is right continuous and increasing.

If x is a point of continuity of F and $\varepsilon > 0$, then there exist r and s rational such that $r < x < s$ and $F(s) - \varepsilon < F(x) < F(r) + \varepsilon$. Then

$$F_{n_j}(x) \geq F_{n_j}(r) \to F(r) > F(x) - \varepsilon$$

and

$$F_{n_j}(x) \leq F_{n_j}(s) \to F(s) < F(x) + \varepsilon.$$

Since ε is arbitrary, $F_{n_j}(x) \to F(x)$.

Since the F_n are tight, there exists M such that $F_n(-M) < \varepsilon$. Then $F(-M) \le \varepsilon$, which implies $\lim_{x \to -\infty} F(x) = 0$. Showing $\lim_{x \to \infty} F(x) = 1$ is similar. Therefore F is in fact a distribution function. □

We conclude by giving an easily checked criterion for tightness.

Proposition A.44 *Suppose there exists $\varphi : [0, \infty) \to [0, \infty)$ that is increasing and $\varphi(x) \to \infty$ as $x \to \infty$. If $a = \sup_n \mathbb{E}\,\varphi(|X_n|) < \infty$, then the sequence $\{X_n\}$ is tight.*

Proof Let $\varepsilon > 0$. Choose M such that $\varphi(x) \ge a/\varepsilon$ if $x > M$. Then

$$\mathbb{P}(|X_n| > M) \le \int \frac{\varphi(|X_n|)}{a/\varepsilon} 1_{(|X_n| > M)} d\mathbb{P} \le \frac{\varepsilon}{a} \mathbb{E}\,\varphi(|X_n|) \le \varepsilon.$$

The conclusion follows. □

In particular, if $\sup_n \mathbb{E}\,|X_n|^2 < \infty$, the sequence $\{X_n\}$ is tight.

A.13 Characteristic functions

We define the *characteristic function* of a random variable X by $\varphi_X(t) = \mathbb{E}\,e^{itx}$ for $t \in \mathbb{R}$.

Note that $\varphi_X(t) = \int e^{itx} \mathbb{P}_X(dx)$. Thus if X and Y have the same law, they have the same characteristic function. Also, if the law of X has a density, that is, $\mathbb{P}_X(dx) = f_X(x)\,dx$, then $\varphi_X(t) = \int e^{itx} f_X(x)\,dx$, so in this case the characteristic function is the same as the definition of the Fourier transform of f_X.

Proposition A.45 $\varphi(0) = 1$, $|\varphi(t)| \le 1$, $\varphi(-t) = \overline{\varphi(t)}$, *and φ is uniformly continuous.*

Proof Since $|e^{itx}| \le 1$, everything follows immediately from the definitions except the uniform continuity. For that we write

$$|\varphi(t + h) - \varphi(t)| = |\mathbb{E}\,e^{i(t+h)X} - \mathbb{E}\,e^{itX}| \le \mathbb{E}\,|e^{itX}(e^{ihX} - 1)| = \mathbb{E}\,|e^{ihX} - 1|.$$

Since $|e^{ihX} - 1|$ tends to zero almost surely as $h \to 0$, the right-hand side tends to zero by dominated convergence. Note that the right-hand side is independent of t. □

Proposition A.46 $\varphi_{aX}(t) = \varphi_X(at)$ *and* $\varphi_{X+b}(t) = e^{itb}\varphi_X(t)$.

Proof The first follows from $\mathbb{E}\,e^{it(aX)} = \mathbb{E}\,e^{i(at)X}$, and the second is similar. □

Proposition A.47 *If X and Y are independent, then*

$$\varphi_{X+Y}(t) = \varphi_X(t)\varphi_Y(t).$$

Proof From the multiplication theorem,

$$\mathbb{E}\,e^{it(X+Y)} = \mathbb{E}\,e^{itX} e^{itY} = \mathbb{E}\,e^{itX}\mathbb{E}\,e^{itY},$$

and we are done. □

Let us look at some examples of characteristic functions.
(1) *Bernoulli*: By direct computation,

$$\varphi_X(t) = pe^{it} + (1 - p) = 1 - p(1 - e^{it}).$$

(2) *Binomial*: Write X as the sum of n independent Bernoulli random variables B_i with parameter p. Thus

$$\varphi_X(t) = \prod_{i=1}^{n} \varphi_{B_i}(t) = [\varphi_{B_i}(t)]^n = [1 - p(1 - e^{it})]^n.$$

(3) *Point mass at a*: $\mathbb{E}\, e^{itX} = e^{ita}$. Note that when $a = 0$, then φ is identically equal to 1.

(4) *Poisson*:

$$\mathbb{E}\, e^{itX} = \sum_{k=0}^{\infty} e^{itk} e^{-\lambda} \frac{\lambda^k}{k!} = e^{-\lambda} \sum \frac{(\lambda e^{it})^k}{k!} = e^{-\lambda} e^{\lambda e^{it}} = e^{\lambda(e^{it}-1)}.$$

(5) *Uniform on $[a, b]$*:

$$\varphi(t) = \frac{1}{b-a} \int_a^b e^{itx} dx = \frac{e^{itb} - e^{ita}}{(b-a)it}.$$

Note that when $a = -b$ this reduces to $\sin(bt)/bt$.

(6) *Exponential*:

$$\varphi(t) = \int_0^{\infty} \lambda e^{itx} e^{-\lambda x}\, dx = \lambda \int_0^{\infty} e^{(it-\lambda)x} dx = \frac{\lambda}{\lambda - it}.$$

(7) *Standard normal*:

$$\varphi(t) = \frac{1}{\sqrt{2\pi}} \int_{-\infty}^{\infty} e^{itx} e^{-x^2/2} dx.$$

This can be done by completing the square and then doing a contour integration. Alternately, $\varphi'(t) = (1/\sqrt{2\pi}) \int_{-\infty}^{\infty} ixe^{itx} e^{-x^2/2} dx$. (Do the real and imaginary parts separately, and use the dominated convergence theorem to justify taking the derivative inside.) Integrating by parts (do the real and imaginary parts separately), $\varphi'(t) = -t\varphi(t)$. The only solution to this differential equation with $\varphi(0) = 1$ is $\varphi(t) = e^{-t^2/2}$.

(8) *Normal with mean μ and variance σ^2*: Writing $X = \sigma Z + \mu$, where Z is a standard normal, then

$$\varphi_X(t) = e^{i\mu t} \varphi_Z(\sigma t) = e^{i\mu t - \sigma^2 t^2/2}. \tag{A.25}$$

(9) *Gamma*. If X has a gamma distribution with parameters λ and r, then its characteristic function is

$$\mathbb{E}\, e^{iuX} = \left(\frac{\lambda}{\lambda - it} \right)^r.$$

Formally, this comes from writing

$$\varphi(t) = \frac{1}{\Gamma(r)} \int_0^{\infty} e^{itx} \lambda e^{-\lambda x} (\lambda x)^{r-1}\, dx = \frac{\lambda^r}{\Gamma(r)} \int_0^{\infty} e^{-(\lambda - it)x} x^{r-1}\, dx$$

and performing a change of variables. To do it properly requires a contour integration around the boundary of the region in the complex plane that is bounded by the positive x axis, the ray $\{(\lambda - it)r : r > 0\}$, $\partial B(0, \varepsilon)$, and $\partial B(0, R)$, and then letting $\varepsilon \to 0$ and $R \to \infty$.

A.14 Uniqueness and characteristic functions

Theorem A.48 *If* $\varphi_X = \varphi_Y$, *then* $\mathbb{P}_X = \mathbb{P}_Y$.

Proof If f is in the Schwartz class, then so is $\widehat{f}$; see Section B.2. We use the Fubini theorem and the Fourier inversion theorem to write

$$\mathbb{E} f(X) = (2\pi)^{-1} \mathbb{E} \left[\int \widehat{f}(u) e^{-iuX} \, du \right] = (2\pi)^{-1} \int \widehat{f}(u) \varphi_X(-u) \, du,$$

and similarly for $\mathbb{E} f(Y)$. Since $\varphi_X = \varphi_Y$, we conclude $\mathbb{E} f(X) = \mathbb{E} f(Y)$. By a limit procedure, we have this equality for all bounded and measurable f, in particular, when f is the indicator of a set. □

The same proof works in higher dimensions: if

$$\mathbb{E} \, e^{i \sum_{j=1}^{n} u_j X_j} = \mathbb{E} \, e^{i \sum_{j=1}^{n} u_j Y_j}$$

for all $(u_1, \ldots, u_n) \in \mathbb{R}^n$, then the joint laws of $(X_1, \ldots, X_n)$ and $(Y_1, \ldots, Y_n)$ are equal. The expression $\mathbb{E} \, e^{i \sum_{j=1}^{n} u_j X_j}$ is called the *joint characteristic function* of $(X_1, \ldots, X_n)$.

The following proposition can be proved directly, but the proof using characteristic functions is much easier.

Proposition A.49 *(1) If X and Y are independent, X is a normal random variable with mean a and variance b^2, and Y is a normal random variable with mean c and variance d^2, then $X + Y$ is normal random variable with mean $a + c$ and variance $b^2 + d^2$.*

(2) If X and Y are independent, X is a Poisson random variable with parameter λ_1, and Y is a Poisson random variable with parameter λ_2, then $X + Y$ is a Poisson random variable with parameter $\lambda_1 + \lambda_2$.

(3) If X and Y are independent random variables, where X has a gamma distribution with parameters λ and r_1 and Y has a gamma distribution with parameters λ and r_2, then $X + Y$ has a gamma distribution with parameters λ and $r_1 + r_2$.

Proof For (1),

$$\varphi_{X+Y}(t) = \varphi_X(t) \varphi_Y(t) = e^{iat - b^2 t^2/2} e^{ict - c^2 t^2/2} = e^{i(a+c)t - (b^2 + d^2)t^2/2}.$$

Now use the uniqueness theorem.

Parts (2) and (3) are proved similarly. □

A.15 The central limit theorem

We need the following estimate on moments.

Proposition A.50 *If $\mathbb{E} |X|^k < \infty$ for an integer k, then φ_X has a continuous derivative of order k and*

$$\varphi_X^{(k)}(t) = \int (ix)^k e^{itx} \mathbb{P}_X(dx).$$

In particular, $\varphi_X^{(k)}(0) = i^k \mathbb{E} X^k$.

Proof   Write

$$\frac{\varphi_X(t+h) - \varphi_X(t)}{h} = \int \frac{e^{i(t+h)x} - e^{itx}}{h} \mathbb{P}(dx).$$

Since $|e^{ihx} - 1| \leq |h| |x|$, the integrand is bounded by $|x|$. Thus if $\int |x| \mathbb{P}_X(dx) < \infty$, we can use dominated convergence to obtain the desired formula for $\varphi_X'(t)$. As in the proof of Proposition A.45, we see $\varphi_X'(t)$ is continuous. We do the case of general k by induction. Evaluating $\varphi_X^{(k)}$ at 0 shows $\varphi_X^{(k)}(0) = i^k \mathbb{E} X^k$.   □

By the above,

$$\mathbb{E} X^2 = -\varphi_X''(0). \tag{A.26}$$

The simplest case of the central limit theorem (CLT) is the case when the X_i's are i.i.d., with mean zero and variance one, $S_n = \sum_{i=1}^n X_i$, and then the CLT says that $S_n/\sqrt{n}$ converges weakly to a standard normal. This is the case we prove.

We need the fact that if w_n are complex numbers converging to w, then $(1+(w_n/n))^n \to e^w$. We leave the proof of this to the reader, with the warning that any proof using logarithms needs to be done with some care, since $\log z$ is a multivalued function when z is complex.

Theorem A.51 *Suppose the X_i's are i.i.d. random variables with mean zero and variance one. Then $S_n/\sqrt{n}$ converges weakly to a standard normal.*

Proof   Since X_1 has finite second moment, then φ_{X_1} has a continuous second derivative by Proposition A.50. By Taylor's theorem,

$$\varphi_{X_1}(t) = \varphi_{X_1}(0) + \varphi_{X_1}'(0)t + \varphi_{X_1}''(0)t^2/2 + R(t),$$

where $|R(t)|/t^2 \to 0$ as $|t| \to 0$. Thus

$$\varphi_{X_1}(t) = 1 - t^2/2 + R(t).$$

Then

$$\varphi_{S_n/\sqrt{n}}(t) = \varphi_{S_n}(t/\sqrt{n}) = (\varphi_{X_1}(t/\sqrt{n}))^n = \left[1 - \frac{t^2}{2n} + R(t/\sqrt{n})\right]^n.$$

Since $t/\sqrt{n}$ converges to zero as $n \to \infty$, we have

$$\varphi_{S_n/\sqrt{n}}(t) \to e^{-t^2/2}.$$

Since $\mathbb{E} S_n^2/n = 1$ for all n, Proposition A.44 tells us that the random variables $S_n/\sqrt{n}$ are tight, and from Theorem A.43, subsequential weak limit points exist. By the preceding paragraph, any weak limit of a subsequence is a normal random variable with mean zero and variance one. Therefore the entire sequence converges weakly to a normal random variable with mean zero and variance one.   □

A.16 Gaussian random variables

A normal random variable is also known as a Gaussian random variable.

Proposition A.52 *If Z is a mean zero normal random variable with variance one and $x \geq 1$, then*

$$\frac{1}{x} e^{-x^2/2} \leq \mathbb{P}(Z \geq x) \leq e^{-x^2/2}.$$

In particular, if $\varepsilon > 0$, there exists x_0 such that

$$\mathbb{P}(Z \geq x) \geq e^{-(1+\varepsilon)x^2/2}$$

if $x \geq x_0$.

Proof For the right-hand inequality,

$$\mathbb{P}(Z \geq x) = \frac{1}{\sqrt{2\pi}} \int_x^\infty e^{-y^2/2} \, dy \leq \int_x^\infty \frac{y}{x} e^{-y^2/2} \, dy = \frac{1}{x} e^{-x^2/2}.$$

The left-hand inequality is left as an exercise. $\square$

Proposition A.53 *If X_n is a normal random variable with mean a_n and variance b_n^2, X_n converges to X weakly, $a_n \to a$, and $b_n \to b \neq 0$, then X is a normal random variable with mean a and variance b^2.*

Proof Since

$$\mathbb{E} X_n^2 = \operatorname{Var} X_n + (\mathbb{E} X_n)^2 = b_n^2 + a_n^2,$$

then $\sup_n \mathbb{E} X_n^2 < \infty$, and the X_n are tight. For each t, the characteristic functions converge:

$$\varphi_X(t) = \lim_{n \to \infty} \varphi_{X_n}(t) = \lim_{n \to \infty} e^{ita_n - t^2 b_n^2/2} = e^{ita - t^2 b^2/2},$$

and the last term is the characteristic function of a normal random variable with mean a and variance b^2. Therefore any weak subsequential limit point of the sequence X_n is a normal random variable with mean a and variance b^2. $\square$

We next prove

Proposition A.54 *If*

$$\mathbb{E} \, e^{i(uX+vY)} = \mathbb{E} \, e^{iuX} \, \mathbb{E} \, e^{ivY} \tag{A.27}$$

for all u and v, then X and Y are independent random variables.

Proof Let X' be a random variable with the same law as X, Y' one with the same law as Y, and so that X' is independent of Y'. (We let $\Omega = [0, 1]^2$, $\mathbb{P}$ a Lebesgue measure, X' a function of the first variable, and Y' a function of the second variable defined as in Proposition A.2.) Then since $e^{iuX'}$ and $e^{ivY'}$ are independent,

$$\mathbb{E} \, e^{i(uX'+vY')} = \mathbb{E} \, e^{iuX'} \, \mathbb{E} \, e^{ivY'}. \tag{A.28}$$

Since X, X' have the same law, $\mathbb{E} \, e^{iuX} = \mathbb{E} \, e^{iuX'}$, and similarly for Y, Y'. Therefore, using (A.27) and (A.28), (X', Y') has the same joint characteristic function as (X, Y). By the uniqueness theorem for characteristic functions, (X', Y') has the same joint law as (X, Y), which implies that X and Y are independent. □

A sequence of random variables $X_1, \ldots, X_n$ is said to be *jointly normal* if there exists a sequence of i.i.d. normal random variables $Z_1, \ldots, Z_m$ with mean zero and variance one and constants b_{ij} and a_i such that

$$X_i = \sum_{j=1}^{m} b_{ij} Z_j + a_i, \qquad i = 1, \ldots, n. \tag{A.29}$$

In matrix notation, $X = BZ + A$. For simplicity, in what follows let us take $A = 0$; the modifications for the general case are easy. The *covariance* of two random variables X and Y is defined to be $\mathbb{E}[(X - \mathbb{E}X)(Y - \mathbb{E}Y)]$. Since we are assuming our normal random variables are mean zero, we can omit the centering at expectations. Given a sequence of mean zero random variables, we can talk about the *covariance matrix*, which is

$$\mathrm{Cov}(X) = \mathbb{E}XX^T,$$

where X^T denotes the transpose of the vector X. In the above case, we see $\mathrm{Cov}(X) = \mathbb{E}[(BZ)(BZ)^T] = \mathbb{E}[BZZ^TB^T] = BB^T$, since $\mathbb{E}ZZ^T = I$, the identity.

Let us compute the joint characteristic function $\mathbb{E}e^{iu^TX}$ of the vector X, where u is an n-dimensional vector. First, if v is an m-dimensional vector,

$$\mathbb{E}e^{iv^TZ} = \mathbb{E}\prod_{j=1}^{m} e^{iv_jZ_j} = \prod_{j=1}^{m} \mathbb{E}e^{iv_jZ_j} = \prod_{j=1}^{m} e^{-v_j^2/2} = e^{-v^Tv/2}$$

using the independence of the Z_j's. Thus

$$\mathbb{E}e^{iu^TX} = \mathbb{E}e^{iu^TBZ} = e^{-u^TBB^Tu/2}.$$

By taking $u = (0, \ldots, 0, a, 0, \ldots, 0)$ to be a constant times the unit vector in the jth coordinate direction, we deduce that X_j is indeed normal, and this is true for each j.

Note that the joint characteristic function of a jointly normal collection of random variables $X = (X_1, \ldots, X_n)$ is completely determined by BB^T, which is the covariance matrix of X. In the case when the X_i's are not mean zero, we can readily check that the joint characteristic function is determined by the covariance matrix together with the vector of means $\mathbb{E}X$. Therefore the joint distribution of a jointly normal collection of random variables is determined by the covariance matrix and the means.

Proposition A.55 *If the X_i are jointly normal and $\mathrm{Cov}(X_i, X_j) = 0$ for $i \neq j$, then the X_i are independent.*

Proof If $\mathrm{Cov}(X) = BB^T$ is a diagonal matrix, then the joint characteristic function of the X_i's factors into the product of the characteristic functions of the X_i's, and so by Proposition A.54, the X_i's will in this case be independent. □

Remark A.56 We note that the analog of Proposition A.53 holds for jointly normal random vectors. That is, if $(X_j^1, \ldots, X_j^n)$ is a jointly normal collection of random variables for each j and each X_j^i converges in probability to X^i and each X_i is nonconstant, then $(X^1, \ldots, X^n)$

is a jointly normal collection of random variables. This follows by looking at the joint characteristic functions as in the proof of Proposition A.53.

We present the multidimensional central limit theorem.

Theorem A.57 *Let $X_j = (X_j^1, \ldots, X_j^d)$ be random vectors taking values in $\mathbb{R}^d$ and suppose the $X_1, X_2, \ldots$ are independent and identically distributed. Suppose $\mathbb{E} X_1^k = 0$ and $\mathbb{E} (X_1^k)^2 < \infty$ for $k = 1, \ldots, d$ and let $C_{k\ell} = \mathbb{E} [X_1^k X_1^\ell]$. If $S_n = \sum_{j=1}^n X_j$, then $S_n / \sqrt{n}$ converges weakly to a jointly normal random vector $Z = (Z^1, \ldots, Z^d)$ where each Z^k has mean zero and the covariance of Z^k and Z^ℓ is $C_{k\ell}$.*

Proof Since

$$\mathbb{E} |S_n|^2 / n = \sum_{j=1}^n \sum_{k=1}^d \mathbb{E} |X_j^k|^2 / n$$

is bounded independently of n, the random vectors $S_n / \sqrt{n}$ are tight, and therefore weak subsequential limit points exist. We need to show that any subsequential limit point is a jointly normal random vector with mean zero and covariance matrix C.

If $u_1, \ldots, u_d \in \mathbb{R}$, then $\sum_{k=1}^d u_k X_j^k$, $j = 1, 2, \ldots$, will be a sequence of i.i.d. random variables with mean zero and variance $\sum_{k,\ell=1}^d u_k u_\ell C_{k\ell}$. By Theorem A.51,

$$\frac{\sum_{j=1}^n \sum_{k=1}^d u_k X_j^k}{\sqrt{n}}$$

converges weakly to a mean zero normal random variable with variance equal to $\sum_{k,\ell=1}^d u_k u_\ell C_{k\ell}$. If we write $S_n = (S_n^1, \ldots, S_n^d)$, then

$$\mathbb{E} \exp \left(i \sum_{k=1}^d u_k S_n^k / \sqrt{n} \right) \to \exp \left(- \sum_{k,\ell=1}^d u_k u_\ell C_{k\ell} / 2 \right).$$

This shows that any subsequential limit point of the sequence $S_n / \sqrt{n}$ has the required law. $\square$

If $(X, Y_1, \ldots, Y_n)$ are jointly normal random variables, then the law of X given $Y_1, \ldots, Y_n$ is also Gaussian.

Proposition A.58 *Suppose $X, Y_1, \ldots, Y_n$ are jointly normal random variables with mean zero. Let A be the $n \times 1$ matrix whose ith entry is $\mathrm{Cov} (X, Y_i)$, B the $n \times n$ matrix whose (i, j)th entry is $\mathrm{Cov} (Y_i, Y_j)$, and Y the $n \times 1$ matrix whose ith entry is Y_i. Suppose B is invertible and let $D = B^{-1} A$. Then for $u \in \mathbb{R}$,*

$$\mathbb{E} [e^{iuX} \mid Y_1, \ldots, Y_n] = e^{iu D^T Y} e^{-(\mathrm{Var} X - A^T B^{-1} A)/2}.$$

In particular, the law of X given $Y_1, \ldots, Y_n$ is that of a normal random variable with mean $D^T Y$ and variance equal to $\mathrm{Var} X - A^T B^{-1} A$.

Proof Note

$$\operatorname{Cov}\left(X - D^T Y, Y_j\right) = \operatorname{Cov}\left(X, Y_j\right) - \sum_{i=1}^{n} D_i \operatorname{Cov}\left(Y_i, Y_j\right)$$

$$= A_j - \sum_{i=1}^{n} D_i B_{ij} = 0,$$

so $X - D^T Y$ is independent of each Y_j. Then

$$\mathbb{E}\left[e^{iuX} \mid Y_1, \ldots, Y_n\right] = e^{iuD^T Y} \mathbb{E}\left[e^{iu(X - D^T Y)} \mid Y_1, \ldots, Y_n\right]$$

$$= e^{iuD^T Y} \mathbb{E}\left[e^{iu(X - D^T Y)}\right]$$

$$= e^{iuD^T Y} \mathbb{E} \, e^{-\operatorname{Var}(X - D^T Y)/2}.$$

To complete the proof, we calculate

$$\operatorname{Var}\left(X - D^T Y\right) = \operatorname{Var} X - 2 \sum_i D_i A_i + \sum_{i,j} D_i B_{ij} D_j$$

$$= \operatorname{Var} X - A^T B^{-1} A,$$

and we are done. $\qquad\square$

Appendix B

Some results from analysis

B.1 The monotone class theorem

The monotone class theorem is a result from measure theory used in the proof of the Fubini theorem.

Definition B.1 $\mathcal{M}$ is a *monotone class* if $\mathcal{M}$ is a collection of subsets of X such that
(1) if $A_1 \subset A_2 \subset \cdots$, $A = \cup_i A_i$, and each $A_i \in \mathcal{M}$, then $A \in \mathcal{M}$;
(2) if $A_1 \supset A_2 \supset \cdots$, $A = \cap_i A_i$, and each $A_i \in \mathcal{M}$, then $A \in \mathcal{M}$.

Recall that an algebra of sets is a collection $\mathcal{A}$ of sets such that if $A_1, \ldots, A_n \in \mathcal{A}$, then $A_1 \cup \cdots \cup A_n$ and $A_1 \cap \cdots \cap A_n$ are also in $\mathcal{A}$, and if $A \in \mathcal{A}$, then $A^c \in \mathcal{A}$.

The intersection of monotone classes is a monotone class, and the intersection of all monotone classes containing a given collection of sets is the smallest monotone class containing that collection.

Theorem B.2 *Suppose $\mathcal{A}_0$ is an algebra of sets, $\mathcal{A}$ is the smallest σ-field containing $\mathcal{A}_0$, and $\mathcal{M}$ is the smallest monotone class containing $\mathcal{A}_0$. Then $\mathcal{M} = \mathcal{A}$.*

Proof A σ-algebra is clearly a monotone class, so $\mathcal{M} \subset \mathcal{A}$. We must show $\mathcal{A} \subset \mathcal{M}$.

Let $\mathcal{N}_1 = \{A \in \mathcal{M} : A^c \in \mathcal{M}\}$. Note $\mathcal{N}_1$ is contained in $\mathcal{M}$, contains $\mathcal{A}_0$, and is a monotone class. Since $\mathcal{M}$ is the smallest monotone class containing $\mathcal{A}_0$, then $\mathcal{N}_1 = \mathcal{M}$, and therefore $\mathcal{M}$ is closed under the operation of taking complements.

Let $\mathcal{N}_2 = \{A \in \mathcal{M} : A \cap B \in \mathcal{M} \text{ for all } B \in \mathcal{A}_0\}$. $\mathcal{N}_2$ is contained in $\mathcal{M}$; $\mathcal{N}_2$ contains $\mathcal{A}_0$ because $\mathcal{A}_0$ is an algebra; $\mathcal{N}_2$ is a monotone class because $(\cup_{i=1}^{\infty} A_i) \cap B = \cup_{i=1}^{\infty} (A_i \cap B)$, and similarly for intersections. Therefore $\mathcal{N}_2 = \mathcal{M}$; in other words, if $B \in \mathcal{A}_0$ and $A \in \mathcal{M}$, then $A \cap B \in \mathcal{M}$.

Let $\mathcal{N}_3 = \{A \in \mathcal{M} : A \cap B \in \mathcal{M} \text{ for all } B \in \mathcal{M}\}$. As in the preceding paragraph, $\mathcal{N}_3$ is a monotone class contained in $\mathcal{M}$. By the last sentence of the preceding paragraph, $\mathcal{N}_3$ contains $\mathcal{A}_0$. Hence $\mathcal{N}_3 = \mathcal{M}$.

We thus have that $\mathcal{M}$ is a monotone class closed under the operations of taking complements and taking intersections. This shows $\mathcal{M}$ is a σ-algebra, and so $\mathcal{A} \subset \mathcal{M}$. $\square$

B.2 The Schwartz class

A function $f : \mathbb{R}^d \to \mathbb{R}$ is in the Schwartz class if f is C^∞ and for each $m, k \geq 0$ and each $i_1, i_2, \ldots, i_k \in \{1, 2, \ldots, d\}$,

$$|x|^m \left| \frac{\partial^k f}{\partial x_{i_1} \cdots \partial x_{i_k}}(x) \right| \to 0$$

as $|x| \to \infty$. (Here $i_1, \ldots, i_k$ need not be distinct.)

Suppose that f is in the Schwartz class. Suppose $m, k \geq 0$ and $i_1, \ldots, i_k$ and $j_1, \ldots, j_n$ are each integers between 1 and d inclusive, and $m_1, \ldots, m_k$ are even positive integers. Let $\widehat{f}$ be the Fourier transform of f:

$$\widehat{f}(u) = \int_{\mathbb{R}^d} e^{iu \cdot x} f(x) \, dx.$$

Then

$$u_{i_1}^{m_1} \cdots u_{i_k}^{m_k} \frac{\partial^{j_1 + \cdots + j_n} \widehat{f}}{\partial u_{j_1} \cdots \partial u_{j_n}}(u)$$

is bounded as a function of u because it is a constant times the Fourier transform of

$$x_{j_1} \cdots x_{j_n} \frac{\partial^{m_1 + \cdots + m_k} f}{\partial x_{i_1}^{m_1} \cdots \partial x_{i_k}^{m_k}},$$

which is in $L^1(\mathbb{R}^d)$ since f is in the Schwartz class. We conclude that $\widehat{f}$ is also in the Schwartz class.

Appendix C

Regular conditional probabilities

Let $\mathcal{E} \subset \mathcal{F}$ be σ-fields, where $(\Omega, \mathcal{F}, \mathbb{P})$ is a probability space. A *regular conditional probability* for $\mathbb{E}[\cdot \mid \mathcal{E}]$ is a map $Q : \Omega \times \mathcal{F} \to [0, 1]$ such that

(1) $Q(\omega, \cdot)$ is a probability measure on $(\Omega, \mathcal{F})$ for each ω;
(2) for each $A \in \mathcal{F}$, $Q(\cdot, A)$ is an $\mathcal{E}$ measurable random variable;
(3) for each $A \in \mathcal{F}$ and each $B \in \mathcal{E}$,

$$\int_B Q(\omega, A) \, \mathbb{P}(d\omega) = \mathbb{P}(A \cap B).$$

$Q(\omega, A)$ can be thought of as $\mathbb{P}(A \mid \mathcal{E})$.

Theorem C.1 *Suppose $(\Omega, \mathcal{F}, \mathbb{P})$ is a probability space, $\mathcal{E} \subset \mathcal{F}$, and Ω is in addition a complete and separable metric space. Then a regular conditional probability for $\mathbb{P}(\cdot \mid \mathcal{E})$ exists.*

Proof Since Ω is a complete and separable metric space, we can embed Ω as a subset of the compact set $I = [0, 1]^{\mathbb{N}}$, where we furnish I with the product topology. Let $\{f_j\}$ be a countable collection of uniformly continuous functions on Ω such that every finite subset of distinct elements is linearly independent and such that $\mathcal{L}_0$, the set of finite linear combinations of the f_j's, is dense in the class of uniformly continuous functions on Ω; let us assume f_1 is identically equal to 1.

For each j, let $g_j = \mathbb{E}[f_j \mid \mathcal{E}]$. (The random variables g_j are only defined up to almost sure equivalence. For each j we select an element g_j from the equivalence class and keep it fixed.) If $r_1, \ldots, r_n$ are rationals with

$$r_1 f_1(\omega) + \cdots + r_n f_n(\omega) \geq 0$$

for all ω, let

$$N(r_1, \ldots, r_n) = \{\omega : r_1 g_1(\omega) + \cdots + r_n g_n(\omega) < 0\}.$$

By the definition of g_j, $\mathbb{P}(N(r_1, \ldots, r_n)) = 0$. Let N_1 be the union of all such $N(r_1, \ldots, r_n)$ with $n \geq 1$, the r_j rational. Then $N_1 \in \mathcal{E}$ and $\mathbb{P}(N_1) = 0$.

Fix $\omega \in \Omega \setminus N_1$. Define a functional L_ω on $\mathcal{L}_0$ by

$$L_\omega(f) = t_1 g_1(\omega) + \cdots + t_n g_n(\omega)$$

if

$$f = t_1 f_1 + \cdots + t_n f_n.$$

We claim L_ω is a positive linear functional. If $f = t_1 f_1 + \cdots + t_n f_n \geq 0$ and $\varepsilon > 0$ is rational, then there exist rationals $r_1, \ldots, r_n$ such that $r_1 f_1 + \ldots + r_n f_n \geq -\varepsilon$ and $|t_i - r_i| \leq \varepsilon$, $i = 1, \ldots, n$, or

$$(r_1 + \varepsilon) f_1 + r_2 f_2 + \cdots + r_n f_n \geq 0.$$

Since $\omega \notin N_1$, then

$$(r_1 + \varepsilon) g_1 + r_2 g_2 + \cdots + r_n g_n \geq 0.$$

Letting $\varepsilon \to 0$, it follows that $t_1 g_1 + \cdots + t_n g_n \geq 0$. This proves that L_ω is positive.

Since $L_\omega(f_1) = 1$, this implies that L_ω is a bounded linear functional, and by the Hahn–Banach theorem L_ω can be extended to a positive linear functional on the closure of $\mathcal{L}_0$. Any uniformly continuous function on Ω can be extended uniquely to $\overline{\Omega}$, the closure of Ω in I, so L_ω can be considered as a positive linear functional on $C(\overline{\Omega})$. By the Riesz representation theorem, there exists a probability measure $Q(\omega, \cdot)$ such that

$$L_\omega(f) = \int f(\omega') Q(\omega, d\omega').$$

The mapping $\omega \to L_\omega(f)$ is measurable with respect to $\mathcal{E}$ for each $f \in \mathcal{L}_0$, hence for all uniformly continuous functions on Ω by a limit argument. If $B \in \mathcal{E}$ and $f = t_1 f_1 + \cdots + t_n f_n$,

$$\int_B \left[\int f(\omega') Q(\omega, d\omega') \right] \mathbb{P}(d\omega) = \int_B L_\omega f(\omega) \, \mathbb{P}(d\omega)$$

$$= \int_B (t_1 g_1 + \cdots + t_n g_n)(\omega) \, \mathbb{P}(d\omega)$$

$$= \int_B \mathbb{E}\left[t_1 f_1 + \cdots + t_n f_n \mid \mathcal{E} \right](\omega) \, \mathbb{P}(d\omega)$$

$$= \int_B f(\omega) \, \mathbb{P}(d\omega)$$

or $\int f(\omega') Q(\omega, d\omega')$ is a version of $\mathbb{E}[f|\mathcal{E}]$ if $f \in \mathcal{L}_0$. By a limit argument, the same is true for all f that are of the form $f = 1_A$ with $A \in \mathcal{F}$.

Let G_{ni} be a sequence of balls of radius $1/n$ (with respect to the metric on Ω) contained in Ω and covering Ω. Choose i_n such that $\mathbb{P}(\cup_{i \leq i_n} G_{ni}) > 1 - 1/(n2^n)$. The set $H_n = \cap_{n \geq 1} \cup_{i \leq i_n} G_{ni}$ is totally bounded; let K_n be the closure of H_n in Ω. Since Ω is complete, K_n is complete and totally bounded, and hence compact, and $\mathbb{P}(K_n) \geq 1 - 1/n$. Hence

$$\mathbb{E}\left[Q(\cdot, \cup_{i=1}^{\infty} K_i); \Omega \setminus N_1 \right] \geq \mathbb{E}\left[Q(\cdot, K_n); \Omega \setminus N_1 \right] = \mathbb{P}(K_n) \geq 1 - (1/n)$$

for each n, or $Q(\omega, \cup_{i=1}^{\infty} K_i) = 1$, a.s. Let N_2 be the null set for which this fails. Thus for $\omega \in \Omega \setminus (N_1 \cup N_2)$, we see that $Q(\omega, d\omega')$ is a probability measure on Ω. For $\omega \in N_1 \cup N_2$, set $Q(\omega, \cdot) = \mathbb{P}(\cdot)$. This Q is the desired regular conditional probability. $\square$

Appendix D

Kolmogorov extension theorem

Suppose S is a metric space. We use $S^{\mathbb{N}}$ for the product space $S \times S \times \cdots$ furnished with the product topology. We may view $S^{\mathbb{N}}$ as the set of sequences $(x_1, x_2, \ldots)$ of elements of S. We use the σ-field on $S^{\mathbb{N}}$ generated by the cylindrical sets. Given an element $x = (x_1, x_2, \ldots)$ of $S^{\mathbb{N}}$, we define $\pi_n(x) = (x_1, \ldots, x_n) \in S^n$.

We suppose we have a Radon probability measure μ_n defined on S^n for each n. (Being a Radon measure means that we can approximate $\mu_n(A)$ from below by compact sets; see Folland (1999) for details.) The μ_n are *consistent* if $\mu_{n+1}(A \times S) = \mu_n(A)$ whenever A is a Borel subset of S^n. The *Kolmogorov extension theorem* is the following.

Theorem D.1 *Suppose for each n we have a probability measure μ_n on S^n. Suppose the μ_n's are consistent. Then there exists a probability measure μ on $S^{\mathbb{N}}$ such that $\mu(A \times S^{\mathbb{N}}) = \mu_n(A)$ for all $A \subset S^n$.*

Proof Define μ on cylindrical sets by $\mu(A \times S^{\mathbb{N}}) = \mu_n(A)$ if $A \subset S^n$. By the consistency assumption, μ is well defined. By the Carathéodory extension theorem, we can extend μ to the σ-field generated by the cylindrical sets provided we show that whenever A_n are cylindrical sets decreasing to $\emptyset$, then $\mu(A_n) \to 0$.

Suppose A_n are cylindrical sets decreasing to $\emptyset$ but $\mu(A_n)$ does not tend to 0; by taking a subsequence we may assume without loss of generality that there exists $\varepsilon > 0$ such that $\mu(A_n) \geq \varepsilon$ for all n. We will obtain a contradiction.

We first want to arrange things so that each $A_n = \pi_n(A_n) \times S^{\mathbb{N}}$. Suppose A_n is of the form

$$A_n = \{(x_1, x_2, \ldots) : (x_1, \ldots, x_{j_n}) \in B_n\},$$

where B_n is a Borel subset of S^{j_n}. We choose $m_n = n + \max(j_1, \ldots, j_n)$. Let $A_0 = S^{\mathbb{N}}$. We then replace our original sequence $A_1, A_2, \ldots$ by the sequence $A_0, \ldots, A_0, A_1, \ldots, A_1, A_2, \ldots, A_2, A_3, \ldots$, where we have m_1 occurrences of A_0, $m_2 - m_1$ occurrences of A_1, $m_3 - m_2$ occurrences of A_2, and so on. Therefore we may without loss of generality suppose $j_n \leq n$. We then have

$$A_n = \{(x_1, x_2, \ldots) : (x_1, \ldots, x_n) \in B_n \times S^{n-j_n}\}.$$

Replacing B_n by $B_n \times S^{j_n - n}$, we may without loss of generality suppose $A_n = \pi_n(A_n) \times S^{\mathbb{N}}$.

We set $\widetilde{A}_n = \pi_n(A_n)$. For each n, choose $\widetilde{B}_n \subset \widetilde{A}_n$ so that $\widetilde{B}_n$ is compact and $\mu(\widetilde{A}_n \setminus \widetilde{B}_n) \le \varepsilon/2^{n+1}$. Let $B_n = \widetilde{B}_n \times \mathcal{S}^{\mathbb{N}}$ and let $C_n = B_1 \cap \ldots \cap B_n$. Hence $C_n \subset B_n \subset A_n$, and $C_n \downarrow \emptyset$, but

$$\mu(C_n) \ge \mu(A_n) - \sum_{i=1}^{n} \mu(A_i \setminus B_i) \ge \varepsilon/2,$$

and $\widetilde{C}_n = \pi_n(C_n)$, the projection of C_n onto $\mathcal{S}^n$, is compact.

We will find $x = (x_1, \ldots, x_n, \ldots) \in \cap_n C_n$ and obtain our contradiction. For each n choose a point $y(n) \in C_n$. The first coordinates of $\{y(n)\}$, namely, $\{y_1(n)\}$, form a sequence contained in $\widetilde{C}_1$, which is compact, hence there is a convergent subsequence $\{y_1(n_k)\}$. Let x_1 be the limit point. The first and second coordinates of $\{y(n_k)\}$ form a sequence contained in the compact set $\widetilde{C}_2$, so a further subsequence $\{(y_1(n_{k_j}), y_2(n_{k_j}))\}$ converges to a point in $\widetilde{C}_2$. Since $\{n_{k_j}\}$ is a subsequence of $\{n_k\}$, the first coordinate of the limit is x_1. Therefore the limit point of $\{(y_1(n_{k_j}), y_2(n_{k_j}))\}$ is of the form (x_1, x_2), and this point is in $\widetilde{C}_2$. We continue this procedure to obtain $x = (x_1, x_2, \ldots, x_n, \ldots)$. By our construction, $(x_1, \ldots, x_n) \in \widetilde{C}_n$ for each n, hence $x \in C_n$ for each n, or $x \in \cap_n C_n$, a contradiction. $\square$

A typical application of this theorem is to construct a countable sequence of independent random variables. We construct $X_1, \ldots, X_n$ as in Proposition A.10. Here $\mathcal{S} = [0, 1]$. Let μ_n be the law of $(X_1, \ldots, X_n)$; it is easy to check that the μ_n form a consistent family. We use Theorem D.1 to obtain a probability measure μ on $[0, 1]^{\mathbb{N}}$. To get random variables out of this, we let $X_i(\omega) = \omega_i$ if $\omega = (\omega_1, \omega_2, \ldots)$.

References

Aldous, D. 1978. Stopping times and tightness. *Ann. Probab.* **6**, 335–40.

Barlow, M. T. 1982. One-dimensional stochastic differential equations with no strong solution. *J. London Math. Soc.* **26**, 335–47.

Bass, R. F. 1983. Skorokhod imbedding via stochastic integrals. *Séminaire de Probabilités XVII*. New York: Springer-Verlag; 221–4.

Bass, R. F. 1995. *Probabilistic Techniques in Analysis*. New York: Springer-Verlag.

Bass, R. F. 1996. The Doob–Meyer decomposition revisited. *Can. Math. Bull.* **39**, 138–50.

Bass, R. F. 1997. *Diffusions and Elliptic Operators*. New York: Springer-Verlag.

Billingsley, P. 1968. *Convergence of Probability Measures*. New York: John Wiley & Sons, Ltd.

Billingsley, P. 1971. *Weak Convergence of Measures: Applications in Probability*. Philadelphia: SIAM.

Blumenthal, R. M. and Getoor, R. K. 1968. *Markov Processes and Potential Theory*. New York: Academic Press.

Bogachev, V. I. 1998. *Gaussian Measures*. Providence, RI: American Mathematical Society.

Boyce, W. E. and DiPrima, R. C. 2009. *Elementary Differential Equations and Boundary Value Problems,* 9th edn. New York: John Wiley & Sons, Ltd.

Chung, K. L. 2001. *A Course in Probability Theory*, 3rd edn. San Diego: Academic Press.

Chung, K. L. and Walsh, J. B. 1969. To reverse a Markov process. *Acta Math.* **123**, 225–51.

Dawson, D. A. 1993. Measure-valued Markov processes. *Ecole d'Eté de Probabilités de Saint-Flour XXI– 1991*. Berlin: Springer-Verlag.

Dellacherie, C. and Meyer, P.-A. 1978. *Probability and Potential*. Amsterdam: North-Holland.

Dudley, R. M. 1973. Sample functions of the Gaussian process. *Ann. Probab.* **1**, 66–103.

Durrett, R. 1996. *Probability: Theory and Examples*. Belmont, CA: Duxbury Press.

Ethier, S. N. and Kurtz, T. G. 1986. *Markov Processes: Characterization and Convergence*. New York: John Wiley & Sons, Ltd.

Feller, W. 1971. *An Introduction to Probability Theory and its Applications*, 2nd edn. New York: John Wiley & Sons, Ltd.

Folland, G. B. 1999. *Real Analysis: Modern Techniques and their Applications*, 2nd edn. New York: John Wiley & Sons, Ltd.

Fukushima, M., Oshima, Y. and Takeda, M. 1994. *Dirichlet Forms and Symmetric Markov Processes*. Berlin: de Gruyter.

Gilbarg, D. and Trudinger, N. S. 1983. *Elliptic Partial Differential Equations of Second Order*, 2nd edn. New York: Springer-Verlag.

Itô, K. and McKean, Jr, H. P. 1965. *Diffusion Processes and their Sample Paths*. Berlin: Springer-Verlag.

Kallianpur, G. 1980. *Stochastic Filtering Theory*. Berlin: Springer-Verlag.

Karatzas, I. and Shreve, S. E. 1991. *Brownian Motion and Stochastic Calculus*, 2nd edn. New York: Springer-Verlag.

Knight, F. B. 1981. *Essentials of Brownian Motion and Diffusion*. Providence, RI: American Mathematical Society.

Kuo, H. H. 1975. *Gaussian Measures in Banach Spaces*. New York: Springer-Verlag.

Lax, P. 2002. *Functional Analysis*. New York: John Wiley & Sons, Ltd.

Liggett, T. M. 2010. *Continuous Time Markov Processes: An Introduction*. Providence, RI: American Mathematical Society.

Meyer, P.-A., Smythe, R. T. and Walsh, J. B. 1972. Birth and death of Markov processes. *Proceedings of the Sixth Berkeley Symposium on Mathematical Statistics and Probability, Vol. III*. Berkeley, CA: University of California Press; 295–305.

Obłój, J. 2004. The Skorokhod embedding problem and its offspring. *Probab. Surv.* **1**, 321–90.

Øksendal, B. 2003. *Stochastic Differential Equations: An Introduction with Applications*, 6th edn. Berlin: Springer-Verlag.

Perkins, E. A. 2002. Dawson–Watanabe superprocesses and measure-valued diffusions. *Lectures on Probability Theory and Statistics (Saint-Flour, 1999)*. Berlin: Springer-Verlag; 125–324.

Revuz, D. and Yor, M. 1999. *Continuous Martingales and Brownian Motion*, 3rd edn. Berlin: Springer-Verlag.

Rogers, L. C. G. and Williams, D. 2000a. *Diffusions, Markov Processes, and Martingales, Vol. 1*. Cambridge: Cambridge University Press.

Rogers, L. C. G. and Williams, D. 2000b. *Diffusions, Markov Processes, and Martingales, Vol. 2*. Cambridge: Cambridge University Press.

Rudin, W. 1976. *Principles of Mathematical Analysis*, 3rd edn. New York: McGraw-Hill.

Rudin, W. 1987. *Real and Complex Analysis*, 3rd edn. New York: McGraw-Hill.

Skorokhod, A. V. 1965. *Studies in the Theory of Random Processes*. Reading, MA: Addison-Wesley.

Stroock, D. W. 2003. *Markov Processes from K. Itô's Perspective*. Princeton, NJ: Princeton University Press.

Stroock, D. W. and Varadhan, S. R. S. 1977. *Multidimensional Diffusion Processes*. Berlin: Springer-Verlag.

Walsh, J. B. 1978. Excursions and local time. *Astérisque* **52–53**, 159–92.

Index

Printed in the United States
By Bookmasters